Regressions in Covariances, Dependencies and Graphs

Multivariate data routinely collected nowadays using modern technological devices display cross-sectional, temporal, and spatial dependence. *Regressions in Covariances, Dependencies and Graphs* emphasizes the phenomenal roles of regression in modeling various dependencies using the twin principles of *parsimony* and *regularization* as a guide. For parsimony, *covariance regression*, mimicking the mean-regression, expresses a covariance matrix or its transform as linear combinations of covariates with the aim of reaching the versatility of the generalized linear models. *Hidden regression* reparametrizes a matrix so as to view its columns as parameters of certain regression models to be estimated iteratively one column at a time via regularized regression. The class of *graphical Lasso algorithms* for sparse graphs and their central roles in the modern high-dimensional data analysis are highlighted. Dimension-reduction through principal component analysis and factor models for multivariate and time series data is illustrated with a particular focus on the role of approximate factor models in the analysis of business and economics data.

The methodologies are illustrated using genuine datasets. At the end of each chapter, practical, ready-to-run R scripts reinforce understanding and hands-on applications. A companion R package `recode` is specifically designed to complement the book's content, featuring real-world and simulated datasets along with a variety of functions to implement and visualize the concepts and results. The book, together with its accompanying R package, helps to bridge the gap between theory and practice, providing the tools one needs to apply advanced and some state-of-the-art statistical methods to real-world scenarios.

Key Features:

- Promotes the regression idea as a unifying framework to model not just the means, but also covariance matrices, graphs and copulas using covariates.
- Highlights the implicit role of Cholesky factor in modeling various dependencies.
- Covers both undirected graphical models and directed graphs for modeling conditional independence structure.
- Bridges the gap between theory and methodology through data examples and exercises in each chapter.
- An R package `(recode)` containing datasets and implementation functions.

Mohsen Pourahmadi is Professor Emeritus of Statistics at Texas A&M University. His research interests are in time series, multivariate and longitudinal data analysis, dealing with dependence all the time.

Aramayis Dallakyan is a statistician and software developer. His research interests lie at the intersection of graphical models, high-dimensional time series, and statistical/machine learning. He earned his Ph.D. in Statistics from Texas A&M University.

CHAPMAN & HALL/CRC
Texts in Statistical Science Series

Recently Published Titles

Analysis of Categorical Data with R, Second Edition
Christopher R. Bilder, Thomas M. Loughin

Applied Nonparametric Statistical Methods, Fifth Edition
Nigel C. Smeeton, Neil H. Spencer, and Peter Sprent

Linear Models with R, Third Edition
Julian J. Faraway

Statistics in Survey Sampling
Jae Kwang Kim

Time Series: A Data Analysis Approach Using R
Robert H. Shumway and David S. Stoffer

Bayesian Statistical Methods: With Applications to Machine Learning
By Brian J. Reich, Sujit K. Ghosh

Experimental Design for Data Science and Engineering
By V. Roshan Joseph

Survival Analysis: Principles and Applications in Clinical Trials and Beyond
Song Yang

Foundations of Bayesian Statistics for Data Scientists: With R and Python
Alan Agresti, Maria Kateri, Ranjini Grove, and Antonietta Mira

A Course in Regression and Smoothing Methods
Zhiqiang Tan

A Course in Large-sample and High-dimensional Theory
Zhiqiang Tan

Markov Chain Monte Carlo
Stochastic Simulation for Bayesian Inference
Dani Gamerman, Hedibert Lopes, and Flávio Bambirra Gonçalves

Introduction to Modern Randomization-Based Design and Analysis for Causal Inference
Tirthankar Dasgupta and Donald B. Rubin

The Analysis of Time Series
An Introduction with R, Eighth Edition
Haipeng Xing and Chris Chatfield

Regressions in Covariances, Dependencies and Graphs
Mohsen Pourahmadi and Aramayis Dallakyan

For more information about this series, please visit: https://www.routledge.com/Chapman--Hall-CRC-Texts-in-Statistical-Science/book-series/CHTEXSTASCI

Regressions in Covariances, Dependencies and Graphs

Mohsen Pourahmadi and Aramayis Dallakyan

CRC Press
Taylor & Francis Group
Boca Raton London New York

CRC Press is an imprint of the
Taylor & Francis Group, an **informa** business
A CHAPMAN & HALL BOOK

Designed Cover Image: Mohsen Pourahmadi and Aramayis Dallakyan

First edition published 2026
by CRC Press
2385 NW Executive Center Drive, Suite 320, Boca Raton FL 33431

and by CRC Press
4 Park Square, Milton Park, Abingdon, Oxon, OX14 4RN

CRC Press is an imprint of Taylor & Francis Group, LLC

ISBN: 978-1-041-06695-8 (hbk)
ISBN: 978-1-041-07114-3 (pbk)
ISBN: 978-1-003-63889-6 (ebk)

DOI: 10.1201/9781003638896

Typeset in CMR12 font
by KnowledgeWorks Global Ltd.

Publisher's Note: This book has been prepared from camera-ready copy provided by the author.

Contents

Preface

This book is about modeling dependencies in multivariate data. Linear and nonlinear (cross-sectional, temporal, spatial) dependencies are prevalent and important in understanding the complex relationships among several variables. While covariances, their inverses and graphs have played significant roles in exploring linear dependencies for a long time, only lately copulas have assumed similar roles in modeling dependence in the presence of nonlinearity, extreme and long-tail behavior.

Our aim is to exploit the idea of regression as a unifying theme in modeling dependencies using covariates and regularization. This is done by mimicking the success of regressions in modeling the mean vector (first moments) and hoping to achieve *almost* the same for the covariance matrix (second moments) of a random vector and eventually its distribution. It is recognized that the road to modeling variances and covariances using covariates is bumpier and more perilous than that for the mean; positivity and large number of parameters are the two major obstacles. There are two ways that regression can help. The first or the *covariance regression* method writes a covariance matrix or its transform as linear combinations of covariates, just as in the mean regression, and strives to reach to generalized linear models (GLM) for covariances. The second, the *hidden regression* idea makes it possible to view columns of a matrix or its transform as parameters of certain regressions and estimate them iteratively one column at a time. This is at the heart of the regularized Gaussian likelihood or the graphical Lasso algorithm which cannot accommodate covariates.

Linear dependence measures like correlations and covariance matrices and their nonlinear counterparts like copulas play a central role in various stages of multivariate data analysis and machine learning. Many fundamental statistical methods like regression analysis, principal component analysis, canonical correlation analysis, linear and quadratic discriminant analysis, cluster analysis, and all their relatives in machine learning and data science require knowledge of the covariance matrix in one form or another. Estimating a covariance matrix and its inverse is a crucial problem in many applications such as functional magnetic resonance imaging, analysis of gene expression arrays, network analysis, risk management and portfolio allocation.

There are many excellent text-books on mean regression, a few on quantile regression and copulas, but virtually none on covariances and depedencies, though Pourahmadi [2013] might be a distant precursor with more focus on high-dimensionality and theory. In the last fifteen years, great progress has been made in developing applicable methodologies using covariates in modeling covariances, graphs and distributions. This book is an attempt to fill the gap as an intermediate-level textbook and a research monograph for students and users in statistics, computer science, business and economics, engineering and physical sciences. A solid background in regression analysis, and familiarity with basic multivariate techniques and matrix algebra at the level of Johnson and Wichern [2008] are expected. With its focus on methodologies, relevant concepts in regularized regressions, matrices and copulas for multivariate data are motivated and brought to life through various examples and real datasets. Each chapter includes practical, ready-to-run R scripts that reinforce understanding and encourage hands-on exploration. The companion R package *recode* has been specially developed for this book. It includes real and simulated datasets, a suite of functions for implementing the methods discussed. Whether you are a student, researcher, or data scientist, the combination of the book, its R scripts, and the *recode* package offers a valuable toolkit to bridge the gap between theory and practice, and to entice the reader to apply modern statistical techniques to real-world problems. The book offers a rich collection of exercises spanning theory, methodology, and hands-on R applications. Their purpose is not only to reinforce the material but also to highlight fresh perspectives and complementary ideas. Some are adapted from current research papers, providing readers with deeper insights that extend beyond the covered topics.

As a research monograph we note that while there are a good number of books covering aspects of estimating sparse covariances and graphs in high-dimensions with a focus on asymptotic guarantees under meticulously stated assumptions, our focus is on presenting certain useful methodologies and state-of-the-art results and to facilitate their implementation. Minimal attention is given to precise statements of the assumptions and proofs of the relevant asymptotic results, but a few illuminating and inspiring proofs of mostly classical theorems are included to encourage and prepare novice readers for the next level of rigor and diving deeper into the current literature while keeping in mind the following nuanced research questions: Why isn't there a bona fide GLM for covariance matrices? Why developing covariance GLM, vine copula distributions, and multivariate distributional regression invariably require an *order* among the variables?

The book has 12 chapters where selection and extent of coverage of topics reflect our interest, comfort and trust in their continued importance in the years to come. The first nine chapters cover background and now standard material in modern multivariate analysis and data sciences. For a one-semester course with a focus on cross-sectional dependence and graphs, we have used selected topics from these chapters. However, with a focus on temporal and spatial dependence selected topics from Chapters 1-9 and 10-12 seem appropriate. The first chapter provides a road map which should be helpful for teaching, self-study and research. The website at rcdg-book.cv provides downloadable R scripts for each chapter. Timely updates, additional resources and announcements will be added at the website. This being our first attempt at writing a text-book with a focus on *dependence modeling using covariates*, surely there is much room for improvement. We would appreciate greatly hearing from readers sharing their finding of errors, inconsistencies and shortcomings or suggesting constructive improvements.

Finally, we would like to thank our families, colleagues and students for their understanding, support and encouragement over many years of this project. Thanks are also due to David Grubbs and his team at Taylor & Francis for their help and timely advice throughout the process.

Mohsen Pourahmadi
Aramayis Dallakyan
College Station, Texas
May 1, 2026

List of Figures

List of Tables

1

Overview

This chapter presents an overview of two broad regression-based techniques in modeling dependencies and covariances. It gives a quick tour of exploiting such techniques in reparameterizing, modeling, and interpreting covariances and graphs of multivariate data. Unconstrained reparameterization of a covariance matrix or its transform is the key in using covariates and regularization. A road map at the end suggests a way to start modeling dependencies using popular methods and R packages by going directly to the relevant sections rather than reading sequentially the whole book.

1.1 Regressions and Dependencies

There are two distinct (direct and indirect) methods where regressions are playing fundamental roles in modeling dependencies (covariances, copulas and graphs). The direct method or *covariance regression* writes regression models for a covariance matrix or any measure of dependence using covariates, while the indirect method or *hidden regression* relies on conditional distributions. For multivariate normal distributions, it regresses each response on a subset of other responses, thus making it possible to view columns of its covariance matrix as regression parameters. These are then regularized and estimated iteratively one column at a time using the coordinate descent algorithm. The most prominent example of this is the popular *graphical Lasso algorithm* (GLasso) for sparse precision matrix or graph estimation, such methods usually cannot handle covariates. Suitable and clever combination of the two methods can bring us very close to the gates of the marvelous world of generalized linear models for dependencies, but not quite there.

1.2 Covariance Regressions

This class of methods, with a much longer history than mean regression, simply writes parametric models for covariance matrices, copulas and eventually the entire distributions in terms of covariates. In what follows, we often simply talk of writing models for a matrix instead of the longer phrases like covariance matrix, inverse covariance (precision) matrix, etc.

DOI: 10.1201/9781003638896-1

Consider a general linear regression model of the form

$$Y = \boldsymbol{X}\boldsymbol{\beta} + \boldsymbol{\epsilon}, \tag{1.1}$$

where $Y = (Y_1, \ldots, Y_p)'$ is the p-vector of responses, $\boldsymbol{X}$ is a known $p \times k$ full rank design matrix, and the error terms $\boldsymbol{\epsilon}$ is a zero-mean random vector with a general $p \times p$ covariance matrix $\boldsymbol{\Sigma} = (\sigma_{ij})$. The price of generality of $\boldsymbol{\Sigma}$ is the quadratic growth of its number of parameters, namely $p(p+1)/2$, and ensuring its positive-definiteness. These make the task of inference about the parameters $(\boldsymbol{\beta}, \boldsymbol{\Sigma})$ a real challenge. The enormous success and popularity of the linear regression models, for the mean $\boldsymbol{\mu} = E(Y)$, can be traced to the quest for *parsimony* by writing the mean vector as linear combinations of the columns of the design matrix $\boldsymbol{X} = (X_1, \ldots, X_k)$:

$$\boldsymbol{\mu} = \boldsymbol{X}\boldsymbol{\beta} = \beta_1 X_1 + \cdots + \beta_k X_k. \tag{1.2}$$

The key advantage of this formulation is in reducing the number of mean parameters from p to a much smaller number k. The background material on least-squares estimation and regularized regression using the Lasso penalty functions is covered in Chapters 2 and 5.

In what follows, it is instructive to view and christen (1.2) as a *vector-on-vector* regression. Then, in developing regression models for covariance matrices, it seems natural to start with a *matrix-on-matrix* regression model

$$\boldsymbol{\Sigma} = \alpha_1 U_1 + \cdots + \alpha_q U_q, \tag{1.3}$$

where the α_i's are the unknown (scalar) covariance parameters, and U_i's are known covariate matrices. This *linear covariance model* (LCM) (1.3) was introduced by Anderson [1973] with U_i's as simple structured matrices such as compound symmetry and banded matrices, (see Chapter 3), with little or no practical relevance to the experiment or the data-generating process.

Of course, as in standard regression models, meaningful interpretation of the parameters α_i's, better understanding and appreciation of the matrix covariates are two interrelated topics of fundamental importance in LCM. Fortunately, there have been some great attempts to deal with the covariate issues and going beyond the LCM by introducing more flexible *covariance regression* models. For example, similarity matrices constructed from a q-vector of covariates $\boldsymbol{x}$ are used in (1.3) [Zou et al., 2017]; Hoff and Niu [2011], Niu and Hoff [2019] have introduced quadratic regression models of the form

$$\Sigma(\boldsymbol{x}) = \mathbf{B}\boldsymbol{x}\boldsymbol{x}'\mathbf{B}',$$

where $\boldsymbol{x}$ is as before and $\mathbf{B}$ is a $p \times q$ matrix parameter. This quadratic model formulation and its implicit connection to latent variables have been further extended by Fox and Dunson [2015] to a nonparametric covariance regression framework within a latent factor model by expressing the columns of the loading matrix and the latent factors in terms of covariates with potentially far-reaching consequences.

In a slightly different direction, a more recent and promising breakthrough in Zhang and Li [2023] and Kim and Zhang [2024] was motivated by the strong need to use genuine covariates in certain genetic applications. Their key idea is to reverse the roles of (α, U) in LCM by writing a *matrix-on-scalar* regression model

$$\boldsymbol{\Sigma}(\boldsymbol{x}) = \mathbf{B}_0 + x_1\mathbf{B}_1 + \cdots + x_q\mathbf{B}_q, \tag{1.4}$$

where $\mathbf{B}_0$ is a positive-definite population-level covariance effect and $\mathbf{B}_i$'s are symmetric matrices representing the effects of the (univariate) covariates x_i's. Such models are more appealing intuitively and a bit easier to work with so far as interpretation of the matrix parameters is concerned. We note that models similar to (1.4) can be written for the inverse covariance or the precision matrix $\boldsymbol{\Sigma}^{-1} = (\sigma^{ij})$.

There is a relatively long and successful history of using log-linear models for variances and covariances to deal with the positivity constraints. However, for covariance matrices there are some drawbacks when it comes to statistical interpretations of the new parameters as discussed in Section 6.1.1.4. Nevertheless, generalized linear models (GLMs) for $\boldsymbol{\Sigma}$ seem to be the most promising direction for introducing covariates and addressing its positive-definiteness, which we present next.

1.3 GLM via Cholesky Decomposition

Thus far the emphasis has been on the role of regression in rendering parsimony in covariances with no regards for its positive-definiteness. Traditionally, in regression analysis, constraints on the mean are dealt with by considering generalized linear models (GLM). In this section we review the limited progress in GLM for covariances.

A promising approach to developing GLM for a covariance matrix is to use the components of its Cholesky decomposition. Interestingly, this direction is closely connected to the autoregressive (AR) models and deeply rooted in time series analysis, where the observations are naturally *ordered* or time-indexed. Variables in a general random vector do not necessarily follow any ordering and this could become a liability. For the time being any generic ordering of the variables can be used to explain the idea.

For a random vector Y with mean zero and covariance matrix $\boldsymbol{\Sigma}$, let $\widehat{Y}_t$ be the linear least-squares predictor of Y_t based on its predecessors $Y_{t-1}, \ldots, Y_1$, $\varepsilon_t = Y_t - \widehat{Y}_t$ be its prediction error with variance $\sigma_t^2 = \mathrm{Var}(\varepsilon_t)$. Then, there are unique scalars ϕ_{tj} $(\phi_{11} = 0)$ so that

$$Y_t = \sum_{j=1}^{t-1} \phi_{tj} Y_j + \varepsilon_t, \; t = 1, \ldots, p. \tag{1.5}$$

Writing this sequence of regressions in matrix form, one arrives at $TY = \boldsymbol{\epsilon}$, where T is a unit lower-triangular matrix with 1's on the diagonal and $-\phi_{tj}$, the negative of regression coefficients, as the (t, j) entry and $\boldsymbol{\epsilon} = (\varepsilon_1, \ldots, \varepsilon_p)'$ with $\mathrm{Cov}(\boldsymbol{\epsilon}) = \mathrm{diag}(\sigma_1^2, \ldots, \sigma_p^2) = D$. Computing the covariance of the linear transformation

$TY = \boldsymbol{\epsilon}$, one obtains the *modified Cholesky decomposition* of the covariance matrix and its inverse:

$$T\boldsymbol{\Sigma}T' = D, \quad \boldsymbol{\Sigma}^{-1} = T'D^{-1}T. \tag{1.6}$$

This can be seen as reparameterizing $\boldsymbol{\Sigma}$ in terms of $(T, \log D)$ with unconstrained entries. An important consequence of such reparameterization or the decompositions (1.6) is that for any estimate of the Cholesky factors (T, D), the estimated covariance matrix is guaranteed to be positive-definite; for more details, see Section 6.1.2.1.

The model (1.5) can be viewed as a regression of a variable on a *subset* of others, a form of hidden regression which is becoming very popular in the literature of spatial data analysis and known as *Veechia approximation* of Gaussian likelihoods.

1.4 Hidden Regressions

The indirect method is newer, less well-known, and somewhat hidden in algorithms scattered in the literature of the last two decades or so. Its mystical premise seems to be: *In the heart of every covariance, there is hidden a regression waiting to be discovered through unconstrained parameterizations.* We present six prominent examples of this appearing in the sparse estimation of columns of a covariance, its inverse, and the associated graphs.

The genesis of hidden regression in modeling dependence in a random vector $Y = (Y_1, \ldots, Y_p)'$ with a probability density f_Y can be traced to the conditional distributions arising from its factorization as

$$f_Y = f_{Y_{-j}} f_{Y_j|Y_{-j}}, \quad j = 1, \ldots, p, \tag{1.7}$$

where Y_{-j} is the original vector with the entry Y_j deleted, and the second factor is the conditional density of Y_j given the rest. For a multivariate normal distribution, this factorization boils down to regressing each response on the others. More precisely, for $Y \sim N_p(\mathbf{0}, \boldsymbol{\Sigma})$ with precision matrix $\boldsymbol{\Sigma}^{-1} = (\sigma^{ij})$, and for clarity we fix $j = p$ in (1.7). Then, the relevant conditional distribution leads to the following linear regression of Y_j on the remaining variables Y_{-p}:

$$Y_p = \sum_{i=1}^{p-1} \beta_i Y_i + \varepsilon_p, \tag{1.8}$$

where the error ε_p is independent of Y_{-p} with

$$\beta_i = -\frac{\sigma^{ip}}{\sigma^{pp}}, \quad \mathrm{Var}(\varepsilon_p) = \frac{1}{\sigma^{pp}}.$$

Evidently, the regression coefficients β_i's are a scale multiple of the entries of the last column of the precision matrix so that they share similar sparsity patterns; see (4.18) and Chapter 4 for more details.

Throughout the book, for $\boldsymbol{Y}_1, \ldots, \boldsymbol{Y}_n$ a sample of size n from Y in (1.1), we use $\boldsymbol{Y}$ to denote the corresponding $n \times p$ data matrix with $\boldsymbol{Y}_i$ as its ith row and $\boldsymbol{y}_j$ its jth column, and emphasize that the vector $\boldsymbol{y}_j$ contains the n sample values of the variable Y_j. Given the data matrix $\boldsymbol{Y}$, it is plausible to estimate the parameters in (1.8) using the least-squares method by minimizing the sum of squares of errors

$$||\boldsymbol{y}_p - \sum_{i=1}^{p-1} \beta_i \boldsymbol{y}_i||^2, \tag{1.9}$$

over $\beta_1, \ldots, \beta_{p-1}$. For high-dimensional data, or when p is large relative to n, one may resort to regularization by adding the Lasso penalty $\lambda \sum_{i=1}^{p-1} |\beta_i|$ to (1.9) where $\lambda > 0$ is a tuning parameter.

As the sparsity of the regressions in (1.8) and that of the precision matrix $\boldsymbol{\Sigma}^{-1}$ are closely linked, it is evident that imposing Lasso-like penalties on these regression problems translates into sparse estimation of the last column of a precision matrix and hence the whole matrix. This is the simplest way to see the central role of the regression idea in sparse estimation of precision matrices or Gaussian graphs.

In fact, aspects of this regression formulation with the Lasso penalty have been used effectively in the last two decades either directly as above or indirectly within the broader likelihood framework. We list here the following six important cases:

1. **Node-wise regression problem:** The representation (1.8) is the key in Meinshausen and Bühlmann [2006]'s nodewise regression or the neighborhood selection (NS) method for sparse estimation of precision matrices. It performs a sequence of Lasso regressions of each node on the others to estimate the columns of a precision matrix, where nonzero estimated regression coefficients indicate the existence of edges in the corresponding graph. The related Sparse PArtial Correlation Estimation (SPACE) algorithm in Peng et al. [2009] deals with certain shortcomings of the NS method by combining the p regressions into a single objective function.

2. **The Iterative Conditional Fitting (ICF):** The ICF algorithm in Chaudhuri et al. [2007] iteratively computes the MLE of a covariance matrix with a prespecified set of zeros, one column at a time using a regression formulation. A somewhat different regression formulation is obtained from a penalized normal likelihood in Hastie et al. [2009, Section 17.3] and solved via an iterative method similar to the GLasso algorithm presented next.

3. **Graphical Lasso (GLasso) Algorithm:** A slightly different reincarnation of the regression idea and Lasso regression appears in developing the fast and popular graphical Lasso (GLasso) algorithm for sparse estimation of precision matrices via penalized Gaussian likelihood [Friedman et al., 2008]. GLasso plays a central role in the literature of high-dimensional precision matrix estimation and graphical model selection and has become the benchmark against which the performances of the newer methods are compared.

Since the GLasso algorithm appears and used frequently in the intervening chapters, here it is worthwhile to take a slight detour and introduce its main ingredients. Recall that for a sample of size n from $Y \sim N_p(\mathbf{0}, \mathbf{\Sigma})$ with precision matrix $\mathbf{\Theta} = \mathbf{\Sigma}^{-1}$ and data matrix $\boldsymbol{Y}$, the sample covariance matrix is given by $\mathbf{S} = \frac{1}{n}\boldsymbol{Y}'\boldsymbol{Y}$. The penalized likelihood method for sparse estimation of the precision matrix solves the convex optimization problem:

$$\min_{\mathbf{\Theta}} -\log\det(\mathbf{\Theta}) + \mathrm{tr}(\mathbf{S}\mathbf{\Theta}) + \lambda\|\mathbf{\Theta}\|_1, \tag{1.10}$$

over the set of positive-definite matrices $\mathbf{\Theta}$, where λ is a tuning parameter and $\|\mathbf{\Theta}\|_1$ is the sum of absolute values of the entries of the matrix. Keeping (1.8) in mind, and to get a sense of its role in developing the iterative GLasso algorithm, we partition $\mathbf{\Theta}$ and $\mathbf{S}$ as

$$\mathbf{\Theta} = \begin{pmatrix} \mathbf{\Theta}_{11} & \boldsymbol{\theta}_{12} \\ \boldsymbol{\theta}_{21} & \theta_{22} \end{pmatrix}, \quad \mathbf{S} = \begin{pmatrix} \mathbf{S}_{11} & \boldsymbol{s}_{12} \\ \boldsymbol{s}_{21} & s_{22} \end{pmatrix}, \tag{1.11}$$

where $\mathbf{\Theta}_{11}$ and $\mathbf{S}_{11}$ are the $(p-1)\times(p-1)$ principal minors of their respective matrices, $\boldsymbol{\theta}_{12}$ and $\boldsymbol{s}_{12}$ are $(p-1)$-vectors and θ_{22}, s_{22} are scalars. Then, using the reparameterization

$$\boldsymbol{\beta} = \boldsymbol{\theta}_{12}, \ \gamma = \theta_{22} - \boldsymbol{\beta}'\mathbf{\Theta}_{11}^{-1}\boldsymbol{\beta}, \tag{1.12}$$

and some basic calculations involving the determinant and inverse of partitioned matrices (Chapter 9) the objective function in (1.10) magically *decouples* into the following sum of two separate convex functions of γ and $\boldsymbol{\beta}$,

$$\min_{\gamma,\boldsymbol{\beta}} \{-\log\gamma + (s_{22}+\lambda)\gamma\} + \{(s_{22}+\lambda)\boldsymbol{\beta}'\mathbf{\Theta}_{11}^{-1}\boldsymbol{\beta} + 2\mathbf{s}_{12}'\boldsymbol{\beta} + 2\lambda\|\boldsymbol{\beta}\|_1\}. \tag{1.13}$$

Consequently, the minimization over γ of the first expression has a neat closed-form solution, and that of the second over $\boldsymbol{\beta}$ is recognized easily as a Lasso regression problem; see Section 2.1.1.

A complete coverage of the GLasso algorithm and many of its extensions and improvements can be found in Chapter 9.

4. **Covariance GLasso (C-GLasso):** In a Gaussian covariance graph, absence of an edge stands for marginal independence rather than conditional independence between variables. The C-GLasso is concerned with sparse estimation of covariance matrices or solving the penalized likelihood

$$\min_{\mathbf{\Sigma}} \log|\mathbf{\Sigma}| + \mathrm{tr}(S\mathbf{\Sigma}^{-1}) + \lambda\|\mathbf{\Sigma}\|_1. \tag{1.14}$$

It is not convex in $\mathbf{\Sigma}$ and is a more challenging computational problem than the GLasso. Surprisingly, Wang [2014] was the first to develop analogs of (1.12) and (1.13) for solving the C-GLasso optimization problem with much improved computational and convergence properties than the earlier proposals in the literature; see Section 9.6.2.

5. **Sparse multivariate regression:** Estimation of the coefficient matrix and precision matrix via jointly penalized Gaussian likelihood in the multivariate regression framework is an important problem. The formulation in Wang [2015] is truly regression-based in spirit and relies on the conditional distribution of each response given the other responses and covariates. Its objective function turns out to be a pseudo-likelihood in the sense of Besag [1975], and its optimization boils down to solving a sequence of regularized regression problems, which can be solved efficiently using the existing software packages; see Section 5.4.2.

6. **Gaussian graphical regression:** Deal with high-dimensional data situations where both the mean and the precision matrix depend on covariates [Zhang and Li, 2023]. The simplest setting corresponds to the multivariate regression $Y \sim N_p(\mathbf{B}\boldsymbol{x}, \Sigma)$ with $\mathbf{B}_{p\times q} = (\mathbf{b}_1, \ldots, \mathbf{b}_q)'$ partitioned into its columns. Then, the parameters $(\mathbf{B}, \boldsymbol{\Sigma}^{-1})$ satisfy a centered and nonlinear version of (1.8):

$$Y_j = \mathbf{b}_j'\boldsymbol{x} + \sum_{k\neq j} \beta_{jk}(Y_k - \mathbf{b}_k'\boldsymbol{x}) + \varepsilon_j,\ j = 1, \ldots, p. \tag{1.15}$$

   The regression parameters in (1.15) can be estimated by imposing a joint Lasso-like penalty on $(\mathbf{B}, \boldsymbol{\Sigma}^{-1})$ where due to nonlinearity, one may alternate between estimating the mean and the precision matrix parameters. When the precision matrix also depends on covariates $\boldsymbol{x}$, one may model $\boldsymbol{\Sigma}^{-1}(\boldsymbol{x})$ as in (1.4). For details, see Section 5.7 and Zhang and Li [2023].

In view of the above cases, it is believable that there is room for hidden regressions to play more serious roles in covariance estimation. In fact, we point out that a bivariate version of (1.8), i.e., regression of any two variables on the rest, plays a central role in developing an *asymptotic normal thresholding* (ANT) method for inference and support recovery in Gaussian graphical models; see Section 9.4.1. A thresholded covariance estimator for a given $\lambda > 0$ and sample covariance $\mathbf{S} = (s_{ij})$ is given by

$$T_\lambda(\mathbf{S}) = \left(s_{ij}\mathbf{1}(|s_{ij}| \geq \lambda)\right),$$

which replaces by zero all small entries with absolute value less than the tuning parameters λ; see Section 8.2. This simple and very useful class of covariance estimators is notorious for not being positive-definite in finite samples, though it is guaranteed to be valid asymptotically under mild conditions. An ingenious idea tied to selecting λ so as to restore positive-definiteness of thresholded covariance estimates without disturbing sparsity in finite samples is presented in Section 8.3.1.

Most of the iterative algorithms mentioned above are computationally demanding and do not scale up well to high-dimensional data, and may not lead to positive-definite covariance estimates. There is a fast *non-iterative* algorithm which directly addresses many of these shortcomings of the iterative methods; see Section 9.8. Using a hidden Cholesky factor, it provides a more universal method for restoring positive-definiteness of a covariance estimate without disturbing its sparsity.

1.5 A Road Map

Unlike regression analysis, there is no standard textbook or manual providing step-by-step guides for covariance or dependence modeling, in spite of great progress in these areas in the last five decades or so. Fortunately, there are semblances of organizations and books for the more focused frameworks of GEE; see Fitzmaurice et al. [2011, Chapter 11] and references therein, mixed-effects models in SAS PROC Mixed [SAS Institute Inc., 2009] and [Pinheiro and Bates, 2000]. The class of LW-shrinkage estimators along with the R package corpcor are very popular in econometrics and biomedical sciences. These provide the simplest gateways to the vast world of modeling correlations and dependencies using parsimonious structured matrices.

The road map in Figure 1.1 is designed to help readers to select, and directly go to the topics of interest to them without studying sequentially the chapters. A novice may consult it when faced for the first time with the daunting task of modeling dependencies in multivariate data. It is developed in analogy with the thought process that an established data analyst may go through when analyzing data calling for regression analysis. While the illustration is for covariance regression, a copula/vine analog can be constructed similarly with minimal effort. We recall that a road map for regression analysis may start with the dichotomy of *parametric* (linear, nonlinear) and *nonparametric* models.

Multivariate data analysis usually starts with computing the sample covariance matrix $\mathbf{S}$, which involves estimating too many parameters even for $p = 10$. Its inverse, which is needed in most classical multivariate methods, does not exist when $p > n$. There are two broad options to resolve these issues. The parametric route for covariance matrices involves *selecting* a simple structured matrix with a few parameters (Sections 3.4 and Example 13). This path has matured to the very popular area of *generalized estimating equations* (GEE) in Section 6.3 with its growing menu of choices for *working correlation* matrices. The next natural step is models which are

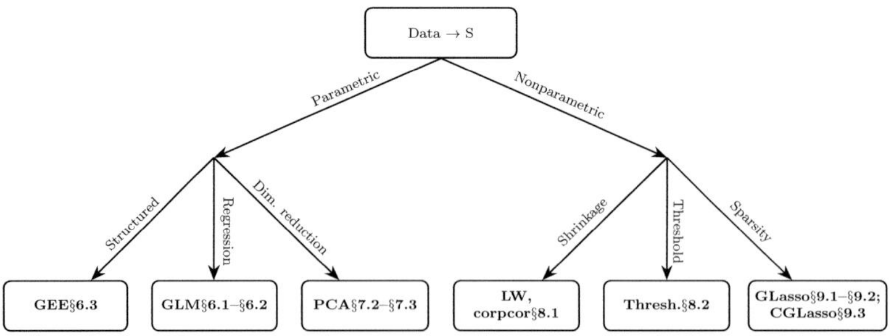

FIGURE 1.1
A road map.

linear combinations of such structured matrices (Section 6.1.1.1) and then move to the GLM (Sections 6.1.1.4 and 6.1.2.1). This requires choosing a suitable link function. Alternatively, one may employ the versatile latent factor models with vast potential for handling covariates (Sections 7.4.3 and 7.3).

The two options in the nonparametric route are: *shrinkage* and *regularization*. For shrinkage, one may appeal to ridge-like or the class of Ledoit-Wolf (LW) shrinkage estimators (Section 8.1.4), which shrink **S** toward a simpler target matrix. The *selection* of a shrinkage target is important in practice. The popular R package corpcor [Schafer and Strimmer, 2005] offers six different target covariance matrices along with analytical formulae for their shrinkage intensities (Section 8.1.5). Regularization or Lasso-type penalties (thresholding, banding, tapering) lead to sparse covariance matrix estimators (Sections 8.2, 8.2.2, 8.3, 8.3.2). The related and more principled penalized normal likelihood leads to the fast and popular graphical Lasso (GLasso) algorithm, which plays a central role in sparse covariance/precision matrix estimation (Chapter 9).

2

Regularized Regression and Thresholding

Regression models are indispensable in the development of statistics, machine learning and modern data science. In this chapter, for high-dimensional multiple linear regression we motivate the need for regularization and focus mostly on the Lasso regression. The central role of thresholding and the coordinate descent algorithm in solving variants of the Lasso regression problems in a computationally efficient manner is motivated through several simple models. The popularity of the Lasso penalty is mostly due to its convexity and the ability to deliver zero estimates of smaller coefficients. We present several (non-convex) alternatives designed to remedy the bias and other drawbacks of Lasso. The scaled Lasso regression has the unique feature of estimating simultaneously the regression coefficients and the error variance. We present a brief overview of the theory of convex optimization which is the backbone of regularized estimation. The R package recode and the hands-on sections toward the end of each chapter provide ample opportunities to get involved with data analysis.

2.1 Least-Squares, Ridge and Regularized Regressions

We start with a vector of observations $\boldsymbol{y} = (y_1, \ldots, y_n)'$, a given $n \times p$ design matrix $\boldsymbol{X}$ and the familiar multiple linear regression model

$$\boldsymbol{y} = \boldsymbol{X}\boldsymbol{\beta} + \boldsymbol{\epsilon}, \tag{2.1}$$

where $\boldsymbol{\beta} = (\beta_1, \ldots, \beta_p)'$ is the vector of regression coefficients, $\boldsymbol{\epsilon}$ is the vector of errors with mean zero and covariance matrix $\sigma^2\mathbf{I}_n$. For a concrete data example of $(\boldsymbol{y}, \boldsymbol{X})$, see Table 2.3.

The least-squares estimator (LSE) of the coefficients $\boldsymbol{\beta} = (\beta_1, \ldots, \beta_p)'$, and its ridge counterpart corresponding to a shrinkage parameter are introduced here, followed by the Lasso (Least Absolute Shrinkage and Selection Operator) regression [Tibshirani, 1996]. The latter encourages selection of a subset of the relevant covariates while simultaneously shrinking the coefficients of less important ones to zero. The Lasso method is powerful and has become the foundation of many complex machine learning algorithms and sparse learning methods [Hastie et al., 2015].

DOI: 10.1201/9781003638896-2

For the standard regression setup, the LSE of the parameter $\boldsymbol{\beta}$ is obtained by minimizing the sum of squares errors

$$Q_{ls}(\boldsymbol{\beta}) = ||\boldsymbol{y} - \boldsymbol{X}\boldsymbol{\beta}||_2^2. \tag{2.2}$$

From here on for a vector $\boldsymbol{y} \in \mathbb{R}^n$, its Euclidean norm is defined by $||\boldsymbol{y}||_2 = \sqrt{y_1^2 + \ldots + y_n^2}$ with the corresponding inner-product $(\boldsymbol{x}, \boldsymbol{y})$ for two vectors. The normal equations for the LSE of the regression coefficients and its explicit solution for $n > p$ are:

$$\boldsymbol{X}'(\boldsymbol{y} - \boldsymbol{X}\hat{\boldsymbol{\beta}}) = 0 \Longrightarrow \hat{\boldsymbol{\beta}} = (\boldsymbol{X}'\boldsymbol{X})^{-1}\boldsymbol{X}'\boldsymbol{y},$$

see Problem 2.2. The vectors of fitted values and residuals, given by $\widehat{\boldsymbol{y}} = \boldsymbol{X}\widehat{\boldsymbol{\beta}}$ and $\boldsymbol{r} = \boldsymbol{y} - \widehat{\boldsymbol{y}}$, are orthogonal to each other in $\mathbb{R}^n$ in the sense that the inner product $(\widehat{\boldsymbol{y}}, \boldsymbol{r}) = 0$. The error variance σ^2 estimated by

$$\widehat{\sigma}^2 = \frac{||\boldsymbol{y} - \boldsymbol{X}\widehat{\boldsymbol{\beta}}||_2^2}{n - p}, \tag{2.3}$$

is an unbiased estimator of its target.

Evidently, computing $\widehat{\boldsymbol{\beta}}$ requires inverting the $\boldsymbol{X}'\boldsymbol{X}$ matrix. However, it is instructive to note that when the columns of $\boldsymbol{X}$ are orthogonal, that is when $\boldsymbol{X}'\boldsymbol{X}$ is a diagonal matrix, then the task of computing $\hat{\boldsymbol{\beta}}$ magically decouples into p univariate (no-intercept) regressions. As a simple example, for a generic covariate $\boldsymbol{x} = (x_1, \ldots, x_n)'$, the task amounts to finding the least-squares estimator of the slope or minimizing $\sum_{i=1}^n (y_i - \beta x_i)^2$, which has the solution

$$\widehat{\beta} = \frac{\sum_{i=1}^n x_i y_i}{\sum_{i=1}^n x_i^2} = \frac{(\boldsymbol{x}, \boldsymbol{y})}{||\boldsymbol{x}||_2}, \tag{2.4}$$

involving the ratio of the inner-product and norm of the two vectors in $\mathbb{R}^n$. Fortunately, reduction to univariate regressions is possible, for any full-rank design matrix, by first successively orthogonalizing its columns and then using the simple formula (2.4); see Problem 2.9. The idea of reducing computations in multiple regression to univariate ones is also at the heart of the coordinate descent algorithm in solving Lasso regression problems; see Example 2.15. Interestingly, this computational advantage is carried over to covariance modeling by reducing it to a sequence of regression problems which is one of the main goals of this book.

At the other extreme, when the columns of the design matrix $\boldsymbol{X}$ are highly correlated, a phenomenon known as the *multicollinearity* in regression analysis, the estimated coefficient vector $\boldsymbol{\beta}$ may not be stable and suffers from high variance. Subset selection and ridge regression are two popular ideas for dealing with such a problem. The first has the aim of selecting a small number of the most relevant variables in the model by penalizing the number of included variables using model complexity penalty criteria like AIC and BIC. This amounts to counting the number of parameters or using the norm $||\beta||_0 = \sum_{i=1}^p I\{\beta_0 \neq 0\}$, with $I(\cdot)$ the indicator function of a set, which is a non-convex, discontinuous function of $\boldsymbol{\beta}$, and a challenge to work with.

However, the ridge regression [Hastie et al., 2015] alternatively aims to balance the variance and bias of the estimators by regularizing the size of the coefficients through their Euclidean norm which is convex and continuous. It is also a great gateway to appreciating the role of the tuning parameter in penalized regressions. For a given tuning parameter $\lambda > 0$, the corresponding ridge estimator is obtained by minimizing the objective function

$$Q_{ridge}(\boldsymbol{\beta}) = \|\boldsymbol{y} - \boldsymbol{X}\boldsymbol{\beta}\|_2^2 + \lambda\|\boldsymbol{\beta}\|_2^2. \tag{2.5}$$

Its estimating equations and the explicit solution are

$$-\boldsymbol{X}'\boldsymbol{y} + (\boldsymbol{X}'\boldsymbol{X})\boldsymbol{\beta} + \lambda \cdot \boldsymbol{\beta} = 0 \Longrightarrow \widehat{\boldsymbol{\beta}}_\lambda = (\boldsymbol{X}'\boldsymbol{X} + \lambda\mathbf{I}_p)^{-1}\boldsymbol{X}'\boldsymbol{y},$$

and in contrast to LSE the solution is unique even when $n < p$; see Problem 2.3.

To get a better feel for the nature of shrinkage and the bias induced by the ridge remedy note that when the columns of $\boldsymbol{X}$ are orthonormal or $\boldsymbol{X}'\boldsymbol{X} = \mathbf{I}_p$, then the ridge solution is simply a scalar multiple of the least squares estimate. More precisely, $\widehat{\boldsymbol{\beta}}_\lambda = \frac{1}{1+\lambda}\widehat{\boldsymbol{\beta}}$, so that it shrinks the least squares estimator as well as the predicted values toward zero or the origin (think of $\lambda \longrightarrow \infty$). That this holds for a general design matrix $\boldsymbol{X}$ can be seen using its singular value decomposition (SVD); see (3.22) and Problem 2.3. The idea that shrinking predicted values pays off in higher dimension was noted and has become very popular in statistics since the 1960's [James and Stein, 1961]. In the context of linear models, shrinking a predictor amounts to finding a scalar c so that the shrunk predictor $\tilde{\boldsymbol{y}} = c\widehat{\boldsymbol{y}}$ has smaller risk, i.e.

$$E\|\boldsymbol{y} - \tilde{\boldsymbol{y}}\|^2 < E\|\boldsymbol{y} - \widehat{\boldsymbol{y}}\|^2. \tag{2.6}$$

To implement the ridge idea, λ as a tuning parameter needs to be selected. There are a number of approaches available in the literature which require computing the solutions $\widehat{\boldsymbol{\beta}}_\lambda$ at different tuning values, the so-called ridge *regularization path*; see Section 2.4. Great computational efficiency in selecting the tuning parameter is achieved by relying on the SVD of the design matrix, in that once the SVD of $\boldsymbol{X}$ is computed, we have the ridge solution for the whole ridge regularization path.

For the fat dataset from our R package *recode* with $n = 252, p = 14$, we present the least-squares, ridge and Lasso estimates of the regression coefficients in Table 2.1. The Lasso and other regularized estimators are introduced in the next few sections.

2.1.1 The Lasso Regression Problem

In this section, we present the general ingredients for introducing the Lasso regression problem, followed by a list of simple and specific examples motivating the soft-thresholding function. These are designed to help readers familiar with the least-squares setup to develop an appreciation for the Lasso penalty and the tools needed to compute the Lasso solution.

In high-dimensional data where the number of parameters is greater than the number of observations or $p > n$, it is reasonable to expect that the models are *sparse*,

TABLE 2.1
fat dataset: Estimates using LS, Ridge, Lasso, thresholded Lasso, SCAD, and MCP.

	LSE	Ridge	Lasso	Thr. Lasso	SCAD	MCP
(Intercept)	-30.365	-15.511	-25.400	-5.168	-29.435	-29.254
age	0.046	0.096	0.047	0.000	0.048	0.048
weight	-0.119	0.005	-0.098	0.000	-0.117	-0.117
height	-0.062	-0.154	-0.066	0.000	-0.048	-0.049
adipos	-0.075	0.303	0.000	0.000	0.000	0.000
neck	-0.316	-0.111	-0.305	-0.523	-0.336	-0.335
chest	0.080	0.153	0.047	0.000	0.069	0.069
abdom	0.867	0.276	0.833	0.789	0.857	0.857
hip	-0.109	0.054	-0.086	0.000	-0.112	-0.114
thigh	0.205	0.108	0.154	-0.138	0.206	0.206
knee	0.025	-0.007	0.000	0.000	0.000	0.000
ankle	0.266	-0.091	0.192	-0.097	0.260	0.260
biceps	0.126	0.033	0.096	0.000	0.119	0.119
forearm	0.457	0.176	0.438	0.404	0.462	0.461
wrist	-1.492	-1.089	-1.452	-1.656	-1.484	-1.486
MSE	13.298	16.973	13.071	14.273	13.165	13.161

or that many of the regression coefficients are negligible. Lasso regression achieves the sparsity goal by using the ℓ_1-norm $\|\boldsymbol{\beta}\|_1 = \sum_{i=1}^{p} |\beta_i|$ as penalty, instead of the Euclidean norm $\|\boldsymbol{\beta}\|_2$ in the ridge regression. While the ridge regression objective function is quadratic, differentiable and convex in $\boldsymbol{\beta}$, the Lasso objective function is neither quadratic nor differentiable (at zero), but is convex with no closed-form solution. For a quick overview of convexity, its relevance in regularized regressions and non-differentiable optimization problems; see Section 2.6.

The Lasso regression problem is that of minimizing the penalized sum of squares of residuals:

$$Q(\boldsymbol{\beta}) = \frac{1}{2}\|\boldsymbol{y} - \boldsymbol{X}\boldsymbol{\beta}\|_2^2 + \lambda\|\boldsymbol{\beta}\|_1, \tag{2.7}$$

where $\lambda > 0$, is a tuning parameter controlling the sparsity of the model. It is evident that for $\lambda = 0$ it reduces to the least-squares estimation so that small penalty values keep the estimators near their LSE. Moreover, with obvious notation the objective function takes the form

$$\frac{1}{2}\boldsymbol{\beta}'\boldsymbol{X}'\boldsymbol{X}\boldsymbol{\beta} - (\boldsymbol{X}'\boldsymbol{y})\boldsymbol{\beta} + \lambda\|\boldsymbol{\beta}\|_1 = \frac{1}{2}\boldsymbol{\beta}'A\boldsymbol{\beta} - \boldsymbol{a}'\boldsymbol{\beta} + \lambda\|\boldsymbol{\beta}\|_1, \tag{2.8}$$

where $A = \boldsymbol{X}'\boldsymbol{X}$ is a positive-definite matrix for a full-rank $\boldsymbol{X}$ and $\boldsymbol{a} = \boldsymbol{X}'\boldsymbol{y}$. For ease of reference, its solution is denoted symbolically by Lasso$(A, \boldsymbol{a}, \lambda)$. By formally differentiating the objective function, one can show that the Lasso solution $\boldsymbol{\beta}$ of (2.7) satisfies the normal equations

$$(\boldsymbol{X}'\boldsymbol{X})\boldsymbol{\beta} - \boldsymbol{X}'\boldsymbol{y} + \lambda \cdot \boldsymbol{\gamma} = A\boldsymbol{\beta} - \boldsymbol{a} + \lambda \cdot \boldsymbol{\gamma} = 0, \tag{2.9}$$

where $\boldsymbol{\gamma} = \text{sign}(\boldsymbol{\beta})$ and the sign function acts componentwise with $\gamma_i = -1$ if $\beta_i < 0$, $\gamma_i = 1$ if $\beta_i > 0$, and $\gamma_i = 0$ otherwise. For a more rigorous derivation of the normal equations, the role of convexity, sub-differential $\boldsymbol{\gamma}$, and KKT conditions; see Lemma 1 in Section 2.6. However, in the interest of simplicity for the time being we rely on (2.9) as the starting point and through a few simple examples we explain and illustrate some remarkable properties of a general Lasso solution.

2.1.2 The Soft-Thresholding Operator

Fundamental to understanding the Lasso solution and computing it is the *soft-thresholding operator*, which turns out to be the minimizer of the simplest version of (2.7):

$$Q(\beta) = \frac{1}{2}(y - \beta)^2 + \lambda|\beta|, \qquad (2.10)$$

for given scalars y and λ.

Since $|\beta|$ is differentiable everywhere except at $\beta = 0$, the (formal) derivative of $Q(\cdot)$ with respect to β is

$$Q'(\beta) = -y + \beta + \lambda \cdot \text{sign}(\beta) = 0, \qquad (2.11)$$

for a more rigorous treatment, see Lemma 1 and Section 2.6. From (2.11) it follows that its explicit solution in terms of y, λ, is

$$\widehat{\beta}(\lambda) = \text{sign}(y)(|y| - \lambda)_+, \qquad (2.12)$$

where $(x)_+ = x$, if $x > 0$ and zero, otherwise, see Problem 2.11.

Surprisingly, this simple closed-form solution already reveals three fundamental characteristics of a general Lasso solution:

1. The Lasso solution is *piecewise linear* in the tuning parameter λ.
2. Increasing the penalty parameter λ will shrink the Lasso estimate $\widehat{\beta}(\lambda)$ toward 0. In fact, as soon as λ exceeds $|y|$, the Lasso estimate $\widehat{\beta}(\lambda)$ becomes zero and will remain so thereafter.
3. The bias, or the difference between the Lasso and the least-squares estimators, is about λ for larger regression coefficients.

Figure 2.1(b) displays the Lasso solution as a function of (y, λ). Note that while the level of sparsity of the estimator is indicated by the width of the horizontal flat line around zero, the dotted line represents the level of bias relative to the least squares estimator. The function (2.12), the building block for solving Lasso regression problems via coordinate descent algorithm, is called the *soft-thresholding operator*, and denoted by

$$S(y, \lambda) = \text{sign}(y)(|y| - \lambda)_+, \qquad (2.13)$$

where y and λ are real and nonnegative numbers, respectively.

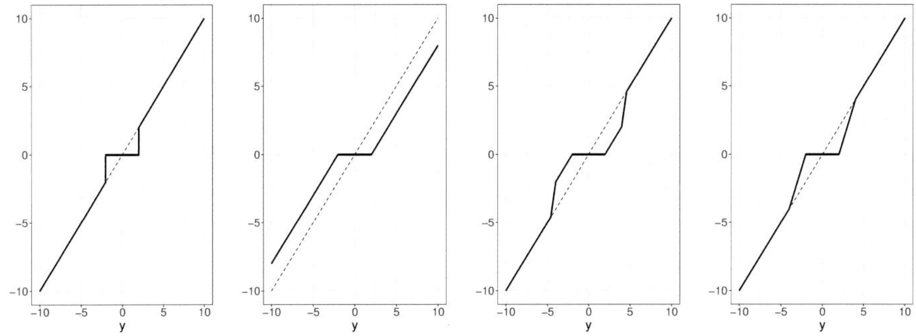

(a) Hard (b) Lasso (c) SCAD (d) MCP

FIGURE 2.1
LSE (dotted), regularized estimators (solid) with $\lambda = 2$ and $\gamma = 3.7$.

There is also the related *hard-thresholding function* at threshold λ defined by

$$H(y, \lambda) = yI(|y| > \lambda),$$

where $I(A)$ is the indicator function of the set A. Note that these two functions are both nonlinear in y, and have in common the *threshold region* $|y| \leq \lambda$ where the signal is estimated to be zero. Figure 2.1(a) illustrates $H(y, \lambda)$.

We note that the solutions of both ridge and Lasso are not invariant under scaling of the data matrix $(\boldsymbol{y}, \boldsymbol{X})$, so it is recommended and quite common to standardize and center the columns of the data matrix before estimation. The centering of the data will remove the intercept term from the model and makes the solution independent of the origin chosen for $\boldsymbol{y}$.

2.1.3 The Coordinate Descent Algorithm

Starting with the simple linear regression with no intercept, the next example further illustrates the central role of the soft-thresholding operator in finding closed-form solutions for some Lasso regression problems. It is also the key to developing a *coordinate descent algorithm* for iteratively solving a general Lasso regression problem.

Example 1 *Suppose* $\boldsymbol{y} = (y_1, \ldots, y_n)'$, $\boldsymbol{X} = (\boldsymbol{x}_1, \ldots, \boldsymbol{x}_n)'$ *and consider minimizing (2.7). For* $\lambda = 0$, *the task is that of finding the least-squares estimate of* β *which is given in (2.4). For simplicity taking* x_i*'s to be normalized so that* $\sum_{i=1}^n x_i^2 = 1$, *for* $\lambda > 0$ *then the Lasso estimate satisfies*

$$Q'(\beta) = -\sum_{1}^{n} x_i(y_i - \beta x_i) + \lambda \cdot sign(\beta) = -\sum_{i=1}^{n} x_i y_i + \beta + \lambda \cdot sign(\beta) = 0.$$

Comparing with (2.11), it follows that the Lasso estimate of the slope parameter β is

$$\widehat{\beta}(\lambda) = S\left(\sum_{i=1}^{n} x_i y_i, \lambda\right), \tag{2.14}$$

where the first argument is $\widehat{\beta}$, the least-squares estimate in a simple linear regression with no intercept.

The setup of Example 1 and its closed-form solution play important roles in solving the sparse multiple regression problem.

The Lasso normal equations (2.9) appears to be similar to that for the ridge regression, however, it does not have a closed-form solution except when the columns of $\boldsymbol{X}$ are orthogonal.

Fortunately, for a general design matrix relying on the *coordinate descent algorithm* the problem can be reduced to a sequence of one-dimensional Lasso problems. The roots of coordinate descent algorithm can be traced back to the Gauss-Seidel method [Seidel, 1874] for solving linear systems. For a comprehensive analysis of its convergence properties, particularly for nondifferentiable objective functions, see Tseng [2001].

The *coordinate descent algorithm* minimizes the objective function (2.7) one β_j at a time while the others are kept fixed. It then cycles through all the parameters until convergence. More precisely, suppose all the values of β_k for $k \neq j$ are held fixed at their current value $\tilde{\beta}_k$, so that $Q(\tilde{\beta})$ can be written as

$$Q(\tilde{\boldsymbol{\beta}}) = \frac{1}{2}\sum_{i=1}^{n}\left(y_i - \sum_{k\neq j} x_{ik}\tilde{\beta}_k - x_{ij}\beta_j\right)^2 + \lambda \sum_{k\neq j} |\tilde{\beta}_k| + \lambda|\beta_j|, \tag{2.15}$$

and viewed as a function of β_j alone. Minimizing $Q(\tilde{\boldsymbol{\beta}})$ with respect to β_j amounts to solving a univariate Lasso problem as in Example 1 with no intercept where the responses are $r_i^{(j)} = y_i - \sum_{k\neq j} x_{ik}\tilde{\beta}_k$. Hence, from (2.14) one obtains the explicit solution

$$\tilde{\beta}_j \leftarrow S\left(\sum_{i=1}^{n} x_{ij} r_i^{(j)}, \lambda\right).$$

Then, the algorithm cycles through all $\beta_j, j = 1, 2, \ldots, p, 1, 2, \ldots, p, \ldots$ until $\tilde{\boldsymbol{\beta}}$ converges.

The more general idea of *block coordinate descent* algorithm where effectively β_j's are replaced by vectors $\boldsymbol{\beta}_j$'s is used widely in various high-dimensional regularized estimation problems. For example, in a multivariate regression setup it might be desirable to shrink to zero certain row/column of the coefficient matrix. Other prominent applications include the graphical Lasso algorithm for sparse estimation of covariance and precision matrices, banding and tapering where one may prefer to shrink to zero certain subdiagonals far away from the main diagonal.

2.1.4 Bias, Consistency and Correlation

Some noteworthy observations about Lasso's bias, consistency and computation and the role of correlations are pointed out in this section.

Although the Lasso penalty is designed to deliver zeros, it has been noticed that in some situations Lasso estimates could be very small but not quite zero. This happens, for example, depending on the correlation structure of the covariates, where even increasing the penalty does not necessarily zero out the noise variables with small coefficients. For example, Weinstein et al. [2022, Figure 1] suggests that false positives of Lasso cannot be eliminated by simply increasing the penalty. For further intuition as why Lasso does not select some true variables for large λ; see Section 2.5.4 and Su et al. [2017, Section 3.7]. The intuitively appealing idea of thresholding initial Lasso estimators [van de Geer et al., 2011] has been shown theoretically to outperform Lasso on variable selection. In addition, Lounici [2008], Meinshausen and Yu [2009] have proved the sign consistency of thresholded Lasso estimator with a proper choice of the second threshold parameter.

Next, selection consistency of Lasso is briefly discussed along with some of the needed conceptual ingredients. Let $S = \{j; \beta_j \neq 0\}$ be the set of nonzero coefficients and $s = |S|$ be its cardinality. The following two conditions on the columns of the design matrix $\boldsymbol{X}$ and magnitude of the non-zero entries in $\boldsymbol{\beta}_S$ are needed for establishing consistency of the nonzero Lasso estimates:

1. The *Irrepresentability Condition*: For some $0 < \theta < 1$,

$$\max_{j \in S^c} |\text{sign}(\boldsymbol{\beta}_S)' (\boldsymbol{X}_S' \boldsymbol{X}_S)^{-1} \boldsymbol{X}_S' X_{.j}| \leq \theta, \tag{2.16}$$

   where $\boldsymbol{X}_S$ is the submatrix of $\boldsymbol{X}$ with the column index in S.

2. The *beta-min condition*,

$$\inf_{j \in S} |\beta_j| > \sqrt{s \log(p)/n}. \tag{2.17}$$

Note that the first condition imposes a great deal of restriction on the size of correlations among the columns of the design matrix, and the second condition demands that the nonzero coefficients be spiked, or stay far away from zero. The condition (2.16) is not easy to check in practice. Nevertheless, when there is evidence that it is not satisfied one could develop a factor model, as in Section 7.3.4, to account for the correlations in the columns of the design matrix.

Define $\widehat{S} = \{j; \widehat{\beta}_j \neq 0\}$ as the set of coefficients estimated as nonzero. Then, under the above two conditions, $\widehat{S}$ is *selection consistent* for S, in the sense that $P(\widehat{S} = S) \to 1$ when n converges to infinity [van de Geer and Bühlmann, 2009]. Note that selection consistency imposes the restriction

$$n > s \log(p), \tag{2.18}$$

on n, p and s.

Regarding computational speed and improvements, from Problem 2.13 it is evident that, for a fixed penalty λ, the Lasso regression is more likely to force to zero the coefficients of those predictors which are less correlated with the response. Furthermore, the Lasso solutions are piecewise linear in λ due to piecewise linearity of the soft-thresholding operator as noted earlier. These observations have motivated a number of insightful and innovative approaches to computing the Lasso solution and improving its performance. The *least angle regression* (LAR) algorithm [Efron et al., 2004], which relies on the correlations among the response and covariates, is closely related to the forward stepwise regression, and the *sure independence screening* (SIS) in Fan and Lv [2008].

Throughout this book, to make the coverage of various methodologies more accessible and relevant we apply them to real datasets as often as possible. The Lasso and other estimates of the regression coefficients for the fat dataset are reported in Table 2.1.

2.2 Improved Lasso Penalty: Reducing the Bias

The Lasso penalty is popular because of its convexity and the ability to deliver some zero estimates. However, it is known to produce biased estimates of the regression coefficients due to the linear increase of the penalty function [Fan and Li, 2001], which shrinks unnecessarily even the largest regression coefficient. Figure 2.9 clearly illustrates the advantage of several (non-convex) alternatives designed to remedy the bias of Lasso. If convinced by the transparent message of this plot you may skip most of this section.

The optimization problem for most of the alternative penalties is that of minimizing:

$$Q(\boldsymbol{\beta}) = \frac{1}{2}Q_{ls}(\boldsymbol{\beta}) + \sum_{j=1}^{p} p_{\lambda_j}(|\beta_j|), \tag{2.19}$$

where $p_{\lambda_j}(\cdot)$ is a sparsity-inducing penalty function, each with its own tuning parameter with Q_{ls} as in (2.2).

A way to correct Lasso's overestimation or bias is to assign different penalties to different coefficients so that the larger coefficients receive smaller weights or penalties. For example, the adaptive Lasso [Zou, 2006] replaces the ℓ_1-penalty in Lasso by a weighted version. More specifically, its objective function is

$$Q(\boldsymbol{\beta}) = \frac{1}{2}Q_{ls}(\boldsymbol{\beta}) + \lambda \sum_{j} \frac{|\beta_j|}{|\widehat{\beta}_j|}, \tag{2.20}$$

where $\widehat{\beta}_j$ is the least-squares estimate of the jth regression coefficient. Its solution, denoted by the vector $\widehat{\boldsymbol{\beta}}_a(\lambda)$, with the subscript a standing for adaptive, has two desirable properties: First when the LS estimator is small or $\widehat{\beta}_j = 0$, then the adaptive

Lasso estimator is also small or $\widehat{\beta}_{a,j}(\lambda) = 0$; second when $\widehat{\beta}_j$ is relatively large, then little shrinkage is done so that the estimator has less bias. Fortunately, for orthogonal design matrices the adaptive Lasso has an explicit solution; see Problem 2.14.

Next, we provide examples of penalty functions with the following three desirable properties [Fan and Li, 2001] so that the resulting estimator enjoys:

1. **Sparsity:** It acts like a thresholding rule, setting small estimated coefficients to zero to accomplish variable selection.
2. **Unbiasedness:** It has low bias, particularly for larger parameter values.
3. **Continuity:** It is continuous in the data to reduce instability in the model prediction.

Example 2 (Smoothly Clipped Absolute Deviation (SCAD)) *It is convenient to start with following generic penalized least-squares objective function:*

$$Q(\beta) = \frac{1}{2}(y - \beta)^2 + p_\lambda(|\beta|), \tag{2.21}$$

with (formal) derivative

$$Q'(\beta) = sign(\beta)\{|\beta| + p'_\lambda(|\beta|)\} - y. \tag{2.22}$$

It turns out that the ensuing penalized least– squares estimator will have the desirable sparsity, unbiasedness and continuity properties [Fan and Li, 2001], respectively, if

$$\begin{aligned} \min_\beta\{|\beta| + p'_\lambda(|\beta|)\} &> 0; \\ p'_\lambda(|\beta|) = 0 \quad &for\ large \quad |\beta|; \\ \operatorname*{argmin}_\beta \left\{|\beta| + p'_\lambda(|\beta|)\right\} &= 0. \end{aligned} \tag{2.23}$$

In a sense, the three properties force the penalty function to be non-differentiable at the origin, and concave over $(0, \infty)$. *To eliminate the bias of penalized estimators one may choose the penalty function to be a constant beyond a second threshold level* $\gamma\lambda$. *As an example, one may consider the following SCAD penalty which is a continuous and differentiable:*

$$p_{\lambda,\gamma}(t) = \lambda \int_0^t \min\{1, \frac{(\gamma\lambda - |y|)_+}{\gamma - 1}\}dy, \tag{2.24}$$

where $\gamma > 2$ *is a constant. Evidently, it corresponds to a quadratic spline function with knots at* λ *and* $\gamma\lambda$, *the value of* $\gamma = 3.7$ *is suggested for practical use in Fan and Li [2001]; see Problem 2.15. Figure 2.1(c) displays the SCAD estimator for the simple regression. Its derivative suggests that SCAD has the same bias as Lasso for small coefficients, but smoothly relaxes this bias as the coefficient increases in magnitude.*

TABLE 2.2
Penalty functions, their derivatives and corresponding estimators. Note the SCAD and MCP penalties are constant after γ, λ.

	$p_{\lambda\gamma}(\lvert\beta\rvert)$	$p'_{\lambda\gamma}(\lvert\beta\rvert)$	Estimator
Hard	$\begin{cases} \frac{\lambda^2-(\lvert\beta\rvert-\lambda)^2}{2} & \text{if } \lvert\beta\rvert < \lambda \\ \frac{\lambda^2}{2} & \text{if } \lvert\beta\rvert \geq \lambda \end{cases}$	$\begin{cases} -\text{sign}(\beta)(\lvert\beta\rvert - \lambda) & \text{if } \lvert\beta\rvert < \lambda, \\ 0 & \text{if } \lvert\beta\rvert \geq \lambda \end{cases}$	$yI(\lvert y\rvert > \lambda)$
Lasso	$\lambda\lvert\beta\rvert$	$\lambda\text{sign}(\beta)$	$\text{sign}(y)(\lvert y\rvert - \lambda)_+$
SCAD	$\begin{cases} \lambda\lvert\beta\rvert & \text{if } \lvert\beta\rvert \leq \lambda, \\ \frac{2\gamma\lambda\lvert\beta\rvert-\beta^2-\lambda^2}{2(\gamma-1)} & \text{if } \lambda < \lvert\beta\rvert < \gamma\lambda \\ \frac{\lambda^2(\gamma+1)}{2} & \text{if } \lvert\beta\rvert \geq \gamma\lambda \end{cases}$	$\begin{cases} \lambda & \text{if } \lvert\beta\rvert \leq \lambda, \\ \frac{\gamma\lambda-\lvert\beta\rvert}{\gamma-1} & \text{if } \lambda < \lvert\beta\rvert < \gamma\lambda, \\ 0 & \text{if } \lvert\beta\rvert \geq \gamma\lambda \end{cases}$	$\begin{cases} \text{sign}(y)(\lvert y\rvert - \lambda)_+ & \text{if } \lvert y\rvert \leq 2\lambda \\ \frac{(\gamma-1)y-\text{sign}(y)\gamma\lambda}{\gamma-2} & \text{if } 2\lambda < \lvert y\rvert \leq \gamma\lambda \\ y & \text{if } \lvert y\rvert > \gamma\lambda \end{cases}$
MCP	$\begin{cases} \lambda\lvert\beta\rvert - \frac{\beta^2}{2\gamma} & \text{if } \lvert\beta\rvert \leq \gamma\lambda, \\ \frac{\gamma\lambda^2}{2} & \text{if } \lvert\beta\rvert > \gamma\lambda \end{cases}$	$\begin{cases} (\lambda - \frac{\lvert\beta\rvert}{\gamma})\text{sign}(\beta) & \text{if } \lvert\beta\rvert \leq \gamma\lambda, \\ 0 & \text{if } \lvert\beta\rvert > \gamma\lambda \end{cases}$	$\begin{cases} \frac{\gamma}{\gamma-1}\text{sign}(y)(\lvert y\rvert - \lambda)_+ & \text{if } \lvert y\rvert \leq \gamma\lambda \\ \lvert y\rvert & \text{if } \lvert y\rvert > \gamma\lambda \end{cases}$

Finally, the general objective function with the SCAD penalty is

$$Q(\boldsymbol{\beta}) = \frac{1}{2}Q_{ls}(\boldsymbol{\beta}) + \sum_{j=1}^{p} p_{\lambda,\gamma}(\beta_j), \tag{2.25}$$

where $p_{\lambda,\gamma}(x)$ is a penalty function with $p_{\lambda,\gamma}(0) = 0$, non-differentiable at zero; see Fan and Li [2001] for more details on the properties of the SCAD penalty function and the related penalized least-squares procedure.

Table 2.2 provides the penalty functions, their derivatives, and form of estimators for most of the estimators considered in this chapter. Figure 2.2 depicts the penalty functions and their derivatives.

Example 3 (Minimax Concave Penalty (MCP)) *Consider the generic penalized least-squares objective function:*

$$Q(\beta) = \frac{1}{2}(y - \beta)^2 + p_{\lambda,\gamma}(\beta), \tag{2.26}$$

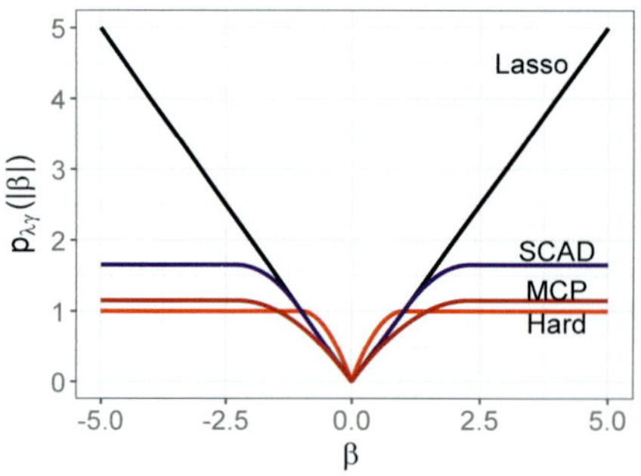

(a) Penalty

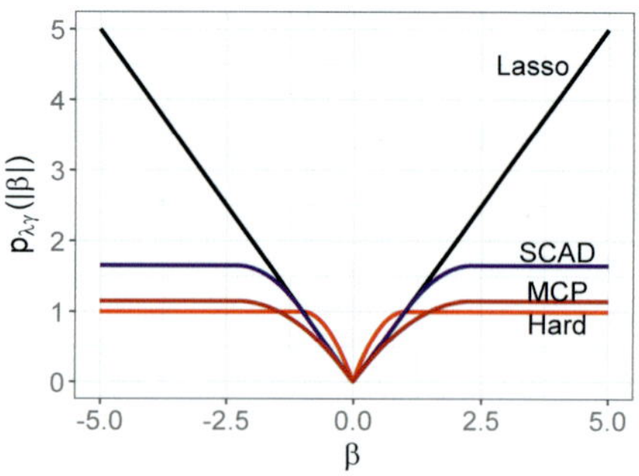

(b) Derivative

FIGURE 2.2
Plot of four different penalty functions $p_{\lambda,\gamma}(|\beta|)$ and their derivatives $p'_{\lambda,\gamma}(|\beta|)$.

where

$$p_{\lambda,\gamma}(\beta) = \lambda \int_0^{|\beta|} \left(1 - \frac{x}{\lambda\gamma}\right)_+ dx,$$

is indexed by the non-convexity parameter $\gamma \in (0, \infty)$. *Note that when* $\gamma = \infty$ *it reduces to the Lasso penalty and* $\gamma = 1_+$ *corresponds to the best subset selection; see Zhang [2010]. In spite of its non-convexity, the minimizer of (2.26) for* $\gamma > 1$ *has the unique closed-form solution:*

$$S_{\lambda,\gamma}(y) = \begin{cases} 0 & \text{if} \quad |y| < \lambda, \\ sign(y)(\frac{|y|-\lambda}{1-\gamma^{-1}}) & \text{if} \quad \lambda < |y| \leq \lambda\gamma, \\ y & \text{if} \quad |y| > \lambda\gamma. \end{cases}$$

Figure 2.1(d) displays the MCP estimator for the generic regression setup. It acts like Lasso and SCAD for the smaller coefficients, but starts its constancy or relaxing the bias much earlier than the SCAD penalty function. Figure 2.3 provides plot of the penalties and their corresponding estimators for SCAD and MCP, and Figure 2.4 illustrates the sample paths of MCP for three different values of γ.

Next, we introduce the fused-Lasso penalty function which unlike the Lasso penalty is not a separable function of components of the regression parameter $\boldsymbol{\beta}$. As the motivation, imagine a regression problem where the vector of regression coefficients or the signal has a natural (time-) order, composed of groups of nearly equal values. It is of interest to formulate the estimation of such a signal or a step function as a regularized regression problem to enforce piece-wise constancy. For example, consider the dataset of daily COVID-19 confirmed cases in Wahltinez et al. [2020], displayed in Figure 2.5. It is evident that the data contains noise, but when the true denoised numbers are expected to be piecewise-constant over contiguous regions, then a new type of penalty function is required to enforce a piecewise-constant structure.

Example 4 (Fused Lasso) *The fused-lasso formulation for a linear regression with* $\boldsymbol{X} = \mathbf{I}_n$ *is that of solving the optimization problem*

$$\min_{\beta} \frac{1}{2} \sum_{i=1}^{n} (y_i - \beta_i)^2 + \lambda \sum_{i=2}^{n} |\beta_i - \beta_{i-1}|. \tag{2.27}$$

Evidently, the fused-lasso penalty encourages neighboring β_i *values to be similar, resulting in a piecewise-constant structure. In Figure 2.5, the fused lasso solution is depicted in red.*

The optimization problem (2.27) is convex and has a well-defined solution. However, the coordinate descent-type algorithm cannot be utilized for solving it as in Lasso regression, since the former penalty is not separable in the parameters and the coordinate descent algorithm can get stuck at a non-optimal point. Algorithms designed to solve such optimization problems include the path algorithm [Hoefling, 2010] and the dual path algorithm [Tibshirani and Taylor, 2012]. To the best of our knowledge,

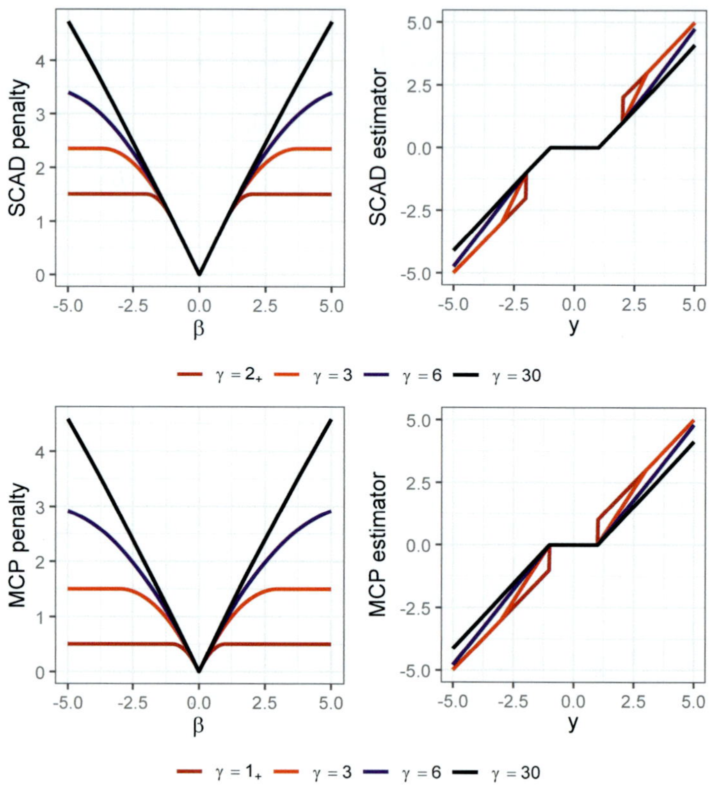

FIGURE 2.3
SCAD and MCP penalties and estimators for different values of γ

the fastest algorithm in Johnson [2013] exploits dynamic programming and requires only $O(n)$ operations.

Note that (2.27) does not impose sparsity on $\boldsymbol{\beta}$ itself. If $\boldsymbol{\beta}$ happens to be sparse, an additional $\lambda_1||\boldsymbol{\beta}||_1$ term can be added to it. Fortunately, it turns out that the solution of such *sparse fused lasso* can be obtained without much additional computational expense by soft thresholding the solution of a particular fused-Lasso [Friedman et al., 2007]. More precisely, for each $i = 1, \ldots, p$, set

$$\hat{\beta}_i(\lambda_1, \lambda_2) = S(\hat{\beta}_i(0, \lambda_2), \lambda_1),$$

where $\hat{\beta}_i(\lambda_1, \lambda_2)$ denotes the solution of the sparse fused lasso with the tuning parameter λ_1 for the ℓ_1 and λ_2 for fused-lasso penalties. In other words, to obtain a solution of the sparse fused-lasso, we estimate $\hat{\beta}$ from (2.27) with $\lambda_1 = 0$ and use soft thresholding for any $\lambda_1 > 0$.

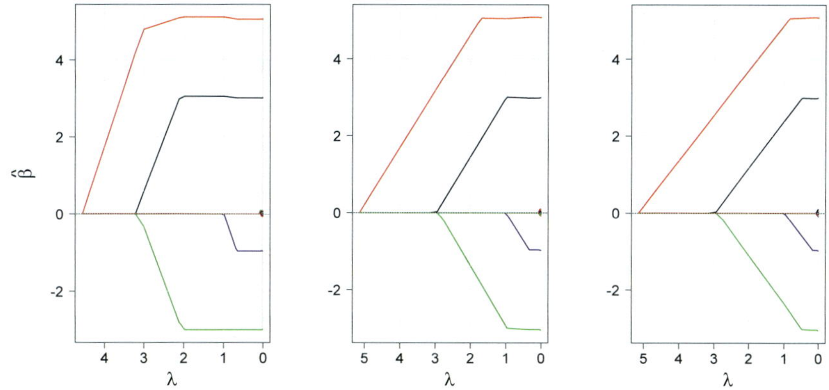

(a) $\gamma = 1_+$ (b) $\gamma = 3$ (c) $\gamma = 6$

FIGURE 2.4
Solution path for MCP for different values of γ.

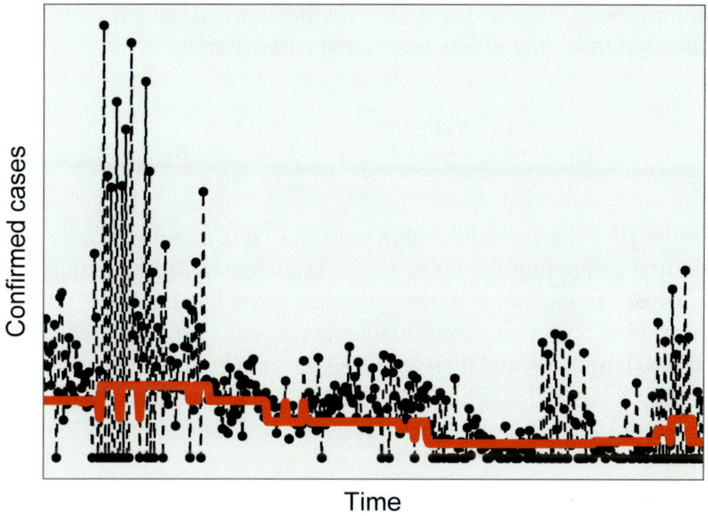

FIGURE 2.5
Fused lasso estimate for daily COVID-19 data.

Another interesting direction to generalize the fused-lasso is to require more smoothness by using a new family of penalties involving the higher-order difference matrix $D^{(k+1)}$:

$$\min_{\beta} \frac{1}{2} \sum_{i=1}^{n} (y_i - \beta_i)^2 + \lambda \| D^{(k+1)} \boldsymbol{\beta} \|_1. \tag{2.28}$$

Note that $k = 0$ leads to a piecewise-constant solution discussed earlier, but using a higher-order difference matrix $D^{(k+1)}$ leads to smoother solutions. In Kim et al. [2009] the interior point algorithm is used to solve the problem for $k = 1$, known as ℓ_1-trend filtering.

2.3 Estimation of the Error Variance

In this section we present various estimators of the error variance σ^2 using the Lasso residuals. Recall that the error variance is a measure of the noise level or the degree to which the response cannot be predicted using the available covariates. Reliable estimation of σ^2 is critically important for quantifying the uncertainty in estimating $\boldsymbol{\beta}$ and constructing confidence intervals for the regression parameters.

While there has been a great deal of work on Lasso estimates of the regression coefficients, estimating the error variance has not received as much attention. We start with the idea of *scaled Lasso* [Sun and Zhang, 2012] which has the unique and fruitful feature of jointly estimating the regression coefficients and the error variance σ^2. The objective function involving all the regression parameters is

$$Q(\boldsymbol{\beta}) = \frac{||\boldsymbol{y} - \boldsymbol{X\beta}||_2^2}{2n\sigma} + \frac{\sigma}{2} + \lambda \sum_{j=1}^{p} \frac{||\boldsymbol{X}_j||}{\sqrt{n}} |\beta_j|, \tag{2.29}$$

where $\boldsymbol{X}_j$ is the jth column of the design matrix $\boldsymbol{X}$. It is minimized over $(\boldsymbol{\beta}, \sigma)$ using iteration, by first estimating the noise level via the mean residual squares, and then scaling the penalty proportional to the estimated noise level.
Note that using this objective function amounts to standardizing each column of the design matrix to length $\sqrt{n}$ and then applying the penalty to the new weighted coefficients [Sun and Zhang, 2012]. They show that the scaled Lasso estimator is equivalent to the square-root Lasso in Belloni et al. [2011]:

$$\tilde{\boldsymbol{\beta}}_{\text{SQRT}} = \underset{\boldsymbol{\beta} \in \mathbb{R}^p}{\operatorname{argmin}} \left(\frac{1}{\sqrt{n}} ||\boldsymbol{y} - \boldsymbol{X\beta}||_2 + 2\lambda ||\boldsymbol{\beta}||_1 \right), \quad \tilde{\sigma}^2_{\text{SQRT}} = \frac{1}{n} ||\boldsymbol{y} - \boldsymbol{X}\tilde{\boldsymbol{\beta}}_{\text{SQRT}}||_2^2. \tag{2.30}$$

It is instructive to note the appearance of $\frac{1}{\sqrt{n}}$ and the disappearance of power 2 in the norm of the errors. Interestingly, the method is also closely related to the penalized likelihood estimator of the error variance in Yu and Bien [2019] presented later in this section.

First, we consider a more direct error variance estimator based on the objective function

$$Q(\boldsymbol{\beta}) = \frac{1}{n} ||\boldsymbol{y} - \boldsymbol{X\beta}||_2^2 + 2\lambda ||\boldsymbol{\beta}||_1, \tag{2.31}$$

which is slightly different from (2.7) in the scalar multipliers. Denote its minimizer and the ensuing residuals by the vectors $\widehat{\boldsymbol{\beta}}_\lambda$ and $\boldsymbol{y} - \boldsymbol{X}\widehat{\boldsymbol{\beta}}_\lambda$, respectively. The first naive error variance estimator is

$$\widehat{\sigma}_1^2 = \frac{1}{n}||\boldsymbol{y} - \boldsymbol{X}\widehat{\boldsymbol{\beta}}_\lambda||_2^2, \tag{2.32}$$

which may underestimate σ^2. Note that for $\lambda = 0$ or close to it, one is in the least-squares setup where it is a well-known and a common practice to correct its bias by replacing n by $n-p$ in the denominator for $n > p$; see (2.3). Such simple and classical bias-correction motivates the alternative and seemingly plausible estimator

$$\widehat{\sigma}_2^2 = \frac{1}{n - \hat{s}_\lambda}||\boldsymbol{y} - \boldsymbol{X}\widehat{\boldsymbol{\beta}}_\lambda||_2^2, \tag{2.33}$$

where $\hat{s}_\lambda$ is the number of nonzero entries in $\widehat{\boldsymbol{\beta}}_\lambda$ for a λ selected using a cross-validation method, see the next section. To study the performance and goodness of such estimators of σ^2 one needs to bring in many ingredients that play fundamental roles in assessing the theoretical properties of Lasso solution so far as variable selection consistency and prediction performance are concerned. Let $S = \{j; \beta_j \neq 0\}$ be the true support set of $\boldsymbol{\beta}$ and $|S|$ be its cardinality or the true number of nonzero regression coefficients. The random variable $\widehat{s}_\lambda$ is an estimator of $|S|$ based on a Lasso solution $\widehat{\boldsymbol{\beta}}_\lambda$. The conditions required for its *selection consistency*, or for the Lasso to recover correctly the support set S with high probability, are much stricter than the conditions needed for it to do well in prediction using the predictor $\boldsymbol{X}\widehat{\boldsymbol{\beta}}_\lambda$; see Section 2.1.4 and van de Geer and Bühlmann [2009]. These depend very much on the selection of λ, and the validity of a procedure employed to declare an entry of $\widehat{\boldsymbol{\beta}}_\lambda$ as a true zero. In short, the random variable $\hat{s}_\lambda$ or the scalar multiplier in (2.33) controls the quality of the estimator, see Reid et al. [2016] for a comprehensive study of the theoretical and computational properties of this error variance estimator.

The penalized normal log-likelihood estimator of σ^2 in Yu and Bien [2019] relies on the following reparameterization of $(\boldsymbol{\beta}, \sigma^2)$:

$$\boldsymbol{\theta} = \frac{\boldsymbol{\beta}}{\sigma^2}, \quad \phi = \frac{1}{\sigma^2}. \tag{2.34}$$

We recall that these are the natural parameters of the normal distribution as a member of the exponential family of distributions [McCullagh and Nelder, 1989]. The negative normal log-likelihood function for the regression model, up to a constant, is

$$L(\boldsymbol{\beta}, \sigma^2|\boldsymbol{X}, \boldsymbol{y}) = \frac{n}{2}\log\sigma^2 + \frac{||\boldsymbol{y} - \boldsymbol{X}\boldsymbol{\beta}||_2^2}{2\sigma^2}, \tag{2.35}$$

where, evidently, for known σ^2, its minimization leads to the familiar least-squares estimators. However, for unknown σ^2, its full minimization is a nonconvex optimization problem, even for a fixed $\boldsymbol{\beta}$; see Problem 2.17. This lack of convexity in σ^2 has more serious computational consequences for estimating the covariance matrix $\boldsymbol{\Sigma}$ of multivariate normal data discussed in Chapter 9.

Fortunately, the proposed reparameterization resolves this major obstacle. In fact, substituting the new parameters in (2.35) and expanding the sum of squares, one arrives at

$$L(\phi^{-1}\boldsymbol{\theta}, \phi^{-1}|\boldsymbol{X}, \boldsymbol{y}) = -\frac{n}{2}\log\phi + \phi\frac{||\boldsymbol{y}||_2^2}{2n} - \frac{1}{n}\boldsymbol{y}'\boldsymbol{X}\boldsymbol{\theta} + \frac{||\boldsymbol{X}\boldsymbol{\theta}||_2^2}{2n\phi}. \tag{2.36}$$

The proposed penalized likelihood estimators are

$$(\widehat{\boldsymbol{\theta}}_\lambda, \widehat{\phi}_\lambda) \in \underset{\phi>0,\boldsymbol{\theta}}{\text{argmin}}\left\{\frac{1}{n}L(\phi^{-1}\boldsymbol{\theta}, \phi^{-1}) + \lambda p(\boldsymbol{\theta}, \phi)\right\}, \tag{2.37}$$

for $p(\boldsymbol{\theta}, \phi)$ a chosen convex penalty function to induce sparsity in θ. Two notable examples of such penalty functions in Yu and Bien [2019] are:

1. $p(\boldsymbol{\theta}, \phi) = ||\theta||_1$,
2. $p(\boldsymbol{\theta}, \phi) = \phi^{-1}||\theta||_1^2$,

and the corresponding estimators are called the *natural* and *organic* Lasso estimators, respectively.

Once a minimizer of (2.37) is found, then solving (2.34) for the original parameters one arrives at the estimators

$$\tilde{\boldsymbol{\beta}}_\lambda = \frac{\widehat{\boldsymbol{\theta}}_\lambda}{\widehat{\phi}_\lambda}, \quad \tilde{\sigma}_\lambda^2 = \frac{1}{\widehat{\phi}_\lambda}. \tag{2.38}$$

How do these estimators relate to the Lasso solution $\widehat{\boldsymbol{\beta}}_\lambda$? Of course, the answer depends on the choice of the penalty function. Fortunately, a clear connection emerges for the *natural* Lasso corresponding to the solution of (2.37) with the penalty $p(\boldsymbol{\theta}, \phi) = ||\boldsymbol{\theta}||_1$.

The following remarkable result in Yu and Bien [2019] shows the minimum value of the Lasso objective function furnishes a hard to beat estimator of the error variance with many desirable properties.

Theorem 1 *The natural lasso estimator* $(\tilde{\boldsymbol{\beta}}_\lambda, \tilde{\sigma}_\lambda^2)$ *in (2.38) with* $(\widehat{\theta}_\lambda, \widehat{\phi}_\lambda)$ *the solution to (2.37) with the penalty* $p(\theta, \phi) = ||\theta||_1$ *are*

$$\tilde{\boldsymbol{\beta}}_\lambda = \widehat{\boldsymbol{\beta}}_\lambda = \underset{\boldsymbol{\beta}\in\mathbb{R}^p}{\text{argmin}}\left(\frac{1}{n}||\boldsymbol{y} - \boldsymbol{X}\boldsymbol{\beta}||_2^2 + 2\lambda||\boldsymbol{\beta}||_1\right), \tag{2.39}$$

and

$$\tilde{\sigma}_\lambda^2 = \widehat{\sigma}_\lambda^2 = \min_{\boldsymbol{\beta}\in\mathbb{R}^p}\left(\frac{1}{n}||\boldsymbol{y} - \boldsymbol{X}\boldsymbol{\beta}||_2^2 + 2\lambda||\boldsymbol{\beta}||_1\right), \tag{2.40}$$

is the minimum value of the Lasso objective function.

From the formula for $\tilde{\sigma}_\lambda^2$, the additive nature of correction for the downward bias of $\widehat{\sigma}_1^2$ is rather clear. This is quite different from the bias-correction for $\widehat{\sigma}_2^2$ which is multiplicative in nature.

Next, we consider the organic Lasso which corresponds to the solution of (2.37) with the penalty function $p(\boldsymbol{\theta}, \phi) = \phi^{-1}||\boldsymbol{\theta}||_1^2$ which is jointly convex in $(\boldsymbol{\theta}, \phi)$. At first, it might appear that solving the organic Lasso problem would require a new algorithm, but Yu and Bien [2019] showed that this is not the case. Indeed, analogous to the natural Lasso, its solution given by

$$\check{\boldsymbol{\beta}}_\lambda = \underset{\boldsymbol{\beta}\in\mathbb{R}^p}{\operatorname{argmin}} \left(\frac{1}{n}||\boldsymbol{y}-\boldsymbol{X}\boldsymbol{\beta}||_2^2+2\lambda||\boldsymbol{\beta}||_1^2\right), \quad \check{\sigma}_\lambda^2 = \min_{\boldsymbol{\beta}\in\mathbb{R}^p} \left(\frac{1}{n}||\boldsymbol{y}-\boldsymbol{X}\boldsymbol{\beta}||_2^2+2\lambda||\boldsymbol{\beta}||_1^2\right), \tag{2.41}$$

can be found without actually solving (2.37). Indeed, computing the organic Lasso estimator amounts to simply solving a penalized least-squares problem where the penalty involves the **square** of the ℓ_1-norm. The connection between the organic Lasso and the square-root Lasso become evident by comparing (2.30) and (2.41) where one squares both the loss and the ℓ_1-penalty.

The Multivariate Case: The responses in linear regression models are usually assumed to be independent with diagonal covariance matrices. However, in application areas like time series, longitudinal studies, spatial and environmental data the observations are correlated and one has to deal with the computational challenges arising from general covariance matrices where the number of parameters grows quadratically with the number of observations.

For multivariate normal data, in analogy with the likelihood approach for estimation of error variance reviewed above, it is common to use the penalized normal log-likelihood

$$\ell(\boldsymbol{\Sigma}) = \log|\boldsymbol{\Sigma}| + \operatorname{tr}(S\boldsymbol{\Sigma}^{-1}) + \lambda||\boldsymbol{\Sigma}||_1, \tag{2.42}$$

where λ is a tuning parameter and $||\boldsymbol{\Sigma}||_1$ is the ℓ_1-norm of the matrix. This objective function is not convex in $\boldsymbol{\Sigma}$, so that it is a challenging computational problem. An iterative block coordinate descent algorithm for computing the solution is given in Wang [2014]; for more details, see Chapter 9.

Interestingly, the idea of scaled Lasso plays a key conceptual role in establishing the asymptotic distribution of the regularized estimators of the entries of precision matrices presented in Chapter 9. Developing the setup and proving analogs of Theorem 1 for the multivariate regression and its error covariance matrix are extremely useful and seem to be open problems.

2.4 Selection of Tuning Parameter: Cross-validation

For Lasso and other regularization methods selection of the tuning parameter λ is of critical importance. It determines the sparsity and performance of the models. A larger value of λ imposes a heavier penalty and produces a sparser model. In the standard linear regression models, one may impose a penalty on the number of selected variables using the AIC or BIC or other data driven methods. For larger p, the idea of subset selection is not computationally feasible, so that continuous penalized methods are preferred.

In this section, our focus is on cross-validation (CV) for tuning parameter selection for penalized regression to improve prediction performance and recovering the true non-zero variables or support recovery. These techniques are not limited to regression problems and slight modifications of them will be applied to regularized covariance and precision matrix estimation in Chapter 9 and other regularized estimation problems. For a recent overview of tuning parameter selection techniques for high-dimensional regression, see Wu and Wang [2020].

K-fold cross-validation is a popular method for selecting the tuning parameters. Its main objective is to find a λ that results in the most accurate model for predicting new independent observations (test dataset) from the same distribution as the original or training dataset. The general idea is, to randomly split the data D into K roughly equal-sized subsets $D_1, \ldots, D_K$ where usually $K = 5$ or 10. For a specific value of λ and fixed k, the remaining $K - 1$ folds $D_1, \ldots, D_{k-1}, D_{k+1}, \ldots, D_K$ are used as the training set to obtain the estimator $\boldsymbol{\theta}(\hat{\lambda}_k)$. Subsequently, the kth fold is used as the validation (test) data, and the corresponding prediction error is computed. For the Lasso, the prediction error on the kth fold is given by $L(D_k, \hat{\boldsymbol{\beta}}(\lambda), \lambda) = \frac{1}{n_{D_k}} \|\boldsymbol{y}_{D_k} - \boldsymbol{X}_{D_k}\hat{\boldsymbol{\beta}}\|_2^2$. This process is then repeated for each of the K folds, using each one as a validation dataset exactly once. The cross-validation error $\mathrm{CV}(\lambda)$ is defined as the average prediction error over all K validation datasets. The optimal λ is the one that achieves the minimum CV value or so-called one-standard-error rule, which selects the smallest value of λ that yields a CV error no more than one standard error above $\mathrm{CV}(\hat{\lambda})$. Typically, the one-standard-error rule yields a more parsimonious model [Hastie et al., 2009, Chapter 7.10] . For Lasso,

$$\mathrm{CV}(\lambda) = \frac{1}{K} \sum_{k=1}^{K} L(D_k, \hat{\boldsymbol{\beta}}(\lambda), \lambda).$$

Figure 2.6 displays the CV curve for the Lasso using the data in Example 5, where the true coefficient vector is $\boldsymbol{\beta} \in \mathbb{R}^{100}$ with only four non-zero elements.

We note that here $\lambda_{\min}$ corresponds to the value that achieves a minimum CV, and $\lambda_{1\mathrm{se}}$ correspond to the one-standard-error rule, respectively. The numbers at the top represent the model size, i.e., the number of estimated non-zero values for the corresponding λ values. In this example, $\lambda_{\min} = 0.04$ and $\lambda_{1\mathrm{se}} = 0.085$, result in model sizes of 22 and 6, respectively.

While cross-validation aims to minimize generalization error, often the goal is to select λ to accurately recover the true set of non-zero elements (variable selection), i.e. the support set $\mathcal{S}_0 = \{j, \beta_j \neq 0, j = 1, \ldots, p\}$. Recent studies on BIC-type selectors show that under certain conditions, they can recover the underlying true model with probability approaching one when combined with nonconvex penalized regression, such as SCAD and MCP. Recall that the classical BIC is defined as:

$$\mathrm{BIC}(\lambda) = -2 \log L(D, \hat{\boldsymbol{\theta}}, \lambda) + d \log n, \tag{2.43}$$

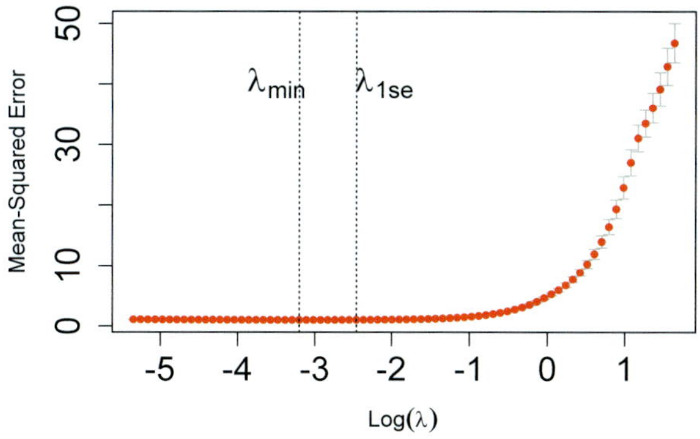

FIGURE 2.6
Cross-validation curve for Lasso regression.

where $\log L(\cdot)$ the log-likelihood function and d is the degrees of freedom. For Lasso, the degrees of freedom $d = |\hat{S}|$, is equal to the size of the set of estimated non-zero elements for a given λ [Tibshirani and Taylor, 2012]. Then, similar to cross-validation, for a given grid of $\lambda_1, \ldots, \lambda_m$ values, one selects λ which achieves a minimum BIC. Another related and commonly used criterion is the extended BIC (EBIC), which is variable selection consistent under some regularity conditions on the dimension p. It has the form

$$\text{EBIC}(\lambda) = -2\log L(\hat{\boldsymbol{\theta}}) + d\log n + 2d\gamma\log p, \tag{2.44}$$

where $\gamma \in [0, 1]$; see Chen and Chen [2008] for more details.

2.5 Hands-on R: Regularized Regressions

In this section, we illustrate and compare the performance of some of the regularized regression methods, first using a simulation example, and then the real dataset fat from the companion recode package provided with this book. The package contains all datasets and custom-written functions used throughout the book. For details, on how to download and install the package, see Appendix A.

Example 5 (Comparing Lasso, SCAD and MCP) *To visualize the difference between Lasso, SCAD, and MCP, we compare solution paths for penalized regressions when using these penalties. Data is generated from a linear regression model with* $p = 100, n = 1000$*, with only the first four elements of the coefficient vector being* $\{5, 3, -3, -1\}$ *non-zero, and the rest is set to zero. A solution path is the plot of the*

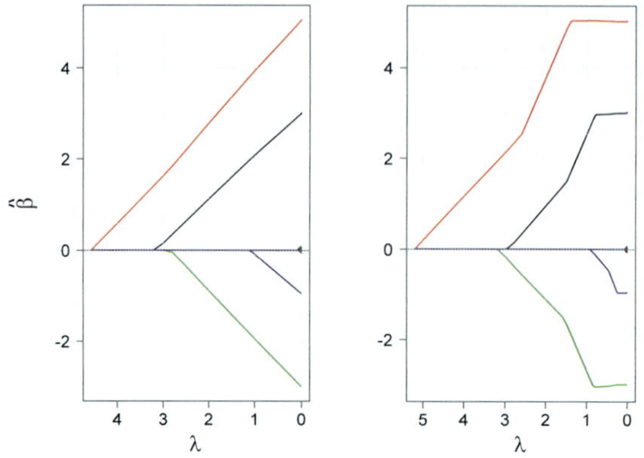

(a) Lasso (b) SCAD (c) MCP

FIGURE 2.7
Solution paths for Lasso, SCAD, and MCP.

estimated coefficient values vs the penalty parameter λ. For example, note that in Figure 2.7 when $\lambda = 5$, all estimated coefficients are zero, and as λ decreases some of the coefficients become active, i.e. non-zero. Moreover, Figure 2.7 reveals that while Lasso introduces bias early in the solution path, that is estimated non-zero coefficients are biased even when λ is small, SCAD and MCP allow the estimated non-zero coefficients to reach larger values for smaller values of λ. In other words, SCAD and MCP are great in bias reduction as expected since they impose less shrinkage on the non-zero coefficients.

The reason for the reversed order of λ values in Figure 2.7 is that some packages implement the pathwise Lasso by starting with its highest value or when all coefficients are 0, see for example Weinstein et al. [2022, Figure 1] .

2.5.1 The Data Example: Ridge, Lasso, SCAD and MCP

We illustrate and compare the performance of Lasso and other regularized estimation methods using the fat dataset [Penrose et al., 1985]. The goal of the study was to predict the percentage of body fat in males, encoded as brozek, using a dataset of $n = 252$ observations and $p = 14$ covariates or predictor variables. These variables include age, weight, height, adiposity index, and 10 body circumference measurements. Table 2.3 reports a small portion of the fat dataset.

It is reasonable to expect that not all of the 14 covariates will affect the percentage of body fat and some of them will have a 0 or very small coefficients. For Lasso and ridge, we use glmnet package and SCAD and MCP are estimated using ncvreg.

We start by importing the required R packages and the data. Since glmnet works only with numerical and quantitative variables, we use model.matrix() function to

TABLE 2.3
Example of fat dataset.

brozek	age	weight	height	adipos	neck	chest	abdom	hip	thigh	knee	ankle	biceps	forearm	wrist
12.60	23	154.25	67.75	23.70	36.20	93.10	85.20	94.50	59.00	37.30	21.90	32.00	27.40	17.10
6.90	22	173.25	72.25	23.40	38.50	93.60	83.00	98.70	58.70	37.30	23.40	30.50	28.90	18.20
24.60	22	154.00	66.25	24.70	34.00	95.80	87.90	99.20	59.60	38.90	24.00	28.80	25.20	16.60
10.90	26	184.75	72.25	24.90	37.40	101.80	86.40	101.20	60.10	37.30	22.80	32.40	29.40	18.20
27.80	24	184.25	71.25	25.60	34.40	97.30	100.00	101.90	63.20	42.20	24.00	32.20	27.70	17.70
20.60	24	210.25	74.75	26.50	39.00	104.50	94.40	107.80	66.00	42.00	25.60	35.70	30.60	18.80

transform qualitative variables into dummy variables. As a good programming practice, before each code block, we provide a concise comment which characterizes the block.

```
## Import required packages
require(recode)
require(glmnet)

data(fat)

## specify the target and feature matrix
y = fat["brozek"]
X <- model.matrix(brozek ~ age + weight +
                    height + adipos +
                    neck + chest +
                    abdom + hip + thigh +
                    knee + ankle +
                    biceps + forearm +
                    wrist, data=fat)[,-1]
```

We split the dataset into training and testing before running any method to avoid the so-called data leakage. The latter may result in an underestimation of the generalization (prediction) error. For each method, the estimated mean squared error (MSE) on the test dataset is used to evaluate prediction performance. The next few lines of code implement data splitting. For reproducibility purposes, it is advisable to set a seed.

```
## Split to train and test datasets
set.seed(123)
sample <- sample(c(TRUE, FALSE), nrow(fat), replace=TRUE, prob=c
    (0.8,0.2))

y_train = y[sample,1]
X_train = X[sample,]

y_test = y[!sample,1]
X_test = X[!sample,]
```

The Least-Squares

We start by implementing the LSE using the lm() function.

```
## Run linear regression
ls <- lm(brozek~age + weight +
           height + adipos +
           neck + chest +
```

```
              abdom + hip + thigh +
              knee + ankle +
              biceps + forearm +
              wrist , data=fat[sample,])
coef(ls)
```

To summarize the model, one can use the summary(ls) function. Here, we used coef() function to print the coefficients. The first column in Table 2.1 reports the results. Next, we estimate the MSE for the test dataset by specifying newdata = fat[!sample,].

```
## Predict and calculate MSE
yhat.ls <-  predict(ls, newdata = fat[!sample,])
mean((yhat.ls - y_test)^2)
[1] 13.29801
```

The Ridge

We use cv.glmnet() function to fit a Ridge regression with 10-fold cross-validation to select the parameter λ. In glmnet(), we specify the option alpha = 0, the default value of 1 for alpha corresponds to the Lasso regression.

```
## Run ridge with 10-fold cross-validation
set.seed(123)
ridge.cv = cv.glmnet(X_train,y_train, alpha = 0)
coef(ridge.cv)
```

The second column in Table 2.1 lists the estimated coefficients. As expected, no coefficient is exactly 0 and they are biased compared to the least- squares estimates. The next few line of codes estimate the MSE for Ridge, which is larger than the MSE for least squares.

```
## Predict and calculate MSE
yhat.ridge <-  predict(ridge.cv, X_test)
mean((yhat.ridge - y_test)^2)
[1] 16.97304
```

The Lasso

We use glmnet() function to fit a Lasso regression and plot the solution path for different tuning parameters λ.

```
## Run lasso
lasso.default <- glmnet(X_train,y_train, standardize = TRUE)
## Plot the solution path
plot(lasso.default)
```

From the solution path in Figure 2.8(a), it is evident that depending on the λ values some coefficients will be exactly 0. To estimate the MSE for a certain value of tuning parameter, for example, $\lambda = 0.5$, we use predict() function.

```
## Predict for specific lambda = 0.5 and calculate MSE
yhat.default = predict(lasso.default , newx = X_test ,
                       s = c(0.50))
mean((yhat.default - y_test)^2)
[1] 16.09849
```

Next, we use cross-validation for the tuning parameter selection and calculate its MSE. By default cv.glmnet() implements a 10-fold cross-validation. The number of folds can be changed by specifying the nfolds option.

```
## Implement 10-fold cross-validation
set.seed(123)
lasso.cv = cv.glmnet(X_train ,y_train)
## Plot the result
plot(lasso.cv)
lambda.cv = lasso.cv$lambda.min
print(lambda.cv)
[1] 0.02259489
```

The plot(lasso.cv) function displays the CV curve in Figure 2.8(b), where the best λ value is around 0.02. We use this value to estimate the MSE.

```
## Predict and calculate MSE
yhat.cv <-  predict(lasso.cv, newx = X_test , s = lambda.cv)
mean((yhat_cv - y_test)^2)
[1] 13.0713
```

Evidently, cross-validation significantly improved the model performance. It is also informative to look at the support of the estimated coefficients.

```
## print coefficients
coef(lasso.cv, s = lambda.cv)
```

The third column in Table 2.1 shows the results. Cross-validated Lasso resulted in two zero coefficients corresponding to knee and adipos.

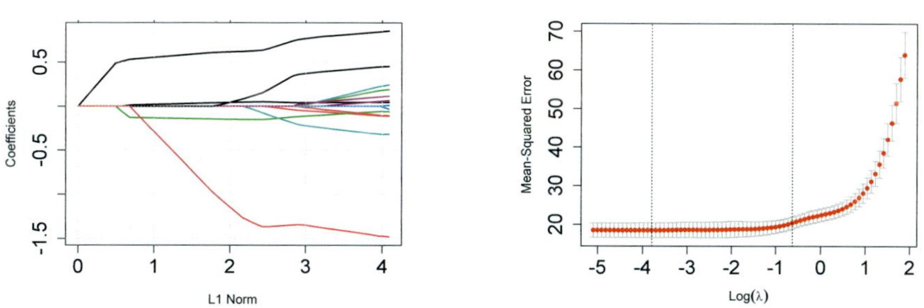

(a) Lasso solution path (b) CV curve for Lasso

FIGURE 2.8
The solution path and CV curve for the Lasso example.

Thresholded Lasso

Next, we illustrate the role of thresholding the Lasso solution in increasing the number of zero estimates. We use cv.threshlasso() function in recode package to implement threshold selection for Lasso using 10-fold cross-validation. Here, the threshold parameter is denoted by τ. Following Zhou [2010] and van de Geer et al. [2011], after thresholding, coefficients for the non-zero variables are estimated using ordinary least squares. It is recommended to choose the sparsity parameter $\lambda \propto \sqrt{\log p/n}$ small, and implement the selection of the threshold τ over a specified grid values [Wang and Allen, 2022]. Here, we use the number of grids equal to 30 from the interval $[\tau_{\min}, \tau_{\max}]$ and $\lambda = 0.01$, where $\tau_{\min} = 0.01$ and $\tau_{\max}$ is chosen such that all coefficients are equal to zero.

```
## Implement threshold-lasso with 10 fold cross-validation
thresh.lasso.cv = cv.threshlasso(X_train, y_train,
                         lambda = 0.01, min.thresh = 0.01, ngrid = 30,
                         nfold = 10, seed = 123)

thresh.lasso =  thresh.lasso.cv$thresh_min
print(thresh.lasso)
[1] 0.1172588
```

The selected τ is approximately 0.117, and the mean squared error (MSE) is 14.27. It is instructive to note that the thresholded Lasso exhibits 8 zero coefficients, whereas the Lasso, SCAD, and MCP methods have only 2 zero coefficients. However, as noted in Bühlmann and van de Geer [2011], the sparser solution comes at the expense of a deteriorated prediction mean squared error.

```
## Estimate MSE
yhat.thresh = predict.threshlasso(thresh.lasso.cv, X_test)
mean((yhat.thresh - y_test)^2)
[1] 14.27333
```

Non-convex penalties

We use ncvreg package, in particular cv.ncvreg() function, for implementing SCAD and MCP using 10-fold cross-validation. The default values for γ in SCAD and MCP are 3.7 and 3, respectively.

```
## Import packages
require(ncvreg)

## Implement SCAD with 10-fold cross-validation
scad.cv = cv.ncvreg(X_train, y_train,
                    penalty = "SCAD", seed = 123)
plot(cv_scad)
lambda.scad = scad.cv$lambda.min
print(lambda.scad)
[1] 0.03769054
```

The selected tuning parameter is approximately equal to 0.04 and the MSE is 13.16, similar to that of the Lasso's.

```
## Estimate MSE
yhat.scad = predict(scad.cv, X_test, lambda = lambda.scad)
mean((yhat.scad - y_test)^2)
[1] 13.16499
```

Interestingly, as can be seen in the relevant column of Table 2.1, the support set for both SCAD and Lasso is the same.

```
## print coefficients
coef(scad.cv)
```

Next we follow similar steps to implement the MCP.

```
## Implement MCP with 10-fold cross-validation
mcp.cv = cv.ncvreg(X_train, y_train,
                   penalty = "MCP", seed = 123)
plot(mcp.cv)
lambda.scad =  mcp.cv$lambda.min
print(lambda.scad)
[1] 0.04041433
## Estimate MSE
yhat.mcp = predict(mcp.cv, X_test, lambda = lambda.scad)
mean((yhat.mcp - y_test)^2)
[1] 13.1612
```

The estimated coefficients in the last column of Table 2.1) are found using

```
## print coefficients
coef(mcp.cv)
```

Overall, for this data, the performance of the three sparse estimation methods are similar with respect to the prediction and support recovery.

Error Variance estimation

We use the testing part of fat dataset and R package natural to compute standard deviations of error using least-squares, naive (2.32), cross-validated (2.33), natural (2.34), and organic methods. The latter method is discussed in Yu and Bien [2019]. For all estimators, we use 3-fold cross-validation.

```
require(natural)
# obtains standard deviation using OLS
sigma_ols = summary(yhat.ls)$sigma
print(sigma_ols)
[1] 4.101745
# obtain standard deviation using organic lasso
olasso_res = olasso_cv(x = X_test, y = y_test,
                       intercept = TRUE, nfold = 3)
sigma_olasso = olasso_res$sig_obj
print(sigma_olasso)
[1] 4.152393
# obtain standard deviation using natural lasso
nlasso_res = nlasso_cv(x = X_test, y = y_test,
```

```
                        intercept = TRUE, nfold = 3)
sigma_nlasso = nlasso_res$sig_obj
print(sigma_nlasso)
[1] 4.603441
# obtain naive standard deviation
sigma_naive = nlasso_res$sig_naive
print(sigma_naive)
[1] 3.766038
# obtain standard deviation using cross-validation
sigma_cv = nlasso_res$sig_df
print(sigma_cv)
[1] 3.983905
```

Note that the error variance estimates are hovering around 3.8, but since the true variance is unknown, there is no guide to decide which estimator is the best. We compare these estimators further in Section 2.5.3 where a semblance of ground truth is constructed for a larger dataset.

2.5.2 Regularization Bias

It is well-known that most regularized estimators are biased [Zhang, 2010]. In this section, we assess and compare the bias of Lasso and MCP using simulation and the data generation process in Example 5 where the true parameter values are known.

We estimate the bias by varying the sample size from 2^7 to 2^{12}. For each sample size, we report the average bias over 100 iterations. To estimate and plot the bias we use the bias.plot() function in the package recode.

```
require(recode)
## Simulation settings
p = 100
n.list = 2^c(7:12)
nsim = 100

## Define coefficients
beta = matrix(0, nrow = p, ncol = 1)
beta[1:4] = c(5,3,-3,-1)

bias.plot(n.list = n.list, beta = beta, nsim = nsim,
    seed = 1234)
```

The results in Figure 2.9 confirm that Lasso remains biased for all samples sizes, but the bias of the MCP estimator vanishes as the sample size grows.

2.5.3 Comparing Error Variance Estimators

In this section, we compare the error variance of Lasso using various estimators discussed in Section 2.3. In particular, we focus on standard deviations of error estimated using naive (2.32), cross-validated (2.33), natural (2.34), and organic methods.

We use part of the Million Song dataset, 10000 observations out of the original 463715, available in R package recode, to compare various methods . The dataset consists of information about 10000 songs, which are mostly Western and range from 1922 to 2011. The goal is to model the release year of a song using p = 90 timbre predictors. Yu and Bien [2019, Section 6] conduct a similar experiment using half of

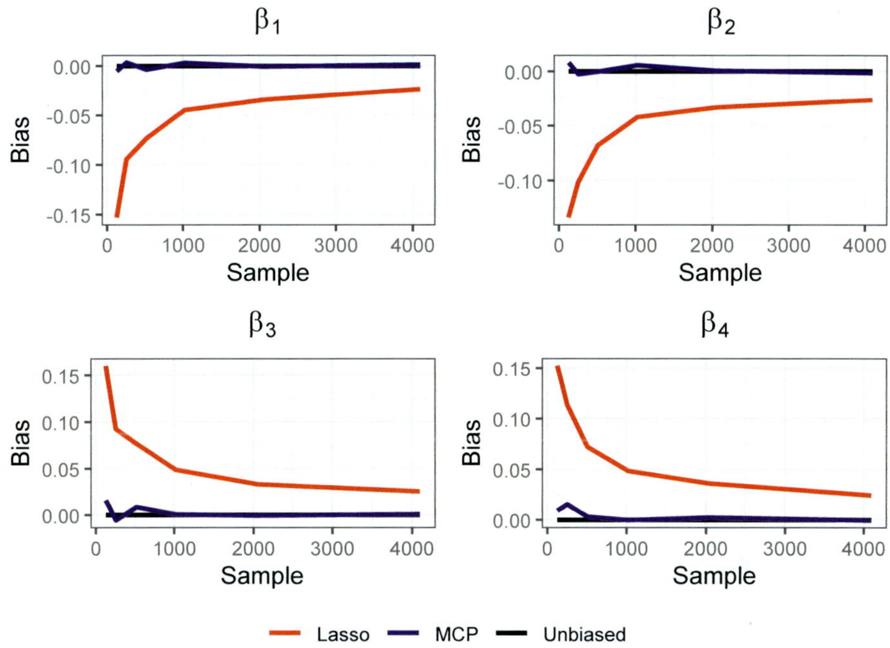

FIGURE 2.9
Bias plot for Lasso and MCP.

the 463715 observations as a training set. They found that all methods perform better than naive. In our case, with small dataset this is not always the case, as can be seen in table 2.4.

We start by importing the required packages and the dataset.

```
require(recode)
require(natural)
data(yearpredictionmsd)
```

Similar to Yu and Bien [2019, Section 6] we randomly select 9000 observations and use the standard deviation estimate of errors obtained from OLS, denoted σ_{ols}, as the ground truth.

```
set.seed(123)
index = sample.int(n = nrow(yearpredictionmsd), size = 9000, replace =
    F)
train = yearpredictionmsd[index,]
test = yearpredictionmsd[-index,]

#obtain standard deviation using ols
ols_result = lm(year~ ., data = train)
sigma_ols = summary(ols_result)$sigma
```

From the remaining data, 1000 observations, that were not previously used, we form a dataset of size $N \in \{50, 100, 150\}$ to estimate the performance of the methods. For each N, we repeat the data selection and error variance estimation 1000 times and report the estimates of the mean squared error $E[(\sigma_{ols}^{-1}\hat{\sigma} - 1)^2]$.

```
test_size = c(50, 100, 150)
n_replicate = 1000
# matrices to save the squared errors
sq.err_olasso = matrix(0, nrow = nsim, ncol = length(test_size))
sq.err_nlasso = matrix(0, nrow = nsim, ncol = length(test_size))
sq.err_naive  = matrix(0, nrow = nsim, ncol = length(test_size))
sq.err_cv     = matrix(0, nrow = nsim, ncol = length(test_size))

for (size in 1:length(test_size)){
  for (rep in 1:n_replicate){
    df = test[sample(test_size[size]) ,]
    x = df[,2:ncol(df)]
    y = df[,1]
    # obtain standard deviation using organic lass
    olasso_res = olasso_cv(x = as.matrix(x), y = y,
                           intercept = TRUE, nfold = 3)
    sigma_olasso = olasso_res$sig_obj
    # obtain standard deviation using natural lasso
    nlasso_res = nlasso_cv(x = as.matrix(x), y = y,
                           intercept = TRUE, nfold = 3)
    sigma_nlasso = nlasso_res$sig_obj
    # obtain naive standard deviation
    sigma_naive = nlasso_res$sig_naive
    # obtain standard deviation using cross-validation
    sigma_cv = nlasso_res$sig_df
    # estimate squared error
    sq.err_olasso[rep, size] = (sigma_olasso/sigma_ols - 1)^2
    sq.err_nlasso[rep, size] = (sigma_nlasso/sigma_ols - 1)^2
    sq.err_naive[rep, size] = (sigma_naive/sigma_ols - 1)^2
    sq.err_cv[rep, size] = (sigma_cv/sigma_ols - 1)^2
  }
}
# Compute mean squared error
mean_sq.err = rbind(colMeans(sq.err_olasso),
                    colMeans(sq.err_nlasso),
                    colMeans(sq.err_naive),
                    colMeans(sq.err_cv)) * 100
```

Table 2.4 reports the results. For $N = 50$ and $N = 100$ natural method performs the best and organic is the best for $N = 150$ case.

2.5.4 Impact of Predictors Correlation on Model Size

Most of the model selection methods considered so far assume that the covariates are cross-sectionally weakly correlated. In this section, we use simulations to examine the impact of strong predictor correlations on model selection or its size. The simulation setting is similar to that in Fan et al. [2020b].

We consider a sparse linear model $\boldsymbol{y} = \boldsymbol{X}\boldsymbol{\beta} + \boldsymbol{\epsilon}$ with no intercept. The coefficients are $\boldsymbol{\beta} = (\beta_1, \ldots, \beta_{10}, \mathbf{0}'_{p-10})'$, where $p = 200$ and the sample size is $n = 100$. The

TABLE 2.4
Mean squared error of noise variance estimation for the Million Song dataset over 1000 replications of the squared error of various methods; each entry is multiplied by 100.

method	$N = 50$	$N = 100$	$N = 150$
organic	0.98	2.70	0.19
natural	0.11	1.03	0.55
naive	0.76	1.74	0.52
cv	0.39	1.24	0.56

10-dimensional non-zero coefficients are drawn from Uniform$[2, 5]$. The predictors $\boldsymbol{X} \sim N_p(\boldsymbol{0}, \boldsymbol{\Sigma})$ where the compound symmetry correlation matrix $\boldsymbol{\Sigma}$ has off-diagonal elements equal to ρ selected from the list $[0, 0.1, 0.2, \ldots, 0.9]$ with a step size of 0.1. For each ρ, we simulate 50 replications and calculate the average model size selected by Lasso, SCAD, and MCP, where the tuning parameter is selected using 10-fold cross-validation.

We start by importing the necessary packages and parameters for the simulation.

```
require(MASS)
require(glmnet)
require(ncvreg)

## Parameters for data generation
n = 100
p = 200
nsim = 50

# List of correlations
rholist = seq(0,9)/10

# Covariance for the error term
Sigma_eps = 0.3 * diag(p)

methods <- c("Lasso", "SCAD", "MCP")
store_size = matrix(0, nrow = length(methods), ncol = length(rholist))
ind = 1
```

The matrix store_size stores the average model sizes for each method and value of ρ. Next, we construct a nested for loop where the outer loop iterates over the values of ρ and the inner loop iterates over the number of simulations and fit Lasso, SCAD, and MCP methods.

```
set.seed(123)

for (rho in rholist){
  Lasso_sel = SCAD_sel = MCP_sel = 0
  for (i in 1:nsim) {

    # Generate true beta
    beta_nz = runif(10, 2, 5)
    beta = c(beta_nz, rep(0,p - 10))
```

```
    # Generate correlated predictors
    Sigma_X = diag(p) * (1 - rho) + rho

    X <- MASS::mvrnorm(n, mu = rep(0, p), Sigma = Sigma_X)

    # Generate response
    y <-  X %*% beta + rnorm(n, 0, Sigma_eps)

    # Get model size
    ## Lasso
    lasso_fit <- cv.glmnet(X, y, alpha = 1, nfolds = 10)
    lambda_selected = lasso_fit$lambda.min
    coef_lasso <- coef(lasso_fit , s = lambda_selected)[-1]
    Lasso_sel <- Lasso_sel + length(which(coef_lasso != 0))

    ## SCAD
    SCAD_fit = cv.ncvreg(X, y, nfolds = 10,
                          penalty = "SCAD")
    lambda_selected = SCAD_fit$lambda.min
    coef_SCAD <- coef(SCAD_fit , s = lambda_selected)[-1]
    SCAD_sel <- SCAD_sel + length(which(coef_SCAD != 0))

    ## MCP
    MCP_fit = cv.ncvreg(X, y, nfolds = 10,
                         penalty = "MCP")
    lambda_selected = MCP_fit$lambda.min
    coef_MCP <- coef(MCP_fit , s = lambda_selected)[-1]
    MCP_sel <- MCP_sel + length(which(coef_MCP != 0))
  }
  store_size[, ind] = c(Lasso_sel , SCAD_sel , MCP_sel) / nsim
  ind = ind + 1
}
```

Figure 2.10 displays the results. As can be seen, Lasso performance deteriorates even when the correlation is small. The deterioration of MCP and SCAD starts around $\rho = 0.5$. Interestingly, while the average model sizes for Lasso and SCAD have a tendency to increase, the average model size for MCP decreases as correlation increases.

The decrease in average model size for MCP decreases as correlation increases is a mysterious and calls for further investigation.

2.6 Convex Optimization

A brief review of topics related to convex functions and their important roles in solving a few regularization problems in statistics is presented in this section. More complete coverage with illuminating examples can be found in Boyd and Vandenberghe [2004].

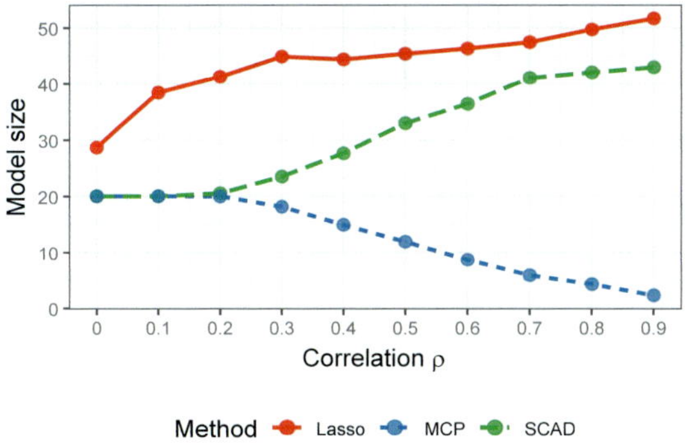

FIGURE 2.10
Impact of correlation on the average model size for Lasso, SCAD, and MCP.

Most of the optimization problems encountered in this book are of the form

$$\begin{aligned} \min_{\boldsymbol{x}} \quad & \ell(\boldsymbol{x}) \\ \text{subject to} \quad & g_i(\boldsymbol{x}) \leq t_i, \quad i = 1, \ldots, m \end{aligned} \tag{2.45}$$

where the functions $\ell, g_i, i = 1, \ldots, m$, are convex. A function g is convex [Boyd and Vandenberghe, 2004] if its domain is a convex set C, i.e., for any $\alpha \in [0, 1]$, $\boldsymbol{x}_1, \boldsymbol{x}_2 \in C$, $\alpha\boldsymbol{x}_1 + (1 - \alpha)\boldsymbol{x}_2 \in C$, and

$$g(\alpha\boldsymbol{x}_1 + (1 - \alpha)\boldsymbol{x}_2) \leq \alpha g(\boldsymbol{x}_1) + (1 - \alpha)g(\boldsymbol{x}_2). \tag{2.46}$$

This definition can be used to show the convexity of the absolute value function $g(x) = |x|$, for $x \in \mathbb{R}$, and the Euclidean norm $g(\boldsymbol{x}) = \|\boldsymbol{x}\|, \boldsymbol{x} \in \mathbb{R}^p, p \geq 1$, (Problem 2.12). Of course, convex objective functions are popular and easier to work with because every local minimum of a convex function is also a global minimum.

Another useful definition of convexity when g is differentiable and the gradient ∇g exists, is

$$g(\boldsymbol{x}_2) \geq g(\boldsymbol{x}_1) + \nabla g(\boldsymbol{x}_1)'(\boldsymbol{x}_2 - \boldsymbol{x}_1), \tag{2.47}$$

for all $\boldsymbol{x}_1, \boldsymbol{x}_2 \in \mathbb{R}^p$.

This alternative definition suggests that, for a convex function, the first-order approximation at any point is a lower bound for the function. Consequently, any point $x \in C$ for which $\nabla g(x) = 0$, is a global minimum. Figure 2.11 illustrates both definitions.

It can be shown that a twice-differentiable function defined on $\mathbb{R}^p$ is convex if and only if its Hessian or its matrix of second derivatives exists at each point of its domain and is nonnegative-definite as a matrix:

$$\nabla^2 f(\boldsymbol{x}) \succeq 0, \tag{2.48}$$

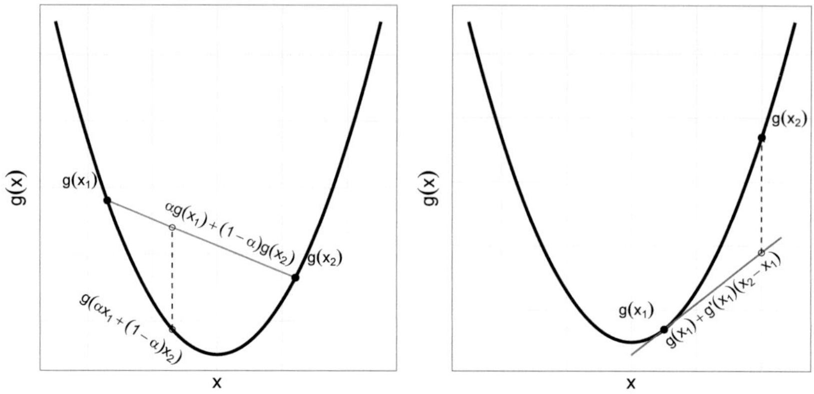

FIGURE 2.11
Two definitions of convex function.

this can be proved using the first-order approximation (2.47); see Problem 2.20. For a function defined on $\mathbb{R}$ this reduces to $f(x)'' \geq 0$, see Boyd and Vandenberghe [2004, Section 3.1.4]. This characterization of convexity is commonly used to check the convexity of the following well-known functions involving positive-definite matrices.

- Negative log-determinant: $g(A) = -\log\det(A)$, for A in the space of positive-definite symmetric matrices [Boyd and Vandenberghe, 2004, Page 644].
- Quadratic forms: $\boldsymbol{x}'A\boldsymbol{x}$ for a positive-definite matrix A, see Example 3.2 in Boyd and Vandenberghe [2004].

Using either definition it follows immediately that sum of any two convex functions is a convex function, the same holds for their linear combinations with nonnegative coefficients. This simple fact is used effectively in the following example dealing with loglikelihood function of several multivariate normal distributions.

Example 6 (Convexity check for Gaussian loglikelihood) *Consider the linear model*

$$\boldsymbol{y} = \boldsymbol{X\beta} + \varepsilon,\ \varepsilon \sim N(0, \sigma^2\boldsymbol{I}_n), \tag{2.49}$$

with the negative loglikelihood function (up to a constant)

$$L(\boldsymbol{\beta}, \sigma^2|\boldsymbol{X}, \boldsymbol{y}) = \frac{n}{2}\log\sigma^2 + \frac{\|\boldsymbol{y} - \boldsymbol{X\beta}\|_2^2}{2\sigma^2}. \tag{2.50}$$

(a) First, we show that the function $f(\sigma^2) = L(\boldsymbol{\beta}, \sigma^2|\boldsymbol{X}, \boldsymbol{y})$ is not convex with respect to σ^2. Note that the first and second derivatives are

$$f'(\sigma^2) = \frac{n}{2\sigma^2} - \frac{\|\boldsymbol{y} - \boldsymbol{X\beta}\|_2^2}{2(\sigma^2)^2}, \quad f''(\sigma^2) = -\frac{n}{2(\sigma^2)^2} + \frac{\|\boldsymbol{y} - \boldsymbol{X\beta}\|_2^2}{(\sigma^2)^3}.$$

Since for some values of σ^2 the second derivative can be negative for a fixed $\boldsymbol{\beta}$, $f(\sigma^2)$ is not convex with respect to the variance σ^2.

(b) *Next, reparameterizing the log-likelihood (2.50) using $\phi = 1/\sigma^2$, which is the inverse variance or the precision parameter, we obtain*

$$f(\phi) = -\frac{n}{2}\log(\phi) + \frac{\|y - X\boldsymbol{\beta}\|_2^2}{2}\phi,$$

with

$$f^{''}(\phi) = \frac{n}{2\phi^2} > 0,$$

so that $f(\phi)$ is convex with respect to the inverse of the variance.

(c) *Finally, for the joint reparameterization $\phi = 1/\sigma^2$, $\boldsymbol{\theta} = \boldsymbol{\beta}/\sigma^2$, the likelihood (2.50) can be written as*

$$L(\phi^{-1}\boldsymbol{\theta}, \phi^{-1}|\boldsymbol{X}, \boldsymbol{y}) = -\frac{n}{2}\log\phi + \phi\frac{\|\boldsymbol{y}\|^2}{2} - \boldsymbol{y}^{'}\boldsymbol{X}\boldsymbol{\theta} + \frac{\|\boldsymbol{X}\boldsymbol{\theta}\|_2^2}{2\phi}. \tag{2.51}$$

Its joint convexity in $(\phi, \boldsymbol{\theta})$ follows from showing that its Hessian matrix is non-negative definite. From Problem 2.21, the Hessian matrix has the form

$$H = \begin{bmatrix} \frac{n}{2\phi^2} + \frac{\|\boldsymbol{X}\boldsymbol{\theta}\|^2}{\phi^3} & -\frac{1}{\phi^2}\boldsymbol{X}^{'}\boldsymbol{X}\boldsymbol{\theta} \\ -\frac{1}{\phi^2}(\boldsymbol{X}^{'}\boldsymbol{X}\boldsymbol{\theta})' & \frac{1}{\phi}\boldsymbol{X}^{'}\boldsymbol{X} \end{bmatrix}, \tag{2.52}$$

with determinant

$$\det(H) = \frac{1}{\phi}\det(\boldsymbol{X}^{'}\boldsymbol{X})\left(\frac{n}{2\phi^2} + \frac{2\|\boldsymbol{X}\boldsymbol{\theta}\|^2}{\phi^3}\right) \geq 0. \tag{2.53}$$

The latter follows since $\boldsymbol{X}^{'}\boldsymbol{X}$ is non-negative definite with $\det(\boldsymbol{X}^{'}\boldsymbol{X}) \geq 0$ and the second term is always positive, so that $H \succeq 0$. .

(d) *(Convexity of regularized Normal Loglikelihood function) Consider the regularized negative loglikelihood*

$$L(\phi^{-1}\boldsymbol{\theta}, \phi^{-1}|\boldsymbol{X}, \boldsymbol{y}) + \lambda p(\boldsymbol{\theta}, \phi),$$

where the regularization function $p(\boldsymbol{\theta}, \phi)$ is convex with respect to $(\phi, \boldsymbol{\theta})$ and $\lambda > 0$. The regularized negative loglikelihood as a sum of two convex functions is also convex.

(e) *More generally, with $\varepsilon \sim N(0, \boldsymbol{\Sigma})$ in (2.49), the negative loglikelihood is of the form (up to a constant)*

$$L(\boldsymbol{\beta}, \boldsymbol{\Sigma}|\boldsymbol{X}, \boldsymbol{y}) = \frac{1}{2}\log\det\boldsymbol{\Sigma} + \boldsymbol{y}'\boldsymbol{\Sigma}^{-1}\boldsymbol{y}.$$

Using matrix derivatives [Petersen and Pedersen, 2008], it can be shown that

$$\frac{\partial L(\boldsymbol{\beta}, \boldsymbol{\Sigma}|\boldsymbol{X}, \boldsymbol{y})}{\partial \boldsymbol{\Sigma} \partial \boldsymbol{\Sigma}} = -\frac{1}{2}\boldsymbol{\Sigma}^{-1} \otimes \boldsymbol{\Sigma}^{-1} + \frac{1}{2}(\boldsymbol{\Sigma}^{-1}\boldsymbol{y}\boldsymbol{y}'\boldsymbol{\Sigma}^{-1}) \otimes \boldsymbol{\Sigma}^{-1},$$

where $\otimes$ is the Kronecker product. This second derivative is not always a non-negative definite matrix; see (a) above. Thus, $L(\boldsymbol{\beta}, \boldsymbol{\Sigma}|\boldsymbol{X}, \boldsymbol{y})$ is not convex with respect to $\boldsymbol{\Sigma}$.

(f) As in (b), by reparameterizing $\boldsymbol{\Theta} = \boldsymbol{\Sigma}^{-1}$, the negative loglikelihood

$$-\frac{1}{2}\log\det\boldsymbol{\Theta} + \boldsymbol{y}'\boldsymbol{\Theta}\boldsymbol{y},$$

as a sum of two convex functions is convex.

In a convex optimization problem, such as (2.45), a vector $\boldsymbol{x}^*$ is feasible if $g_i(\boldsymbol{x}^*) \leq t_i$ for $i = 1, \ldots, m$. Moreover, a necessary and sufficient condition for a feasible vector $\boldsymbol{x}^*$ to achieve a global optimum is that

$$\nabla\ell(\boldsymbol{x}^*)'(\boldsymbol{x} - \boldsymbol{x}^*) \geq 0,$$

for all feasible points $\boldsymbol{x}$, for ℓ differentiable. For an unconstrained optimization problem, i.e., $m = 0$, the necessary and sufficient condition reduces to the first-order condition

$$\nabla\ell(\boldsymbol{x}^*) = 0. \tag{2.54}$$

Example 7 (Lasso and ridge as convex optimization problems) *The Lasso problem in (2.7) can be reformulated as*

$$\begin{aligned} \min_{\boldsymbol{\beta}} \quad & \frac{1}{2}||\boldsymbol{y} - \boldsymbol{X}\boldsymbol{\beta}||^2, \\ \textit{subject to} \quad & ||\boldsymbol{\beta}||_1 \leq t, \end{aligned} \tag{2.55}$$

which is a convex optimization problem. It is instructive to note that there exists a one-to-one correspondence between λ in (2.7) and the bound t. Here, a smaller t limits the sum of the absolute values of the parameter estimates. The formulation (2.7) is known as the Lagrangian form of (2.55). A similar representation holds for the ridge regression (2.5), but instead of ℓ_1 norm, the ℓ_2 norm is used as the constraint.

For a general optimization problem, as in (2.45), its Lagrangian form is defined as

$$L(\boldsymbol{x}, \lambda) = \ell(\boldsymbol{x}) + \sum_{i=1}^{m} \lambda_i g_i(\boldsymbol{x}).$$

Here, we assume that $t_i = 0$, and $\lambda_i \geq 0$ is referred to as the dual variables or the Lagrange multipliers. The Lagrange dual function is the minimum value of the Lagrangian over $\boldsymbol{x}$:

$$f(\lambda) = \min_{\boldsymbol{x}} L(\boldsymbol{x}, \lambda),$$

so that it is a lower bound on the optimal value of (2.45), i.e., $f(\lambda) \leq \ell(x^*)$. The problem of maximizing $f(\lambda)$ subject to $\lambda \geq 0$ is known as a Lagrange *dual problem* and the original problem (2.45) is referred to as a *primal problem*. The former is always a convex optimization problem, even if the primal is not convex.

A closely related object to dual functions is the so-called *conjugate functions*. The conjugate function ℓ^* of a function ℓ is defined as

$$\ell^*(\boldsymbol{y}) = \max_{\boldsymbol{x}}(\boldsymbol{y}'\boldsymbol{x} - \ell(\boldsymbol{x})) \tag{2.56}$$

where the maximum is over the domain of ℓ. We note that ℓ^* is always a convex function, even if ℓ is not. The conjugate function can be considered as the maximum gap between the linear function $\boldsymbol{y}'\boldsymbol{x}$ and $\ell(\boldsymbol{x})$. If ℓ is differentiable, then this maximum gap occurs at $\boldsymbol{y} = \nabla f(\boldsymbol{x})$. For details on the relationship between the dual and conjugate functions, see Problem 2.19.

It turns out that for a convex primal problem, under Karush-Kuhn-Tucker (KKT) conditions listed below, the gap between the optimal value of the primal (2.45) and its dual disappears, i.e., $f(\boldsymbol{\lambda}^*) = \ell(\boldsymbol{x}^*)$. In other words, KKT conditions provide a necessary and sufficient conditions for optimality. For differentiable ℓ and g_i, the KKT conditions are

$$\begin{aligned} g_i(\boldsymbol{x}) &\leq 0, \quad i = 1, \ldots, m \\ \lambda_i &\geq 0, \quad i = 1, \ldots, m \\ \lambda_i g_i(\boldsymbol{x}) &= 0, \quad i = 1, \ldots, m \\ \nabla\ell(\boldsymbol{x}) + \sum_{i=1}^{m} \lambda_i \nabla g_i(\boldsymbol{x}) &= 0 \end{aligned} \tag{2.57}$$

The third condition is known as complementary slackness, which assures that the ith optimal Lagrange multiplier is zero if the ith constraint is inactive, i.e., $g_i(\boldsymbol{x}) < 0$.

Up to this point, it was assumed that ℓ and g_i's are differentiable. However, most problems we encounter contain a non-differentiable component. For example, $|x|$ or $||\boldsymbol{\beta}||_1$ in Lasso is not differentiable at the origin. For such problems, the first-order condition (2.54) and the KKT conditions (2.57) are not applicable. Fortunately, for convex functions there are more general optimality conditions involving the notions of subgradient and subdifferential.

For a convex function g, a vector $\boldsymbol{\gamma}$ is said to be a subgradient of g at $\boldsymbol{x}$ if for all $\boldsymbol{y}$

$$g(\boldsymbol{y}) \geq g(\boldsymbol{x}) + \boldsymbol{\gamma}'(\boldsymbol{y} - \boldsymbol{x}). \tag{2.58}$$

For a given point $\boldsymbol{x}$, the set of all subgradients is called its subdifferential, and denoted by $\partial g(\boldsymbol{x})$. If g is differentiable, then subdifferential contains only a single vector, which is the gradient $\nabla g(\boldsymbol{x})$. Figure 2.12 illustrates examples of subgradients for a convex function g. At the point x_1 on the left, the function is differentiable and $\gamma_1 = \nabla g(x_1)$. On the other hand, at point x_2, the function is not differentiable and the subdifferential is a set.

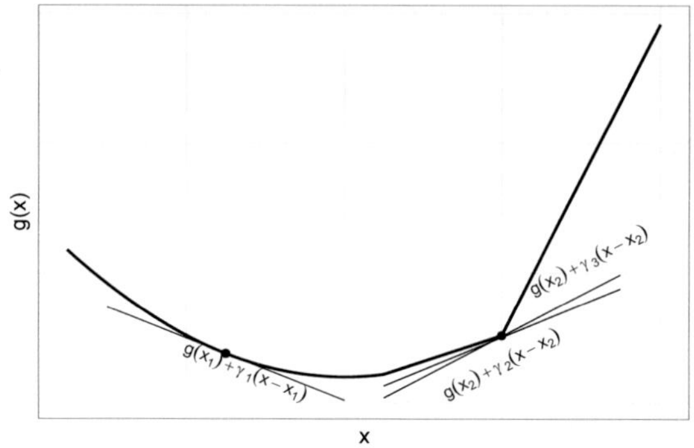

FIGURE 2.12
Examples of subgradients for a convex function g.

Exploiting the subdifferential, the fourth KKT condition in (2.57) can be modified to

$$0 \in \partial \ell(\boldsymbol{x}) + \sum_{i=1}^{m} \lambda_i \partial g_i(\boldsymbol{x}). \tag{2.59}$$

Example 8 (Subgradient of the absolute value function) *Consider the univariate function $f(\beta) = |\beta|$, which is differentiable at all points $\beta \in \mathbb{R}$, except at the origin $\beta = 0$. Its subgradient $\gamma = sign(\beta)$ is unique when $\beta \neq 0$, but for $\beta = 0$ from the definition of subgradient in (2.58), for any $\beta' \neq 0$ one arrives at*

$$|\beta'| \geq \gamma\beta'.$$

This is satisfied if and only if $\gamma \in [-1, 1]$, so that the subdifferential is the interval $\partial f(0) = [-1, 1]$. This derivation can be extended to the ℓ_1 norm $f(\boldsymbol{\beta}) = \|\boldsymbol{\beta}\|_1 = \sum_{i=1}^{p} |\beta_i|$, with the subdifferential

$$\partial f(\boldsymbol{\beta}) = \{\boldsymbol{\gamma} \in \mathbb{R}^p : \gamma_i = sign(\beta_i) \text{ if } \beta_i \neq 0,\ \gamma_i \in [-1, 1] \text{ if } \beta_i = 0\}.$$

The next lemma exploits the modified KKT conditions for the subdifferential to characterize the Lasso solution.

Lemma 1 (Characterization of the Lasso Solution) *For the Lasso regression (2.7), let the gradient of its quadratic (differentiable) component be denoted by the vector*

$$G(\boldsymbol{\beta}) = -\boldsymbol{X}'(\boldsymbol{y} - \boldsymbol{X}\boldsymbol{\beta}),$$

with $G_j(\cdot)$ its jth entry. Then, from (2.59) for a given λ, a necessary and sufficient condition for $\boldsymbol{\beta}$ to be the minimizer of (2.7) is that

$$0 \in \partial Q(\boldsymbol{\beta}) = G(\boldsymbol{\beta}) + \lambda\boldsymbol{\gamma}.$$

More explicitly, for $j = 1, \ldots, p$,

$$G_j(\boldsymbol{\beta}) = -sign(\beta_j)\lambda \quad if \quad \beta_j \neq 0,$$
$$|G_j(\boldsymbol{\beta})| \leq \lambda \quad if \quad \beta_j = 0.$$

2.7 Problems

2.1 Consider a simple linear model with the full-rank $n \times 2$ design matrix $\boldsymbol{X}$ with columns labeled as $\mathbf{1}_n, \boldsymbol{x}$ where $\boldsymbol{x} = (x_1, \ldots, x_n)'$.

(a) Compute $\boldsymbol{X}'\boldsymbol{X}$, its inverse and the least squares estimators of β_0 and β_1.

(b) Express $\hat{\beta}_1$ as a linear combination of the observations y_i's and show that

$$\text{var}(\hat{\beta}_1) = \frac{\sigma^2}{\sum_{i=1}^n (x_i - \bar{x})^2}.$$

Discuss a strategy for reducing the variance of the estimator of the slope. Do you prefer the sample variance of the x_i's to be large or small?

(c) For a $\lambda > 0$, repeat (a) and (b) for the ridge estimators of the intercept and slope.

2.2 (a) Consider the quadratic form $Q(\boldsymbol{\beta}) = (\boldsymbol{y} - \boldsymbol{X}\boldsymbol{\beta})'(\boldsymbol{y} - \boldsymbol{X}\boldsymbol{\beta})$, which is the sum of squares of errors in the general linear regression model. Show that $\boldsymbol{X}'(\boldsymbol{y} - \boldsymbol{X}\widehat{\boldsymbol{\beta}}) = \mathbf{0}$, and

$$Q(\boldsymbol{\beta}) = Q(\widehat{\boldsymbol{\beta}}) + (\widehat{\boldsymbol{\beta}} - \boldsymbol{\beta})'\boldsymbol{X}'\boldsymbol{X}(\widehat{\boldsymbol{\beta}} - \boldsymbol{\beta}) \geq Q(\widehat{\boldsymbol{\beta}}).$$

(b) Use the identity in (a) to conclude that $Q(\boldsymbol{\beta})$ is minimized at the least square estimator $\hat{\boldsymbol{\beta}}$.

2.3 (a) Show that the minimizers of $Q_{ls}(\boldsymbol{\beta})$ and $Q_{ridge}(\boldsymbol{\beta})$ satisfy their respective stated normal equations.

(b) Find $E(\widehat{\boldsymbol{\beta}})$ and $E(\widehat{\boldsymbol{\beta}}_\lambda)$, discuss their biases or lack thereof.

(c) Show that $\text{Cov}(\widehat{\boldsymbol{\beta}}) = \sigma^2(\boldsymbol{X}'\boldsymbol{X})^{-1}$, compute $\text{Cov}(\widehat{\boldsymbol{\beta}}_\lambda)$, and compare the two.

(d) Show that the vectors of least squares fitted values and residuals have the following representations

$$\widehat{\boldsymbol{y}} = \boldsymbol{X}\widehat{\boldsymbol{\beta}} = \boldsymbol{H}\boldsymbol{y}, \ \boldsymbol{y} - \widehat{\boldsymbol{y}} = (\boldsymbol{I} - \boldsymbol{H})\boldsymbol{y},$$

where $\boldsymbol{H} = \boldsymbol{X}(\boldsymbol{X}'\boldsymbol{X})^{-1}\boldsymbol{X}'$ is the so-called hat matrix. Discuss their orthogonality (uncorrelatedness) using these representations.

(e) Show that $\boldsymbol{H}$ is a projection matrix, that is $\boldsymbol{H}^2 = \boldsymbol{H}$, and consequently the vectors of fitted values and residuals are orthogonal.

2.4 For $\boldsymbol{X}$ an $n \times p$ matrix of rank $r \leq \min\{n, p\}$, it is known that there exist matrices $\mathbf{U}, \mathbf{V}$ and $\mathbf{D}$ giving its singular value decomposition (SVD):

$$\boldsymbol{X} = \mathbf{UDV}' = \sum_{k=1}^{r} d_k \boldsymbol{u}_k \mathbf{v}_k', \tag{2.60}$$

where the columns of $\mathbf{U} = (\boldsymbol{u}_1, \ldots, \boldsymbol{u}_r), \mathbf{V} = (\mathbf{v}_1, \ldots, \mathbf{v}_r)$ are orthonormal, and the diagonal entries of $\mathbf{D} = \text{diag}(d_1, \ldots, d_r)$ are ordered nonnegative numbers, i.e., $d_1 \geq d_2 \cdots \geq d_r > 0$, see Chapter 3].

(a) Use the SVD of $\boldsymbol{X}$ and show that the ridge estimator of $\boldsymbol{\beta}$ and the corresponding fitted values are

$$\widehat{\boldsymbol{\beta}}_\lambda = \mathbf{V}(\mathbf{D}'\mathbf{D} + \lambda\mathbf{I})^{-1}\mathbf{D}'\mathbf{U}'\boldsymbol{y} = \sum_{d_j>0} \frac{d_j^2}{d_j^2 + \lambda}(\boldsymbol{u}_j, \boldsymbol{y})\mathbf{v}_j,$$

and

$$\widehat{\boldsymbol{y}}_\lambda = \sum_{d_j>0} \frac{d_j^2}{d_j^2 + \lambda}(\boldsymbol{u}_j, \boldsymbol{y})\boldsymbol{u}_j.$$

Thus, once the SVD of $\boldsymbol{X}$ is computed, we have the ridge solution and fitted values for all values of λ.

(b) For $n > p$ and $\boldsymbol{X}$ full rank, show that

$$\text{Bias}(\widehat{\boldsymbol{\beta}}_\lambda) = \sum_{i=1}^{p} \mathbf{v}_i \frac{\lambda}{\lambda + d_i^2}(\mathbf{v}_j, \boldsymbol{\beta}),$$

so that the bias is a linear combination of the columns of $\mathbf{V}$, and its every component increases for larger λ.

(c) For $n > p$ and $\boldsymbol{X}$ full rank, show that

$$\text{Cov}(\widehat{\boldsymbol{\beta}}_\lambda) = \sigma^2 \sum_{i=1}^{p} \frac{d_i^2}{(\lambda + d_i^2)^2}\mathbf{v}_j\mathbf{v}_j'.$$

2.5 Given the standard regression data $(\boldsymbol{X}, \boldsymbol{y})$ and a $\lambda > 0$, consider the augmented data $(\boldsymbol{X}_a, \boldsymbol{y}_a)$ where

$$\boldsymbol{X}_a = (\boldsymbol{X}', \sqrt{\lambda}\mathbf{I}_p)', \ \boldsymbol{y}_a = (\boldsymbol{y}', \mathbf{0}_p')'.$$

Write the normal equations for the least-squares estimator of $\boldsymbol{\beta}$ for the augmented data $(\boldsymbol{X}_a, \boldsymbol{y}_a)$, and note that they are the same as the normal equations for the ridge estimator of the original data $(\boldsymbol{X}, \boldsymbol{y})$. Discuss the role of λ in the augmented data and ridge estimators when it goes to zero or infinity. Would the results hold when $\mathbf{0}_p$ in $\boldsymbol{y}_a$ is replaced by $\mathbf{1}_p$?

2.6 Consider estimating the regression parameters in the standard linear model $\boldsymbol{y} = \boldsymbol{X\beta} + \boldsymbol{\epsilon}$, with the $n \times p$ design matrix $\boldsymbol{X}$, by minimizing the quadratic criterion

$$Q(\boldsymbol{\beta}) = (\boldsymbol{y} - \boldsymbol{X\beta})'\mathbf{W}(\boldsymbol{y} - \boldsymbol{X\beta}) + (\boldsymbol{\beta} - \boldsymbol{\beta}_0)'\Delta(\boldsymbol{\beta} - \boldsymbol{\beta}_0),$$

where $\mathbf{W}, \Delta$ are non-negative definite matrices and $\boldsymbol{\beta}_0$ is a given target vector. Note that this criterion is rather general as seen in the following:

(a) For $\Delta = 0$, it reduces to the weighted least-squares (WLS) criterion. Derive the normal equations and state all conditions needed for it to have a unique solution. Write down the solution.

(b) For $\mathbf{W} = \mathbf{I}_n, \boldsymbol{\beta}_0 = \mathbf{0}_p$, it reduces to a generalized ridge criterion since for $\Delta = \lambda\mathbf{I}_p$ it is the familiar ridge criterion. Derive the normal equations, solve it and compare the solution to the standard ridge shrinkage estimator.

(c) Derive the normal equations for minimizing the general $Q(\boldsymbol{\beta})$, and solve it. State any conditions needed for a unique solution.

(d) Discuss in detail the connections between this rather general weighted ridge estimator with the special cases in (a)–(c) with a focus on Δ and $\boldsymbol{\beta}_0$, see [van Wieringen, 2023, Chapter 3] .

2.7 As another example of the criterion in the previous problem, consider estimating the regression parameters in the standard linear model $\boldsymbol{y} = \boldsymbol{X\beta} + \boldsymbol{\epsilon}$, with the $n \times p$ design matrix $\boldsymbol{X}$, by minimizing

$$Q(\boldsymbol{\beta}) = (\boldsymbol{y} - \boldsymbol{X\beta})'(\boldsymbol{y} - \boldsymbol{X\beta}) + \lambda \sum_{j=2}^{p} (\beta_j - \beta_{j-1})^2,$$

for a $\lambda > 0$.

(a) Writing $\sum_{j=2}^{p} (\beta_j - \beta_{j-1})^2 = \boldsymbol{\beta}'\Delta\boldsymbol{\beta}$, find the explicit form of the matrix Δ.

(b) Find the eigenvalues and rank of the matrix Δ, and decide whether this quadratic penalty function is strictly convex.

2.8 Show that the derivative of the penalty function in (2.24) is

$$p'_{\lambda,\gamma}(y) = \lambda I(|y| \leq \lambda) + \frac{(\gamma\lambda - |y|)_+}{\gamma - 1} I(|y| > \lambda).$$

2.9 Consider the linear model $\boldsymbol{y} = \boldsymbol{X\beta} + \boldsymbol{\epsilon}$ with the full-rank $n \times (p+1)$ design matrix $\boldsymbol{X}$ with columns labelled as $\boldsymbol{X}_0, \boldsymbol{X}_1, \ldots, \boldsymbol{X}_p$. Set $z_0 = \boldsymbol{X}_0$, and for $j \geq 1$, let

$$\hat{\boldsymbol{X}}_j = \sum_{i=0}^{j-1} b_{ij} z_i, \quad z_i = \boldsymbol{X}_i - \hat{\boldsymbol{X}}_i,$$

be the best linear predictor (projection) of $\boldsymbol{x}_j$ based on the linear subspace generated by $\boldsymbol{X}_0, \boldsymbol{X}_1, \ldots, \boldsymbol{X}_{j-1}$, and its prediction error (co-projection). Show that

(a) (Gram-Schmidt Procedure) the coefficients b_{ij}'s and prediction errors (residuals) are given by

$$b_{ij} = \frac{(z_i, \boldsymbol{X}_j)}{||z_i||^2}, \; z_j = \boldsymbol{X}_j - \sum_{i=0}^{j-1} b_{ij} z_i.$$

(b) (QR Decomposition of $\boldsymbol{X}$) the second equality in (a), after rearranging the terms, can be written in matrix form as

$$\boldsymbol{X} = \mathbf{ZB},$$

where $\mathbf{Z}$ is the $n \times (p+1)$ matrix with the orthogonal vector z_j as its jth column and $\mathbf{B} = (b_{ij})$ is an upper-triangular matrix. With $\mathbf{D} = \text{diag}(||z_0||, \ldots, ||z_p||)$, setting $Q = \mathbf{ZD}^{-1}$ amounts to orthonormalizing the columns of $\mathbf{Z}$ so that $Q'Q = \mathbf{I}$, and

$$\boldsymbol{X} = \mathbf{ZD}^{-1}\mathbf{DB} = QR,$$

where R is a $(p+1) \times (p+1)$ upper-triangular matrix. (Aptly called the QR decomposition of the matrix $\boldsymbol{X}$.)

(c) (Computing $\hat{\boldsymbol{\beta}}$, fitted values and residuals) using the QR decomposition in (b) the least squares estimator and the fitted values are computed easily by

$$\hat{\boldsymbol{\beta}} = R^{-1}Q'\boldsymbol{y}, \; \hat{\boldsymbol{y}} = QQ'\boldsymbol{y}.$$

The ease in computation is due to the upper-triangularity of R.

2.10 (Added-variable plot) Added-variable plots are helpful in visually assessing the effect of each covariate, after adjusting for the effects of other predictors. Imagine adding a new variable z to the linear model (2.1) with a full-rank $n \times (p+1)$ design matrix $\boldsymbol{X}$, namely

$$\boldsymbol{y} = \boldsymbol{X}\boldsymbol{\beta} + \alpha z + e.$$

Evidently, the coefficient α measures the effect of z on $\boldsymbol{y}$ after adjusting for the effect of $\boldsymbol{X}$ on $\boldsymbol{y}$. Show that α satisfies the no-intercept simple linear model

$$e_y = \alpha e_z + e^*,$$

where e_y and e_z are the residuals of regressing $\boldsymbol{y}$ and z on $\boldsymbol{X}$, respectively, and e^* is the errors. The scatter plot of e_y versus e_z is called the added-variable plot for z. It is the plot of residuals after removing the effect of $\boldsymbol{X}$ on $\boldsymbol{y}$ and z.

2.11 Show that the function in (2.12) solves the relevant normal equation.

2.12 A simple geometric way to get a feel why Lasso penalty is more likely to produce more zeros than the Ridge penalty, plot in $\mathbb{R}^2$ the following two sets:

$$M_1 = \{(x_1, x_2) \in \mathbb{R}^2 : |x_1| + |x_2| \leq 1\}, \; M_2 = \{(x_1, x_2) \in \mathbb{R}^2 : x_1^2 + x_2^2 \leq 1\}.$$

(a) Compare the smoothness and shapes of their boundaries.

(b) Discuss the situation in $\mathbb{R}^3$ and higher dimensions.

(c) Discuss the situation for the sets

$$M_q = \{(x_1, x_2) \in \mathbb{R}^2 : |x_1|^q + |x_2|^q \leq 1\},\ q = 0, 0.5.$$

- Use the definition of the convex function and triangle inequality to show that every norm is a convex function.

2.13 Show that for an orthonormal design matrix $\boldsymbol{X} = (\boldsymbol{x}_1, \ldots, \boldsymbol{x}_p)$, i.e., $\boldsymbol{X}'\boldsymbol{X} = I_p$, the Lasso solution decouples into p separate simple Lasso regressions with closed-form solutions for $i = 1, \ldots, p$:

$$\widehat{\beta}_i(\lambda) = \text{sign}(\boldsymbol{x}_i'\boldsymbol{y})(\boldsymbol{x}_i'\boldsymbol{y} - \lambda)_+ = S(\boldsymbol{x}_i'\boldsymbol{y}, \lambda),$$

and in vector form

$$\widehat{\boldsymbol{\beta}}(\lambda) = \text{sign}(|\boldsymbol{X}'\boldsymbol{y}|)(|\boldsymbol{X}'\boldsymbol{y}| - \lambda)_+ = S(|\widehat{\boldsymbol{\beta}}|, \lambda),$$

where

$$\widehat{\boldsymbol{\beta}} = \boldsymbol{X}'\boldsymbol{y} = \left(\boldsymbol{x}_1'\boldsymbol{y}, \ldots, \boldsymbol{x}_p'\boldsymbol{y}\right)',$$

is the least-squares estimator.

2.14 Show that for an orthogonal design matrix the solution of the adaptive Lasso is of the form

$$\widehat{\beta}_{a,j}(\lambda) = \text{sign}(\widehat{\beta}_j)\left(|\widehat{\beta}_j| - \frac{\lambda}{|\widehat{\beta}_j|}\right)_+,\ j = 1, \ldots, p,$$

which involves the so-called *adaptive thresholding function*

$$S_a(x, \lambda) = \text{sign}(x)\left(|x| - \frac{\lambda}{|x|}\right)_+.$$

2.15 For the commonly used ℓ_q-penalty function $p(\beta) = |\beta|^q$ show that

(a) for $q = 1$ or the ℓ_1 penalty or Lasso does not have the unbiasedness property,

(b) for $q > 1$ it does not satisfy the sparsity condition,

(c) for $0 \leq q < 1$ the continuity condition is not satisfied.

Thus, so far as SCAD is concerned the ℓ_q penalty function does not satisfy simultaneously the conditions for sparsity, unbiasedness and continuity for any q.

2.16 Show that the function $\mathbf{S}_{\lambda,\gamma}(y)$ is the minimizer of the MCP objective function under the stated restriction on γ.

2.17 (a) Show that the function in (2.35) is convex in $\boldsymbol{\beta}$ for a fixed σ^2, but nonconvex in σ^2 for a fixed $\boldsymbol{\beta}$.

(b) Consider a reparameterization of the log-likelihood in (2.35) using

$$\rho = \sigma^{-1}, \quad \gamma = \frac{\boldsymbol{\beta}}{\sigma}.$$

Show that the reparameterized objective function

$$-\log \rho + \frac{1}{2n}||\rho \boldsymbol{y} - \boldsymbol{X}\gamma||_2^2 + 2\lambda||\gamma||_1,$$

is jointly convex in (ρ, γ) [Stadler et al., 2010].

2.18 Show that for the normal linear regression model the negative log-likelihood function in (2.35) is not jointly convex in $(\boldsymbol{\beta}, \sigma)$, but the reparameterized log-likelihood function (2.36) is jointly convex in $(\boldsymbol{\theta}, \phi)$.

2.19 For the optimization problem

$$\begin{aligned} &\min_x \quad \ell(x), \\ &\text{subject to} \quad x = 0, \end{aligned} \tag{2.61}$$

find the dual function $f(\lambda)$ and show that $f(\lambda) = \ell^*(-\lambda)$, where ℓ^* is the conjugate function in (2.56).

2.20 Use the definition of convexity in (2.46) to prove that a twice differentiable function $g(x)$, $x \in \mathbb{R}$ is convex on an interval if and only if its second derivative $g''(x)$ is nonnegative on that interval. Hint: Apply the Mean Value Theorem twice to obtain an expression for the second derivative $f''(\cdot)$, then use (2.46) to prove nonnegativity.

2.21 (a) Show that the Hessian matrix of (2.51) with respect to $(\phi, \boldsymbol{\theta})$ has the form (2.52). Recall that for the function $f(x, y)$, its Hessian matrix is

$$H = \begin{bmatrix} \frac{\partial f(x,y)}{\partial x^2} & \frac{\partial f(x,y)}{\partial x \partial y} \\ \frac{\partial f(x,y)}{\partial y \partial x} & \frac{\partial f(x,y)}{\partial y^2} \end{bmatrix}.$$

(b) Show that the determinant of (2.52) is equal to 2.53. Hint: For a block matrix

$$\det \begin{bmatrix} A & B \\ C & D \end{bmatrix}$$

is equal to

$$\det(D)\det(A - BD^{-1}C)$$

2.22 Consider the Boston housing dataset from the R package MASS that contains housing values in 506 Boston suburbs with 13 predictors. The goal of this exercise is to use sparse regression techniques to predict the median house price (medv) using the 13 predictors.

(a) Plot the histogram of medv, using function hist() and observe that the distribution is right skewed. Take the logarithmic transformation to fix the right skewness.

(b) Split the original data into train 80% and test 20% datasets.

(c) Use cv.glmnet() function from the glmnet package to fit a Lasso regression with 10-fold cross-validation and predict the log of median house prices on test dataset.

(d) Repeat the previous step but using cv.ncvreg() function from the ncvreg package.

3

Covariances and Dependencies

The notion of dependence between two or more random variables is at the heart of statistics and modern data sciences with applications in almost every area of human endeavors. Historically, much focus has been placed on linear dependence where the correlation coefficient $\rho \in [-1, 1]$ between two random variables is the most common and popular concept. A covariance/correlation matrix carries information about the pairwise dependence of the components of a random vector. However, information on the nature of dependence among more than two variables as in regressions and graphs are hidden in their inverse, eigenvalues-eigenvectors (spectral decomposition) and the Cholesky factors among others. Some basic properties of covariance matrices with a focus on the positive-definiteness are presented in this chapter, along with a few examples of structured covariance matrices with one or two parameters. While nonlinear dependence is prevalent in practice, there are noticeably fewer popular nonlinear tools available. The notion of bivariate copula, a distribution of two random variables on $[0, 1]^2$ with uniform marginals seems to be the most formidable counterpart of the correlation coefficient ρ. The role of vine copulas (trees) in construction of multivariate copulas using only bivariate copulas as building blocks is reviewed briefly.

3.1 Covariance Matrices

The covariance matrix of a random vector, its basic properties along with its characterization as a positive-definite matrix are presented in this section.

For $Y = (Y_1, \ldots, Y_p)'$ a p-dimensional random vector with the *mean vector*

$$\boldsymbol{\mu} = E(Y) = (\mu_1, \ldots, \mu_p)', \ \mu_i = E(Y_i). \tag{3.1}$$

Its *covariance matrix* is defined as the $p \times p$ matrix

$$\Sigma = E(Y - \boldsymbol{\mu})(Y - \boldsymbol{\mu})' = (\sigma_{ij}), \tag{3.2}$$

where $\sigma_{ij} = E(Y_i - \mu_i)(Y_j - \mu_j)$ is the covariance between the two random variables Y_i and Y_j for $i \neq j$, and σ_{ii} is simply the variance of Y_i. Two distinct random variables Y_i and Y_j are said to be *uncorrelated* or *orthogonal* if their covariance $\sigma_{ij} = 0$. It is evident that if entries of the p-vector Y are pairwise uncorrelated, then its covariance

DOI: 10.1201/9781003638896-3

matrix is a diagonal matrix with $\sigma_{ii} > 0$ as its diagonal entries and written compactly as

$$\mathbf{\Sigma} = \text{diag}(\sigma_{11}, \ldots, \sigma_{pp}).$$

A covariance matrix is characterized by its *positive-definiteness*, a deep mathematical concept, constraining all its entries simultaneously. The definition (3.2) suggests that like the variance of a random variable, the covariance matrix of a random vector is "nonnegative". More precisely, for any vector $\mathbf{c} = (c_1, \ldots, c_p)'$ the variance of the linear combination $\mathbf{c}'Y = \sum_{i=1}^{p} c_i Y_i$:

$$\mathbf{c}'\mathbf{\Sigma}\mathbf{c} = \sum_{i=1}^{p}\sum_{j=1}^{p} c_i c_j \sigma_{ij} = E\left[\mathbf{c}'(Y - \boldsymbol{\mu})\right]^2 = \text{Var}\left[\mathbf{c}'(Y - \boldsymbol{\mu})\right] \geq 0, \tag{3.3}$$

is nonnegative. Motivated by this, we say a symmetric matrix $\mathbf{\Sigma}$ is *nonnegative definite* or *positive semi-definite* (psd) if it satisfies (3.3). It is said to be *positive-definite* (pd) if the inequality (3.3) is strict for all non-zero vectors $\mathbf{c}$.

Note that if the equality in (3.3) holds for a nonzero vector $\mathbf{c} = \mathbf{c}_0 \neq 0$, then

$$\mathbf{c}_0'(Y - \boldsymbol{\mu}) = \sum_{i=1}^{p} c_{0,i}(Y_i - \mu_i) = 0, \tag{3.4}$$

so that the random variables $Y_1, \ldots, Y_p$ are *linearly dependent*. To avoid such redundancies, in the sequel we often work with the positive-definite covariance matrices.

Although the definition (3.3) looks deceptively simple, verifying it for all nonzero $\boldsymbol{c}$ is a computationally and/or analytically challenging task. However, a few specifically chosen $\boldsymbol{c}$'s lead to some interesting insights into the world of positive-definite matrices:

1. If $\mathbf{\Sigma}$ is positive-definite, then all its diagonal entries must be nonnegative:

$$\sigma_{ii} \geq 0, \quad i = 1, \ldots, p.$$

   This follows from (3.3) by choosing $\boldsymbol{c}$ to have 1 as its ith entry for $i = 1, \ldots, p$, and the rest zeros. It is hard to believe the converse that nonnegativity of the diagonal entries implies positive-definiteness of the matrix; see Problem 3.1. Perhaps, one needs to examine some larger submatrices, as discussed next.

2. If $\mathbf{\Sigma}$ is positive-definite, then all 2×2 symmetric matrices with entries $\sigma_{ii}, \sigma_{ij}, \sigma_{jj}, i \neq j$, are positive-definite. Moreover, it follows that

$$|\sigma_{ij}| \leq \sqrt{\sigma}_{ii}\sqrt{\sigma}_{jj},\ i \neq j. \tag{3.5}$$

   See Problem 3.2 for a possible converse and extension to larger submatrices.

3. $\mathbf{\Sigma}$ is positive-definite, if and only if all its *leading principal minors*, that is for all $k \times k$ matrices

$$\mathbf{\Sigma}_{kk} = \begin{pmatrix} \sigma_{11} & \cdots & \sigma_{1k} \\ \vdots & \ddots & \vdots \\ \sigma_{k1} & \cdots & \sigma_{kk} \end{pmatrix} > 0, \quad k = 1, \ldots, p, \tag{3.6}$$

are positive-definite. This follows from (3.3) for vectors $\boldsymbol{c}$ having arbitrary first k entries for $k = 1, \ldots, p$, and the rest zeros.

Throughout this book we use fairly standard matrix operations and notation. For example, for a matrix A, A' stands for its transpose; for square matrices A and B, $\mathrm{tr}(A)$ and $\det A$ or $|A|$ stand for its trace (sum of diagonal values) and determinant, respectively, where $\mathrm{tr}(AB)$=$\mathrm{tr}(BA)$ and $|AB| = |A||B|$, and A^{-1} stands for the inverse of a square, invertible matrix A. The book Golub and Van Loan [1996] provides an excellent coverage of topics in matrix algebra with a focus on computational issues, see Abadir and Magnus [2005] for topics with statistical and econometrical relevance.

Linear transformations of one or several random vectors are common in statistics and machine learning. The impact of such transformations on their dependence can be studied readily using the basic definitions (3.1) and (3.2). In fact, for Y, a $p \times 1$ random vector with covariance matrix $\mathrm{Cov}(Y)$, and suitable (nonrandom) matrices A, B, and vectors $\boldsymbol{c}$ and $\boldsymbol{d}$, following the line of computation in (3.3) one obtains

$$\mathrm{Cov}(AY + \boldsymbol{b},\ CY + \boldsymbol{d}) = A\mathrm{Cov}(Y)C^{'}, \tag{3.7}$$

so that adding constants to random variables does not impact their linear dependence, but multiplying will. The result suggests ways to construct more elaborate covariance matrices from the simpler ones like the identity or diagonal matrices or to reduce more general covariance matrices to such simple forms through various matrix decompositions such as spectral and Cholesky decompositions. For example, taking $A = C$ and starting from a Y with the identity covariance matrix, it follows that AA' is a genuine covariance matrix for any matrix A. In fact, all covariance matrices turn out to have this form. In other words, a matrix $\boldsymbol{\Sigma}$ is a covariance matrix, if and only if it is of the form

$$\boldsymbol{\Sigma} = AA', \tag{3.8}$$

for a matrix A. It can be taken to be a symmetric matrix (symmetric square root) or a lower triangular matrix, usually referred to as the asymmetric square root of $\boldsymbol{\Sigma}$. Proof of the existence of A relies on the spectral decomposition theorem of symmetric matrices discussed later.

Whatever the form of A, (3.8) suggests the existence of vectors $\boldsymbol{u}_1, \ldots, \boldsymbol{u}_p$ in $\mathbb{R}^p$, think of the rows of A, such that

$$\sigma_{ij} = \boldsymbol{u}_i'\boldsymbol{u}_j = (\boldsymbol{u}_i, \boldsymbol{u}_j), \tag{3.9}$$

so that entries of a positive-definite matrix are realized as the inner-product of certain vectors from an inner-product space. Consequently, using (3.9) one can reinterpret (3.5) as the Cauchy-Schwarz inequality:

$$|(\boldsymbol{u}_i, \boldsymbol{u}_j)|^2 \leq (\boldsymbol{u}_i, \boldsymbol{u}_i)(\boldsymbol{u}_j, \boldsymbol{u}_j),\ \boldsymbol{u}_i, \boldsymbol{u}_j \in \mathbb{R}^p, \tag{3.10}$$

with equality, if and only if $\boldsymbol{u}_i = c\boldsymbol{u}_j$ for a scalar c.

In view of (3.9), the matrix $\boldsymbol{\Sigma} = (\sigma_{ij})$ is sometimes called the *Gram matrix* of the vectors $\boldsymbol{u}_1, \ldots, \boldsymbol{u}_p$, and written as $\boldsymbol{\Sigma} = G(\boldsymbol{u}_1, \ldots, \boldsymbol{u}_p)$. In general, the result holds when the inner-product $(\cdot,\cdot)$ is formed for suitable vectors (functions) in any inner product spaces. It plays a central role in establishing positive-definiteness of matrices, construction of the reproducing kernel Hilbert spaces [Fan et al., 2020c, Section 2.7] which is popular in modern machine learning. For a proof of spectral representation of stationary processes and the role of unitary operators, etc., see Pourahmadi [2001].

Given a symmetric matrix, a great deal of ingenuity is needed to establish its positive-definiteness. Computing its eigenvalues is not a viable option as it is computationally expensive for large numerical matrices and analytically challenging for non-numerical matrices. The next two examples illustrate ways to establish positive-definiteness of certain matrices.

Example 9 (Compound Symmetry) *The compound symmetry (CS) covariance matrix*

$$\boldsymbol{\Sigma} = \sigma^2 R = \sigma^2 \left[(1-\rho)I + \rho \mathbf{1}_p \mathbf{1}_p'\right]$$

has only two parameters σ^2 and ρ. We use the abbreviation CS(σ^2, ρ) in the sequel to emphasize these parameters.

A direct and instructive method of establishing the positive-definiteness of the CS matrix is to construct a random vector Y having such a $\boldsymbol{\Sigma}$ as its covariance matrix. Let us start with defining the random variables

$$Y_i = X + \varepsilon_i, \quad i = 1, \dots, p, \tag{3.11}$$

where X and ε_i's are independent random variables with means zero and $Var(X) = \sigma_x^2$ with $Var(\boldsymbol{\epsilon}_i) = \sigma_e^2$. Then, with $Y = (Y_1, \dots, Y_p)', \boldsymbol{\epsilon} = (\varepsilon_1, \dots, \varepsilon_p)'$, the vector form of (3.11) is

$$Y = \mathbf{1}_p X + \boldsymbol{\epsilon}, \tag{3.12}$$

and computing its covariance matrix one arrives at

$$Cov(Y) = \sigma_x^2 \mathbf{1}_p \mathbf{1}_p' + \sigma_e^2 I_p = \sigma^2[(1-\rho)I_p + \rho \mathbf{1}_p \mathbf{1}_p'], \tag{3.13}$$

which is of the form CS(σ^2, ρ) with

$$\sigma^2 = \sigma_x^2 + \sigma_e^2, \quad \rho = \frac{\sigma_x^2}{\sigma_x^2 + \sigma_e^2}.$$

Here it is instructive to think of the random variable X as a common factor (signal) shared by all the p variables. For example, some genetic characteristics are common among the siblings in a family due to shared parents, and for this reason the parameter ρ is called the *heritability index.*

Example 10 *For any positive scalars $\alpha_1, \dots, \alpha_p$, show that the matrix $A = \left(\frac{1}{\alpha_i + \alpha_j}\right)$ is positive-definite.*

We first show that it is a Gram matrix. In fact, choosing the functions $f_i(x) = x^{\alpha_i - 1/2}, i = 1, \ldots, p$, *in* $L^2[0, 1]$ *with the usual inner product, using basic integration it can be verified that*

$$\frac{1}{\alpha_i + \alpha_j} = (f_i, f_j) = \int_0^1 f_i(x) f_j(x) dx.$$

In the special case of $\alpha_i = i$, *for all* i, *the matrix* A *is known as the Hilbert matrix.*

For more interesting examples of the above type and discussions on the role of the Grammian matrices, see Bhatia [2006]. Several popular examples of positive-definite functions defined on $\mathbb{Z}^2$ of interest in spatial data analysis can be found in Chapter 12.

3.2 Three Matrix Decompositions

In this section, we briefly discuss three of the most important matrix decompositions, namely the spectral, singular value and Cholesky decompositions. The eigen (eigenvalue-eigenvector) decomposition or the spectral decomposition of a square matrix is a powerful tool in matrix computation and analysis. For rectangular matrices, a similar role is played by its *singular value decomposition* (SVD). Applications of these decompositions, along with the Cholesky decomposition of a covariance matrix to statistics, computation and dimension-reduction are explored further in the next chapter.

A deeper appreciation of the notions of eigenvalues and eigenvectors is crucial for understanding the spectral decomposition and applications in statistics. A $p \times p$ square matrix A is said to have an *eigenvalue* λ, with a corresponding *eigenvector* $\boldsymbol{x} \neq 0$, if

$$A\boldsymbol{x} = \lambda \boldsymbol{x}, \quad \text{or} \quad (A - \lambda I_p)\boldsymbol{x} = 0. \tag{3.14}$$

For this homogeneous linear system of equations to have a nonzero vector $\boldsymbol{x}$ as its solution, the eigenvalue λ must be chosen so that the matrix $(A - \lambda I_p)$ is singular or its determinant is zero, i.e., $|A - \lambda I| = 0$. This is a polynomial equation of degree p in λ with possibly complex-valued roots. In what follows, we normalize $\boldsymbol{x}$ so that it has unit length, namely $\boldsymbol{x}'\boldsymbol{x} = 1$ and denote such a generic normalized eigenvector by $\mathbf{e}$.

Next, we illustrate in great details the process of computing the eigenvalues and eigenvectors for two simple asymmetric and symmetric matrices.

Example 11 *(Real and Complex Eigenvalues)*

(a) The eigenvalues of the 2×2 *matrix* $A = \begin{pmatrix} 1 & 9 \\ -1 & 1 \end{pmatrix}$ *satisfy the quadratic equation*

$$|A - \lambda I| = \begin{vmatrix} 1 - \lambda & 9 \\ -1 & 1 - \lambda \end{vmatrix} = (1 - \lambda)^2 + 9 = 0.$$

Its roots $\lambda = 1 \pm 3i$, with $i = \sqrt{-1}$, are complex-valued, this is not unusual since the matrix A is not symmetric.

(b) The eigenvalues of the 2×2 *symmetric matrix*

$$A = \begin{pmatrix} 1 & 3 \\ 3 & 1 \end{pmatrix},$$

are $\lambda_1 = 4$ *and* $\lambda_2 = -2$ *which are real-valued and ordered from the largest to the smallest. The normalized eigenvector corresponding to* $\lambda_1 = 4$ *is* $\mathbf{e}_1 = \left(\frac{1}{\sqrt{2}}, \frac{1}{\sqrt{2}}\right)'$ *and that corresponding to* $\lambda_2 = -2$ *is* $\mathbf{e}_2 = \left(-\frac{1}{\sqrt{2}}, \frac{1}{\sqrt{2}}\right)'$. *Interestingly,* $\mathbf{e}_1$ *and* $\mathbf{e}_2$ *are* orthogonal *to each other, that is,* $\mathbf{e}_1'\mathbf{e}_2 = 0$ *and hence the* 2×2 *matrix* $P = (\mathbf{e}_1, \mathbf{e}_2)$ *is an* orthogonal matrix, *namely* $PP' = P'P = I_2$. *Moreover, routine computation shows that the matrix P diagonalizes A, namely,*

$$P'AP = \Lambda = \begin{pmatrix} 4 & 0 \\ 0 & -2 \end{pmatrix}, \tag{3.15}$$

where the diagonal entries of Λ *are precisely the ordered eigenvalues of A.*

It turns out that these properties of the eigenvalues and eigenvectors of the 2×2 symmetric matrix A hold for all symmetric matrices. In fact, (3.15) is a precursor or an implicit statement of the fundamental spectral decomposition of symmetric matrices.

To state the spectral decomposition, as motivated by the previous numerical example, we start with A a $p\times p$ symmetric matrix with p pairs of real (ordered, distinct) eigenvalues and (orthonormal) eigenvectors $(\lambda_1, \mathbf{e}_1)$, $(\lambda_2, \mathbf{e}_2), \ldots, (\lambda_p, \mathbf{e}_p)$, with

$$\lambda_1 \geq \lambda_2 \geq \cdots \geq \lambda_p, \quad \Lambda = \text{diag}(\lambda_1, \ldots, \lambda_p), \quad P = (\mathbf{e}_1, \mathbf{e}_2, \ldots, \mathbf{e}_p),$$

where Λ is *diagonal* and P is an orthogonal matrix, namely $PP' = P'P = I$. The *spectral decomposition* states that any symmetric matrix A can be diagonalized by an orthogonal matrix P:

$$A = P\Lambda P' = \lambda_1\mathbf{e}_1\mathbf{e}_1' + \lambda_2\mathbf{e}_2\mathbf{e}_2' + \cdots + \lambda_p\mathbf{e}_p\mathbf{e}_p'. \tag{3.16}$$

That the matrix A is decomposed as a sum of rank-one matrices $\mathbf{e}_i\mathbf{e}_i'$'s with decreasing coefficients λ_i's suggests truncation of the sum as a simple and natural way to approximate the matrix A by lower rank matrices. In fact, this idea is at the heart of principal component analysis (PCA), factor analysis, principal component regression and other dimension-reduction techniques in high-dimensional data analysis and machine learning.

Another powerful consequence of the spectral decomposition of a symmetric matrix is the characterization of its positive-definiteness in terms of the positivity of its eigenvalues: A symmetric matrix A is *positive-definite*, if and only if all its eigenvalues are positive.

Furthermore, the matrices P, Λ in (3.16) being orthogonal and diagonal, respectively, renders computing various powers of a symmetric matrix A rather easy and transparent. Indeed, it follows from (3.16) that $A^2 = AA = P\Lambda P'P\Lambda P' = P\Lambda^2 P'$, and $A^m = P\Lambda^m P'$, for any positive integer m where for a diagonal matrix $\Lambda^m = \text{diag}(\lambda_1^m, \ldots, \lambda_n^m)$. Evidently, the same argument works for noninteger and negative powers. For example, if all the eigenvalues of the matrix A are nonnegative, then for $m = \frac{1}{2}$, setting

$$A^{1/2} = \sqrt{\lambda_1} e_1 e_1' + \cdots + \sqrt{\lambda_n} e_n e_n' = P\Lambda^{1/2} P',$$

it follows that

$$A^{1/2} A^{1/2} = P\Lambda^{1/2} P' P\Lambda^{1/2} P' = P\Lambda P' = A,$$

so that $A^{1/2}$ is, indeed, the *symmetric square root* of the matrix A. Similarly, the *inverse* of a nonsingular symmetric matrix A is given by

$$A^{-1} = \sum_{i=1}^{p} \lambda_i^{-1} e_i e_i' = P\Lambda^{-1} P'.$$

There are other functions of a matrix like the (matricial) logarithm of a covariance matrix which play important roles in modeling covariances using covariates. We use the spectral decomposition of a symmetric matrix to define and study its properties in Section 3.5.

There is an interesting and fruitful connection between the eigenvalues and maximum of quadratic forms involving a positive-definite matrix A. Evidently, from (3.14) the definition of the eigenvalue-eigenvector it follows that for a nonzero eigenvector $\boldsymbol{x}$,

$$\frac{\boldsymbol{x}' A \boldsymbol{x}}{\boldsymbol{x}' \boldsymbol{x}} = \lambda,$$

which connects the eigenvalues to ratios of certain quadratic forms. In fact, maximizing the numerator quadratic form subject to $\boldsymbol{x}'\boldsymbol{x} = 1$ and using the idea of Lagrange multiplier leads to

$$\frac{\partial}{\partial \boldsymbol{x}} \{\boldsymbol{x}' A \boldsymbol{x} - \lambda(\boldsymbol{x}'\boldsymbol{x} - 1)\} = 2(A\boldsymbol{x} - \lambda \boldsymbol{x}) = 0 \longrightarrow A\boldsymbol{x} = \lambda \boldsymbol{x}, \tag{3.17}$$

identifying $\boldsymbol{x}$ as the eigenvector corresponding to the largest eigenvalue λ.

Next, a more rigorous proof of this is given using the spectral decomposition and the following generalized Cauchy-Schawrz inequality.
For any $p \times p$ positive-definite matrix A and vectors $\boldsymbol{u}, \boldsymbol{v} \in \mathbb{R}^p$,

$$(\boldsymbol{u}'\mathbf{v})^2 \leq (\boldsymbol{u}' A \boldsymbol{u})(\mathbf{v}' A^{-1} \mathbf{v}), \tag{3.18}$$

where the equality holds, if and only if $\boldsymbol{u} = cA^{-1}\mathbf{v}$ for a scalar c. Thanks to the existence of $A^{1/2}$ and its inverse for a positive-definite matrix, the proof follows by writing $\boldsymbol{u}'\mathbf{v} = (\boldsymbol{u}' A^{1/2})(A^{-1/2}\mathbf{v})$, and applying the Cauchy-Schawrz inequality (3.10); see Problem 3.19.

Next, for A a $p \times p$ positive-definite matrix with ordered eigenvalues and normalized eigenvectors $(\lambda_i, \mathbf{e}_i)$ and the spectral decomposition (3.16), we show that

$$\max_{x \neq 0} \frac{\boldsymbol{x}' A \boldsymbol{x}}{\boldsymbol{x}' \boldsymbol{x}} = \lambda_1, \tag{3.19}$$

with the maximum attained for $\boldsymbol{x} = \mathbf{e}_1$.

Start with the spectral decomposition $A = P\Lambda P'$, set $\boldsymbol{y} = P'\boldsymbol{x}$ and note that $\boldsymbol{x}'\boldsymbol{x} = \boldsymbol{y}'\boldsymbol{y}$. The result then follows from

$$\begin{aligned} \boldsymbol{x}'A\boldsymbol{x} &= \boldsymbol{x}'A^{1/2}A^{1/2}\boldsymbol{x} = \boldsymbol{x}'P\Lambda^{1/2}P'P\Lambda^{1/2}P'\boldsymbol{x} \\ &= \boldsymbol{y}'\Lambda\boldsymbol{y} = \sum_{i=1}^{p} \lambda_i y_i^2 \le \lambda_1 \sum_{i=1}^{p} y_i^2 = \lambda_1 \boldsymbol{x}'\boldsymbol{x}, \end{aligned} \tag{3.20}$$

and the equality is achieved for $\boldsymbol{y} = P'\boldsymbol{e}_1 = (1, 0, \ldots, 0)'$ corresponding to the choice of $\boldsymbol{x} = \boldsymbol{e}_1$. Next, we give the outline of proving

$$\max_{\boldsymbol{x} \perp \mathbf{e}_1, \ldots, \mathbf{e}_k} \frac{\boldsymbol{x}'A\boldsymbol{x}}{\boldsymbol{x}'\boldsymbol{x}} = \lambda_{k+1}, \tag{3.21}$$

and the maximum is attained for $\boldsymbol{x} = \mathbf{e}_{k+1}, k = 1, 2, \ldots, p-1$, where the symbol $\boldsymbol{x} \perp \boldsymbol{y}$ stands for "$\boldsymbol{x}$ is orthogonal to $\boldsymbol{y}$". The key idea is to note that

$$\boldsymbol{x} = P\boldsymbol{y} = y_1\boldsymbol{e}_1 + \cdots + y_p\boldsymbol{e}_p \perp \boldsymbol{e}_1, \ldots, \boldsymbol{e}_k,$$

implies that $y_i = 0, i \le k$, and proceed as in (3.20); see Problem 3.18.

Other functions of the eigenvalues like their products and ratios appear in various applications. For example, the *condition number* of a positive-definite matrix defined as λ_1/λ_p or the ratio of its largest and smallest eigenvalues, is used commonly as a measure of ease of computing of inverses, determinants, etc.

The *singular value decomposition* (SVD) of a rectangular matrix is a natural generalization of the spectral decomposition of a square symmetric matrix. It plays important roles in low-rank approximation of matrices, solving regression problems with a rectangular design matrix $\boldsymbol{X}$ and connections with the eigendecomposition and principal component analysis. For $\boldsymbol{X}$ an $n \times p$ matrix of rank $r \le \min\{n, p\}$, there exist matrices $\mathbf{U}$, $\mathbf{V}$ and $\mathbf{D}$ such that

$$\boldsymbol{X} = \mathbf{UDV}' = \sum_{k=1}^{r} d_k \boldsymbol{u}_k \mathbf{v}_k', \tag{3.22}$$

where the columns of $\mathbf{U} = (\boldsymbol{u}_1, \ldots, \boldsymbol{u}_r), \mathbf{V} = (\mathbf{v}_1, \ldots, \mathbf{v}_r)$ are orthonormal, and the diagonal entries of $\mathbf{D} = \mathrm{diag}(d_1, \ldots, d_r)$ are ordered nonnegative numbers: $d_1 \ge d_2 \cdots \ge d_r > 0$. The columns of $\mathbf{U}$ and $\mathbf{V}$ are called the *left and right-singular vectors* of $\boldsymbol{X}$, respectively, and the diagonal entries of $\mathbf{D}$ are the corresponding *singular values* [Golub and Van Loan, 1996].

The form of the decomposition $\boldsymbol{X}$ as a weighted sum of rank-1 matrices with deceasing weights suggests that for any integer $l \le r$, the matrix

$$\boldsymbol{X}^{(l)} = \sum_{k=1}^{l} d_k \boldsymbol{u}_k \mathbf{v}_k', \tag{3.23}$$

obtained by truncating the sum in (3.22), is the closest rank-l approximation to $\boldsymbol{X}$ in the Frobenius norm in (3.27) (Eckart-Young Theorem), i.e.,

$$\boldsymbol{X}^{(l)} = \operatorname*{argmin}_{\boldsymbol{X}^* \in \mathcal{M}_{(l)}} ||\boldsymbol{X} - \boldsymbol{X}^*||_F^2,$$

where $\mathcal{M}_{(l)}$ is the set of all $n \times p$ matrices of rank l.

Assuming the existence of the SVD of X, how does one compute the $\mathbf{v}_i$'s, $\boldsymbol{u}_i$'s and d_i's? Amazingly, the simple answer lies in noting that

$$X'X = \mathbf{VDU'UDV'} = \mathbf{VD}^2\mathbf{V}', \tag{3.24}$$

so that the columns of $\mathbf{V}$ are the normalized eigenvectors of $X'X$ corresponding to the eigenvalues d_i^2. Then, the $\boldsymbol{u}_i$'s are computed from

$$X\mathbf{v}_i = d_i\boldsymbol{u}_i,\ i = 1, \ldots, r, \tag{3.25}$$

which can be shown to be the eigenvectors of XX'. The $\mathbf{v}_i$'s and $\boldsymbol{u}_i$'s have proved to be useful in statistics as the directions of the principal components and the (normalized) principal component score vectors of the data matrix X, see Chapter 7.

These results are particularly useful in high-dimensional computation and data analysis. For example, for X a 10000×5 matrix it reduces the expensive task of computing the eigen-decomposition of the 10000×10000 matrix XX' to that of the 5×5 matrix $X'X$. This is also central to the low-rank plus sparse approximations of covariance matrices as $\frac{1}{n}X'X$ is, indeed, the sample covariance matrix of the column-centered data; see Problem 3.16.

Cholesky decomposition of covariance matrices is ubiquitous in statistics, computer science and computational linear algebra. The standard Cholesky decomposition of a positive-definite matrix encountered in some optimization techniques and matrix computation [Golub and Van Loan, 1996] is of the form

$$A = CC', \tag{3.26}$$

where $C = (c_{ij})$ is a unique lower-triangular matrix with positive diagonal entries. Given a $p \times p$ positive-definite matrix A, finding its Cholesky factor amounts to solving $p(p+1)/2$ quadratic equations in the entries of C; see Problem 3.11. Statistical relevance or interpretation of the entries of C for a covariance matrix is not clear in its present form, but a slight modification of it, discussed in Chapter 4, will reveal its deep connection to certain regression parameters.

3.3 Norms and Proximity of Matrices

The accuracy of matrix approximations is usually assessed using various matrix norms. In this section we present a few commonly used matrix norms. The choice of a norm or loss function for matrix approximation varies according to specific applications and needs.

For a matrix $A = (a_{ij})$, its squared Frobenius norm is defined by

$$\|A\|_F^2 = \mathrm{tr}(A'A) = \sum\sum a_{ij}^2, \tag{3.27}$$

which is the sum of squares of all its entries. The choice of the norm for covariance matrix estimation critically depends on the norm used to measure the error.

For instance, matrix estimation under the Frobenius norm is essentially a vector estimation problem. However, under spectral (operator) norm, introduced next, the matrix estimation is highly non-additive with respect to the (two-way) entries of the matrix.

For A a $p \times p$ matrix and $0 \le r, s \le \infty$, we define a general class of norms by

$$||A||_{(r,s)} = \max\{||Ax||_s \,;\, x \in \mathbb{R}^p,\ ||x||_r = 1\} \tag{3.28}$$

where $||x||_r = (|x_1|^r + \cdots + |x_p|^r)^{1/r}$ is the rth-norm of the vector x. Consequently,

$$||Ax||_s \le ||A||_{(r,s)} ||x||_r. \tag{3.29}$$

The case of $r = s = 2$ is of particular interest, where $||A|_{(2,2)} = ||A||_2$ is simply the *operator* or the *spectral* norm of A, which is also denoted and defined more directly as

$$||A||^2 = \lambda_{\max}(A'A),$$

where $\lambda_{\max}(\cdot)$ is the largest magnitude of the eigenvalues of the indicated matrix. When A is a symmetric matrix, then $||A|| = \max_{i \le p} |\lambda_i|$.

It is plausible that matrices that are close in some sense should have eigenvalues and eigenvectors that are also close. This statement is made more precise by using some commonly used norms for matrices. The spectral decomposition which writes a symmetric matrix A as the sum of rank-1 matrices with decreasing weights suggests a natural strategy to approximate A by lower rank matrices. In particular, establishing consistency of covariance estimators under the spectral norm is more natural and useful in statistics as it implies consistency of the eigenvalues and eigenvectors. Indeed, the following results due to Weyl [1912], and Davis and Kahan [1970] are the key in linking the closeness of the eigenstructures of two matrices to the operator norm of their difference.

Theorem 2 *Let $\{\lambda_i\}, \{\widehat{\lambda}_i\}, \{e_i\}$, and $\{\widehat{e}_i\}$ be the eigenvalues and associated eigenvectors of the two $p \times p$ matrices $\boldsymbol{\Sigma}, \widehat{\boldsymbol{\Sigma}}$, respectively, where the eigenvalues are in descending order. Then*

(a) (Weyl's Theorem): $|\widehat{\lambda}_i - \lambda_i| \le ||\widehat{\boldsymbol{\Sigma}} - \boldsymbol{\Sigma}||$.

(b) (Davis and Kahan sinθ Theorem): If $\widehat{e}_i' e_i \ge 0$, then

$$||\widehat{e}_i - e_i|| \le \frac{\sqrt{2}||\widehat{\boldsymbol{\Sigma}} - \boldsymbol{\Sigma}||}{\min(|\widehat{\lambda}_{i-1} - \lambda_i|, |\lambda_i - \widehat{\lambda}_{i+1}|)}.$$

From this theorem it is immediate that if the sequence $||\widehat{\boldsymbol{\Sigma}_n} - \boldsymbol{\Sigma}|| \to 0$, then both the *eigenvalues* and *eigenvectors* of $\boldsymbol{\Sigma}_n$ converge to those of $\boldsymbol{\Sigma}$. In particular, consistency or convergence of a covariance matrix estimator in the operator norm guarantees the convergence of their principal components loadings [Johnstone and Lu, 2009b].

3.4 Structured Covariance Matrices

In addition to maintaining positive-definiteness, another challenge in estimation of covariance matrices is its large number of parameters which grows quadratically in p. A popular remedy is to restrict attention to structured covariance matrices with 2 or 3 parameters. A few examples of commonly used structured covariance matrices such as compound symmetry (CS), autoregression (AR) are introduced in this section. Such covariance matrices are used often in time series, longitudinal data analysis, engineering, finance and economics, just to name a few.

Example 9 cont. *The compound symmetry (CS) correlation matrix*

$$R = \left[(1-\rho)I + \rho \mathbf{1}_p \mathbf{1}_p'\right]$$

or CS(ρ, 1) has only one parameter ρ on the off-diagonal entries. Its positive-definiteness controlled by ρ can be checked directly by verifying that for any vector **c**,

$$\mathbf{c}' R \mathbf{c} = (1-\rho)\sum_{i=1}^{p}(c_i - \bar{c})^2 + n\bar{c}^2[1 + (p-1)\rho], \tag{3.30}$$

where $\bar{c}$ is the average of the entries of **c**; *see Problem 3.8. Then, a sufficient condition for it to be positive-definite is that*

$$1 + (p-1)\rho \geq 0, \quad \text{or} \quad -(p-1)^{-1} \leq \rho \leq 1.$$

In fact, one can directly verify that when ρ is positive, its largest eigenvalue is

$$\lambda_1 = 1 + (p-1)\rho,$$

with the associated eigenvector

$$\mathbf{e}_1 = \frac{1}{\sqrt{p}}(1, \ldots, 1)' = \frac{1}{\sqrt{p}}\mathbf{1}_p,$$

and the remaining $(p-1)$ eigenvalues are

$$\lambda_2 = \lambda_3 = \cdots = \lambda_p = 1 - \rho,$$

so that the eigenvalues are all nonnegative only if $1 \geq \rho > -(p-1)^{-1}$.

Three noteworthy features of CS matrices which serve as the prime example and motivation for following topics of central importance in this book are:

1. **Spiked Covariance:** The largest eigenvalue of a CS matrix goes to infinity as $p \to \infty$, while the rest is bounded. This property makes it one of the most natural examples of the class of *spiked covariance matrices*, see section 7.1.3 and

Johnstone [2001], where a few of the eigenvalues grow to infinity and the rest remains bounded. This class plays a key role in PCA and studying the asymptotic properties of estimators of eigenvectors in PCA and factor analysis. What are the other concrete examples of spiked covariance matrices? Are there generalizations of the CS model (3.11) that are spiked? See Problem 3.9.

2. **Factor Models:** A generalization of (3.11) is the following *low-rank plus sparse (diagonal)* decomposition of a covariance matrix:

$$\Sigma = LL' + \Psi, \tag{3.31}$$

where L is a $p \times r$ ($r < p$) loading matrix and Ψ is a diagonal matrix. For $r = 1$ and the obvious choice of (L, Ψ) it reduces to the CS matrix. Such matrices, including the CS, do not have zero entries and are not sparse at all. However, they are ideal models for situations calling for non-sparse matrices.

3. **Long Memory:** In CS matrices the faraway off-diagonal entries, namely $\sigma_{|i-j|}$, is a constant and does not go to zero as $|i - j| \to \infty$. In other words, the corresponding process has *long memory*.

   The next example provides just the opposite of long-memory processes where its faraway off-diagonal entries drops off to zero geometrically. This is a highly desirable property of covariance matrices in statistics and physical sciences, mimicking dependence in nature where faraway measurements over time and space are less correlated than those near each other.

Example 12 (The AR(1) Covariance Structure) *It is defined simply as*

$$\Sigma = \sigma^2(\rho^{|i-j|}), \tag{3.32}$$

where $|\rho| < 1$ and $\sigma^2 > 0$. However, it is not easy to show directly that it is positive-definite. Fortunately, it arises as the covariance matrix of a process defined by

$$Y_t = \sum_{k=0}^{\infty} \rho^k \varepsilon_{t-k}, \tag{3.33}$$

where $\{\varepsilon_t\}$ is a sequence of i.i.d. random variables with $E(\varepsilon_t) = 0$ and $Var(\varepsilon_t) = \sigma_\varepsilon^2$. In fact, it can be shown that the random vector $Y = (Y_1, \ldots, Y_p)'$ has Σ as its covariance matrix, namely

$$Cov(Y_i, Y_j) = \sigma^2 \rho^{|i-j|},$$

where $\sigma^2 = \sigma_\varepsilon^2(1 - \rho^2)^{-1}$. Furthermore, the process $\{Y_t\}$ satisfies the difference equation

$$Y_t = \varepsilon_t + \sum_{k=1}^{\infty} \rho^k \varepsilon_{t-k} = \varepsilon_t + \rho \sum_{k=0}^{\infty} \rho^k \varepsilon_{t-1-k} = \rho Y_{t-1} + \varepsilon_t.$$

Processes satisfying such simple difference equations are called the autoregressive process *of order one, AR(1) for short.*

The next class subsumes CS and AR(1) covariance matrices and is at the heart of stationary time series and spatial data. It is characterized by the constancy of its entries along all (sub)diagonals where p or the number of parameters grows linearly with the size of the matrix.

Example 13 (Toeplitz Matrices) *Given a sequence $\{\gamma_k; k \in \mathbb{Z}\}$ of complex-numbers with $\gamma_k = \bar{\gamma}_{-k}$, and a positive integer p, the corresponding $p \times p$ Toeplitz matrix is defined by*

$$T_p = (\gamma_{i-j})_{i,j=1}^{p}.$$

We note that such a matrix is conjugate-symmetric ($T_p = T_p^{*'}$). Its positive-definiteness for a fixed p, requires that for any nonzero vector $\boldsymbol{c} = (c_1, \ldots, c_p)'$ with complex entries, it must hold that

$$\boldsymbol{c}^* T_p \boldsymbol{c} = \sum_{i,j=1}^{p} c_i \bar{c}_j \gamma_{i-j} \geq 0. \tag{3.34}$$

Often in modern high-dimensional statistics and machine learning, it is desired to allow p to grow gradually and approach infinity. In this situation (3.34) must hold for all $p \geq 1$ which amounts to imposing restriction on the whole sequence $\{\gamma_k; k \in \mathbb{Z}\}$. Now, it is instructive to view this sequence as a function on the set $\mathbb{Z}$ of all integers. This brings us to the more general notion of *positive-definite functions* on the more general domains (groups) $\mathbb{Z}^d$ and $\mathbb{R}^d$, d a positive integer, and their characterizations involving Fourier transform of certain nonnegative functions or measures. Here, we focus on functions $\gamma(\cdot)$ defined on the set of all integers; it is positive-definite if (3.34) holds for all $p \geq 1$, and all $\boldsymbol{c} = (c_1, \ldots, c_p)'$ with complex entries.

Theorem 3 *A function $\gamma(\cdot)$ defined on the set of all integers is positive-definite, if and only if there exists a distribution function $F(\cdot)$ so that*

$$\gamma(k) = \int_{-\pi}^{\pi} exp(ik\omega) dF(\omega), \quad k = 0, \pm 1, \ldots. \tag{3.35}$$

Moreover, when $F(\cdot)$ is differentiable, its nonnegative derivative $f(\cdot)$ can be expressed using Fourier inversion formula as

$$f(\omega) = \sum_{k=-\infty}^{\infty} \gamma(k) \exp(-ik\omega), \tag{3.36}$$

provided that $\sum_{k=-\infty}^{\infty} |\gamma(k)| < \infty$. The function f, called the spectral density function, plays a central role in the study of stationary stochastic processes with the autocovariance function $\gamma(\cdot)$, studied in Chapter 11.

The function f is also of interest because of its close connection with the asymptotic behavior of the eigenvalues of T_p for p large. We note that the eigenvalues of T_p are hard to compute even for the AR(1) covariance matrix, but for p large and under some smoothness conditions on f, the eigenvalues of T_p mimick closely the values of $f(\omega), \omega \in [-\pi, \pi]$ as described in the following fundamental *Szegö distribution theorem*. For an elementary proof and an extension, see Trench [2014].

Theorem 4 *Let $a \leq f(\omega) \leq b$ be a continuous function and $\lambda_{1p} \leq \ldots \leq \lambda_{pp}$ be the ordered eigenvalues of its associated $p \times p$ Toeplitz matrix T_p. Then,*

(a) $a \leq \lambda_{1p} \leq \ldots \leq \lambda_{pp} \leq b$, for all $p \geq 1$.

(b) $\lim_{p \longrightarrow \infty} \lambda_{1p} = a$ and $\lim_{p \longrightarrow \infty} \lambda_{pp} = b$.

(c) For any continuous function $g(\cdot)$ defined on $[a, b]$, it holds that

$$\lim_{p \longrightarrow \infty} \frac{1}{p} \sum_{i=1}^{p} g(\lambda_{ip}) = \frac{1}{2\pi} \int_{-\pi}^{\pi} g(f(\omega)) d\omega.$$

For example, choosing $g(x) = x$, it follows that the average of eigenvalues of T_p converges to the integral of f, with similar conclusions for functions $g(x) = x^r$, for a nonnegative power r. In view of the increasing importance of spiked covariance matrices in developing asymptotic framework for high-dimensional principal component analysis and factor models (see Section 7.1.3), the theorem provides valuable information on the distribution of eigenvalues of growing Toeplitz matrices. On another front, applications of the theorem to prediction theory of stochastic processes, signal processing and information theory are reviewed in Gray [2006] with another elementary proof.

The class of *banded matrices* where subdiagonals faraway from the main diagonal are zero plays important roles in statistics, numerical linear algebra and other applications. The simplest example is the tri-diagonal matrix

$$\mathbf{\Sigma} = \begin{pmatrix} \gamma_0 & \gamma_1 & 0 & \cdots & 0 \\ \gamma_1 & \gamma_0 & \gamma_1 & \ddots & \vdots \\ \vdots & \ddots & \ddots & \ddots & 0 \\ \vdots & & \ddots & \ddots & \gamma_1 \\ 0 & \cdots & 0 & \gamma_1 & \gamma_0 \end{pmatrix}, \tag{3.37}$$

for given γ_0, γ_1. Its eigenvalues and eigenvectors are given by

$$\lambda_k = \gamma_0 + 2\gamma_1 \cos k\theta, \;\; k = 1, \ldots, p$$

and

$$\mathbf{e}_k = (\sin k\theta, \sin 2k\theta, \ldots, \sin pk\theta)',$$

where $\theta = \pi/(p+1)$, for a neat derivation of these see Kazdan [2006]. For explicit inverse of banded Toeplitz matrices, see Trench [1985].

3.5 Log of a Covariance Matrix

In this section we show that the spectral decomposition of a symmetric matrix plays a central role in defining the log of a matrix. Like log-linear models for variances, the (matricial) logarithm of a covariance matrix plays important roles in removing the positive-definiteness constraint and hence modeling covariances using covariates [Chiu et al., 1996] and correlation matrices.

For a function $f(\cdot)$ with the Taylor expansion

$$f(x) = \sum_{k=0}^{\infty} f^{(k)}(0)\frac{x^k}{k!},$$

and a symmetric matrix A with the spectral decomposition (3.16), one defines formally its matrix version by

$$f(A) = \sum_{k=0}^{\infty} f^{(k)}(0)\frac{A^k}{k!} = P(\sum_{k=0}^{\infty} f^{(k)}(0)\frac{\Lambda^k}{k!})P' = Pf(\Lambda)P', \tag{3.38}$$

where $f(\Lambda) = \text{diag}(f(\lambda_1), \ldots, f(\lambda_p))$ is a diagonal matrix. Now, for $f(x) = e^x$, using (3.38), and a $p \times p$ symmetric matrix A, its matrix exponential e^A is defined by the power series expansion

$$e^A = I + \frac{A}{1!} + \frac{A^2}{2!} + \cdots,$$

where using the spectral decomposition of A, computing the power A^k simplifies and the whole task reduces to

$$e^A = P\text{diag}(e^{\lambda_1}, \ldots, e^{\lambda_p})P'.$$

The function of interest to us here is $f(x) = \log x$. Using its Taylor's expansion, the *logarithm of a matrix* A, denoted by $\log A$, is defined using (3.38) by

$$\log A = P(\log \Lambda)P'. \tag{3.39}$$

For a diagonal matrix Λ with positive diagonal entries, its logarithm is simply the diagonal matrix of $\log \lambda_i$'s:

$$\log \Lambda = \text{diag}(\log \lambda_1, \ldots, \log \lambda_p).$$

Evidently, from (3.39) the logarithm of a positive-definite matrix is simply a symmetric matrix with unconstrained entries, so that the log transformation removes the notorious positive definiteness constraint on a covariance matrix; see Problem 3.22. In fact, it is evident that a matrix A is positive-definite, if and only if $\log A$ is a real symmetric matrix.

The inverse transformation, namely going from $\log \boldsymbol{\Sigma}$ to the covariance matrix follows the same principle. However, entries of the covariance matrix turn out to be complicated functions of those of $\log \boldsymbol{\Sigma}$ and their statistical interpretation becomes harder; see Problem 3.21.

Example 14 (Extracting $\boldsymbol{\Sigma}$ from $\log \boldsymbol{\Sigma}$ for CS) *Consider the case*

$$\log \boldsymbol{\Sigma} = (\alpha - \beta) I_p + \beta J_p,$$

where J_p is the $p \times p$ matrix of one's. From Example 3.10, the eigenvalues of $\log \boldsymbol{\Sigma}$ are $\alpha + (p-1)\beta$ with the normalized eigenvector $\mathbf{e}_1 = \frac{1}{\sqrt{p}} 1_p$, and the other eigenvalues of $\boldsymbol{\Sigma}$ are $\alpha - \beta$ with multiplicity $p-1$. Thus, with $\Lambda = diag[\alpha + (p-1)\beta, \alpha - \beta, \ldots, \alpha - \beta]$ we have

$$\exp(\Lambda) = e^{\alpha - \beta} I_p + (e^{p\beta} - 1) diag(1, 0, \ldots, 0)$$

and

$$\boldsymbol{\Sigma} = P \exp(\Lambda) P' = e^{\alpha - \beta} I_p + p^{-1}(e^{p\beta} - 1) J_p$$

has exactly the same form as $\log \boldsymbol{\Sigma}$ [Chiu et al., 1996].

3.6 Correlations and Copulas

One of the simplest decompositions of a covariance matrix $\boldsymbol{\Sigma}$ involves separation of the standard deviations and correlations of the underlying random variables which amounts to working with the standardized random variables or removing heterogeneity. For multivariate distributions there is a similar separation of marginal distributions and the dependence structure called its *copula*; a multivariate distribution with $U(0, 1)$ marginals [Joe, 2014]. This section provides a brief introduction to copulas as an attempt to highlight a sense of similarity between correlation matrices and multivariate copulas in quantifying linear and nonlinear dependencies.

We note that for any random vector $Y = (Y_1, \ldots, Y_p)'$ its covariance matrix can decomposed as

$$\boldsymbol{\Sigma} = DRD, \tag{3.40}$$

where $D = \text{diag}(\sqrt{\sigma_{11}}, \ldots, \sqrt{\sigma_{pp}})$ is a diagonal matrix of standard deviations of the variables and $R = (\rho_{ij})$ is the *correlation matrix* of Y with

$$\rho_{ij} = \text{Corr}(Y_i, Y_j) = \frac{\sigma_{ij}}{\sqrt{\sigma_{ii}\sigma_{jj}}}.$$

Often, it is useful to think of the matrix R as the covariance matrix of the standardized random variables $(Y_i - \mu_i)/\sqrt{\sigma_{ii}}$, $i = 1, \ldots, p$. From (3.5), the correlation coefficient between any two random variables is less than one in absolute value, that is $|\rho_{ij}| \leq 1$. While for $p = 2$ checking positive-definiteness of R is easy, it is not the case for $p > 2$. An encyclopedic review of the developments of the concept of correlation and its evolutions over the last 150 years can be found in Zorich [2024].

3.6.1 Some Nonparametric Measures of Dependence

Measuring dependence between random variables is of central importance in statistics with applications in financial risk modeling, environmental prediction, genomics, reliability theory and network analysis. In these areas marginal distributions often exhibit heavy-tailed behavior which standard Pearson product-moment correlation coefficient may fail to capture effectively.

The classical nonparametric measures of dependence usually rely on the similarity or concordance of the ranks of the observations. Important examples include Kendall's τ and Spearman's ρ, etc. These are widely used due to their robustness and invariance under monotonic transformations, see Czado and Nagler [2022] and Joe [2014]. The more recent distance covariance (dCov) and distance correlation (dCor) measures have further broadened the scope of nonparametric dependence detection, particularly for nonlinear and nonmonotonic relationships. We recall that distance covariance in Székely et al. [2007] is a measure of dependence between two real random variables X and Y. Its popularity is due to its desirable property that the measure is zero, if and only if the two random variables are independent, and positive otherwise, see Leyder et al. [2025] and references therein for a recent review of distance covariance and improvements from the robustness perspectives.

3.6.2 Copulas

The popular idea of separation of variances and correlations of a covariance matrix (second moments of a distribution) can be effectively extended to a whole multivariate distribution. Indeed, this is done through copula construction which allows separate marginal distributions for each variable coupled with a simpler multivariate distribution capturing the dependence structure. Instead of the standardization of the variables in (3.3), the key player here is the familiar probability integral transform which transforms a general random variable to the uniform distribution on $[0, 1]$. More precisely, if the random variable $X \sim F$, then $F(X) \sim U[0, 1]$. For a random vector, its analog is made possible by the Sklar [1959] theorem.

Theorem 5 *For a p-variate continuous random vector $Y = (Y_1, \ldots, Y_p)'$, its joint distribution $F(\cdot)$ with marginals $F_i, i = 1, \ldots, p$, can be written as*

$$F(y_1, \ldots, y_p) = C(F_1(y_1), \ldots, F_p(y_p)), \tag{3.41}$$

where $C(\cdot)$ is a unique p-variate distribution on the hypercube $[0, 1]^p$ with uniform marginals, called its copula distribution. Furthermore, if F is absolutely continuous then its density is given by

$$f(y_1, \ldots, y_p) = c(F_1(y_1), \ldots, F_p(y_p)) f_1(y_1) \cdots f_p(y_p), \tag{3.42}$$

where $c(\cdot)$ is the copula density of the copula distribution $C(\cdot)$ and f_i is the density of the ith random variable.

Evidently, the Theorem 5 shows how to write the joint distribution in terms of the marginal distributions and the copula distribution. Conversely, when the joint and the marginal distributions are known, the copula can be recovered by

$$C(u_1, \ldots, u_p) = F(F_1^{-1}(u_1), \ldots, F_p^{-1}(u_p)), \tag{3.43}$$

where F_i^{-1} is the quantile function or the inverse of F_i. To explain the construction we provide a few examples of copulas or the dependence structure.

1. *Independence Copula:* The simplest example of multivariate copula is

   $$C(u_1, \ldots, u_p) = u_1 \cdots u_p,$$

   which corresponds to p independent random variables with a flat contour plot and no parameters. Copulas with more parameters, denoted by θ, are introduced next.

2. *Gaussian Copula:* Using the general form (3.43), it is given by

   $$C(u_1, \ldots, u_p; R) = \Phi_p(\Phi^{-1}(u_1), \ldots, \Phi^{-1}(u_p)), \tag{3.44}$$

   where Φ_p is the distribution function of a $p-$variate normal random vector with mean zero and correlation matrix R, and Φ is the distribution function of a univariate $N(0, 1)$. It has the correlation matrix R as its parameters where the number of parameters grows quadratically in p, but for the bivariate Gaussian copula there is only one parameter $\theta = \rho$, namely the correlation coefficient between the two variables. In this case, the copula takes the form

   $$C(u_1, u_2; \theta) = \int_{-\infty}^{\Phi^{-1}(u_1)} \int_{-\infty}^{\Phi^{-1}(u_2)} f(x_1, x_2; \theta) dx_1 dx_2,$$

   where f is the bivariate density of two correlated standard normal random variables with correlation coefficient ρ. In general, to reduce the number of parameters in the Gaussian copula one may need to restrict R to be structured like the compound symmetry correlation matrix with only one parameter, see Section 3.4 for more examples.

   The t copula corresponding to a p-variate t-distribution with ν degrees of freedom can be defined similarly. Both the Gaussian and t copulas belong to a larger family known as the *elliptical copula*.

3. *Archimedean copula:* This class of copulas is defined as

   $$C(u_1, \ldots, u_p; \theta) = \psi(\psi^{-1}(u_1) + \cdots + \psi^{-1}(u_p)) \tag{3.45}$$

   for suitable univariate functions ψ. For example, the Gumbel-Hougaard copula corresponds to

   $$C_{GH}(u_1, \ldots, u_p; \theta) = \exp(-(\sum_{i=1}^{p}(-\log u_i)^{\theta})^{1/\theta}), \tag{3.46}$$

with a single parameter θ. We note that for $\theta = 1$ it reduces to the independence copula, and when $\theta \to \infty$, it reduces to

$$C_{co}(u_1, \ldots, u_p) = \min\{u_1, \ldots, u_p\}, \tag{3.47}$$

usually referred to as the *comonotonicity copula.*

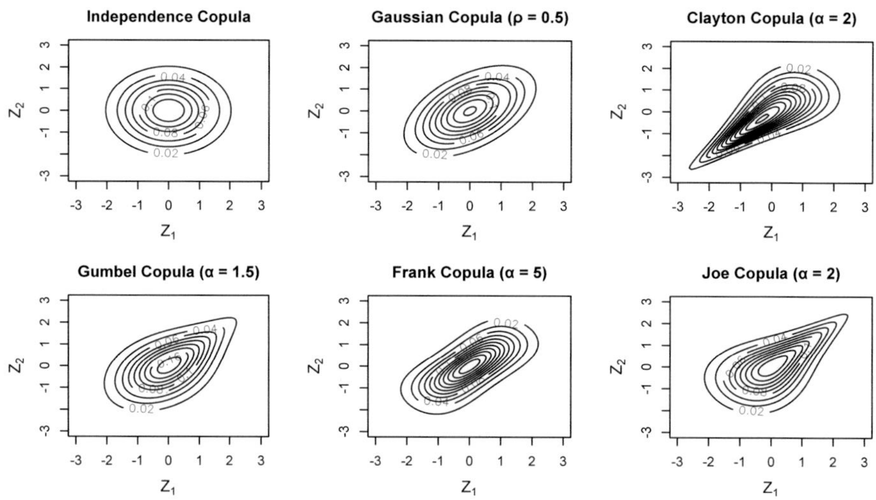

FIGURE 3.1
Contour plots of different bivariate copulas with z-scale.

While scatterplots are popular for assessing linear relationships between two variables, bivariate contour plots play a similar role of assessing the nature and type of nonlinear dependence in the bivariate copula setting. For visual comparison of bivariate contour plots it is advisable to rely on marginally normalized contour plots. Between the copula scale $(U_1, U_2) = (F_1(X_1), F_2(X_2))$ and the z-scale $(Z_1, Z_2) = (\Phi^{-1}(X_1), \Phi^{-1}(X_2))$ where Φ is the cdf of the standard normal, the z-scale is preferred in practice. Interestingly, just as in linear dependence, appearance of circle and elliptic contour plots are indicators of independence and Gaussian copulas, respectively. Figure 3.1 displays the contour plots for some of the popular examples of copulas presented earlier. Availability of computationally tractable estimation and model selection procedures is one of the main reasons for popularity of copula models in various application areas [Czado and Nagler, 2022].

Unfortunately, multivariate copulas like elliptical (the Gaussian and t-copula) and Archimedean (Clayton and Gumbel copulas) though capture nonlinear dependence, they are not flexible so far as symmetry and tail dependence are concerned, see Figure 3.1. In fact, the Archimedean copulas and some other classical copulas are symmetric (permutation-invariant) functions. In this sense they are the copula analogs of the compound symmetry covariance matrix, and hence have limited flexibility. It is then natural to search for a principled method to expand the list of available copulas enjoying greater flexibility.

In this direction, the idea of vine copulas is quite effective in constructing flexible multivariate copulas using only bivariate copulas as the building blocks. The first such copula construction in terms of distribution functions was given in Joe [1996], only a few years later Bedford and Cooke [2001, 2002] developed them in terms of densities and within a framework that identifies all possible constructions.

Factorization of the joint density of a random vector $Y = (Y_1, \ldots, Y_p)'$ as the product of conditional densities is crucial for construction of vine copulas. Of course, it depends on the ordering of the random variables in Y, to see the role of such ordering we note that for $p = 2$ there are only two possible ordering: $(Y_1, Y_2)', (Y_2, Y_1)'$. For the first one, the corresponding factorization is

$$f(y_1, y_2) = f_{1|2}(y_1|y_2) f_2(y_2),$$

and the conditional density $f_{1|2}$ can be expressed in terms of the bivariate copula as

$$f_{1|2}(y_1|y_2) = c_{12}(F_1(y_1), F_2(y_2)) f_1(y_1), \tag{3.48}$$

where the (conditional) copula term is called a *pair copula*. For the formula for the corresponding conditional distribution, see Problem 3.24. In higher dimensions the number of possible factorizations is bound to increase as seen in the next example. We follow the recent review paper by Czado and Nagler [2022] in the rest of this section, it provides a comprehensive and balanced coverage of developments in copula modeling.

Example 15 *For $p = 3$ the starting point of vine copula constructions using bivariate copulas is the generic factorization*

$$f(y_1, y_2, y_3) = f_{3|12}(y_3|y_1, y_2) f_{2|1}(y_2|y_1) f_1(y_1), \tag{3.49}$$

where from here on $F_{j|D}$ or $f_{j|D}$ stands for the conditional distribution or density of Y_j given the random variables with indices in D. Our first order of business is to show that the first factor in (3.49) can be expressed in terms of $f_{13|2}(y_1, y_3|y_2)$ and $c_{13;2}(\cdot, \cdot; y_2)$, which stands for the bivariate copula density associated with the conditional density of (Y_1, Y_3) given $Y_2 = y_2$. In fact, viewing it as a "bivariate density" and using (3.42) we have

$$f_{13|2}(y_1, y_3|y_2) = c_{13;2}(F_{1|2}(y_1|y_2), F_{3|2}(y_3|y_2); y_2) f_{1|2}(y_1|y_2) f_{3|2}(y_3|y_2). \tag{3.50}$$

For the conditional density $f_{3|12}$, using the identity in (3.48) one obtains

$$f_{3|12}(y_3|y_1, y_2) = c_{13;2}(F_{1|2}(y_1|y_2), F_{3|2}(y_3|y_2); y_2) f_{3|2}(y_3|y_2). \tag{3.51}$$

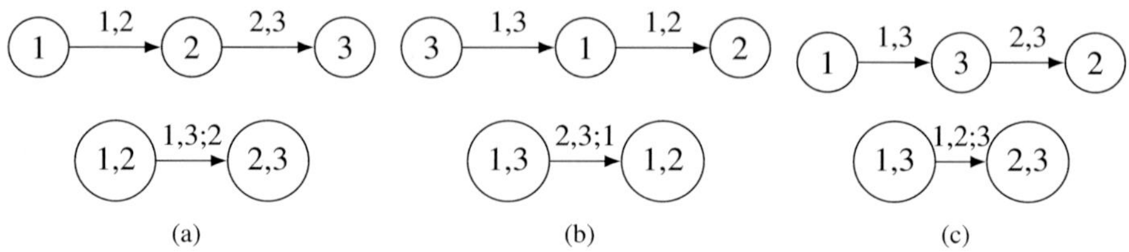

FIGURE 3.2
Graphical representation of the three possible pair copula constructions corresponding to the three variable orderings: (a) $c_{23} - c_{12} - c_{13;2}$, (b) $c_{13} - c_{12} - c_{23;1}$, and (c) $c_{13} - c_{23} - c_{12;3}$.

Repeated applications of (3.48) give

$$f_{2|1}(y_2|y_1) = c_{12}(F_1(y_1), F_2(y_2))f_2(y_2), \tag{3.52}$$

and

$$f_{3|2}(y_3|y_2) = c_{32}(F_3(y_3), F_2(y_2))f_3(y_3). \tag{3.53}$$

Substituting (3.50)–(3.53) into (3.49) one obtains the following pair copula decomposition of a trivariate probability density function

$$\begin{aligned} f(y_1, y_2, y_3) = {} & c_{13;2}(F_{1|2}(y_1|y_2), F_{3|2}(y_3|y_2); y_2) \\ & \times c_{23}(F_2(y_2), F_3(y_3))c_{12}(F_1(y_1), F_2(y_2)) \\ & \times f_1(y_1)f_2(y_2)f_3(y_3), \end{aligned} \tag{3.54}$$

as the product of only bivariate copula densities, conditional distribution functions and marginal densities.

We note that this decomposition is not unique, and there are two other possible decompositions based on permuting or reordering the variables the results of which are shown in Figure 3.2 using path graphs.

In spite of the nonuniqueness, the grand plan is decomposition of a multivariate density involving conditional bivariate copula terms of the form $c_{ij;k}(\cdot, \cdot; y_k)$ called a *pair copula*. This insistence on using pair copulas is perhaps inspired by the use of partial correlations in deriving alternative representation of multivariate normal distributions, but in general it could lead to the need to check validity of the final function as a probability density.

For managing the parameters in the decomposition and their eventual estimation, it is common to suppress the dependence on the specific conditioning value like y_k, the action aptly called the *simplifying assumption*. Then, introducing the copula parameter vector $\boldsymbol{\theta} = (\boldsymbol{\theta}_{12}, \boldsymbol{\theta}_{23}, \boldsymbol{\theta}_{12;3})'$, the simplified pair copula construction leads to

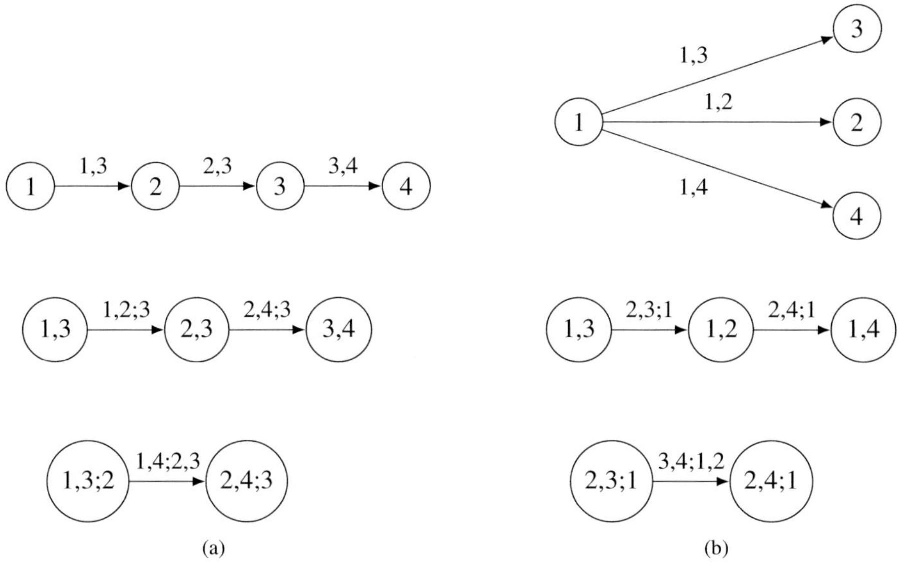

FIGURE 3.3
Graphical representation of two types of regular vine structures for $p = 4$.

$$\begin{aligned} f(y_1, y_2, y_3; \boldsymbol{\theta}) = {} & c_{13;2}(F_{1|2}(y_1|y_2), F_{3|2}(y_3|y_2); \boldsymbol{\theta}_{13;2}) \\ & \times c_{23}(F_2(y_2), F_3(y_3); \boldsymbol{\theta}_{23}) c_{12}(F_1(y_1), F_2(y_2); \boldsymbol{\theta}_{12}) \\ & \times f_1(y_1) f_2(y_2) f_3(y_3), \end{aligned} \tag{3.55}$$

where $c_{ij;k}(\cdot, \cdot; \boldsymbol{\theta}_{ij;k})$'s are arbitrary parametric bivariate copula densities. Consequently, (3.55) gives the construction of a trivariate probability density function using only bivariate copulas. In particular, it reduces to a bona fide trivariate copula density when the marginals are the uniform distributions on $[0, 1]$. For $p = 3$, there are three possible pair copula constructions shown in Figure 3.2.

Figure 3.3 shows graphs of the two types of regular vine structures in four dimensions corresponding to two decompositions. We note the graph in (a) is composed of path graph at each level, but the one (b) has a star graph where one node is connected to all the other nodes. Furthermore, the number of such constructions is 24 [Czado and Nagler, 2022] for $p = 4$, and their numbers increase superexponentially in p. In higher dimensions the simplifying assumption is expected to impose more severe constraints on the joint distribution.

Remark 1 *The ordering of the variables and the ensuing technical challenges for checking the validity of constructions of multivariate copulas like (3.55) are reminders of similar issues that one may encounter in dealing with the positive-definiteness of covariance matrices. For instance, checking validity of a proto-covariance matrix* *constructed from 2×2 and higher-order submatrices, see (3.6), and the implicit need for ordering the variables in Cholesky-based modeling and estimation of covariance matrices. A reason for such similarity or duality is to keep in mind that a zero-mean multivariate normal distribution is uniquely identified by its covariance matrix. In this case, factorization of the joint density in terms of conditional distributions relates to partitioning its covariance matrix, see Theorem 7.*

Fortunately, there is a way to organize all valid factorizations by relying on a graphical object called *vine* on a set of nodes, developed by Bedford and Cooke [2001, 2002]. Indeed, a vine simply consists of linked sequence of trees where the edges in one tree become the nodes of the next.

More precisely, a vine on a set V_1 with p elements is a sequence of trees, denoted by

$$\mathcal{V} = \{T_1 = (V_1, E_1), \ldots, T_{p-1} = (V_{p-1}, E_{p-1})\},$$

such that

(i) T_1 has vertices V_1 and edges E_1 with the cardinality $|E_1| = p - 1$;

(ii) for each integer $k \in \{2, \ldots, p-1\}$, the set of vertices of the level-k tree T_k is $V_k = T_{k-1}$;

(iii) for any edge $e = \{(a, b)\}$ in E_k where $a = \{a_1, \ldots, a_{k+1}\}$ and $b = \{b_1, \ldots, b_{k+1}\}$ are subsets of V_1, one must have $|a\Delta b| = 2$, where Δ is the symmetric difference operator.

The *proximity condition* in (iii) states that two vertices can be connected only if there are exactly two uncommon elements. For an edge e connecting nodes a and b, the conditioning set consists of their common variables $S_e = a \cap b$, and the conditioned set is a pair of variables (a_e, b_e) such that $a \setminus S_e = \{a_e\}$ with a similar definition for b_e. Evidently, each edge $e = (a_e, b_e|S_e)$ then represents a conditional bivariate distribution.

The next example relies on the conditional independence of several variables given another to introduce the important class of factor-copula models connecting copulas and latent factor models in Chapter 7.

Example 16 *Consider the uniform random variables $Y_1, \ldots, Y_p$ which are conditionally independent given a single variable $X \sim U(0, 1)$. Then, the joint distribution of $Y = (Y_1, \ldots, Y_p)'$ for a $y = (y_1, \ldots, y_p)' \in (0, 1)^p$ is given by the copula*

$$C(y) = \int_0^1 \Pi_{j=1}^p P(Y_j \leq y_j|X = x)dx = \int_0^1 \Pi_{j=1}^p C_{j|X}(y_j|x)dx, \qquad (3.56)$$

where $C_{j|X}(y_j|x)$ is the conditional distribution of Y_j given $X = x$. Their corresponding joint density is

$$c(y) = \int_0^1 \Pi_{j=1}^p c_{jX}(y_j, x)dx,$$

where $c_{jX}(y_j, x)$ is the joint density of the indicated variables. Thus, the joint distribution of Y or its multivariate copula is completely determined by the p bivariate copulas C_{jX}, $j = 1, \ldots, p$, for the pair of variables (Y_j, X).

Factor-copulas in Krupskii and Joe [2013] are parametric copula models suitable for describing the dependence structure of high-dimensional variables via conditional independence given a few latent factors, They show that that the bivariate marginals in factor copula constructions inherit the tail dependence properties of the linking copulas C_{jk}'s. For example, in the one-factor model (3.56), C_{jk} may have lower (upper) tail dependence if both C_{jX} and C_{kX} have lower (upper) tail dependence. Thus, choosing the bivariate linking copulas suitably, the ensuing factor-copula models can account for lower and/or upper tail dependence and tail asymmetry.

More general models of the type in (3.56) can be introduced by partitioning the set $\{1, \ldots, p\}$ into G non-overlapping groups $J_1, \ldots, J_G$ with group sizes $p_g, g = 1, \ldots, G$, and $Y = (Y'_{J_1}, \ldots, Y'_{J_G})'$. Let the variables X_j, $j = 1, \ldots, G$ be the latent $U(0, 1)$ random variables with the G-variate copula $C_{1:G}$. For each group G the dependence structure is determined by p_g bivariate linking copulas, one for each (Y_j, X_g), $j \in J_g$. The joint density of Y for all $y \in \mathbb{R}^p$ is then given by

$$C(y) = \int_{(0,1)^G} \Pi_{g=1}^{G} \Pi_{j \in J_g} c_{jX}(y_j, x_g) dC_{1:G}(x_1, \ldots, x_G). \tag{3.57}$$

The class of distributions in (3.56)–(3.57) called factor-copula models, extends classical Gaussian factor models by addressing some of their limitations. In particular, it provides more flexibility with respect to tail dependence. The recent work by Li et al. [2025] provides an application of vine copulas in the time series setup for the analysis data for flood insurance losses.

The preceding coverage barely touches the rapidly growing literature on modeling dependence using copulas and vine copulas. The book by Czado [2019] provides a lucid and comprehensive coverage of various topics of interest including estimation, model selection and data analysis using bivariate and pair copulas. For a deeper and theoretical aspects, the book by Joe [2014] provides an excellent coverage of copulas and their roles in modeling nonlinear dependencies in multivariate data.

3.6.3 Hands-on R: Generating Samples Using Copulas

In this section, we use copulas to generate simulated joint samples with certain marginal distributions and dependence structures.

Suppose we are interested in assessing the flood risk and examine the joint behavior of the amount of rainfall and river flow rate. It is expected that heavy rainfall usually leads to high river flow but this relationship, in general, is non-linear and non-Gaussian. We start by importing fictional floodrisk dataset from the R package recode and the required packages. The data contains two variables with 500 observations. Figure 3.4(left) helps to visualize the dependence.

```
## import required packages
library(recode)
```

```
library(fitdistrplus)
library(copula)

data("floodrisk")
rainfall <- floodrisk[,1]
riverflow <- floodrisk[,2]
```

Next, we transform the variables into uniform margins. First, we estimate the marginal distributions by assuming gamma distribution for the rainfall and log-normal for the rainflow. Then use the estimated parameters to obtain the uniform marginals.

```
## estimate marginal distributions
# Fit gamma to rainfall
fit_rain <- fitdist(rainfall, "gamma")
# Fit lognormal to river flow
fit_flow <- fitdist(riverflow, "lnorm")

# Transform to uniform
U <- pgamma(rainfall, shape = fit_rain$estimate["shape"], rate = fit_
    rain$estimate["rate"])
V <- plnorm(riverflow, meanlog = fit_flow$estimate["meanlog"], sdlog =
      fit_flow$estimate["sdlog"])

# Combine into a pseudo-sample
UV <- cbind(U, V)
```

We use Gumbel-Hougaard copula, defined in (3.46) to fit the copula to data.

```
# Choose a copula (e.g., Gumbel for upper tail dependence)
cop <- gumbelCopula(dim = 2)
# Fit the copula to the data
fit_cop <- fitCopula(cop, UV, method = "ml")
summary(fit_cop)
# Extract fitted copula with parameter
fitted_cop <- fit_cop@copula
```

Finally, we simulate new sample from the copula and use the estimated parameters from the gamma and log-normal distributions to transform the simulated sample back to original scale using inverse CDFs.

```
# Simulate new pairs from the copula
sim_uv <- rCopula(500, fitted_cop)
# Transform back to original scale using inverse CDFs
sim_rainfall <- qgamma(sim_uv[,1], shape = fit_rain$estimate["shape"],
      rate = fit_rain$estimate["rate"])
sim_riverflow <- qlnorm(sim_uv[,2], meanlog = fit_flow$estimate["
    meanlog"], sdlog = fit_flow$estimate["sdlog"])
```

Figure 3.4(right) plots the simulated sample.

3.6.4 Distributional and Copula Regressions

An ambitious class of models known as *distributional regression* has the goal of expressing the full distribution of a response variable in terms of covariates. A less ambitious task is quantile regression which models each quantile of the distribution

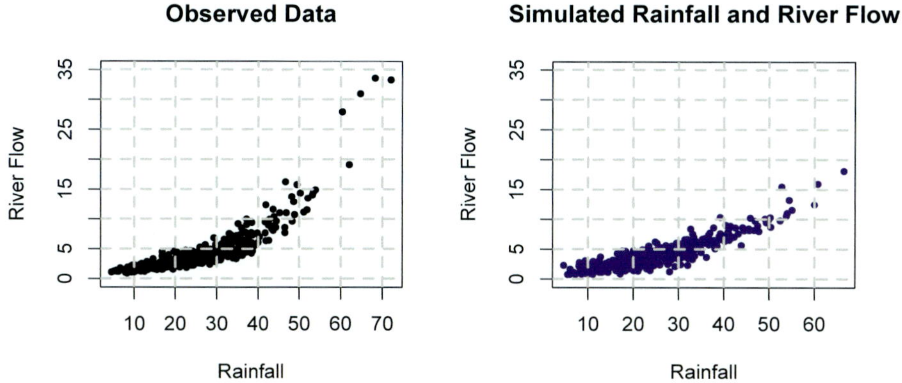

FIGURE 3.4
Observed and simulated sample from the flood risk dataset.

separately, however, fitting a separate model for each quantile is a computational challenge and extensions to multivariate responses are nontrivial. This section provides a brief review of the progress in distributional regressions, and highlights the need for *ordering* the variables for multivariate response.

For a univariate response a simple approach to distributional regression is to link each parameter of a distribution to covariates. Examples of univariate response distributions include count data [Klein et al., 2015] and geoadditive hazards regression [Kneib and Fahrmeir, 2007]. Distributional regressions for multivariate responses are more challenging, though great progress is made in the bivariate cases through copulas. Important examples include Klein et al. [2015] who considered bivariate Gaussian and t-distributions based on the variance-correlation decomposition, Marra and Radice [2017]'s framework where bivariate responses are modeled via copulas with continuous margins, then all distributional parameters are modeled additively. Other relevant contributions are the covariate-dependent copula approaches of Vatter and Nagler [2018] and Hans et al. [2023] dealing with four- and two-dimensional responses, respectively. Two great review papers in this area are Klein [2024] and Kneib et al. [2023].

The class of multivariate conditional transformation models (MCTM) in Klein et al. [2022] makes minimal distributional assumptions on the response vector. The key player is an unknown, bijective, strictly monotonically increasing transformation function $h : \mathbb{R}^p \to \mathbb{R}^p$, which maps the p-vector Y, with unknown distribution, to a set of p independent and identically distributed standard normal random variables $Z_j, j = 1, \ldots, p$:

$$h(Y) = (h_1(Y), \ldots, h_p(Y)) \sim Z = (Z_1, \ldots, Z_p). \tag{3.58}$$

Indeed, it is known [Hothorn et al., 2018] that for a response vector Y with an absolutely continuous distribution there exists a unique h to satisfy (3.58). Moreover, the result holds even when the standard normal random variables Z_i's are replaced by other i.i.d. random variables having an absolutely continuous distribution with a log-concave density function. The choice of normality of the transformed response $h(Y)$ allows its dependence structure to be that of a Gaussian copula.

The MCTM model is rather general and possibly cumbersome to work with so far as interpretation and computational tractability are concerned. A simpler sequential (triangular) parameterization of h that leads to more manageable and interpretable models. For this, first we assume that the jth component of the function h depends only on the first j elements of its argument Y: $h_j(Y) = h_j(Y_1, \dots, Y_j)$. Second, it is assumed these functions are linear combinations of certain univariate functions $\tilde{h}_j : \mathbb{R} \to \mathbb{R}$, so that

$$h_j(Y_1, \dots, Y_j) = \lambda_{j1}\tilde{h}_1(Y_1) + \dots + \lambda_{jj}\tilde{h}_j(Y_j),$$

where each $\tilde{h}_j$ increases strictly monotonically and $\lambda_{jj} > 0$, $j = 1, \dots, p$, to ensure the bijectivity of h. Because the last coefficient, λ_{jj}, cannot be separated from the marginal transformation function $\tilde{h}_j(Y_j)$, for sake of identifiability we set $\lambda_{jj} = 1$. Thus, the recursive parameterization of the transformation function h takes the simpler form

$$h_j(Y_1, \dots, Y_j) = \lambda_{j1}\tilde{h}_1(Y_1) + \dots + \tilde{h}_j(Y_j). \tag{3.59}$$

In summary, the simplified MCTM is characterized by a set of marginal transformations $\tilde{h}_j(Y_j)$, $j = 1, \dots, p$, each acting on only a single component of the vector Y, where the coefficients form the unit lower triangular matrix $\Lambda = (\lambda_{ij})$. Of course, this simplification carries the hefty cost of requiring an ordering of the elements in Y which is quite reminiscent of that in the Cholesky decomposition and vine copula constructions. Gioia et al. [2024] and Muschinski et al. [2022] consider distributional regression with multivariate Gaussian responses. They propose non-parametric modeling of the elements of Cholesky-based parameterizations. Klein [2024] provides an excellent review of distributional regressions and software.

3.7 Problems

3.1 Consider the 2×2 symmetric matrices

$$A = \begin{pmatrix} 1 & 2 \\ 2 & 1 \end{pmatrix}, \quad B = \begin{pmatrix} 2 & 1 \\ 1 & 2 \end{pmatrix}.$$

Show that the first with positive diagonal entries is not a covariance matrix while the second is.

3.2 (a) Show that if $\mathbf{\Sigma} = (\sigma_{ij})$ is positive-definite, then all its 2×2 symmetric submatrices matrices

$$\begin{pmatrix} \sigma_{ii} & \sigma_{ij} \\ \sigma_{ij} & \sigma_{jj} \end{pmatrix}, \quad i \neq j,$$

are positive-definite. Is the converse true?

(b) Show that the entries of the 2×2 matrix in (a) must satisfy the inequality

$$|\sigma_{ij}| \leq \sqrt{\sigma}_{ii}\sqrt{\sigma}_{jj}, \; i \neq j. \tag{3.60}$$

(c) State and prove an analog of (a) for all $k \times k$ submatrices of $\mathbf{\Sigma}$ for $k = 3, \ldots, p$.

3.3 Consider the symmetric matrix

$$A = \begin{pmatrix} \alpha & \beta \\ \beta & \gamma \end{pmatrix},$$

where α, β and γ are real numbers.

(a) Show that the eigenvalues and the normalized eigenvectors of A are

$$\lambda_1 = \tfrac{1}{2}(\alpha + \gamma - \sqrt{\Delta}), \quad \mathbf{e}_1 = d_1^{-1}(-\beta, \alpha - \lambda_1)',$$

$$\lambda_2 = \tfrac{1}{2}(\alpha + \gamma + \sqrt{\Delta}), \quad \mathbf{e}_2 = d_2^{-1}(-\beta, \alpha - \lambda_2)'$$

where

$$\Delta = (\alpha - \gamma)^2 + 4\beta^2, \; d_1^2 = \sqrt{\Delta}\,(\alpha - \lambda_1), \; d_2^2 = \sqrt{\Delta}\,(\lambda_2 - \alpha).$$

(b) For $\alpha = \gamma = 1$ and $\beta = \rho$, a correlation coefficient, express the eigenvalues and eigenvectors of A in terms of ρ.

(c) (Equal Eigenvalues): Show that a 2×2 real symmetric matrix has equal eigenvalues if and only if the matrix is diagonal with equal diagonal entries.

3.4 For the asymmetric matrix A in Example 11 (a) find its normalized eigenvectors, form the matrix P and compute PAP'. Is it diagonal?

3.5 A symmetric matrix $A = (a_{ij})$ is said to be *diagonally dominant* if

$$|a_{ii}| > \sum_{j=1, j\neq i}^{p} |a_{ij}|, \text{ for } i = 1, \ldots, p.$$

It is known that diagonally dominant matrices are positive-definite (Hadamard theorem). Check whether the following matrix is diagonally dominant and hence positive-definite:

$$\Sigma_1 = \begin{pmatrix} 1.0 & 0.2 & 0.5 \\ 0.2 & 1.0 & 0.4 \\ 0.5 & 0.4 & 1.0 \end{pmatrix}.$$

How about the matrix

$$\Sigma_2 = \begin{pmatrix} 1.0 & 0.9 & 0.7 \\ 0.9 & 1.0 & 0.4 \\ 0.7 & 0.4 & 1.0 \end{pmatrix}?$$

Is it diagonally dominant? Positive-definite?

3.6 For a scalar x consider the matrix

$$A(x) = \begin{pmatrix} 1.0 & x & 0.6 \\ x & 1.0 & 0.7 \\ 0.6 & 0.7 & 1.0 \end{pmatrix}.$$

Find the range of values of x so that the matrix is diagonally dominant and hence positive-definite. (This is a simple example of *positive-definite matrix completion problem* where x stands for a missing or unknown entry).

3.7 The *Schur product* or entry-wise product of two square matrices $A = (a_{ij}), B = (b_{ij})$ of the same size is the matrix $A \circ B = (a_{ij}b_{ij})$. Show that if A and B are positive-definite matrices, then so is their Schur product $A \circ B$.

Hint: Take two independent zero-mean normal random vectors $X = (X_1, \ldots, X_p)'$ and $Y = (Y_1, \ldots, Y_p)'$ with A and B as their covariance matrices, respectively. Then define the random vector $Z = (X_1Y_1, \ldots, X_pY_p)'$ and show that $Cov(Z) = A \circ B$.

3.8 For a CS matrix verify the identity (3.30). Show that the other normalized eigenvectors of a CS covariance matrix are given by

$$\mathbf{e}_k = (\frac{1}{\sqrt{(k-1)k}}, \ldots, \frac{1}{\sqrt{(k-1)k}}, \frac{-(k-1)}{\sqrt{(k-1)k}}, 0, \ldots, 0), \; k = 2, \ldots, p.$$

3.9 The (Huynh-Feldt) matrix given below is more general than the compound symmetry matrix

$$\mathbf{\Sigma} = \sigma^2(\alpha I_p + \mathbf{a}\mathbf{1}_p' + \mathbf{1}_p\mathbf{a}'),$$

where $\mathbf{a} = (a_1, \ldots, a_p)'$ and α is a scalar chosen so that $\mathbf{\Sigma}$ is positive-definite.

(a) Show that the matrix

$$\mathbf{a}\mathbf{1}_p' + \mathbf{1}_p\mathbf{a}',$$

has two nonzero eigenvalues

$$\mathbf{1}_p'\mathbf{a} \pm \sqrt{p\mathbf{a}'\mathbf{a}},$$

and the rest are zero, see Wolfinger [1996]. Which eigenvalue is negative?

(b) Show that the matrix $\mathbf{\Sigma}$ is positive-definite if

$$\alpha > \mathbf{1}_p'\mathbf{a} - \sqrt{p\mathbf{a}'\mathbf{a}}.$$

3.10 The (Schatten) m-norm of a $p \times p$ matrix A with the singular values $\lambda_i, i = 1, \ldots, p$, is defined as the vector ℓ_m-norm of its spectra:

$$||A||_{m,S} = (\sum_{i=1}^{p} |\lambda_i|^m)^{1/m}. \tag{3.61}$$

(a) Show that when $m = 2$, it is the same as the Frobenious norm.

(b) When $m = 1$, it is called the *nuclear* norm of A and is usually used as a penalty function for low-rank approximation of matrices. Discuss why the nuclear norm should be preferred to Frobenious norm when the goal is low-rank approximation and not sparsity.

3.11 For the matrix

$$A = \begin{pmatrix} 2 & 1 \\ 1 & 2 \end{pmatrix}$$

show that its unique Cholesky factor with positive diagonal entries is given by

$$C = \begin{pmatrix} \sqrt{2} & 0 \\ \frac{1}{\sqrt{2}} & \frac{\sqrt{3}}{\sqrt{2}} \end{pmatrix}.$$

3.12 (a) With the matrix A as above, find the eigenvalues, eigenvectors and the spectral decomposition of the 3×3 partitioned matrix

$$\mathbf{A} = \begin{pmatrix} A & 0_2 \\ 0_2' & 1 \end{pmatrix}.$$

(b) Find the Cholesky decomposition (factor) of $\mathbf{A}$.

(c) Note that the matrix $\mathbf{A}$ is block-diagonal. Describe how the spectral and Cholesky decompositions of block diagonal matrices relate to the corresponding decompositions of its diagonal blocks.

3.13 (Recursive Computation of the Cholesky Factor) Let $\mathbf{\Sigma}$ be a $p \times p$ positive-definite matrix partitioned as

$$\mathbf{\Sigma} = \begin{pmatrix} \mathbf{\Sigma}_{11} & \boldsymbol{\sigma}_{12} \\ \boldsymbol{\sigma}_{21} & \sigma_{22} \end{pmatrix} = CC',$$

where $\sigma_{22} > 0$ is a scalar and C its Cholesky factor partitioned as

$$C = \begin{pmatrix} C_{11} & 0 \\ c_{21} & c_{22} \end{pmatrix},$$

with C_{11} the $(p-1) \times (p-1)$ Cholesky factor of Σ_{11}.

(a) Show that the last row of the Cholesky factor can be computed via

$$c_{21} = \boldsymbol{\sigma}_{21}(C_{11}')^{-1}, \quad c_{22} = (\sigma_{22} - \boldsymbol{\sigma}_{21}\mathbf{\Sigma}_{11}^{-1}\boldsymbol{\sigma}_{12})^{1/2} = (\sigma_{22} - c_{21}c_{21}')^{1/2}.$$

Devise a plan for computing recursively the rows of Cholesky factor of a given positive-definite matrix.

(b) Apply (a) to compute the Cholesky factor of the matrix $\mathbf{A}$ in Problem 3.12.

(c) Use induction to prove existence of Cholesky decomposition of a symmetric, positive-definite matrix.

3.14 Consider the two 4×4 matrices from Khare et al. [2024],

$$\boldsymbol{\Sigma} = \begin{pmatrix} 3 & 1 & 0 & 1 \\ 1 & 3 & 1 & 0 \\ 0 & 1 & 3 & 2 \\ 1 & 0 & 2 & 3 \end{pmatrix}, \quad C = \begin{pmatrix} 1.732 & 0 & 0 & 0 \\ 0.577 & 1.633 & 0 & 0 \\ 0 & 0.612 & 1.620 & 0 \\ 0.577 & 0.204 & 1.312 & 0.951 \end{pmatrix}.$$

(a) Is the matrix $\boldsymbol{\Sigma}$ a Hadamard matrix? Is it positive-definite?

(b) Compute the matrix CC' and compare it with $\boldsymbol{\Sigma}$. Are they equal?

(c) Use the recursive method of the previous problem to compute the Cholesky factor of $\boldsymbol{\Sigma}$, and compare it with the lower triangular matrix C given above. You may use the *chol* function in R to compute the Cholesky factor of $\boldsymbol{\Sigma}$. Note that the number of zero entries in the lower triangular part of C is smaller than that in $\boldsymbol{\Sigma}$.

(d) Let $\boldsymbol{\Sigma}_{11}$ be the leading 3×3 principal minor of $\boldsymbol{\Sigma}$. Find its Cholesky factor and compare the number of zeros in its lower triangular part with that of $\boldsymbol{\Sigma}_{11}$.

The question of matching sparsity in the lower triangular part of C and $\boldsymbol{\Sigma}$ is of central importance in the graphical model literature, see Khare et al. [2014, 2024].

3.15 (a) Consider the $p \times q$ matrix $\boldsymbol{X} = c\boldsymbol{x}\boldsymbol{y}'$ where $\boldsymbol{x}$ and $\boldsymbol{y}$ are p- and q-vectors of length one and c is a scalar. Find the singular value and vectors of $\boldsymbol{X}$ when $c = 1, c = -2$.

(b) Find the spectral decompositions of $\boldsymbol{X}'\boldsymbol{X}$ and $\boldsymbol{X}\boldsymbol{X}'$.

(c) Find the eigenvalues, singular values, singular vectors and SVD of the matrix

$$X = \begin{pmatrix} 0 & 1 & 0 & 0 \\ 0 & 0 & 2 & 0 \\ 0 & 0 & 0 & 3 \\ 0 & 0 & 0 & 0 \end{pmatrix}.$$

3.16 (a) Show that for any $n \times p$ matrix $\boldsymbol{X}$, the nonzero eigenvalues of $\boldsymbol{X}\boldsymbol{X}'$ and $\boldsymbol{X}'\boldsymbol{X}$ are the same and the corresponding eigenvectors are related to each other. More precisely, if $\mathbf{v}$ is an eigenvector of the former matrix corresponding to a nonzero eigenvalue λ, then $\boldsymbol{X}'\mathbf{v}$ is the corresponding eigenvector of the latter.

(b) Let $\boldsymbol{X}$ be a $n \times p$ matrix with orthogonal columns. What are the eigenvalues and eigenvectors of $\boldsymbol{X}'\boldsymbol{X}$ and $\boldsymbol{X}\boldsymbol{X}'$?

(c) Let $\boldsymbol{X}$ be the design matrix for the fat dataset in Chapter 2. Use a computer program and compute the SVD of $\boldsymbol{X}$ and the spectral decomposition of $\boldsymbol{X}'\boldsymbol{X}$. Plot the ordered eigenvalues of $\boldsymbol{X}'\boldsymbol{X}$ against their rank (index) $i = 1, \ldots, p$. Do you see any shape or pattern?

3.17 Verify and fill in the details in (3.17).

3.18 Eigenvalues as maximum of quadratic forms: Let A be a $p \times p$ positive-definite matrix with ordered eigenvalues and normalized eigenvectors $(\lambda_i, \mathbf{e}_i)$. Using the spectral decomposition of A show that

(a) $\max_{\boldsymbol{x} \neq 0} \frac{\boldsymbol{x}'A\boldsymbol{x}}{\boldsymbol{x}'\boldsymbol{x}} = \lambda_1$, and the maximum is attained for $\boldsymbol{x} = \mathbf{e}_1$.

(b) $\max_{\boldsymbol{x} \perp \mathbf{e}_1, \ldots, \mathbf{e}_k} \frac{\boldsymbol{x}'A\boldsymbol{x}}{\boldsymbol{x}'\boldsymbol{x}} = \lambda_{k+1}$, and the maximum is attained for $\boldsymbol{x} = \mathbf{e}_{k+1}, k = 1, 2, \ldots, p-1$.

(c)

$$\min_{\boldsymbol{x} \neq 0} \frac{\boldsymbol{x}'A\boldsymbol{x}}{\boldsymbol{x}'\boldsymbol{x}} = \lambda_p,$$

with the minimum attained for $\boldsymbol{x} = \mathbf{e}_p$.

3.19 Complete the proof of the generalized Cauchy-Schwarz inequality (3.18) using the provided hint.

3.20 Verify that for Y_t in (3.33),

$$\mathrm{Cov}(Y_i, Y_j) = \sigma_\varepsilon^2 (1-\rho^2)^{-1} \rho^{i-j} = \sigma^2 \rho^{i-j},$$

where $\sigma^2 = \sigma_\varepsilon^2 (1-\rho^2)^{-1}$.

3.21 Set $\log \boldsymbol{\Sigma} = A$, where A is the matrix in Problem 3.3.

(a) Show that the entries of the 2×2 covariance matrix $\boldsymbol{\Sigma}$ are given by

$$\sigma_{11} = \frac{1}{2\sqrt{\Delta}} \exp\left(\tfrac{\alpha+\gamma}{2}\right) \{\sqrt{\Delta}\, s^+ - (\alpha-\gamma)s^-\},$$

$$\sigma_{22} = \frac{1}{2\sqrt{\Delta}} \exp\left(\tfrac{\alpha+\gamma}{2}\right) \{\sqrt{\Delta}\, s^+ + (\alpha-\gamma)s^-\},$$

and

$$\sigma_{12} = \sigma_{21} = \frac{\beta}{2\sqrt{\Delta}} \exp\left(\frac{\alpha+\gamma}{2}\right) s^-,$$

where

$$s^{\pm} = \exp\left(\frac{\sqrt{\Delta}}{2}\right) \pm \exp\left(-\frac{\sqrt{\Delta}}{2}\right).$$

(b) Show that $\log A$ for the matrix A is

$$A = \begin{pmatrix} 2 & 1 \\ 1 & 2 \end{pmatrix}, \quad \log A = \frac{\log 3}{2} \begin{pmatrix} 1 & 1 \\ 1 & 1 \end{pmatrix}.$$

so that while A has full-rank two, its log is singular with rank one.

3.22 For ρ a correlation coefficient taking values in $(-1, 1)$, show that

$$\log\begin{pmatrix} \rho & 1 \\ 1 & \rho \end{pmatrix} = \frac{1}{2}\begin{pmatrix} \log(1-\rho^2) & \log\frac{1+\rho}{1-\rho} \\ \log\frac{1+\rho}{1-\rho} & \log(1-\rho^2) \end{pmatrix}.$$

Identify the range of values of the entries of the matrix in the right-hand side and compare them to those of the matrix on the left.

3.23 Consider the partitioned matrices

$$\mathbf{A} = \begin{pmatrix} \mathbf{A}_{11} & \mathbf{A}_{12} \\ \mathbf{A}_{21} & \mathbf{A}_{22} \end{pmatrix}, \ \mathbf{B} = \begin{pmatrix} \mathbf{I} & \mathbf{0} \\ -\mathbf{A}_{21}\mathbf{A}_{11}^{-1} & \mathbf{I} \end{pmatrix},$$

where $\mathbf{A}_{11}$ is invertible and $\mathbf{B}$ is a unit lower triangular with determinant equal to 1.

(a) (Determinant of partitioned matrices) Show that

$$\mathbf{BAB}' = \begin{pmatrix} \mathbf{A}_{11} & \mathbf{0} \\ \mathbf{0} & \mathbf{A}_{22} - \mathbf{A}_{21}\mathbf{A}_{11}^{-1}\mathbf{A}_{12} \end{pmatrix},$$

and consequently

$$|\mathbf{A}| = |\mathbf{A}_{11}||\mathbf{A}_{22} - \mathbf{A}_{21}\mathbf{A}_{11}^{-1}\mathbf{A}_{12}|.$$

The matrix $\mathbf{A}_{22.1} = \mathbf{A}_{22} - \mathbf{A}_{21}\mathbf{A}_{11}^{-1}\mathbf{A}_{12}$, which appears frequently in statistics and linear algebra, is usually called the *Schur complement* of $\mathbf{A}_{22}$ in $\mathbf{A}$.

(b) (Inverse of partitioned matrices) Show that

$$\mathbf{A}^{-1} = \begin{pmatrix} \mathbf{A}_{11}^{-1} + \mathbf{A}_{11}^{-1}\mathbf{A}_{12}\mathbf{A}_{22.1}^{-1}\mathbf{A}_{21}\mathbf{A}_{11}^{-1} & -\mathbf{A}_{11}^{-1}\mathbf{A}_{12}\mathbf{A}_{22.1}^{-1} \\ -\mathbf{A}_{22.1}^{-1}\mathbf{A}_{21}\mathbf{A}_{11}^{-1} & \mathbf{A}_{22.1}^{-1} \end{pmatrix}.$$

3.24 For the bivariate random vector $(Y_1, Y_2)'$ with the joint density factorization

$$f(y_1, y_2) = f_{1|2}(y_1|y_2)f_1(y_1),$$

show that the conditional distribution corresponding to $f_{1|2}(y_1|y_2)$ can be expressed in terms of the bivariate copula as

$$F_{1|2}(y_1|y_2) = \frac{\partial}{\partial F_2(y_2)} C_{12}(F_1(y_1), F_2(y_2)).$$

3.25 For a random vector $Y = (Y_1, \ldots, Y_p)'$ with continuous marginal distributions and $g_i(\cdot), i = 1, \ldots, p$, univariate continuous and strictly increasing functions, show that the vectors Y and $(g_1(Y_1), \ldots, g_p(Y_p))'$ share the same copula function.

4

Hidden Regressions

For a random vector $Y = (Y_1, \dots, Y_p)'$ with a probability density function f_Y, the roots of the regression idea for modeling its dependence can be traced to the standard factorization of its probability density function

$$f_Y = f_{Y_{-j}} f_{Y_j|Y_{-j}}, \; j = 1, \dots, p, \tag{4.1}$$

where Y_{-j} is the original vector with the entry Y_j removed, and the second factor is the conditional density of Y_j given the rest. For a multivariate normal distribution the ensuing conditional distributions have profound implications for revealing a host of hidden linear regressions and various factorizations of its covariance matrix when Y_{-j} is replaced by a subset of its entries. Interestingly, such factorizations break the "two-way, row-column" character of a matrix into simpler factors related to regressions and can be estimated more readily. The most prominent examples are regression interpretation of columns of a precision matrix and the Cholesky decomposition with important roles in modeling covariances using covariates with guaranteed positive-definiteness. In this chapter, we present regularization of hidden regressions derived from multivariate normal conditional distributions using Lasso-type penalties. Regularization of a few other hidden regressions arrived at using other means than the conditional distributions are reviewed.

4.1 Multivariate Normal Distributions

In this section some basic properties of a multivariate normal random vector are reviewed briefly. These include its closure under linear transformations, conditional distributions, maximum likelihood estimation (MLE) of the mean and covariance parameters, and the dependence structure of the entries of the sample covariance matrix. For more details on other standard properties of multivariate normal distributions the reader may refer to Johnson and Wichern [2008] and Anderson [2003].

The probability density function of a multivariate normal random vector $Y \sim N_p(\boldsymbol{\mu}, \boldsymbol{\Sigma})$ is given by

$$f(\boldsymbol{y}) = \frac{1}{(2\pi)^{p/2}|\boldsymbol{\Sigma}|^{1/2}} \exp\{-\frac{1}{2}(\boldsymbol{y} - \mu)'\boldsymbol{\Sigma}^{-1}(\boldsymbol{y} - \mu)\}, \; \boldsymbol{y} \in \mathbb{R}^p, \tag{4.2}$$

DOI: 10.1201/9781003638896-4

provided that $\mathbf{\Sigma}$ is invertible or positive-definite. For singular covariance matrices the density is not defined, though it is supported on lower-dimensional subspaces. There are two alternative definitions of multivariate normality which obviates the need for determinant of $\mathbf{\Sigma}$ and its inverse. The first one is in terms of characteristic function of Y:

$$\phi_Y(\boldsymbol{t}) = \exp(i\boldsymbol{\mu}'\boldsymbol{t} - \frac{1}{2}\boldsymbol{t}'\mathbf{\Sigma}\boldsymbol{t}), \boldsymbol{t} \in \mathbb{R}^p,$$

and the second, which we prefer, involves univariate normality of all linear combinations $\boldsymbol{c}'Y$:

Theorem 6 *A random vector $Y \sim N_p(\boldsymbol{\mu}, \mathbf{\Sigma})$, if and only if for all $\boldsymbol{c} \in \mathbb{R}^p$,*

$$\boldsymbol{c}'Y = \sum_{i=1}^{p} c_i Y_i \sim N(\boldsymbol{c}'\boldsymbol{\mu}, \boldsymbol{c}'\mathbf{\Sigma}\boldsymbol{c}).$$

This characterization of multivariate normality implicitly confirms the inherent preservation of normality under linear transformation: If $Y \sim N_p(\boldsymbol{\mu}, \mathbf{\Sigma})$, then

$$AY + \boldsymbol{b} \sim N_q(A\boldsymbol{\mu} + \boldsymbol{b}, A\mathbf{\Sigma}A'),$$

for any $q \times p$ matrix A and q-vector $\boldsymbol{b}$.

4.1.1 Maximum Likelihood Estimation

For a sample $\boldsymbol{Y}_1, \ldots, \boldsymbol{Y}_n$ of size n from a multivariate normal Y, we denote the corresponding data matrix by the $n \times p$ matrix $\boldsymbol{Y}$ with $\boldsymbol{Y}_i$ as its ith row and use $\boldsymbol{y}_j$ to denote its jth column. This convention obviates the need for using double indices to label the rows and columns of a data matrix. For data from a normal distribution with parameters $(\boldsymbol{\mu}, \mathbf{\Sigma})$, the parameters are usually estimated by maximizing the likelihood function

$$L(\mu, \mathbf{\Sigma}|\text{data}) = \frac{1}{(2\pi)^{np/2}|\mathbf{\Sigma}|^{n/2}} \exp\left\{-\frac{1}{2}\sum_{i=1}^{n}(\boldsymbol{Y}_i - \boldsymbol{\mu})'\mathbf{\Sigma}^{-1}(\boldsymbol{Y}_i - \boldsymbol{\mu})\right\}, \quad (4.3)$$

over the parameter space. The maximum likelihood estimators (MLE) of these parameters are

$$\bar{\boldsymbol{Y}} = \frac{1}{n}\sum_{i=1}^{n}\boldsymbol{Y}_i, \quad \mathbf{S} = \frac{1}{n}\sum_{i=1}^{n}(\boldsymbol{Y}_i - \bar{\boldsymbol{Y}})(\boldsymbol{Y}_i - \bar{\boldsymbol{Y}})', \quad (4.4)$$

see Anderson [2003]. It is easy to show that the sample mean is unbiased for $\boldsymbol{\mu}$, but the MLE $\mathbf{S}$ is biased for $\mathbf{\Sigma}$ which can be corrected by replacing the n in the denominator by $n - 1$. In the sequel we refer to $\mathbf{S}$ in (4.4) as the sample covariance matrix.

Evidently, the entries of $\mathbf{S}$ are correlated and heterogeneous. The variance and dependence nature of the entries of the sample covariance matrix $\mathbf{S}$ play important roles in selecting threshold parameters and other aspects of sparse estimation of a covariance matrix and its inverse. Next, we present explicit formulae for these for a sample from normal distributions. Under normality, the sample covariance matrix $\mathbf{S}$ has a Wishart distribution with $n-1$ degrees of freedom, which is the matrix generalization of the familiar chi-squared distribution for the sample variance of a sample from univariate $N(\mu, \sigma^2)$:

$$\frac{(n-1)S^2}{\sigma^2} \sim \chi^2_{n-1}.$$

The fact that this is realized as the sum of squares of $n-1$ standard normal random variables motivates a more direct definition of Wishart distributions. A $p \times p$ random matrix $\boldsymbol{W}$ defined by

$$\boldsymbol{W} = \sum_{i=1}^{n} \boldsymbol{Z}_i \boldsymbol{Z}_i' \sim W_p(n, \boldsymbol{\Sigma}), \tag{4.5}$$

for independent $\boldsymbol{Z}_i \sim N_p(\mathbf{0}, \boldsymbol{\Sigma})$, is said to have a Wishart distribution with degrees of freedom n and the scale (covariance) matrix $\boldsymbol{\Sigma}$. In this case, the random matrix $\boldsymbol{U} = \boldsymbol{W}^{-1}$ is said to have an *inverse Wishart* distribution where it can be shown that

$$E(\boldsymbol{U}) = \frac{\boldsymbol{\Sigma}^{-1}}{n-p-1}, \tag{4.6}$$

provided that $\boldsymbol{\Sigma}$ is invertible and $n > p+1$.

The succinct and closed-form expressions for dependence among and heterogeneity of entries of the sample covariance matrix $\mathbf{S}$ involve the Wishart distribution, the vec operator and Kronecker's product of two matrices. We recall that for a $p \times p$ matrix A, the operator $\text{vec}(A)$ transforms the matrix to a long column vector with p^2 entries by simply stacking up the first column, then second and so on. For a symmetric matrix A, $\text{vech}(A)$ is a shorter column vector with $p(p+1)/2$ entries discarding the upper diagonal entries of A. The duplication matrix D_p is a matrix which transforms $\text{vech}(A)$ into $\text{vec}(A)$ for any $p \times p$ symmetric matrix A, and defined by

$$D_p \text{vech}(A) = \text{vec}(A),$$

see Problem 4.2. With these notation it is known [Abadir and Magnus, 2005] that

$$\text{Cov}(\text{vech}(\mathbf{S})) = \frac{2}{n} D_p^+ (\boldsymbol{\Sigma} \otimes \boldsymbol{\Sigma}) D_p^+, \tag{4.7}$$

where $D_p^+ = (D_p' D_p)^{-1} D_p'$ is the Moore-Penrose inverse of D_p and $\otimes$ stands for the Kronecker's product of the two indicated matrices.

4.1.2 Conditional Distributions and Regressions

This section highlights the role of the conditional distribution of a multivariate normal random vector in revealing hidden regressions and the Cholesky decomposition of its covariance matrix.

For a p-dimensional partitioned random vector $Y = (Y_{(1)}', Y_{(2)}')'$ of dimensions p_1 and p_2, respectively, we partition its mean vector, covariance and precision matrices as

$$\boldsymbol{\mu} = \begin{pmatrix} \boldsymbol{\mu}_1', & \boldsymbol{\mu}_2' \end{pmatrix}', \boldsymbol{\Sigma} = \begin{pmatrix} \boldsymbol{\Sigma}_{11} & \boldsymbol{\Sigma}_{12} \\ \boldsymbol{\Sigma}_{21} & \boldsymbol{\Sigma}_{22} \end{pmatrix}, \boldsymbol{\Sigma}^{-1} = \begin{pmatrix} \boldsymbol{\Sigma}^{11} & \boldsymbol{\Sigma}^{12} \\ \boldsymbol{\Sigma}^{21} & \boldsymbol{\Sigma}^{22} \end{pmatrix}. \tag{4.8}$$

When Y has a multivariate normal distribution with positive-definite covariance matrix, then the following well-known result provides valuable information about the conditional distribution of $Y_{(2)}$ given $Y_{(1)}$.

Theorem 7 *If* $Y \sim N_p(\boldsymbol{\mu}, \boldsymbol{\Sigma})$ *is partitioned as above, then the conditional distribution of* $Y_{(2)}$ *given* $Y_{(1)}$ *is normal*

$$Y_{(2)}|Y_{(1)} \sim N_{p_2}(\boldsymbol{\mu}_{2.1}, \boldsymbol{\Sigma}_{22.1}), \tag{4.9}$$

where

$$\boldsymbol{\mu}_{2.1} = \boldsymbol{\mu}_2 - \boldsymbol{\Sigma}_{21}\boldsymbol{\Sigma}_{11}^{-1}(Y_{(1)} - \boldsymbol{\mu}_1),\ \boldsymbol{\Sigma}_{22.1} = \boldsymbol{\Sigma}_{22} - \boldsymbol{\Sigma}_{21}\boldsymbol{\Sigma}_{11}^{-1}\boldsymbol{\Sigma}_{12}. \tag{4.10}$$

This theorem clearly shows the *linearity* of the conditional mean of $Y_{(2)}$ given $Y_{(1)}$ or revealing a hidden regression in the covariance matrix. More importantly the matrix of regression coefficients, namely $\boldsymbol{\Sigma}_{21}\boldsymbol{\Sigma}_{11}^{-1}$, involves entries of the covariance and precision matrices. The conditional covariance $\boldsymbol{\Sigma}_{22\cdot1}$ in (4.10), called the *Schur complement* of $\boldsymbol{\Sigma}_{11}$ in $\boldsymbol{\Sigma}$, appears frequently in linear algebra and statistics. Its appearance implicitly signals the existence of a (hidden) regression setup.

It turns out normality of the distribution is not required for some of the above results to hold. Next, we derive (4.10) using only the second moments of the distribution and the notion of orthogonality without requiring Y to be normally distributed. For simplicity Y is taken to be a mean zero random vector, and we find interesting relationships among certain blocks of $\boldsymbol{\Sigma}$ and $\boldsymbol{\Sigma}^{-1}$ by considering the linear least-squares regression (prediction) of $Y_{(2)}$ based on $Y_{(1)}$. More precisely, setting

$$Y_{(2)} = \mathbf{B}Y_{(1)} + \boldsymbol{\epsilon}, \tag{4.11}$$

where $\mathbf{B}$ is the $p_2 \times p_1$ (population) regression coefficient matrix, so that the vector of residuals $\boldsymbol{\epsilon} = Y_{(2)} - \mathbf{B}Y_{(1)}$, is uncorrelated with $Y_{(1)}$, namely $\boldsymbol{\epsilon} \perp Y_{(1)}$ or

$$\mathbf{0} = \text{Cov}(Y_{(2)} - \mathbf{B}Y_{(1)}, Y_{(1)}) = \text{Cov}(Y_{(2)}, Y_{(1)}) - \mathbf{B}\text{Cov}(Y_{(1)}, Y_{(1)}).$$

Then, it follows that $\mathbf{B}$ and the prediction error covariance matrix satisfy

$$\mathbf{B} = \boldsymbol{\Sigma}_{21}\boldsymbol{\Sigma}_{11}^{-1}, \tag{4.12}$$

and

$$\text{Cov}(\boldsymbol{\epsilon}) = \text{Cov}(Y_{(2)} - \mathbf{B}Y_{(1)}, Y_{(2)}) = \boldsymbol{\Sigma}_{22} - \boldsymbol{\Sigma}_{21}\boldsymbol{\Sigma}_{11}^{-1}\boldsymbol{\Sigma}_{12}, \tag{4.13}$$

where the expression in the right-hand side of (4.13) is the same as $\boldsymbol{\Sigma}_{22\cdot1}$ in (4.10).

Next, we re-express the matrix of regression coefficients $\mathbf{B}$ and the covariance matrix of the residuals $\boldsymbol{\Sigma}_{22\cdot1}$ in terms of the corresponding blocks of the precision matrix. This plays a central role in relating the zero entries of precision matrices to those of the related regression coefficients, or *connecting sparsity of the regression coefficients to that of the columns of the precision matrix.*

The key step is to realize that the vector of residuals $\boldsymbol{\epsilon}$ is a linear transformation of $Y_{(1)}$ and $Y_{(2)}$, and one may use it to write

$$\begin{pmatrix} Y_{(1)} \\ \boldsymbol{\epsilon} \end{pmatrix} = \begin{pmatrix} \boldsymbol{I}_{p_1} & \mathbf{0} \\ -\mathbf{B} & \boldsymbol{I}_{p_2} \end{pmatrix} \begin{pmatrix} Y_{(1)} \\ Y_{(2)} \end{pmatrix} = TY, \tag{4.14}$$

where T is seen to be a block lower-triangular matrix with the identity blocks down the diagonal and $-\mathbf{B}$, the negative of the regression coefficients, in the $(2, 1)$ block. Let $D = \text{diag}(\boldsymbol{\Sigma}_{11}, \boldsymbol{\Sigma}_{22\cdot 1})$ be the block-diagonal covariance matrix of the random vector in the left-hand side of (4.14), then computing covariance of this linear transformation it follows that

$$T\boldsymbol{\Sigma}T' = D,$$

and the precision matrix is given by

$$\boldsymbol{\Sigma}^{-1} = T'D^{-1}T. \tag{4.15}$$

Important consequences of multiplying out the partitioned matrices in the right-hand side of this identity and matching with the $(2, 2)$ and $(1, 2)$ blocks of $\boldsymbol{\Sigma}^{-1}$ it follows that

$$\mathbf{B} = \boldsymbol{\Sigma}_{21}\boldsymbol{\Sigma}_{11}^{-1} = -(\boldsymbol{\Sigma}^{22})^{-1}\boldsymbol{\Sigma}^{21}, \tag{4.16}$$

and

$$\boldsymbol{\Sigma}_{22\cdot 1} = (\boldsymbol{\Sigma}^{22})^{-1}. \tag{4.17}$$

There is a flurry of activities emanating from these two amazing representations of the multivariate regression parameters in terms of subblocks of the precision matrix as discussed in the next section. They are the key to connecting, the now standard, sparse estimation of regression coefficients to the more challenging problem of sparse estimation of precision matrices.

The idea of progressively orthogonalizing (Gram-Schmidt) the entries of Y leading to (4.15) is at the heart of deriving the modified Cholesky decomposition of a covariance matrix with an eye on the regression-related interpretability of the entries of the Cholesky factors (T, D) in Section 4.3.

4.2 Regressions in Precision Matrices

For regression-inspired interpretation of entries of the precision matrix, we use the partition $(Y'_{(1)}, Y'_{(2)})'$ and focus on the special choices of $p_2 = 1, 2$, for the number of entries in $Y_{(2)}$. They lead to powerful results directly connecting the regression coefficients $\mathbf{B}$ and the prediction error covariance $\boldsymbol{\Sigma}_{22\cdot 1}$ to the individual entries of the precision matrix $\boldsymbol{\Sigma}^{-1}$. In particular, writing the variance-correlation decomposition as $\boldsymbol{\Sigma}^{-1} = (\sigma^{ij}) = DRD$, it is shown that the entries of $\boldsymbol{\Sigma}^{-1}$, R, D have direct statistical interpretations in terms of the partial correlations, regression coefficients and variance of predicting a variable given the *rest*.

Theorem 8 *(Precision Matrix, Regression, Partial Correlation).*

Let $Y = (Y_1, \ldots, Y_p)'$ *be a zero-mean random vector with a precision matrix* $\boldsymbol{\Sigma}^{-1} = (\sigma^{ij})$*. Then,*

(a) For any fixed i,

$$Y_i = \sum_{j \neq i} \beta_{i,j} Y_j + \varepsilon_i, \tag{4.18}$$

where

$$\beta_{i,j} = -\frac{\sigma^{ij}}{\sigma^{ii}}, \quad j \neq i, \tag{4.19}$$

and

$$Var(\varepsilon_i) = Var(Y_i | Y_j, j \neq i) = \frac{1}{\sigma^{ii}}, \quad i = 1, \ldots, p. \tag{4.20}$$

(b) The partial correlation coefficient between any two random variables Y_i *and* $Y_j, i \neq j$*, adjusted for the linear effects of the remaining* $p-2$ *variables, is given by*

$$\rho^{ij} = -\frac{\sigma^{ij}}{\sqrt{\sigma^{ii}\sigma^{jj}}}. \tag{4.21}$$

Consequently, the regression coefficients have a "symmetric representation" *in terms of the partial correlations as*

$$\beta_{ij} = \rho^{ij}\sqrt{\frac{\sigma^{jj}}{\sigma^{ii}}}. \tag{4.22}$$

Proof. (a) In (4.11) take $p_2 = 1$, and for a fixed i set $Y_{(2)} = Y_i$, and $Y_{(1)} = (Y_1, \ldots, Y_{i-1}, Y_{i+1}, \ldots, Y_p)' = Y_{-i}$. Then, $\mathbf{B}$ is a row vector with $(p-1)$ entries β_{ij}'s and (4.18) holds. Define $\widehat{Y}_i$ as the linear least-squares predictor of Y_i based on the rest Y_{-i}, and set $\varepsilon_i = Y_i - \widehat{Y}_i$, $d_i^2 = \text{Var}(\varepsilon_i)$ as its prediction error and prediction error variance, respectively. Then, $\boldsymbol{\Sigma}_{22\cdot 1}$ is a scalar, called the *partial variance* of Y_i given the rest. Moreover, it follows immediately from (4.16)–(4.17) that the regression coefficients of Y_i on Y_{-i}, are given by (4.19) and (4.20). This shows that σ^{ij}, the (i, j) entry of the precision matrix, is up to a scalar, the coefficient of variable j in the multiple regression of variable i on the rest. As such each $\beta_{i,j}$ is an unconstrained real number with $\beta_{j,j} = 0$, but $\beta_{i,j}$ is not necessarily symmetric in (i, j).

(b) For $p_2 = 2$, take $Y_{(2)} = (Y_i, Y_j)$, and $Y_{(1)} = Y_{-(ij)}$ comprising the remaining variables. Then, it follows from (4.17) that the covariance matrix between Y_i and Y_j after eliminating the linear effects of the others is

$$\boldsymbol{\Sigma}_{22\cdot 1} = \begin{pmatrix} \sigma^{ii} & \sigma^{ij} \\ \sigma^{ij} & \sigma^{jj} \end{pmatrix}^{-1} = \Delta^{-1} \begin{pmatrix} \sigma^{jj} & -\sigma^{ij} \\ -\sigma^{ij} & \sigma^{ii} \end{pmatrix},$$

where $\Delta = \sigma^{ii}\sigma^{jj} - (\sigma^{ij})^2$. Thus, the correlation coefficient in $\boldsymbol{\Sigma}_{22\cdot 1}$ is, indeed, the *partial correlation coefficient* between Y_i and Y_j adjusted for the linear effect of others as in (4.21). Moreover, from (4.19) and (4.21) it follows that the regression coefficients have a *symmetric representation* in terms of the partial correlations as in (4.22).

Remark 2 *1. The representation (4.22) confirms that $\boldsymbol{\Sigma}^{-1}$ and the matrix $\Pi = (\rho^{ij})$ of the partial correlations share the same sparsity patterns. This observation is the basis for the Peng et al. [2009]'s SPACE (Sparse PArtial Correlation Estimation) algorithm which imposes a Lasso penalty on the off-diagonal entries of the symmetric matrix of partial correlations Π, thus obviates the need to impose the symmetry constraint.*

2. Computing all partial correlations using regression residuals or Schur complement involves inverting $\binom{p}{2}$ matrices of order $(p-2) \times (p-2)$, which is computationally challenging for high dimensional data. By contrast, the alternative formula in (4.21) requires inverting the $p \times p$ matrix $\boldsymbol{\Sigma}$ only once.

4.3 Hidden Regressions in the Cholesky Factors

The role of regression and statistical interpretation of the entries of the standard Cholesky factor are somewhat hidden in $\boldsymbol{\Sigma} = CC'$, with $C = (c_{ij})$ a lower triangular matrix. In this section, we show that by standardizing its columns or shifting attention to the *modified Cholesky decomposition* (MCD) of the covariance matrix reveals a fruitful connection to a sequence of one-sided regressions in contrast to the two-sided regressions in (4.18). Alternatively, one could rely on orthogonalization in (4.14) to arrive directly at the modified Cholesky decomposition and hence similar regression-based interpretations.

We start with the diagonal matrix $D_1 = \text{diag}(c_{11}, \ldots, c_{pp})$ and write

$$\boldsymbol{\Sigma} = CD_1^{-1}D_1D_1D_1^{-1}C' = LDL', \tag{4.23}$$

where $L = CD_1^{-1}$ and $D = D_1^2$ are unit lower-triangular and diagonal matrices. Note that L is obtained from C by dividing the entries of its ith column by c_{ii}. The factorization (4.23), called the *modified Cholesky decomposition* of $\boldsymbol{\Sigma}$, also admits the following two equivalent forms

$$T\boldsymbol{\Sigma}T' = D, \quad \boldsymbol{\Sigma}^{-1} = T'D^{-1}T, \tag{4.24}$$

where $T = L^{-1}$ is a unit lower-triangular matrix.

The statistical relevance and meaning of the non-redundant entries of (T, D) in (4.24) will emerge using the idea of linear prediction and assuming an *order* among the variables in a random vector Y with mean zero and a positive-definite covariance matrix $\boldsymbol{\Sigma}$. Let $\widehat{Y}_t$ be the linear least-squares predictor of Y_t based on its predecessors (past) $Y_{t-1}, \ldots, Y_1$, and $\varepsilon_t = Y_t - \widehat{Y}_t$ be its prediction error with variance $\sigma_t^2 = \mathrm{Var}(\varepsilon_t)$. Then, one may write

$$\widehat{Y}_t = \sum_{j=1}^{t-1} \phi_{tj} Y_j,$$

for some scalars ϕ_{tj}'s leading to the following p regression models:

$$Y_t = \sum_{j=1}^{t-1} \phi_{tj} Y_j + \varepsilon_t, \quad t = 1, \ldots, p. \tag{4.25}$$

For $t = 1$, by convention $Y_1 = \varepsilon_1$.

Next, we show how to express the regression coefficients $\phi_t = (\phi_{t1}, \ldots, \phi_{t,t-1})'$ in (4.25) and σ_t^2 in terms of the given covariance matrix $\boldsymbol{\Sigma}$. For a fixed t, $2 \leq t \leq p$, let $\boldsymbol{\Sigma}_t$ be the $(t-1) \times (t-1)$ leading principal minor of $\boldsymbol{\Sigma}$ and $\tilde{\boldsymbol{\sigma}}_t$ be the column vector composed of the first $t-1$ entries of the tth column of $\boldsymbol{\Sigma}$. Then, with $Y_{(1)} = (Y_1, \ldots, Y_{t-1})', Y_{(2)} = Y_t$ from (4.12)–(4.13), it follows that

$$\phi_t = \boldsymbol{\Sigma}_t^{-1} \tilde{\boldsymbol{\sigma}}_t, \sigma_t^2 = \sigma_{tt} - \tilde{\boldsymbol{\sigma}}_t' \boldsymbol{\Sigma}_t^{-1} \tilde{\boldsymbol{\sigma}}_t. \tag{4.26}$$

It is instructive to connect the regression parameters ϕ_{tj}'s to the matrix T in (4.24). Rewriting the regressions (4.25) in matrix form as

$$\Phi Y = \boldsymbol{\epsilon},$$

for a matrix Φ where its non-redundant entries are -ϕ_{tj} and $\boldsymbol{\epsilon} = (\varepsilon_1, \ldots, \varepsilon_p)'$, it follows from the uniqueness of the factors in (4.24) that $\Phi = T$; see Problem 4.6. Consequently, the nonredundant entries of T are, indeed, the regression coefficients and those of D are the prediction error variances in a successive regression scheme, and the pair (T, D) offers a new reparameterization of the covariance matrix. A natural question is whether they are unconstrained parameters. Fortunately, the answer is in the affirmative as presented next.

Note that since the ϕ_{ij}'s in (4.26) are simply the regression coefficients computed from an unstructured covariance matrix, these coefficients along with $\log \sigma_t^2$ are unconstrained parameters. Thus, one may employ linear models with covariates, nonparametric and Bayesian methods, etc. in modeling them; see Section 6.4.2 and Pourahmadi [2013, Section 3.9]. An important consequence of the decomposition (4.24) is that for any estimate $(\widehat{T}, \widehat{D})$ of the Cholesky factors, the estimated covariance and the precision matrix $\widehat{\boldsymbol{\Sigma}}^{-1} = \widehat{T}'\widehat{D}^{-1}\widehat{T}$ are guaranteed to be positive-definite. From (4.25) it is evident that, the successive regression scheme reduces the task of modeling a covariance matrix to that of a sequence of p varying-coefficient and varying-order regression models.

An alternative Cholesky decomposition of the form

$$\boldsymbol{\Sigma} = D_1 \tilde{L} \tilde{L}' D_1, \quad \tilde{L} = D_1^{-1} C,$$

is given in Chen and Dunson [2003] where $\tilde{L}$ is obtained from C by dividing the entries of its ith row by c_{ii}. It has proved useful for joint variable selection in the linear mixed-effects models and modeling correlation matrices [Pourahmadi, 2007a].

Of course, the success of modeling covariance via modified Cholesky decomposition hinges critically on the existence or choice of an order among the variables. This is not an issue for time series and longitudinal data, but is a challenge for general multivariate data. Fortunately, there are several heuristic and principled methods to deal with the order problem. For example, in Rajaratnam and Salzman [2013] the goal is to order variables with the smallest sum of p variances in the sequential regression. Other ideas include Park and Kim [2020]'s Gaussian Structural Equation Model (GSEM), and the PC-algorithm in Kalisch and Bühlmann [2007] which estimates the complete partial directed acyclic graph (CPDAG) of the data to find an equivalant class of orders consistent with the graph, see Chapter 10. The most notable from computational standpoint which searches for the sparsest Cholesky factor is the Reverse Cuthill-Mckee (RCM) method [Davis, 2006], and it is implemented in the chol function in R and will be used as the default method of ordering variables in this book.

4.4 Penalized Hidden Regressions

In the search for sparsity, the hidden regressions revealed in terms of the columns of the precision matrix and rows of the Cholesky factor can be regularized using the Lasso or related penalties. The general plan is to reduce sparse estimation of matrices to that of solving a sequence of regularized regression problems. The alternative penalized likelihood approach is covered in Chapter 9, though, references are made to it in this chapter from time to time.

4.4.1 Regularized Nodewise Regression

The regression in (4.18) is the starting point of Meinshausen and Bühlmann [2006]'s nodewise selection (NS) idea of reducing the sparse precision matrix estimation problem to that of solving p separate Lasso regressions one for each column of $\boldsymbol{\Theta} = \boldsymbol{\Sigma}^{-1}$.

Let $\boldsymbol{y}_i$ be the ith column of the $n \times p$ data matrix $\boldsymbol{Y}$, and $\boldsymbol{Y}_{-i}$ be the $\boldsymbol{Y}$ after removing its ith column. Then, the regularized estimator of $\boldsymbol{\beta}_i$, the vector of regression coefficients in (4.18), is given by

$$\widehat{\boldsymbol{\beta}}_i = \underset{\boldsymbol{\beta}_i}{\text{argmin}} ||\boldsymbol{y}_i - \boldsymbol{Y}_{-i}\boldsymbol{\beta}_i||^2 + \lambda_i ||\boldsymbol{\beta}_i||_1, \ i = 1, \dots, p. \tag{4.27}$$

Indeed, such regularized *nodewise regressions* will encourage zero coefficients and hence a sparser inverse covariance or a sparser graph. The following fundamental equivalancies are the key to keeping track of the zeros for the sparse precision matrix, edge selection and partial correlations: For distinct j, k,

$$\beta_{j,k} \neq 0 \;\&\; \beta_{k,j} \neq 0 \Leftrightarrow (j,k) \;\&\; (k,j) \in E \Leftrightarrow \theta_{j,k} \neq 0 \Leftrightarrow \rho_{j,k|V\setminus\{j,k\}} \neq 0,$$

where E is the set of graph edges and $\rho_{j,k|V\setminus\{j,k\}}$ is the partial correlation coefficient between Y_j and Y_k after adjusting for the effects of others. Even though, regularized nodewise regressions encourage sparsity, the estimated precision matrix is neither symmetric nor positive-definite. Fortunately, there are simple and useful remedies based on OR/AND rules to yield a symmetric estimator, see for example (4.33).

Inspired by the approach in Meinshausen and Bühlmann [2006], the idea of penalized regression has been used increasingly in several settings listed below dealing with sparse estimation of covariance and precision matrices as well as multivariate regressions. In what follows to keep the notation simple, we fix the column/variable index at $i = p$ in (4.27) and focus on the regression of the last variable Y_p on $Y_1, \ldots, Y_{p-1}$, with regression coefficients $\boldsymbol{\beta} = (\beta_1, \ldots, \beta_{p-1})'$. Once an algorithm is developed for computing $\boldsymbol{\beta}$, it will be applied repeatedly to columns $p-1, p-2, \ldots, 2, 1$, by moving each column to the last position. This convention will be followed when solving the optimization problem corresponding to the following applications:

1. Nodewise selection (NS) algorithm in Meinshausen and Bühlmann [2006] and the sparse PArtial Correlation Estimation (SPACE) algorithm in Peng et al. [2009].

2. Iterative Conditional Fitting (ICF) algorithm in Chaudhuri et al. [2007] for computing MLE of covariance matrices with a prespecified set of zeros.

3. Graphical Lasso (GLasso) algorithm in Friedman et al. [2008] for sparse estimation of precision matrix via penalized likelihood.

4. Covariance Glasso (C-GLasso) algorithm for sparse estimation of covariance matrix via penalized likelihood [Wang, 2014].

5. Sparse estimation of multivariate regression coefficient matrix and precision matrix via jointly penalized likelihood [Wang, 2015].

We note that in the NS algorithm the regression models are fitted for each variable separately and consequently it suffers from two major drawbacks. First, it does not take into account the intrinsic symmetry of the precision and partial correlation matrices, and second the same penalty parameter is used for all p Lasso regressions.

4.4.2 The SPACE Algorithm

The Sparse PArtial Correlation Estimation (SPACE) algorithm in Peng et al. [2009] deals simultaneously with the two shortcomings of the NS algorithm by minimizing the objective function

$$\frac{1}{2}\sum_{i=1}^{p} w_i||\boldsymbol{y}_i - \sum_{j\neq i}\rho^{ij}\sqrt{\frac{\theta_{jj}}{\theta_{ii}}}\boldsymbol{y}_j||^2 + \lambda\sum_{i<j}|\rho^{ij}|, \tag{4.28}$$

over the partial correlations ρ^{ij} where w_i are some known nonnegative weights and $\boldsymbol{y}_i$ is the sample data from the ith variable or the ith column of the data matrix $\boldsymbol{Y}$. Note that while Meinshausen and Bühlmann [2006] in (4.27) uses p separate linear regressions to estimate the neighborhood of one node at a time, the SPACE algorithm merges all p linear regressions into a single least-squares problem. Furthermore, the variables associated to each regression are weighted differently according to w_i's. Since it estimates the partial correlations simultaneously, the symmetry of the estimated precision matrix is ensured.

It is fruitful to connect SPACE and penalized regression in general to penalized likelihood estimation. For this we consider the following optimization problem, inspired by the SPACE idea above, which minimizes

$$Q_0(\Theta) = -\frac{n}{2}\sum_{i=1}^{p}\log\theta_{ii} + \frac{1}{2}\sum_{i=1}^{p} w_i||\boldsymbol{y}_i - \sum_{j\neq i}\rho^{ij}\sqrt{\frac{\theta_{jj}}{\theta_{ii}}}\boldsymbol{y}_j||^2 + \lambda\sum_{i<j}|\rho^{ij}|, \quad (4.29)$$

over the partial correlations ρ^{ij}'s and θ_{ii}'s. Note that fixing the diagonal entries θ_{ii}'s, it reduces to (4.28). From (4.19)–(4.22) governing the relations among $\beta_{i,j}, \theta_{i,j}, \rho^{ij}$, and imposing the penalty on the entries of the precision matrix $\boldsymbol{\Theta}$ instead of the partial correlations, a modified SPACE objective function can be reformulated in terms of the entries of the precision matrix as minimizing

$$Q_1(\boldsymbol{\Theta}) = -\frac{n}{2}\sum_{i=1}^{p}\log\theta_{ii} + \frac{1}{2}\sum_{i=1}^{p}\theta_{ii}||\boldsymbol{y}_i + \sum_{j\neq i}\frac{\theta_{jj}}{\theta_{ii}}\boldsymbol{y}_j||^2 + \lambda\sum_{i<j}|\theta_{ij}|, \quad (4.30)$$

over $\boldsymbol{\Theta}$. This objective function turns out to be convex; for more details and extensions, see Problem 9.8. In fact, it is part of the regression-based CONvex CORrelation selection methoD (CONCORD) method in Khare et al. [2014] which unifies about five penalized precision matrix estimation proposals in the literature and leads to a powerful coordinate descent algorithm for minimizing this objective function involving a covariance matrix in a manner similar to that for Lasso regression in Section 2.1.3.

4.4.3 Constrained L_1-Minimization for Inverse Matrix Estimation

This section provides a brief description of another example of hidden regression unearthed algebraically from a particular sparse precision matrix estimation method proposed in Cai et al. [2011].

The CLIME estimator is defined as the solution of the following convex optimization problem

$$\text{minimize } ||\boldsymbol{\Theta}||_1 \quad \text{subject to} \quad ||\boldsymbol{S\Theta} - \boldsymbol{I}||_\infty \leq \lambda, \quad (4.31)$$

where λ is a tuning parameter and $||\cdot||_1, ||\cdot||_\infty$ are the element-wise ℓ_1- and ℓ_∞-norm, respectively, of the precision matrix [Cai et al., 2011]. The connection to regression is not immediate at all, but the motivation for the curious constraint comes from the form of the gradient of the penalized normal likelihood, see (9.2), and observing that

$$||\boldsymbol{\Theta}^{-1} - \mathbf{S}||_\infty \leq \lambda.$$

Due to the presence of matrix inverse, the feasible set of solutions can be complicated, however, multiplying it by $\mathbf{\Theta}$ leads to the simpler form (4.31). Now using some algebraic manipulations it can be shown that the optimization problem is equivalent to the following p separate regularized regression problems:

$$\text{minimize } ||\boldsymbol{\beta}_{.i}||_1 \quad \text{subject to} \quad ||\mathbf{S}\boldsymbol{\beta}_i - \boldsymbol{e}_i||_\infty \leq \lambda, \tag{4.32}$$

where $\boldsymbol{\beta}_i \in \mathbb{R}^p$ is the ith column of $\mathbf{\Theta}$ and $\boldsymbol{e}_i$ is the standard unit vector with 1 in the ith position and 0 elsewhere; see Problem 4.9. The reformulation in (4.32) is known as the Dantzig selector in the regression analysis for the ith node where its solution can be obtained by solving the corresponding linear programming problem [Candes and Romberg, 2006].

Like the NS estimator, the CLIME solution is neither symmetric nor positive-definite. The final CLIME estimator is usually obtained by symmetrizing the solution of the optimization problem, namely replacing the entries (i, j) and (j, i) by the one corresponding to the smaller magnitude of the estimates of these two entries. More precisely, the solution $(\hat{\theta}^1_{ij})$ is symmetrized using

$$\hat{\theta}_{ij} = \hat{\theta}^1_{ij} I(|\hat{\theta}^1_{ij}| \leq |\hat{\theta}^1_{ji}|) + \hat{\theta}^1_{ji} I(|\hat{\theta}^1_{ji}| \leq |\hat{\theta}^1_{ij}|), \quad i \neq j. \tag{4.33}$$

As for positive-definiteness of such symmetric estimator, Theorem 1 in Cai et al. [2011] shows that the final CLIME estimator is positive-definite with high probability under suitable conditions when both n and p grow to infinity. This sort of asymptotic guarantee of positive-definiteness is quite common in the high-dimensional covariance matrix estimation literature which is not so reassuring for finite sample situations. Fortunately, lately there has been attempts to select the tuning parameters to ensure positive-definiteness of thresholded covariance estimators in finite sample situations; see Section 8.3.1. In this connection, we note that Bonat and Jørgensen [2016] model directly the covariance matrix using covariates and tune the optimizer to avoid generating indefinite matrices.

4.4.4 Regularized Covariance Estimation

In this section we employ a simpler (moment-based) reincarnation of the regression idea; for two other instances of this; see (6.20) and Section 7.1.6.

We exploit the fact that the sample covariance matrix is an unbiased estimator of the population covariance matrix:

$$E(\mathbf{S}) = \mathbf{\Sigma}.$$

This amounts to viewing $\mathbf{S}$ as the "response vector" and the "population" covariance matrix as its mean. Then, one may rewrite the above identity involving $\mathbf{S}$ as simple intercept-only regression models

$$s_{ij} = \sigma_{ij} + \varepsilon_{ij}, i \geq j, \tag{4.34}$$

where ε_{ij}'s are mean-zero random errors which are definitely *correlated*.

We use the matrix vec and vech operations and rewrite these $p(p+1)/2$ models as a single linear model where the non-redundant entries of $\mathbf{S}$ and $\mathbf{\Sigma}$ form the response and parameter vectors, respectively. More specifically, using the identity design matrix, we arrive at

$$s = \text{vech}(\mathbf{S}) = \boldsymbol{I}_{p(p+1)/2}\text{vech}(\mathbf{\Sigma}) + \boldsymbol{\epsilon} = \boldsymbol{I}_{p(p+1)/2}\boldsymbol{\sigma} + \boldsymbol{\epsilon}, \tag{4.35}$$

where the errors are zero-mean and correlated random variables with $\text{Cov}(\boldsymbol{\epsilon}) = V$, see (4.7), and its mean parameters $\boldsymbol{\sigma}$ are constrained. In short, unlike the standard linear models, the simple regression formulation (4.35) suffers from two drawbacks: first, the parameters in $\boldsymbol{\sigma}$ are constrained to ensure that $\mathbf{\Sigma}$ is a positive-definite matrix, and second the errors are correlated. There are variety of approaches for estimating $\mathbf{\Sigma}$ in the model (4.35) including (penalized) least-squares and maximum likelihood estimation.

For the moment it is helpful to assume that the error covariance matrix V is a known positive-definite matrix with $V^{1/2}$ as its square-root. Then, pre-multiplying both sides of the model by $V^{-1/2}$ leads to a new linear model

$$V^{-1/2}s = V^{-1/2}\boldsymbol{\sigma} + V^{-1/2}\boldsymbol{\epsilon}, \tag{4.36}$$

with the response vector $V^{-1/2}s$, design matrix $V^{-1/2}$ and the uncorrelated errors $V^{-1/2}\boldsymbol{\epsilon}$. This removes the second drawback and brings the estimation problem a step closer to the framework of ordinary least-squares theory, keeping in mind that the new mean parameters are still constrained.

For sparse estimation of a covariance matrix $\mathbf{\Sigma}$ it is intuitively reasonable to separate and penalize differently its diagonal and off-diagonal entries. Thus, we let $\boldsymbol{\sigma}_{(d)} \in \mathbb{R}^p$ and $\boldsymbol{\sigma}_{(o)} \in \mathbb{R}^{p(p-1)/2}$ be the vectors of its diagonal and off-diagonal entries, respectively, with similar notation for the diagonal and off-diagonal entries of $\mathbf{S}$. Given an estimator $\widehat{V}^{-1/2}$ of $V^{-1/2}$, a novel sparse estimator of $\mathbf{\Sigma}$ proposed in Kim and Gaynanova [2025] solves the following penalized linear regression

$$\widehat{\boldsymbol{\sigma}} = \arg\min_{\boldsymbol{\sigma}} \frac{1}{2n}||\widehat{V}^{-1/2}s - \widehat{V}^{-1/2}\boldsymbol{\sigma}||^2 + \lambda p(\boldsymbol{\sigma}_{(o)}) \quad \text{subject to} \quad \boldsymbol{\sigma}_{(d)} = s_{(d)}, \tag{4.37}$$

where $\lambda > 0$, is a tuning parameter and $p(\cdot)$ is a penalty function on the off-diagonal entries. Popular choices of the penalty functions are the Lasso, SCAD and MCP introduced in Chapter 2. The constraint $\boldsymbol{\sigma}_{(d)} = s_{(d)}$ which fixes the diagonal entries of $\mathbf{\Sigma}$ to that of the sample covariance matrix and does not penalize them is becoming a fairly common phenomenon in high-dimensional covariance estimation. Among other benefits, it reduces the number of parameters to be estimated by p and the bias of estimating the diagonal entries of $\mathbf{\Sigma}$. We note that once a sparse estimator of $\widehat{\boldsymbol{\sigma}}_{(o)}$ is obtained, the resulting covariance estimate denoted by $\mathbf{\Sigma}(\widehat{\boldsymbol{\sigma}}_{(o)}, s_{(d)})$ is also sparse.

Interestingly, the simplest choice of $\widehat{V} = \boldsymbol{I}_{p(p+1)/2}$ leads to the following familiar and well-known estimators of $\boldsymbol{\sigma}$ depending on the choice of the penalty function:

1. Hard-thresholding corresponds to $p(\boldsymbol{\sigma}_{(o)}) = ||\boldsymbol{\sigma}_{(o)}||_0$, counting the number of non-zero entries of $\boldsymbol{\sigma}_{(o)}$, see Bickel and Levina [2008b].

2. Soft-thresholding corresponds to $p(\boldsymbol{\sigma}_{(o)}) = ||\boldsymbol{\sigma}_{(o)}||_1$ where (4.37) decouples into several simple Lasso problems of the form (2.10), see also (8.20).

Unfortunately, the positive-definiteness of $\boldsymbol{\Sigma}(\widehat{\boldsymbol{\sigma}}_{(o)}, \boldsymbol{s}_{(d)})$ cannot be guaranteed for finite samples. To resolve this shortcoming, Kim and Gaynanova [2025] adds additional constraints like $\boldsymbol{\Sigma} \succ 0$ to (4.37). However, this comes at an extra computational cost since then the popular coordinate descent algorithm helpful in solving the Lasso regression problem is not applicable. For more details and the computational algorithm for the case of a more general non-diagonal matrix V, see Kim and Gaynanova [2025].

4.5 Problems

4.1 For $Y \sim N_p(\boldsymbol{\mu}, \boldsymbol{\Sigma})$ where $\boldsymbol{\Sigma}$ is the CS(ρ, 1) matrix, find the expressions for the conditional mean and covariance matrix given in Theorem 7. Are all the required square submatrices and their inverses belong to the class of CS matrices? If so, identify their corresponding (ρ, σ^2) parameters.

4.2 (a) For a symmetric 3×3 matrix $A = (a_{i,j})$, write the vectors vec(A) and vech(A) and the duplication matrix D_3. What are the dimensions of these three matrices for a general p?

(b) For $p = 3$ and arbitrary sample size n, write the explicit form of all the entries of Cov(vech($\mathbf{S}$)) in (3) in terms of the entries of $\boldsymbol{\Sigma}$. In particular, is it true that Var(s_{ii}), $i = 1, 2, 3$, are equal?

(c) What is the Cov(s_{12}, s_{23})?

(d) Write the explicit form of all the entries of Cov(vech($\mathbf{S}$)) when $\boldsymbol{\Sigma} = \text{CS}(\rho, \sigma^2)$.

4.3 Show that the normal loglikelihood function corresponding to (4.3) is

$$\log \det(\boldsymbol{\Theta}) - \text{tr}(\mathbf{S}\boldsymbol{\Theta}),$$

up to some constants where $\boldsymbol{\Theta} = \boldsymbol{\Sigma}^{-1}$.

4.4 The Kullback-Leibler divergence between two probability distributions with density functions $f_1(\boldsymbol{x})$ and $f_2(\boldsymbol{x})$ with respect to the Lebesgue measure on $\mathbb{R}^p$ is defined as

$$D_{KL}(f_1, f_2) = \int \log \frac{f_1(\boldsymbol{x})}{f_2(\boldsymbol{x})} f_1(\boldsymbol{x}) d\boldsymbol{x}.$$

Show that, if $f_i, i = 1, 2$ are two p-variate zero-mean normal distributions with covariance matrices $\boldsymbol{\Sigma}_i, i = 1, 2$, then the KL divergence reduces to

$$2D_{KL}(f_1, f_2) = \text{tr}(\boldsymbol{\Sigma}_2^{-1}\boldsymbol{\Sigma}_1) + \log|\boldsymbol{\Sigma}_2| - \log|\boldsymbol{\Sigma}_1| - p. \quad (4.38)$$

This is known as the popular *entropy loss function* in the covariance estimation literature going back to Stein [1975].

4.5 Starting from the $n \times p$ normal data matrix $\boldsymbol{X}$ from $N_p(\boldsymbol{\mu}, \boldsymbol{\Sigma})$, consider the linear function

$$Y = AXB,$$

where A and B are non-random matrices of dimensions $m \times n$ and $p \times q$, respectively.

(a) For $A = n^{-1}1_n'$ and $B = I_q$, describe the random vector $\boldsymbol{Y}$ in terms of sample averages and state its distribution.

(b) Show that the rows of $\boldsymbol{Y}$ are independent, if and only if

(i) $A1 = a1$ for some scalar a, or $B'\boldsymbol{\mu} = \boldsymbol{0}$, and

(ii) $AA' = bI$ for some scalar b, or $B'\boldsymbol{\Sigma}B = \boldsymbol{0}$.

(c) When both (i) and (ii) are satisfied, show that $\boldsymbol{Y}$ is a normal data matrix from a q-variate normal distribution. What are the mean vector and covariance matrix of this distribution?

4.6 Write (4.25) in the matrix form $\Phi Y = \boldsymbol{\epsilon}$, and conclude that $T = \Phi$, where T is defined in (4.24).

4.7 Show the columns of the Cholesky factor C in $\boldsymbol{\Sigma} = CC'$ can be computed using the following recursions

$$c_{mm} = \sqrt{\sigma_{mm} - \sum_{k=1}^{m-1} c_{mk}^2}, \quad c_{jm} = \frac{1}{c_{mm}}\left(\sigma_{jm} - \sum_{k=1}^{m-1} c_{mk}c_{jk}\right), \quad j > m$$

(a) Use the above recursions to compute the standard Cholesky factor of the matrix

$$\Sigma = \begin{pmatrix} 1.0 & 0.2 & 0.5 \\ 0.2 & 1.0 & 0.4 \\ 0.5 & 0.4 & 1.0 \end{pmatrix}.$$

(b) Apply the formulas in (4.26) to obtain the corresponding regression coefficients and prediction error variances (modified Cholesky decomposition).

4.8 (Schur complement and regression) Let $\boldsymbol{\Sigma}$ be the covariance matrix of a random vector Y, with $\boldsymbol{\Sigma}_{11}$ its $(p-1) \times (p-1)$ leading principal minor, $\boldsymbol{\sigma}_{12}$ the vector of the first $(p-1)$ entries and σ_{22} the last entry of its last column. Consider

$$\boldsymbol{\beta} = \boldsymbol{\sigma}_{12}, \quad \sigma_{22.1} = \gamma = \sigma_{22} - \boldsymbol{\sigma}_{12}'\boldsymbol{\Sigma}_{11}^{-1}\boldsymbol{\sigma}_{12},$$

as a reparameterization of its last column.

(a) Show that by applying the block matrix inversion formula to $\mathbf{\Sigma}$ in (3.23) one obtains

$$\mathbf{\Sigma}^{-1} = \begin{pmatrix} \mathbf{\Sigma}_{11}^{-1} + \mathbf{\Sigma}_{11}^{-1}\boldsymbol{\beta}\boldsymbol{\beta}'\mathbf{\Sigma}_{11}^{-1}\gamma^{-1} & -\mathbf{\Sigma}_{11}^{-1}\boldsymbol{\beta}\gamma^{-1} \\ -\boldsymbol{\beta}'\mathbf{\Sigma}_{11}^{-1}\gamma^{-1} & \gamma^{-1} \end{pmatrix}.$$

Discuss the connection and interpretation of $\gamma, \boldsymbol{\beta}$ as the regression parameters when regressing Y_p on the $(Y_{p-1}, \ldots, Y_1)'$.

(b) (Schur complement and the likelihood function) For $Y = (Y_{(1)}', Y_{(2)}')' \sim N_p(\mathbf{0}, \mathbf{\Sigma})$, with suitable partition of its covariance matrix, show that

$$Y'\mathbf{\Sigma}^{-1}Y = (Y_{(2)} - \mathbf{\Sigma}_{21}\mathbf{\Sigma}_{11}^{-1}Y_{(1)})'\mathbf{\Sigma}_{22.1}^{-1}(Y_{(2)} - \mathbf{\Sigma}_{21}\mathbf{\Sigma}_{11}^{-1}Y_{(1)}) + Y_{(1)}'\mathbf{\Sigma}_{11}^{-1}Y_{(1)},$$

and hence the loglikelihood of Y decomposes

$$\ell_Y(\mathbf{\Sigma}) = \ell_{Y_{(2)}|Y_{(1)}}(\mathbf{\Sigma}_{21}, \mathbf{\Sigma}_{22.1}) + \ell_{Y_{(1)}}(\mathbf{\Sigma}_{11}).$$

as the sum of the marginal distribution of $Y_{(1)}$ and the conditional distribution of $Y_{(2)}$ given $Y_{(1)}$.

(c) Define the pseudo-variables $\boldsymbol{Z}_1 = \mathbf{\Sigma}_{11}^{-1}Y_{(1)}$ and show that the first term in (b) reduces to

$$Q(\mathbf{\Sigma}_{21}) = (Y_{(2)} - \mathbf{\Sigma}_{21}Z_1)'\mathbf{\Sigma}_{22.1}^{-1}(Y_{(2)} - \mathbf{\Sigma}_{21}Z_1),$$

which can be viewed as the weighted sum of squares of residuals when $Y_{(2)}$ is regressed on the vector of pseudo-predictors Z_1 with $\mathbf{\Sigma}_{21}$ as the unknown regression coefficient matrix.

(d) Using the simple partition $Y = (Y_{(1)}', Y_p)'$ and the earlier notation show that the decomposition in (b) reduces to

$$\frac{1}{\gamma}(Y_p - \boldsymbol{\beta}'Z_1)'(Y_p - \boldsymbol{\beta}'Z_1) + Y_{(1)}'\mathbf{\Sigma}_{11}^{-1}Y_{(1)},$$

with the corresponding decomposition of the likelihood function. In particular, its conditional loglikelihood

$$\ell_{Y_p|Y_{(1)}}(\boldsymbol{\beta}, \gamma) = -\log\gamma - \frac{1}{\gamma}(Y_p - \boldsymbol{\beta}'Z_1)^2,$$

is readily recognized as that of the linear regression of Y_p on Z_1 with $\boldsymbol{\beta}$ as the vector of regression coefficients and the error variance γ.

4.9 Show that the optimization problem (4.31) can be written as

$$\min \sum_{i=1}^{p} ||\beta_{.i}||_1, \text{ subject to } \quad \max_i ||\mathbf{S}\beta_{.i} - e_i||_\infty \leq \lambda,$$

and use it to establish the equivalence between CLIME and the p regression problems in (4.32).

5

Multivariate Regressions and Graphs

Multivariate data are routinely collected in many application areas such as e-commerce, financial econometrics, engineering, health and biomedical sciences, physical and social sciences. The interest lies usually in predicting q responses using a set of p covariates which requires estimating the pq parameters in the regression coefficient matrix, a challenging task when both p and q are large. In this chapter, we review the class of general linear models which subsume all well-known regression models. After pointing out some shortcomings of least-squares estimation, we explore the role of shrinkage and regularizing simultaneously the coefficient matrix and the $q \times q$ covariance (precision) matrix of the response vector. The most promising method seems to be the regularized conditional pseudo-likelihood formulation [Wang, 2015]. This is a truly regression-based method which relies on the conditional distribution of a response given other responses and all the covariates. We finish the chapter with the newer and important topic of *graph regressions* where the graph structure or the precision matrix is linked to individual or subject-level covariates.

5.1 General Linear Regression Models

We start with a broad class of linear models with correlated errors which subsumes multiple and multivariate regression models, time series and spatial models among others. The goal is to highlight the role of a covariance matrix or dependence among the observations in regression and prediction. A reader familiar with this topic may skip this section.

In many application areas including e-commerce, time series, spatial and spatial-temporal data, the responses are often viewed as a long dependent vector. They are analyzed within a general linear regression model of the form

$$\boldsymbol{y} = \boldsymbol{X}\boldsymbol{\beta} + \boldsymbol{\epsilon}, \tag{5.1}$$

where $\boldsymbol{y}$ is the n-vector of observations, $\boldsymbol{X}$ is a known $n \times k$ full rank design matrix, and the error term $\boldsymbol{\epsilon}$ is a zero-mean random vector. The term *general* emphasizes the generality of the $n \times n$ covariance matrix $\boldsymbol{\Sigma}$ or the correlation structure of the

DOI: 10.1201/9781003638896-5

vector of observations. The price of generality of the dependence structure $\mathbf{\Sigma}$ is the quadratic growth of its number of parameters which is $n(n+1)/2$, and its positive-definiteness. These make the task of simultaneous estimation of the parameters $(\boldsymbol{\beta}, \mathbf{\Sigma})$ a real challenge even for a univariate response as encountered in time series and spatial data analysis.

Historically, one controls the growing number of covariance parameters using the *parsimony* idea or using *structured covariance matrices* $\mathbf{\Sigma}(\theta)$, with a known functional form and a few unknown parameters; see Section 3.4. Furthermore, clever interplay between block-diagonal covariance matrices and carefully chosen design matrices give rise to a host of familiar and parsimonious classes of linear regression models, like heteroscedastic, seemingly unrelated regression (SUR), multivariate regression models, and regression with time series errors [Kariya and Kurata, 2004]. After a brief review of the interplay between the pair $(\boldsymbol{X}, \mathbf{\Sigma})$ in this section, we discuss least-squares estimation, shrinkage and regularization for estimating $\boldsymbol{\beta}$ and $\mathbf{\Sigma}$.

5.1.1 Structured Covariances and Design Matrices

For correlated and heterogeneous data, a simple and natural way to start and expand the least-squares estimation theory to go beyond the familiar $\mathbf{\Sigma} = \sigma^2 I_n$ is to include one additional dependence parameter θ:

$$\mathbf{\Sigma} = \sigma^2 V(\theta), \quad V(\theta)^{-1} = I_n + v_n(\theta)C, \quad \theta \in \Theta \subset \mathbb{R}, \tag{5.2}$$

where C is a known symmetric matrix and $v_n(\cdot)$ is a continuous real-valued function so that $\mathbf{\Sigma}$ is positive-definite.

Since C is symmetric and known, without loss of generality, it can be replaced with a diagonal matrix. In fact, using the orthogonal matrix P from its spectral decomposition $P'CP = \Lambda$, the transformed model $P'\boldsymbol{y} = P'X\beta + P'\boldsymbol{\epsilon}$, is a linear regression model with a heterogeneous diagonal covariance structure:

$$\mathrm{Cov}(P'\boldsymbol{\epsilon}) = \sigma^2 P'V(\theta)P = \sigma^2 V_0(\theta), \; V_0(\theta)^{-1} = I_n + v_n(\theta)\Lambda,$$

where its diagonal elements given by $1 + \lambda_j v_j(\theta)$ vary with j.

Several notable examples of structured covariance matrices are presented in Section 3.4. The most important example, the compound symmetry (CS) covariance matrix is discussed next.

Compound Symmetry (Equi-correlated, Uniform, Exchangeable,...) is the $n \times n$ matrix of the form

$$\mathbf{\Sigma} = \sigma_e^2 \begin{pmatrix} 1 & \rho & \cdots & \rho \\ \rho & 1 & \cdots & \rho \\ \vdots & \vdots & \ddots & \vdots \\ \rho & \rho & \cdots & 1 \end{pmatrix}, \tag{5.3}$$

where $\sigma_e^2 > 0$ and $\theta = \rho \in (-\frac{1}{n-1}, 1)$ ensures its positive-definiteness. It is a member of the class (5.2) since

$$\mathbf{\Sigma} = \sigma^2 V(\theta), \quad V(\theta)^{-1} = I_n + \lambda_n(\theta)\mathbf{1}_n\mathbf{1}_n',$$

with

$$\sigma^2 = \sigma_e^2(1-\theta), \ \ \lambda_n(\theta) = -\frac{\theta}{1+(n-1)\theta},$$

and $\mathbf{1}_n = (1, \ldots, 1)'$ is an n-vector of ones.

It turns out that adding a few extra parameters, choosing block-diagonal covariances and some special design matrices in (5.1) give rise to some of the most interesting and important classes of linear regression models encountered in various applications.

5.1.2 Heteroscedastic Regression Models

In the standard regression models and least-squares theory, the observations are assumed to be homogeneous or have equal variances. When this is not the case, then it seems plausible to partition or regroup $\boldsymbol{y}$ into p nearly homogeneous groups as $\boldsymbol{y} = (\boldsymbol{Y}_1', \ldots, \boldsymbol{Y}_p')'$, where $\boldsymbol{Y}_i$ is an $n_i \times 1$ vector and $n = \sum_{i=1}^{p} n_i$, with p different variances so that

$$\boldsymbol{\Sigma} = \text{block diag}(\sigma_1^2 I_{n_1}, \ldots, \sigma_p^2 I_{n_p}).$$

Compatible partitioning of the rows of $\boldsymbol{X}_{n\times k} = (\boldsymbol{X}_1', \ldots, \boldsymbol{X}_p')'$, $\boldsymbol{\epsilon} = (\boldsymbol{\epsilon}_1', \ldots, \boldsymbol{\epsilon}_p')'$ suggests a heteroscedastic regression model for the data composed of p separate multiple linear regression models

$$\boldsymbol{Y}_i = \boldsymbol{X}_i\boldsymbol{\beta}_i + \boldsymbol{\epsilon}_i, \ \ \text{Cov}(\boldsymbol{\epsilon}_i) = \sigma_i^2 I_{n_i}, \ \ i = 1, \ldots, p, \tag{5.4}$$

where $\boldsymbol{X}_i$ is $n_i \times k$. In spite of the subscript i, here we note that the $\boldsymbol{\beta}_i$'s are of the same dimension and equal to a common $\boldsymbol{\beta}$, but this changes in the next section.

5.1.3 Seemingly Unrelated Regression (SUR) Models

While the p homogeneous groups in heteroscedastic regression models (5.4) were implicitly assumed to be uncorrelated, the SUR models [Zellner, 1962] allow them to be correlated with specific block structures for $\boldsymbol{X}$ and $\boldsymbol{\Sigma}$.

Let $\boldsymbol{y} = (\boldsymbol{Y}_1', \ldots, \boldsymbol{Y}_p')'$ be partitioned as above, but with $n_1 = \ldots = n_p = m, n = mp$, so that

$$\boldsymbol{X}_{n\times k} = \text{block diag}(\boldsymbol{X}_1, \ldots, \boldsymbol{X}_p), \ \boldsymbol{\beta} = (\boldsymbol{\beta}_1', \ldots, \boldsymbol{\beta}_p')', \ \boldsymbol{\epsilon} = (\boldsymbol{\epsilon}_1', \ldots, \boldsymbol{\epsilon}_p')',$$

where $\boldsymbol{X}_i$ and $\boldsymbol{\beta}_i$ are $m \times k_i$ and $k_i \times 1$, respectively, with $\sum_{i=1}^{p} k_i = k$. Dependence among the groups is introduced by setting $\boldsymbol{\Sigma} = \Sigma_p \otimes I_m = (\sigma_{ij} I_m)$ where $\Sigma_p = (\sigma_{ij})$ is a $p \times p$ covariance matrix and $\otimes$ stands for the Kronecker product of two matrices. Compared to (5.4), this model is equivalent to the following p regression models

$$\boldsymbol{Y}_i = \boldsymbol{X}_i\boldsymbol{\beta}_i + \boldsymbol{\epsilon}_i, \ \ \text{Cov}(\boldsymbol{\epsilon}_i, \boldsymbol{\epsilon}_j) = \sigma_{ij} I_m, \ \ i, j = 1, \ldots, p. \tag{5.5}$$

However, unlike (5.4) they are linked to each other via their correlated errors. The model (5.5) is usually referred to as a *p-equation SUR model.*

Using matrix notation, the model (5.5) can be rewritten as a *multivariate linear regression* model of the form:

$$Y = X\mathbf{B} + \mathbf{E},$$

where

$$Y_{m\times p} = (Y_1, \ldots, Y_p), \;\; X_{m\times k} = (X_1, \ldots, X_p),$$

$$\mathbf{E}_{m\times p} = (\epsilon_1, \ldots, \epsilon_p), \; E(\mathbf{E}) = \mathbf{0}, \; \text{Cov}\,(\text{vec}(\mathbf{E})) = I_m \otimes \Sigma_p,$$

and

$$\mathbf{B}_{k\times p} = \text{block diag}(\boldsymbol{\beta}_1, \ldots, \boldsymbol{\beta}_p).$$

In fact, the block-diagonality of **B** reveals that SUR models are restricted multivariate regression models. However, when

$$k_1 = \ldots = k_p = k_0, \; X_1 = \ldots = X_p = X_0,$$

then the model is a bona fide multivariate linear regression model

$$Y = X_0\mathbf{B}_0 + \mathbf{E}, \; \mathbf{B}_0 = (\boldsymbol{\beta}_1, \ldots, \boldsymbol{\beta}_p),$$

with absolutely no restrictions on $\mathbf{B}_0$.

5.2 Multivariate Data and Regression

In this section, we review some standard methods for estimating the parameters of multivariate regression models. In a search for parsimonious models, as in multiple regression, one may start with the classical variable selection methods like the stepwise selection and AIC. However, due to the discrete nature of such procedures [Breiman, 1996] they are unstable in the sense that small perturbations in the data may lead to very different estimators or predictors.

For the sample data $(\boldsymbol{x}_i, \boldsymbol{Y}_i), i = 1, \ldots, n$, where $\boldsymbol{Y}_i = (Y_{i1}, \ldots, Y_{iq})'$ is the vector of responses and $\boldsymbol{x}_i = (x_{i1}, \ldots, x_{ip})'$ the predictors, a multivariate linear regression model relates the responses to the predictors via

$$\boldsymbol{Y}_i = \mathbf{B}'\boldsymbol{x}_i + \boldsymbol{\epsilon}_i, \; i = 1, \ldots, n. \tag{5.6}$$

Here **B** is a $p \times q$ regression coefficient matrix and $\boldsymbol{\epsilon}_i = (\varepsilon_{i,1}, \ldots, \varepsilon_{i,q})'$ is the vector of errors for the ith subject. In what follows, we make the standard assumption that $\boldsymbol{\epsilon}_1, \ldots, \boldsymbol{\epsilon}_n$ are $i.i.d \sim N_q(\mathbf{0}, \boldsymbol{\Sigma})$ with a $q \times q$ covariance matrix $\boldsymbol{\Sigma} = (\sigma_{ij})$. Note that when $q = 1$, this reduces to the familiar multiple linear regression models where **B** is then the p-vector of regression coefficients, see Chapter 2.

5.2.1 Least-Squares and Shrinkage Estimates

For estimation, it is convenient to write the model in matrix notation. Let $\boldsymbol{Y}$ denote the $n \times q$ data matrix where its ith row is $\boldsymbol{Y}_i'$, $\boldsymbol{X}$ is the $n \times p$ predictor matrix where its ith row is $\boldsymbol{x}_i'$, and let $\mathbf{E}$ denote the $n \times q$ random error matrix where its ith row is $\boldsymbol{\epsilon}_i'$. Then, the multivariate regression model in matrix form is

$$\boldsymbol{Y} = \boldsymbol{X}\mathbf{B} + \mathbf{E}. \tag{5.7}$$

For simplicity, in this chapter we assume that $\boldsymbol{X}$ is of full-rank, the columns of $\boldsymbol{X}$ and $\boldsymbol{Y}$ have been centered and hence the intercept terms are omitted. The ordinary least-squares (ols) estimator of $\mathbf{B}$, obtained by solving the optimization problem

$$\min_{\mathbf{B}} \operatorname{tr}\left[(\boldsymbol{Y} - \boldsymbol{X}\mathbf{B})'(\boldsymbol{Y} - \boldsymbol{X}\mathbf{B})\right], \tag{5.8}$$

is given by

$$\widehat{\mathbf{B}}_{ols} = (\boldsymbol{X}'\boldsymbol{X})^{-1}\boldsymbol{X}'\boldsymbol{Y},$$

see Problem 5.1. Curiously, it decouples as the least-squares estimators of the coefficients of q separate multiple regression, one for each response. This amounts to ignoring the correlations and such an estimator might be inferior to jointly estimating all columns of $\mathbf{B}$ by accounting for the dependence in the responses [Anderson, 2003, Section 8.2]. A possible explanation for this unusual phenomenon is that the model (5.7) is equivalent to q separate multivariate regressions, see Problem 5.2, so that one might as well start by applying univariate regression techniques separately to each response variable.

5.2.2 C&W Shrinkage of the Predictor

It is somewhat unsettling that the fitted (predicted) values $\widehat{\boldsymbol{Y}}_{ols} = \boldsymbol{X}\widehat{\mathbf{B}}_{ols}$ ignores the correlations in $\boldsymbol{Y}$ or lets the precious correlations go to waste. In this situation, pooling predicted values or borrowing strength from other responses seems like a plausible remedy which has been studied using various statistical tools and ideas including the *Curds and Whey (C&W)* shrinkage method in Breiman and Friedman [1997] and references therein. It employs the canonical coordinates obtained from the *canonical correlation analysis* (CCA) of the two sets of variables in X, Y. We present briefly the C&W shrinkage here; for more details on CCA is given in Section 5.8.

One of the most popular shrinkage-type predictor for multivariate regression, called the Curds and Whey (C&W) method [Breiman and Friedman, 1997]. Its Curd part fits a univariate regression model for each response against all the covariates, and the Whey part refits each response against these fitted values. The end action is to shrink the least-squares predictor of the multivariate response by

$$\tilde{\boldsymbol{Y}} = \widehat{\boldsymbol{Y}}_{\text{ols}}\boldsymbol{M}, \tag{5.9}$$

where the $q \times q$ shrinkage matrix is of the form $\boldsymbol{M} = \mathbf{U}\Lambda\mathbf{U}^{-1}$. The rows of the matrix $\mathbf{U}$ are the sample canonical coordinates $\boldsymbol{u}_k$ and $\Lambda = \operatorname{diag}(\lambda_1^2, \ldots, \lambda_m^2)$ is a diagonal matrix of squares of the values of the canonical correlations. Section 5.6.2 presents numerical performance of this shrinkage predictor relative to regularized methods for a dataset.

5.3 Regularizing Regression Coefficients

Similar to multiple regressions, in multivariate regressions when p and/or q are larger than n, the matrix $\mathbf{B}$ cannot be estimated uniquely since $\boldsymbol{X}$ is not of full column rank, and $\mathbf{S}$ the sample covariance matrix is not invertible. In these situations, the traditional estimators for $\mathbf{B}$ and $\boldsymbol{\Sigma}$ with pq and $q(q+1)/2$ parameters, respectively, have rather poor computational and statistical performances and they are not suitable for use in prediction and inference. Thus, one must seek suitable alternative estimators by assuming sparsity or relying on *regularization*. The idea of regularization is even more appealing in the multivariate regression setup as there are more parameters to be estimated and more directions and natural ways to introduce penalty functions on the two matrices of parameters.

The goal of regularized multivariate regression is that of solving the optimization problem:

$$\underset{\mathbf{B}}{\operatorname{argmin}} \left\{ \operatorname{tr}\left[(\boldsymbol{Y} - \boldsymbol{X}\mathbf{B})'(\boldsymbol{Y} - \boldsymbol{X}\mathbf{B}) \right] + \lambda p(\mathbf{B}) \right\}, \tag{5.10}$$

where $p(\mathbf{B})$ is often a matrix norm and λ is a tuning parameter. Popular examples include the ℓ_∞-norm penalty [Turlach et al., 2005], and the ℓ_2-norm penalty, $p(\mathbf{B}) = \sum_{j,k} b_{jk}^2$, in which case (5.10) has a closed-form solution recognized as the ridge-regression solution. The ℓ_1-norm $p(\mathbf{B}) = \sum_{j,k} |b_{jk}|$ leads to the Lasso estimate:

$$\widehat{\mathbf{B}} = \underset{\mathbf{B}}{\operatorname{argmin}} \left\{ \operatorname{tr}\left[(\boldsymbol{Y} - \boldsymbol{X}\mathbf{B})'(\boldsymbol{Y} - \boldsymbol{X}\mathbf{B}) \right] + \lambda \sum_{j,k} |b_{jk}| \right\}. \tag{5.11}$$

Historically, regularization has been applied separately either on the regression coefficient matrix $\mathbf{B}$ or the precision matrix $\boldsymbol{\Theta} = \boldsymbol{\Sigma}^{-1}$, but not on both simultaneously. Most often a penalty is imposed only on the regression coefficient matrix $\mathbf{B}$ while ignoring the $q(q+1)/2$ parameters in $\boldsymbol{\Sigma}$. The covariance-regularized regression method [Witten and Tibshirani, 2009] assumes multivariate normality of the $(p+q)$-dimensional random vector (Y, X), but only regularizes the inverse covariance of the predictors in X to perform variable selection.

5.4 Joint Regularization

In this section, we present joint regularization of $(\mathbf{B}, \boldsymbol{\Theta})$ aiming to provide sparse estimators of both matrices in a systematic manner. The multivariate regression with covariance estimation (MRCE) method in Rothman et al. [2010b] penalizes the normal log-likelihood $\ell(\mathbf{B}, \boldsymbol{\Theta})$. While MRCE's focus is on improved estimation of $\mathbf{B}$ for the task of prediction, the method in Yin and Li [2011] aims to improve the estimation of the precision matrix after removing the effects of the covariates on the mean which has evolved into the *graph regressions* method.

5.4.1 Jointly Regularized Likelihood Function

Assuming normality of the data in the multivariate linear regression, the optimization problem to solve jointly over $(\mathbf{B}, \mathbf{\Theta})$ is

$$\underset{\mathbf{B},\mathbf{\Theta}}{\operatorname{argmin}} \left\{ \operatorname{tr}\left[\frac{1}{n}(\boldsymbol{Y} - \boldsymbol{X}\mathbf{B})\mathbf{\Theta}(\boldsymbol{Y} - \boldsymbol{X}\mathbf{B})'\right] - \log|\mathbf{\Theta}| \right. \\ \left. + \lambda_1 \sum_{j' \neq j} |\theta_{jj'}| + \lambda_2 \sum_{j,k} |b_{jk}| \right\} \tag{5.12}$$

where $\mathbf{\Theta} = (\theta_{jj'}) = \mathbf{\Sigma}^{-1}$ is the precision matrix and λ_1, λ_2 are two tuning parameters. The Lasso penalty on $\mathbf{B}$ introduces sparsity in the regression coefficients matrix, improves the prediction performance and possibly facilitates interpretation of the fitted model. Note that the Lasso penalty on the precision matrix involves only its off-diagonal entries.

The objective function in (5.12) is biconvex in $(\mathbf{B}, \mathbf{\Theta})$ in that it is convex in one when the other matrix is fixed. Any algorithm for it minimization is unstable when $p \geq n$, mostly due to the objective function (5.12) not being jointly convex in $(\mathbf{B}, \mathbf{\Theta})$ and the term $\log|\mathbf{\Theta}|$ dominating the other terms. In fact, if $p \geq n$ some diagonal entries of $(\boldsymbol{Y} - \boldsymbol{X}\mathbf{B})(\boldsymbol{Y} - \boldsymbol{X}\mathbf{B})'$ could be zero, and consequently the corresponding diagonal entries of $\mathbf{\Theta}$ could become very large without changing the objective function. Such an estimate of the precision matrix with large diagonal entries is not desirable in applications as it leads to very small residual variances in regression of the corresponding responses; see (4.20). Thus, instead of alternating between solving for $\mathbf{B}$ and $\mathbf{\Theta}$ and dealing with the associated computational issues, the plug-in method in Lee and Liu [2012] avoids such computational problems. In addition, it allows the use of prior modeling methods to screen the variables or reduce dimensions and then use MRCE or joint estimation.

5.4.2 Regularized Pseudo-Likelihood

The penalized normal likelihood formulations in the previous section estimate the two parameter matrices of the multivariate regression using alternating (two-stage) methods. These are computationally expensive and do not guarantee convergence to the global optimum of the objective function. By contrast, the alternative formulation in Wang [2015], which we present next, is truly regression-based in spirit in that it relies on the conditional distribution of each response given the other responses and covariates. Its objective function turns out to be a pseudo-likelihood in the sense of Besag [1975], and its optimization boils down to solving q regularized (hidden) regression problems, which can be solved efficiently, and even in parallel if so desired, using the existing software packages.

From the normality of the response vector and the multivariate linear regression model (5.7), it follows that for $i = 1, \ldots, q$, the conditional distribution of $\boldsymbol{y}_i$, the ith column of the data matrix $\boldsymbol{Y}$, given others and the covariates has also a multivariate normal distribution. Formally, it is given by

$$\boldsymbol{y}_i | (\boldsymbol{X}, \boldsymbol{Y}_{-i}) \sim N_n(\boldsymbol{X}\boldsymbol{\beta}_i + (\boldsymbol{Y}_{-i} - \boldsymbol{X}\mathbf{B}_{-i})\boldsymbol{\gamma}_i, \sigma_{ii.-i}\boldsymbol{I}_n), \tag{5.13}$$

where $\boldsymbol{Y}_{-i}$ and $\mathbf{B}_{-i}$ are, respectively, the data matrix and the regression coefficient matrix with their ith column removed, and

$$\boldsymbol{\gamma}_i = \boldsymbol{\Sigma}_{-i,-i}^{-1}\boldsymbol{\Sigma}_{-i,i} = -\boldsymbol{\Theta}_{-i,i}/\theta_{i,i}, \quad \sigma_{ii.-i} = \sigma_{ii} - \boldsymbol{\Sigma}'_{-i,i}\boldsymbol{\Sigma}_{-i,-i}^{-1}\boldsymbol{\Sigma}_{-i,i}.$$

We recall that $\boldsymbol{\Sigma}_{-i,-i}$ stands for $\boldsymbol{\Sigma}$ with its ith row and column removed, and $\boldsymbol{\Sigma}_{-i,i}$ is the ith column of $\boldsymbol{\Sigma}$ with its ith entry removed. Note that since $\theta_{i,i}$ is always positive, the sparsity of $(\mathbf{B}, \boldsymbol{\Theta})$ is shared with those of $(\boldsymbol{\beta}_i', \boldsymbol{\gamma}_i')'$ for $i = 1, \ldots, q$.

Indeed, the model (5.13) for a given $\mathbf{B}_{-i}$, is a larger linear regression model where regression coefficients are defined by the vector $(\boldsymbol{\beta}_i', \boldsymbol{\gamma}_i')'$. Thus, their sparse estimation may proceed by solving the following sequence of q regularized regressions:

$$\begin{aligned}&\underset{\boldsymbol{\beta}_i, \boldsymbol{\gamma}_i}{\operatorname{argmin}} \left\{ ||\boldsymbol{y}_i - \boldsymbol{X}\boldsymbol{\beta}_i - (\boldsymbol{Y}_{-i} - \boldsymbol{X}\mathbf{B}_{-i})\boldsymbol{\gamma}_i||_2^2 + \lambda_1 p_1(\boldsymbol{\beta}_i) + \lambda_2 p_2(\boldsymbol{\gamma}_i) \right\}, \\ &i = 1, \ldots, q, \end{aligned} \tag{5.14}$$

where λ_1, λ_2 are the tuning parameters and $p_1(\cdot)$ and $p_2(\cdot)$ are two sparsity-inducing convex penalty functions. The setup can be seen as a generalized version of the node-wise selection in Meinshausen and Bühlmann [2006] for sparse estimation of a precision matrix, to which it reduces when $\boldsymbol{X} = \mathbf{0}$ and $\lambda_1 = 0$. Consequently, a sparse estimate of $\boldsymbol{\Theta}$ is neither symmetric nor positive-definite.

In practice, one may replace $\mathbf{B}_{-i}$ with a consistent initial estimate $\widehat{\mathbf{B}}_{-i}^{(0)}$. For example, a natural and simple initial estimate $\widehat{\mathbf{B}}_{-i}^{(0)}$ can be constructed from fitting separate Lasso regressions ignoring the correlation structure. Then, the penalized conditional multivariate regression problem reduces to solving

$$\underset{\mathbf{B}, \boldsymbol{\Gamma}}{\operatorname{argmin}} \sum_{i=1}^{q} \left(||\boldsymbol{y}_i - \boldsymbol{X}\boldsymbol{\beta}_i - (\boldsymbol{Y}_{-i} - \boldsymbol{X}\widehat{\mathbf{B}}_{-i}^{(0)})\boldsymbol{\gamma}_i||_2^2 + \lambda_1 p_1(\boldsymbol{\beta}_i) + \lambda_2 p_2(\boldsymbol{\gamma}_i) \right). \tag{5.15}$$

Grouping together the parameters and data in $\boldsymbol{\phi} = (\mathbf{B}', \boldsymbol{\Gamma}')'$, $\tilde{\boldsymbol{X}}_i = [\boldsymbol{X}, \boldsymbol{Y}_{-i} - \boldsymbol{X}\hat{\mathbf{B}}_{-i}^{(0)}]$, and $\boldsymbol{\phi}_i$ denoting the ith column of $\boldsymbol{\Phi}$, then (5.15) can be written as

$$\arg\min_{\boldsymbol{\phi}} \sum_{i=1}^{q} \left(||\boldsymbol{y}_i - \tilde{\boldsymbol{X}}_i \boldsymbol{\phi}_i||_2^2 + \lambda p(\boldsymbol{\phi}_i) \right),$$

which is a Lasso regression problem and jointly convex in $(\mathbf{B}, \boldsymbol{\Gamma})$. Thus, a coordinate descent algorithm can be used to solve it and the convergence is guaranteed for each $1 \leq i \leq q$. Here, for simplicity, we have used a single penalty λ, but since $p(\boldsymbol{\phi}_i)$ is separable into $p_1(\boldsymbol{\beta}_i)$ and $p_2(\boldsymbol{\gamma}_i)$, it is straightforward to generalize it to $\boldsymbol{\lambda} = [\lambda_1, \lambda_2]$.

As for the penalty functions in (5.15), one may use Lasso, adaptive Lasso and MCP from Chapter 2. Using the adaptive Lasso penalty for the joint estimation of the sparse multivariate regression coefficient matrix and the sparse precision matrix gives rise to the aMCR algorithm [Wang, 2015]. Simulation studies and real data analysis reveal its improved performance. Moreover, under standard assumptions, the asymptotic selection consistency and normality of the regression coefficient matrix are established when dimensions of the covariates and number of responses diverge.

It should be noted that adding $\sum_{i=1}^{q} \log \sigma^2_{ii.-i}$ to the objective function

$$\sum_{i=1}^{q} ||\boldsymbol{y}_i - \boldsymbol{X}\boldsymbol{\beta}_i - (\boldsymbol{Y}_{-i} - \boldsymbol{X}\widehat{\mathbf{B}}_{-i}^{(0)})\boldsymbol{\gamma}_i||_2^2,$$

the estimation criterion becomes closely related to the pseudo-likelihood procedure in Besag [1975]. This connection is helpful in establishing various asymptotic properties of the resulting estimators.

5.5 Covariate-Adjusted Precision Matrix Estimation (CAPME)

An attractive and more direct alternative to the regularized likelihood method for sparse estimation of the regression coefficient and precision matrices in the context of multivariate regression is the CAPME algorithm in Cai et al. [2012]. It is a natural extension of CLIME in Section 4.4.3 and does not assume normality of the data.

When covariates are present, the CAPME method [Cai et al., 2012] formulates two separate Dantzig selector formulations for $\mathbf{B}$ and $\boldsymbol{\Theta}$ and amounts to solving two convex programs. As a two-stage method, it starts with estimating the regression coefficient matrix using a Dantzig selector first, and then estimating the precision matrix using the regression residuals $\widehat{\boldsymbol{\epsilon}} = \boldsymbol{Y} - \boldsymbol{X}\widehat{\mathbf{B}}$ as input to the CLIME method in Section 4.4.3. For more details the readers are encouraged to consult the original paper.

A powerful tuning-free methodology in Chen et al. [2016], for the joint estimation of the regression coefficients and the precision matrix, has the surprising property that the estimator for each entry of the precision matrix is asymptotically normal and efficient. They employ this result for support recovery by edge-wise adaptive (non-uniform) thresholding of the precision matrix via a two-step procedure. Its first step is a tuning-free scaled Lasso regression to obtain an estimator $\widehat{\mathbf{B}}$ of $\mathbf{B}$ and the residuals as $\widehat{\boldsymbol{\epsilon}} = \boldsymbol{Y} - \boldsymbol{X}\widehat{\mathbf{B}}$. The second step, applies the tuning-free scaled Lasso regression for the standard Gaussian graphical model to estimate each entry θ_{ij} of $\boldsymbol{\Theta}$ treating $\widehat{\boldsymbol{\epsilon}}$ as the data; see Section 9.4.1.

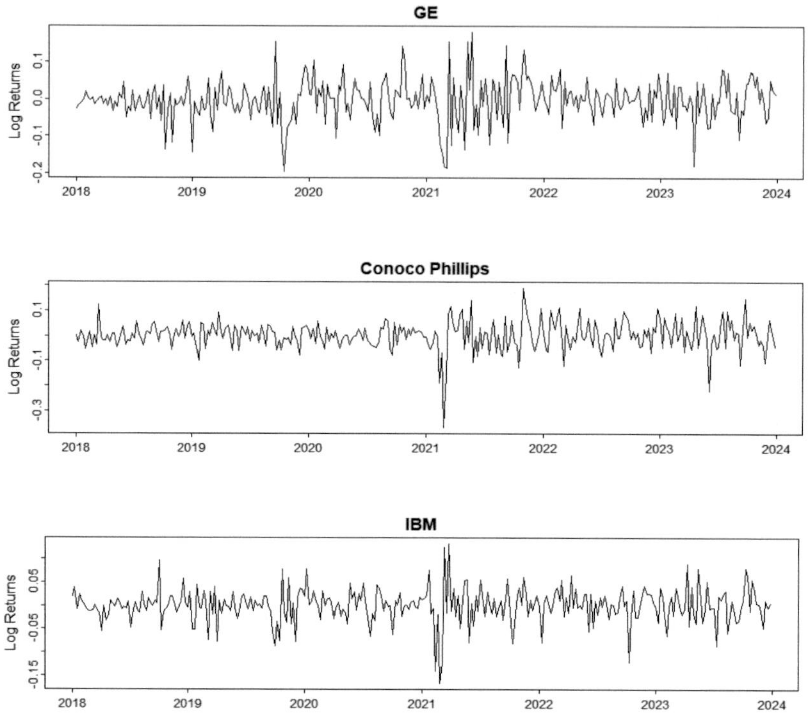

FIGURE 5.1
Time series plots for GE, Conoco Phillips, and IBM stocks.

5.6 Hands-on R: Regression and Prediction

In this section, we illustrate applications of multivariate regression, various estimation methods and compare the forecast performance of the MRCE and CAPME methods using two real datasets.

5.6.1 Predicting Asset Returns

We use the weekly log-returns of eight US stocks (Walmart, Exxon, Ford, GE, Conoco Phillips, Citigroup, IBM, and AIG) from 2017 to 2022. The dataset, downloaded from Yahoo Finance, consists of 314 observations and is available in the R package recode. Figure 5.1 shows the time series plots (observations versus their time indices) of the first three stocks where reasonable similarities are seen in the dependence and volatility of the returns even though these stocks are from different sectors of the economy.

Let $\boldsymbol{Y}_t$ be the vector of log returns of the eight stocks at time (week) t which should be viewed and analyzed as time series data, see Chapter 11. However, for forecasting and related purposes, there is a long tradition in econometrics to capture the dynamic (temporal dependence) of the data by suitably lagging and stacking the data so that

$$Y = (Y_2', Y_3', \ldots, Y_n')',$$

is viewed as the response values and

$$X = (Y_1', Y_2', \ldots, Y_{n-1}')',$$

as the covariates/predictors in a multivariate regression model for (Y, X),

$$Y = X\mathbf{B} + \mathbf{E}.$$

In fact, this is an instance of the vector autoregression of order one, with **B** as its transition matrix, see Chapter 11.

To assess the predictive performance of the model, we split the data and retain the log-returns of all stocks for the last 26 weeks in 2022 as the test set and the remaining 288 observations as the training set. The model parameters are estimated using the training set, while the predictive performance is measured by the average mean squared error over the test set for each stock.

We start by importing the data and splitting it into the training and test datasets.

```
library(recode)
data(stocks)
X = stocks$X
Y = stocks$Y

X_train = X[1: (nrow(X) - 26) ,]
Y_train = Y[1: (nrow(Y) - 26) ,]

X_test = X[(nrow(X) - 25) : nrow(X) ,]
Y_test = Y[(nrow(Y) - 25) : nrow(Y) ,]
```

For MRCE, the tuning parameters are selected from the set $\Lambda = \{2^x : x = -25, -24, \ldots, 10\}$ using 5-fold cross-validation. Next, we show implementation of the MRCE algorithm, and note that different partitioning of the data in the cross-validation would lead to slightly different results, here we fix a partition and present the results for illustration.

```
set.seed(1234)
library(MRCE)
lam1.vec=rev(2^seq(from=-25,to=0,by=1))
lam2.vec=rev(2^seq(from=-25,to=0,by=1))
cvfit=mrce(Y=Y_train,X = X_train, lam1.vec = lam1.vec,
            standardize = TRUE,
            lam2.vec = lam2.vec, method = "cv", kfold= 5)
## Print estimated coefficient matrix
print(cvfit$Bhat)
```

The CAPME algorithm has two tuning parameters α and τ; see Section 5.5 and Cai et al. [2012]. For α we use a logarithmic spaced grid which spans from $\alpha_{\min} = 1e{-}6$ to $\alpha_{\max} = \max(\mathrm{Cov}(X)) * (1-1/n) * \max(\mathrm{Cov}(Y) * (1-1/n) * \sqrt{4 * \log(p)^2/n}$, for $n = 288$, and $p = 8$. The latter grid is the default in the R package capme.

```
library(capme)
cv.capme.obj<-cv.capme(loss = "likelihood",fold=5,x=X_train,
            y=Y_train, nlambda = 50, ntau = 25,
            linsolver.Gamma="simplex",
            linsolver.Omega="simplex")
lambdaopt<-cv.capme.obj$lambdaopt
tauopt<-cv.capme.obj$tauopt
capme.obj<-capme(x=X_train,y= Y_train,lambda=lambdaopt,
            tau=tauopt, linsolver.Gamma="simplex",
            linsolver.Omega="simplex")
print(capme.obj$Gammalist)
```

The estimated coefficient matrix **B** for MRCE and CAPME are reported in Figure 5.2. Both matrices capture significantly different non-zero (black squares) patterns. For example, for MRCE, the relevant predictors of log-returns for AIG at week t are Walmart, Ford, CPhi, IBM, Citi, and AIG at week $t-1$. We present the average squared prediction error for each company over the last 26 observations in 2022 in Table 5.1. The results using aMCR with MCP penalty, the separate Lasso, Ridge, Curd and Whey and LSE estimates are also included for comparison. The results for Curd & Whey are not included in Table 5.1 due to its comparatively poor performance, as previously observed in Rothman et al. [2010a]. The aMCR algorithm is described in Section 5.4.2 and the aMCR() function is available in the R package recode. For each company and method, the best result is highlighted. It is evident from the results in the Table 5.1 that the Sep. Lasso and aMCR are performing the best, followed by CAPME.

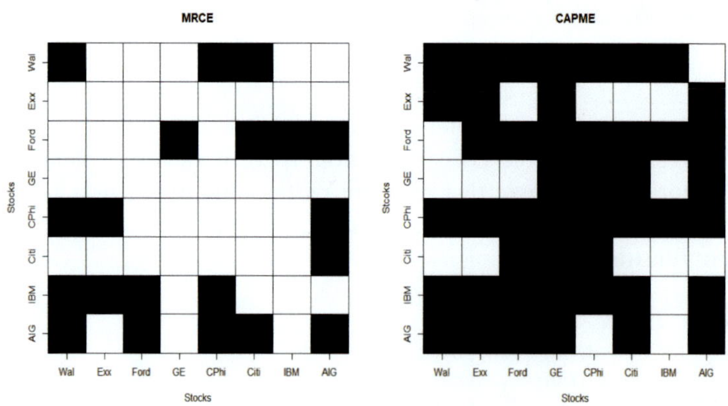

FIGURE 5.2
Estimated coefficient matrices for MRCE and CAPME algorithms. The black squares indicate non-zero values.

TABLE 5.1
Average squared prediction error for each company $\times 10^3$ based on 26 points.

	MRCE	CAPME	aMCR	Sep. Lasso	Sep. Ridge	Sep. LSE
Walmart	0.82	0.82	**0.80**	**0.80**	**0.80**	9.50
Exxon	2.78	2.79	**2.75**	2.77	2.78	33.31
Ford	4.40	4.41	**4.39**	4.41	4.43	50.37
GE	2.59	2.60	**2.56**	2.71	2.72	31.68
CPhi	3.73	3.66	3.72	**3.61**	3.62	43.50
Citigroup	1.90	**1.81**	1.85	**1.81**	**1.81**	22.48
IBM	1.37	**1.30**	1.34	**1.30**	**1.30**	15.10
AIG	1.57	**1.51**	1.55	**1.51**	1.52	22.74

5.6.2 A Chemometrics Example

In this section, we apply aMCR and separate Lasso, the top performers of the previous section, to the chemometrics dataset chemo, available in R package recode and borrowed from Skagerberg et al. [1992]. The goal of the study in Skagerberg et al. [1992] was to demonstrate the use of multivariate regression techniques to perform quality control and statistical process control.

The dataset consists of $n = 56$ observations, each with $p = 22$ predictor variables and $q = 6$ responses. The data are taken from a simulation of a low density tubular polyethylene reactor. The predictors consist of temperatures measured at equal distances along the reactor together with the wall temperature of the reactor and the feed rate. The six responses are the output characteristics of the produced polymers:

- Y_1: number-average molecular weight
- Y_2: weight-average molecular weight
- Y_3: frequency of long chain branching
- Y_4: frequency of short chain branching
- Y_5: content of vinyl groups
- Y_6: content of vinylidene groups

This data was analyzed in Breiman and Friedman [1997], and as a baseline we also include their results from C&W method [Breiman and Friedman, 1997, Table 2] . We start by importing the required R packages and the dataset. Because the distributions of the values of all of the response variables appear to be highly skewed to the right, we use the logarithms of their corresponding values.

```
library(recode)
data(chemo)
# Obtain the dataset
X = chemo$X
y = chemo$y
log_y = log(y)
```

Figure 5.3 displays the correlation matrix of the log-transformed responses. Evidently, the first two responses Y_1 and Y_2 and Y_4, Y_5, Y_6 form two strongly correlated groups. However, the correlation of Y_3 with the others is comparably weak. We scale both matrices X and log_Y to have mean 0 and variance 1 using the scale() function. We also compute the vector of possible λ values for tuning through 10-fold cross-validation and matrices aMCR.mse and lasso.mse and save the results.

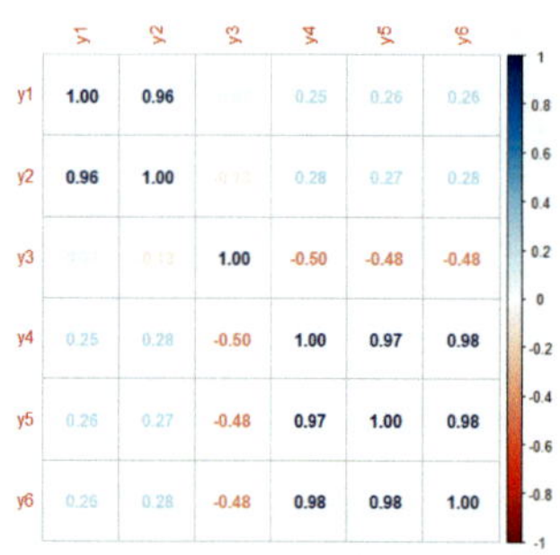

FIGURE 5.3
Correlation matrix of responses for the Chemometrics example.

```
log_y = scale(log(y))
X = scale(X)
## lambda vector
lam1.vec= rev(2^seq(from=-25,to=0,by=0.5))
## Matrix to store the results
n = nrow(log_y)
q = ncol(log_y)
aMCR.mse = matrix(0, nrow = n, ncol = q)
lasso.mse = matrix(0, nrow = n, ncol = q)
```

The predictive accuracy of each method was assessed using leave-one-out cross-validation. Specifically, the predictive model was fitted on 55 out of the 56 observations, and the squared error was calculated for the observation left out. This process was repeated 56 times, each time excluding a different observation, and the resulting 56 squared errors were then averaged.

```
set.seed(1234)
for (i in 1:n){
  X_train = X[-i,]
  Y_train = log_y[-i,]
  X_test = X[i,]
  Y_test = log_y[i,]
  ## Run aMCR
  cv.aMCR <- aMCR(Y = Y_train, X = X_train, lam.vec = lam1.vec)
  ## Predict aMCR
  Yhat.aMCR = cv.aMCR$mu_hat + X_test %*% cv.aMCR$B_hat
  aMCR.mse[i,] = (Y_test - Yhat.aMCR)^2

  ## Run seperate Lasso and predict
  for(j in 1:ncol(Y_train)){
```

```
    cv.lasso = cv.glmnet(x = X_train, y = Y_train[,j], lambda = lam1.
    vec)
    lasso.mse[i, j] = (Y_test[j] - predict(cv.lasso, newx=X_test))^2
  }
}
```

Table 5.2 reports the predictive squared error for each response. As can be seen, aMCR delivers the best performance for Y_1 and Y_2 and C&W is the best for the other 4 responses. On average, however, aMCR and Sep. Lasso perform similarly.

TABLE 5.2
Predictive squared error for the chemometrics example for each response.

	aMCR	Sep. Lasso	C&W
Y_1	**0.16**	0.18	0.55
Y_2	**0.22**	0.25	0.81
Y_3	0.31	0.33	**0.25**
Y_4	0.34	0.17	**0.11**
Y_5	0.30	0.40	**0.19**
Y_6	0.29	0.27	**0.15**
Average	0.29	**0.27**	0.34

5.7 Graph Regressions

Gaussian graphical models are widely used in many areas, but the progress in linking the graph structure or the precision matrix to subject-level covariates has been rather slow, and focused mostly on the mean as in Section 5.5. In this section we follow the recent and novel approach in Zhang and Li [2023] where both the mean and the precision matrix of the underlying Gaussian distribution are modeled using covariates. Sparse estimation of the regression parameters is pursued via a group Lasso penalty.

The need for covariates in graphs is motivated by applications in genetic where graph structures are known to depend on individuals' characteristics, leading to the notion of heterogeneous or subject-specific graphical regressions. For example, in gene expression networks, covariates such as genetic variants, clinical and environmental factors, may affect both the expression levels of individual genes and the co-expression relationships among them. Genetic variants that alter co-expression relationships are referred to as co-expression quantitative trait loci (QTL), and identifying them is of great scientific interest [Wang et al., 2012b, van der Wijst et al., 2018].

Following Zhang and Li [2023], we begin within the standard multivariate regression setting $Y \sim N_p(\mathbf{B}\boldsymbol{x}, \boldsymbol{\Sigma})$ where $\mathbf{B}_{p\times q} = (b_1, \ldots, b_q)$ with b_i its ith column and the parameters $(\mathbf{B}, \boldsymbol{\Sigma}^{-1})$ enter the regression:

$$Y_j = b_j'\boldsymbol{x} + \sum_{k\neq j} \beta_{jk}(Y_k - b_k'\boldsymbol{x}) + \epsilon_j,\ j = 1, \ldots, p, \tag{5.16}$$

and ϵ_j is independent of $\boldsymbol{Y}_{-j} = \{Y_k; k \neq j\}$ if and only if $\beta_{jk} = -\frac{\sigma^{jk}}{\sigma^{jj}}$ with $\mathrm{Var}(\epsilon_j) = 1/\sigma^{jj}$; see (4.18)–(4.19) and Problem 5.6. In addition, when the precision matrix or the regulatory network also depends on the covariates $\boldsymbol{x}$, Zhang and Li [2023] replaces the LCM (1.3) by the *matrix-on-scalar* regression

$$\boldsymbol{\Sigma}^{-1}(\boldsymbol{x}) = \mathbf{B}_0 + \sum_{h=1}^{q} \mathbf{B}_h x_h. \tag{5.17}$$

Its novelty is in using $\mathbf{B}_h$ matrices as the regression parameters for the scalar covariates x_h's. A key advantage of such precision matrix regression models is the ease of interpretation of its matrix coefficients. In fact, here $\mathbf{B}_0$ captures the population level regulatory network and $\mathbf{B}_h$ characterizes the effect of covariate x_h. In what follows, the σ^{jj}'s are assumed not to depend on the covariate $\boldsymbol{x}$ or the error variances do not vary with the covariates.

Of course, due to the linearity of (5.17) it is desirable to know when such a matrix is positive-definite and whether its parameters are identifiable. Fortunately, conditions to ensure that the precision matrix is diagonally dominant and hence positive-definite are given in Zhang and Li [2023] and Problem 3.5.

When both the mean and precision matrix depend on the covariate $\boldsymbol{x}$, one may estimate their respective parameters via the following sequence of *Gaussian graphical regressions* where for each node $j = 1, \dots, p$,

$$Y_j = b_j'\boldsymbol{x} + \sum_{k\neq j} \beta_{jk0}(Y_k - b_k'\boldsymbol{x}) + \sum_{k\neq j}\sum_{h=1}^{q} \beta_{jkh} x_h \times (Y_k - b_k'\boldsymbol{x}) + \epsilon_j, \tag{5.18}$$

with $\beta_{jkh} = -[\mathbf{B}_h]_{jk}/\sigma^{jj}$ for all (j, k, h). A pleasant feature of the model (5.18) is the inclusion of the interaction terms between the nodes in $\boldsymbol{Y}_{-j}$ and the covariate $\boldsymbol{x}$. Setting $Z = Y - \mathbf{B}\boldsymbol{x} = (Z_1, \dots, Z_p)'$, and rewriting (5.18)

$$Z_j = \sum_{k\neq j} \beta_{jk0} Z_k + \sum_{k\neq j}\sum_{h=1}^{q} \beta_{jkh} x_h Z_k + \epsilon_j, \tag{5.19}$$

gives a clear rational for requiring the diagonal elements of the precision matrix not to depend on the covariates. Indeed, this can be seen as the variance of Z_j being homoscedastic after removing effects of $\boldsymbol{x}$, $\boldsymbol{Z}_{-j}$ and their interactions which is a familiar assumption in the literature of linear models.

Given observations $(\boldsymbol{Y}, \boldsymbol{X})$ for n subjects and assuming for the moment that $\mathbf{B}$ is known, with $Z = \boldsymbol{Y} - \mathbf{B}\boldsymbol{X}$ we denote its jth column by the vector z_j and the sample values of the hth covariate by the vector $\boldsymbol{x}_h$. Then, the sample version of (5.19) is

$$z_j = \sum_{k\neq j} \beta_{jk0} z_k + \sum_{k\neq j}\sum_{h=1}^{q} \beta_{jkh} \boldsymbol{x}_h \odot z_k + \epsilon_j, \tag{5.20}$$

where $\boldsymbol{x}_h \odot z_k$ is the component-wise product of two n-vectors, and $\epsilon_j \sim N(\boldsymbol{0}, 1/\sigma^{jj}\boldsymbol{I}_n)$. Next, we rewrite the above as a standard linear regression model.

For this, let $\boldsymbol{\beta}_{jh} = (\beta_{j1h}, \ldots, \beta_{jph})' \in \mathbb{R}^p$ be the hth group of parameters measuring the effects of covariate $\boldsymbol{x}_h$ on the edges of node j, where for $h = 0$ it could be interpreted as the population level effects on the edges of a node j. Defining the vector $\boldsymbol{\beta}_j = (\boldsymbol{\beta}_{j0}, \boldsymbol{\beta}_{j1}, \ldots, \boldsymbol{\beta}_{jq})'$, then (5.20) takes the more compact and familiar form

$$z_j = \mathbf{W}_{-j}\boldsymbol{\beta}_j + \boldsymbol{\epsilon}_j, \tag{5.21}$$

where $\boldsymbol{W}_{-j} = (z_1, \boldsymbol{x}_1 \odot z_1, \ldots, \boldsymbol{x}_1 \odot z_q, \ldots, \boldsymbol{x}_q \odot z_{j-1}, z_{j+1}, \boldsymbol{x}_1 \odot z_{j+1}, \ldots, \boldsymbol{x}_q \odot z_p)$ is its design matrix.

For estimation of the regression coefficients $\boldsymbol{\beta}_j$, consider the squared error loss function

$$\ell_j(\boldsymbol{\beta}_j, \mathcal{D}) = \frac{1}{2n}||z_j - \mathbf{W}_{-j}\boldsymbol{\beta}_j||_2^2, \; j = 1, \ldots, p.$$

For sparse estimation, one may impose simultaneously an element-wise sparsity by penalizing $\|\boldsymbol{\beta}_j\|_1$ and a group sparsity via penalizing $\sum_{h=1}^q \|\boldsymbol{\beta}_{jh}\|_2$. Such simultaneous regularization is tantamount to assuming that $\boldsymbol{\beta}_j$ is *group sparse*, in the sense that only a few covariates may impact edges, the so-called *effective covariates*. Moreover, *element-wise sparsity* of $\boldsymbol{\beta}_j$ means that effective covariates may influence only a few edges. Note that $\boldsymbol{\beta}_{j0}$ is excluded from the group sparsity constraint (but not in the element-wise sparsity), as it determines the population level regulatory network. Fortunately, the desire for such simultaneous sparsity is well-supported by genetic studies [van der Wijst et al., 2018].

Thus, for sparse estimation of $\boldsymbol{\beta}_j$ it is advisable to consider the objective function

$$\ell_j(\boldsymbol{\beta}_j, \mathcal{D}) + \lambda||\boldsymbol{\beta}_j||_1 + \lambda_g \sum_{h=1}^{q} ||\boldsymbol{\beta}_{j,h}||_2, \tag{5.22}$$

where λ and λ_g are nonnegative tuning parameters. For unknown **B**, we may replace it by its regularized estimator. The combined penalty in (5.22) is known as *the sparse group lasso* penalty. It being convex, can be optimized by using the existing gradient descent algorithms for sparse group lasso. Since the solution is not guaranteed to be a symmetric matrix, a post-processing step similar to (4.33) can be used to symmetrize it. In addition to lack of positive-definiteness, there are other theoretical challenges arising from regressing both the means and the precision matrices on covariates. The two parameters λ, λ_g can be jointly selected via K-fold cross-validation.

For directed graphs and in a Bayesian framework, Niu et al. [2023] develops models for the underlying heterogeneous multivariate observations where the covariates may enter the scene implicitly/nonlinearly. Their first key step is to group/cluster observations, leading to subgroups with distinct covariance matrices. This is done via product partition models, the novelty of such covariate-dependent Gaussian graphical model is in allowing graphs to vary with covariates so that observations whose covariates are similar share a similar undirected graph. Their modeling framework seems to be in line with those in Fox and Dunson [2015] who considered non-linear covariate effects, and Franks [2022] considering high dimensional responses and in Alakus et al. [2023] using random forests.

5.8 Canonical Correlation Analysis (CCA)

In this section we present the basic ingredients for canonical correlation analysis of two zero-mean random vectors $Y = (Y_1, \dots, Y_q)'$ and $X = (X_1, \dots, X_p)'$. It can be viewed as an attempt to revive multivariate regression by fully engaging the correlations or the covariance matrix of the response vector. Indeed, it subsumes the familiar notions of correlation coefficient ρ between two random variables ($p = q = 1$) and the multiple correlation coefficient R^2 when $q = 1$ and $p > 1$. It helps to recall that R^2 for $q = 1$ is the square of the maximum of $\text{Corr}(Y, \mathbf{v}'X)$ over $\mathbf{v} \in \mathbb{R}^p$ or linear combinations of the entries of X.

For two random vectors or when both p and q are bigger than one, CCA can be viewed as converting the task of the multivariate regression of Y on X into a number of univariate regressions so that a linear combination $\mathbf{v}'X$ of the covariates is the best predictor of $\boldsymbol{u}'Y$. In words, one searches for a pair of vectors $\boldsymbol{u}, \mathbf{v}$ so that the corresponding linear combinations enjoy the maximum correlation.

More precisely, in the CCA of two zero-mean random vectors $Y = (Y_1, \dots, Y_q)'$ and $X = (X_1, \dots, X_p)'$ with covariance matrices $\boldsymbol{\Sigma}_{11}$ and $\boldsymbol{\Sigma}_{22}$ and the cross-covariance matrix $\boldsymbol{\Sigma}_{12} = \text{Cov}(Y, X)$, one searches for linear combinations $U = \boldsymbol{u}'Y$ and $V = \mathbf{v}'X$ having the largest correlation coefficient [Hotelling, 1935]. In other words, one searches for $\boldsymbol{u}$ and $\mathbf{v}$, so that U is best predicted by V or that

$$\text{Corr}(\boldsymbol{u}'Y, \mathbf{v}'X), \tag{5.23}$$

is maximized.

Since $\text{Cov}(U, V) = \boldsymbol{u}'\boldsymbol{\Sigma}_{12}\mathbf{v}$, it is plausible that the singular values and vectors of the $q \times p$ cross-covariance matrix $\boldsymbol{\Sigma}_{12}$ of the pair (Y, X) might play important roles in computing the quantities of interest. In fact, it turns out that the key players are the ordered eigenvalues λ_k, $k = 1, \dots, m = \min\{p, q\}$, and the corresponding normalized eigenvectors of the matrix

$$A = \boldsymbol{\Sigma}_{11}^{-1/2}\boldsymbol{\Sigma}_{12}\boldsymbol{\Sigma}_{22}^{-1}\boldsymbol{\Sigma}_{21}\boldsymbol{\Sigma}_{11}^{-1/2} = BB', \tag{5.24}$$

where $B = \boldsymbol{\Sigma}_{11}^{-1/2}\boldsymbol{\Sigma}_{12}\boldsymbol{\Sigma}_{22}^{-1/2}$ is the "standardized" matrix $\boldsymbol{\Sigma}_{12}$. In fact, when both $\boldsymbol{\Sigma}_{11}$ and $\boldsymbol{\Sigma}_{22}$ are identity matrices, B reduces to $\boldsymbol{\Sigma}_{12}$.

Next, we illustrate and highlight some of the key ideas and ingredients of maximizing the correlation leading to (5.24). First, note that

$$\text{Var}(U) = \boldsymbol{u}'\boldsymbol{\Sigma}_{11}\boldsymbol{u}, \quad \text{Var}(V) = \mathbf{v}'\boldsymbol{\Sigma}_{22}\mathbf{v}, \quad \text{Cov}(U, V) = \boldsymbol{u}'\boldsymbol{\Sigma}_{12}\mathbf{v},$$

and

$$\text{Corr}(U, V) = \frac{\boldsymbol{u}'\boldsymbol{\Sigma}_{12}\mathbf{v}}{\sqrt{(\boldsymbol{u}'\boldsymbol{\Sigma}_{11}\boldsymbol{u})(\mathbf{v}'\boldsymbol{\Sigma}_{22}\mathbf{v})}}. \tag{5.25}$$

Then, we rewrite the terms in the denominator using $\boldsymbol{\Sigma}_{ii}^{1/2}$, the square root of $\boldsymbol{\Sigma}_{ii}$, as

$$\boldsymbol{u}'\boldsymbol{\Sigma}_{11}\boldsymbol{u} = \boldsymbol{u}'\boldsymbol{\Sigma}_{11}^{1/2}\boldsymbol{\Sigma}_{11}^{1/2}\boldsymbol{u} = \mathbf{a}'\mathbf{a}, \quad \mathbf{v}'\boldsymbol{\Sigma}_{22}\mathbf{v} = \mathbf{v}'\boldsymbol{\Sigma}_{22}^{1/2}\boldsymbol{\Sigma}_{22}^{1/2}\mathbf{v} = \mathbf{b}'\mathbf{b},$$

where $\mathbf{a} = \mathbf{\Sigma}_{11}^{1/2}\boldsymbol{u}$ and $\mathbf{b} = \mathbf{\Sigma}_{22}^{1/2}\mathbf{v}$. Writing the numerator in terms of $\mathbf{a}, \mathbf{b}$ and applying the Cauchy-Schwarz inequality one obtains

$$\begin{aligned} \boldsymbol{u}'\mathbf{\Sigma}_{12}\mathbf{v} = (\mathbf{a}'\mathbf{\Sigma}_{11}^{-1/2}\mathbf{\Sigma}_{12}\mathbf{\Sigma}_{22}^{-1/2})\mathbf{b} &\le (\mathbf{a}'\mathbf{\Sigma}_{11}^{-1/2}\mathbf{\Sigma}_{12}\mathbf{\Sigma}_{22}^{-1/2}\mathbf{\Sigma}_{21}\mathbf{\Sigma}_{11}^{-1/2}\mathbf{a})^{1/2}(\mathbf{b}'\mathbf{b})^{1/2} \\ &= (\mathbf{a}'A\mathbf{a})^{1/2}(\mathbf{b}'\mathbf{b})^{1/2}. \end{aligned} \tag{5.26}$$

Let λ_1 be the largest eigenvalue of the symmetric matrix A in (5.24) with the corresponding normalized eigenvector $\boldsymbol{e}_1$. Then, it follows that

$$\mathbf{a}'A\mathbf{a} \le \lambda_1\mathbf{a}'\mathbf{a}, \tag{5.27}$$

and the equality is achieved if $\mathbf{a} = \boldsymbol{e}_1$. Furthermore, the equality in (5.26) holds if $\mathbf{b} = c\mathbf{\Sigma}_{22}^{-1/2}\mathbf{\Sigma}_{21}\mathbf{\Sigma}_{11}^{-1/2}\mathbf{a}$ for a scalar c. Putting these pieces together, from (5.26) it follows that

$$\max \operatorname{Corr}(U, V) = \lambda_1^{1/2},$$

and the maximum is attained for

$$\boldsymbol{u}_1 = \mathbf{\Sigma}_{11}^{-1/2}\boldsymbol{e}_1, \quad \mathbf{v}_1 = c\mathbf{\Sigma}_{22}^{-1/2}\mathbf{\Sigma}_{21}\mathbf{\Sigma}_{11}^{-1/2}\boldsymbol{e}_1, \tag{5.28}$$

where the arbitrary constant c is chosen so that the correlation is positive.

We refer to the pair $(\boldsymbol{u}_1, \mathbf{v}_1)$ of vectors in (5.28) as the *first pair of canonical coordinates*, and the corresponding pair of random variables (U_1, V_1) as the *first pair of canonical variates* with the correlation coefficient $\rho_1 = \lambda_1^{1/2}$. The subsequent pairs of canonical coordinates $(\boldsymbol{u}_k, \mathbf{v}_k)$, variates (U_k, V_k) and the corresponding maximum correlations $\rho_k = \lambda_k^{1/2}$ for $k = 2, \ldots, m = \min\{p, q\}$ follows steps similar to those in Section 7.1 for finding the second and higher order PCs.

Given data $(\boldsymbol{X}_i, \boldsymbol{Y}_i)$, $i = 1, \ldots, n$, one may compute the sample canonical variates and correlations by replacing the population covariance matrices by their sample counterparts in (5.24). For p and q fixed and smaller than the sample size n, the procedure produces consistent estimators of the population canonical coordinates and correlations for n going to infinity. For more details and a readable account of the CCA methodologies, geometrical interpretation of the canonical coordinates coupled with numerous data examples and illustrations, see Johnson and Wichern [2008]. However, when p and q exceed n, the sample covariances are not invertible and one cannot proceed using (5.24). In such situations it is common to impose sparsity constraint on the canonical coordinates or assume that most of their entries are zero. Consequently, the correlation between two random vectors can be explained using a small set of features or coordinates. This effectively reduces the dimensionality and improves the interpretability of the results.

When and under what conditions are the vectors of canonical coordinates sparse? The following recent and remarkable result due to Chen et al. [2013] sheds considerable light on the question: The $(\boldsymbol{u}_k, \mathbf{v}_k)$'s maximize the correlation in (5.25) if and only if the matrices $(\mathbf{\Sigma}_{11}, \mathbf{\Sigma}_{22}, \mathbf{\Sigma}_{12})$ satisfy

$$\mathbf{\Sigma}_{12} = \mathbf{\Sigma}_{11}\left(\sum_{k=1}^{r} \lambda_k \boldsymbol{u}_k \mathbf{v}_k'\right)\mathbf{\Sigma}_{22}, \tag{5.29}$$

where $r = \text{rank}(\boldsymbol{\Sigma}_{12})$ and $(\boldsymbol{u}_k, \mathbf{v}_k)$'s are orthonormal with respect to $\boldsymbol{\Sigma}_{11}$ and $\boldsymbol{\Sigma}_{22}$, namely

$$\boldsymbol{u}_i'\boldsymbol{\Sigma}_{11}\boldsymbol{u}_j = \delta_{ij}, \quad \mathbf{v}_i'\boldsymbol{\Sigma}_{22}\mathbf{v}_j = \delta_{ij},$$

for δ_{ij} the Kronecker's delta function with values one when $i = j$, and zero otherwise. An immediate and deep consequence of the transparent characterization in (5.29) is that the canonical coordinates are sparse, if and only if $\boldsymbol{u}_k, \mathbf{v}_k$'s are. Thus, the sparsity assumption is made explicit in this probabilistic model.

Closely related to CCA is the idea of *reduced rank regression* [Anderson, 1951] which simply performs least-squares estimation in (5.8) subject to the rank constraint, $\text{rank}(\mathbf{B}) = r$, for some $r \leq \min(p, q)$. It turns out that this constraint forces the least-squares method to use effectively the correlations and cross-correlations among the responses and covariates. In fact, its solution is related to the CCA and the singular-value decomposition (SVD), see Hastie et al. [2009, Chapter 3]. Another good exposition of this connection is given in Aldrin [2000]. Unfortunately, both methods work well in the low dimensional setup, and do not deal well with the challenges of growing data dimensions.

5.9 Problems

5.1 The expression $Q(\mathbf{B}) = \text{tr}(\boldsymbol{Y} - \boldsymbol{X}\mathbf{B})'(\boldsymbol{Y} - \boldsymbol{X}\mathbf{B})$ is the sum of squares and cross-products of errors in the multivariate regression model with the fixed (non-random) design matrix $\boldsymbol{X}$.

(a) Show that $\boldsymbol{X}'(\boldsymbol{Y} - \boldsymbol{X}\widehat{\mathbf{B}}) = \mathbf{0}$, where $\widehat{\boldsymbol{B}} = \widehat{\boldsymbol{B}}_{ols}$. Describe its geometrical meaning so far as orthogonality of the residuals to the column space of $\boldsymbol{X}$ is concerned.

(b) Use the result in (a) to establish the identity

$$Q(\boldsymbol{B}) = Q(\widehat{\boldsymbol{B}}) + \text{tr}(\widehat{\boldsymbol{B}} - \boldsymbol{B})'\boldsymbol{X}'\boldsymbol{X}(\widehat{\boldsymbol{B}} - \boldsymbol{B}),$$

and that $Q(\boldsymbol{B})$ is minimized at $\hat{\mathbf{B}} = \widehat{\boldsymbol{B}}_{ols}$.

5.2 (a) Show that the multivariate regression model in (5.7) is equivalent to the following q (column-wise) regression models:

$$Y_{\cdot i} = \boldsymbol{X}\mathbf{b}_{\cdot i} + \boldsymbol{e}_{\cdot i},\ i = 1, \ldots, q,$$

where for a matrix A, $A_{\cdot i}$ stands for its ith column. Are these q models uncorrelated?

(b) Find the least squares estimator of $\mathbf{b}_{\cdot i}$ in (a) and compare it with the ith column of $\widehat{\mathbf{B}}_{ols}$.

(c) Show that $Q(\mathbf{B})$ in the previous problem can be rewritten as

$$Q(\mathbf{B}) = \sum_{i=1}^{q} (Y_{\cdot i} - \boldsymbol{X}\mathbf{b}_{\cdot i})'(Y_{\cdot i} - \boldsymbol{X}\mathbf{b}_{\cdot i}).$$

5.3 Substitute the LCM-type model (5.17) into (5.16) and write the details of deriving (5.19)–(5.21).

5.4 Let $(Y', X')'$ be a $(q+p)$-dimensional zero-mean random vector with the following partitioned positive-definite covariance matrix

$$\boldsymbol{\Sigma} = \begin{pmatrix} \boldsymbol{\Sigma}_{11} & \boldsymbol{\Sigma}_{12} \\ \boldsymbol{\Sigma}_{21} & \boldsymbol{\Sigma}_{22} \end{pmatrix},$$

where $\boldsymbol{\Sigma}_{11}$ is a $q \times q$ matrix.

(a) For $p = q = 2$ and $\boldsymbol{\Sigma}$ a $\mathrm{CS}(\rho, \sigma^2)$ matrix find the canonical correlations and canonical coordinates between Y and X.

(b) Repeat (a) for general p and q with $q \leq p$.

5.5 Show that the objective function in (5.12) for $\boldsymbol{\Theta} = \boldsymbol{\Theta}_0$ is convex $\mathbf{B}$ and can be reduced to the following Lasso regression problem

$$\arg\min_{\beta} \left\{ \|\tilde{\boldsymbol{Y}} - \tilde{\boldsymbol{X}}\boldsymbol{\beta}\|^2 + \lambda_2 \sum_{j=1}^{pq} |\beta_j| \right\},$$

where $\boldsymbol{\beta}$ is the vector of length pq given by

$$(b_{11}, b_{21}, \ldots, b_{p1}, \ldots, b_{1q}, b_{2q}, \ldots, b_{pq})',$$

and

$$\boldsymbol{X}_i = I_{q\times q} \otimes \boldsymbol{x}_i',$$

where $\otimes$ is the Kronecker product and $I_{q\times q}$ is the identity matrix, and

$$\tilde{\boldsymbol{Y}} = \frac{1}{\sqrt{n}} (\boldsymbol{Y}_1' L', \boldsymbol{Y}_2' L', \ldots, \boldsymbol{Y}_n' L')',$$

which is of length qn

$$\tilde{\boldsymbol{X}} = \frac{1}{\sqrt{n}} (\boldsymbol{X}_1' L', \ldots, \boldsymbol{X}_n' L')',$$

and L is the $q \times q$ upper triangular matrix from the Cholesky decomposition $\boldsymbol{\Theta}_0 = L'L$.

5.6 Starting from (4.19) prove the statement in (5.16).

5.7 (a) Repeat the analysis of the log-returns data in Section 5.6.1, this time by retaining the log-returns of all stocks for the last 30 weeks in 2022 as the test set and the remaining 284 observations as the training set.

(b) Repeat the analysis in (a) by using a finer partition for the tuning parameter α in the CAPME algorithm. For example, splitting each interval into two equal parts.

5.8 Use the data from Section 5.6.2 and

(a) note that in Figure 5.3 the last three variables have very weak correlations with others. Plot the histogram and Q-Q plot of the last three responses.

(b) Compute and plot the correlation matrix of the untransformed data as in Figure 5.3. Compare the correlation plots of the original and transformed data.

(c) Plot and compare the scatterplot matrices of the original and the transformed data. Use the discussions and plots in Section 3.6.3 as guide to suggest plausible pair copula for each scatterplot.

(d) Find the canonical variates between the first two responses $(Y_1, Y_2)'$ and the last three, namely $(Y_4, Y_5, Y_6)'$.

(e) Find the canonical variates between all the responses Y and covariates in X.

5.9 Consider the psych dataset in package recode. This hypothetical dataset contains 600 observations and six variables. The psychological variables are locus of control (locus_of_control), self-concept (self_concept), and motivation (motivation). The academic variables are standardized tests scores in reading (read), writing (write), and science (science). Similar to the data example in Section 5.6.2, use the psychological variables as responses and 10-fold cross-validation to apply aMCR algorithm and separate LSE to obtain predictive accuracy for both methods.

6

Covariance Regressions

For a response vector Y and a multivariate predictor X, the (mean) regression or modeling $\mu(x) = E(Y|X = x)$ as a function of covariates is a great success story in statistics. There is also a rich literature on heterogeneous regression models allowing the error variances to vary with predictors [Carroll and Ruppert, 1988]. However, there is a dearth of cogent and flexible statistical methodologies for modeling $\mathbf{\Sigma}(x) = \text{Cov}(Y|X = x)$ using covariates or *covariance regression*. In this chapter we review the progress in covariance regression following the path of developments in regression, going through the phases of linear, log-linear and generalized linear models (GLM). Indeed, inspired by the success of GLM for the first moment, it is plausible to write a covariance GLM (C-GLM) as

$$g(\mathbf{\Sigma}) = \alpha_1 U_1 + \cdots + \alpha_q U_q, \tag{6.1}$$

for a *link* function $g(\cdot)$ and a nonnegative integer q, where α_i's are the unknown scalar parameters and U_i's are some known symmetric basis matrices (covariates). Some popular choices of the link functions are the identity, inverse and matrix logarithm. However, at present there is no bona fide C-GLM due to lack of link functions satisfying the C-GLM wishlist in Remark 3, though the search goes on. A broad class of real-valued convex differentiable functions $g(\cdot)$ leads to the *linear entropic covariance* models. Model estimation and generalized estimating equations are briefly reviewed and the chapter ends with three case studies.

6.1 Covariance Model Formulation

A covariance matrix captures the correlations among the responses and expressing these in terms of covariates is highly desirable. In addition to providing insights into the nature of linear dependencies, it could serve as an important diagnostic for violation of the usual assumption of equality of covariances or homoscedasticity in regression analysis with significant impact on inferences. In analogy with the regression GLM, the success of GLM for covariance matrices hinges on finding unconstrained and statistically meaningful reparameterizations of covariance matrices (suitable link functions or transformations). This is usually resolved using various matrix decompositions including the spectral and Cholesky decompositions and the variance-correlation separation.

DOI: 10.1201/9781003638896-6

The positive-definiteness of a covariance matrix is the major stumbling block in its modeling using covariates. This constraint, perhaps due to the "two-way" structure of a matrix, is considerably more complicated to deal with than the range constraint on the individual entries of a mean vector. Thus, any genuine covariance reparameterization or transformation must involve *all its entries simultaneously*, in contrast to term-wise transformations like log, logit, etc. for a mean vector. Once a link function or a suitable transformation and a list of covariates are identified, in principle, the 3-stage iterative modeling procedure (*model formulation, estimation* and *diagnostic checking*) popular in regression and time series analysis could be implemented for covariance matrices.

Three loosely defined classes of transformations working on the whole, aspects of covariance matrices, and the whole distribution of the response vector are identified and reviewed in this section.

6.1.1 Modeling the *Whole* Covariance Matrix

Virtually all early GLM and regularization methods work on the whole covariance matrix as the object of modeling. This seems to be in line with the development of mean regression where the whole mean vector is written as linear combinations of like (basis) vectors or columns of a design matrix (vector-on-vector regression).

6.1.1.1 Linear Covariance Models

The identity and inverse link functions in (6.1) are the simplest transformations of covariance matrices. They lead to a rather general class of *linear covariance models* (LCM):

$$\boldsymbol{\Sigma}^{\pm 1} = \alpha_1 U_1 + \cdots + \alpha_q U_q, \tag{6.2}$$

where U_i's are known symmetric basis matrices (covariates) and α_i's are the unknown parameters [Anderson, 1973]. For simplicity we restrict discussion to $\boldsymbol{\Sigma}$. The vector $\boldsymbol{\alpha} = (\alpha_1, \ldots, \alpha_q)' \in \mathbb{R}^q$ is restricted so that the matrix is positive-definite. Furthermore, it is assumed that there is at least a vector $\boldsymbol{\alpha}$ such that $\boldsymbol{\Sigma}$ is positive-definite. We may write $\boldsymbol{\Sigma}(U)$ to emphasize the covariates, even though at this point they are simply some mathematical constructs and are not genuine from experimental perspectives. It is instructive to view (6.2) as a matrix-on-matrix regression.

Remarkably, the model (6.2) is general and flexible enough to include all $p \times p$ covariance matrices. Indeed, for $q = p(p+1)/2$, any covariance matrix has the LCM representation:

$$\boldsymbol{\Sigma} = (\sigma_{ij}) = \sum_{i=1}^{p} \sigma_{ii} e_i e_i' + \sum_{i=1}^{p} \sum_{j>i}^{p} \sigma_{ij}(e_i e_j' + e_j e_i'), \tag{6.3}$$

where now the α_i's are simply the unknown entries of the covariance matrix and e_i's are the standard basis for $\mathbb{R}^p$ with 1 in the ith position and 0 elsewhere. For future applications, it is instructive to keep in mind that the set of matrices $\{e_i e_j' + e_j e_i'; i \geq$

$j\}$ forms an orthogonal basis for $\mathcal{S}$, the space of all $p \times p$ symmetric matrices. For $p = 2$, (6.3) reduces to the simple form

$$\boldsymbol{\Sigma} = \sigma_{11}\begin{pmatrix} 1 & 0 \\ 0 & 0 \end{pmatrix} + \sigma_{22}\begin{pmatrix} 0 & 0 \\ 0 & 1 \end{pmatrix} + \sigma_{12}\begin{pmatrix} 0 & 1 \\ 1 & 0 \end{pmatrix}.$$

In spite of its simplicity, LCM suffers from the following three drawbacks:

1. The constraint coefficients make the task of estimation rather difficult.

2. Though the covariates U_i's are assumed known, in practice, they are hard to come by in a given experimental context. In fact, it is seldom the case that one collects matrix-valued covariates directly from experiments. Nevertheless, simple *structured (fixed) matrices* such as compound symmetric and autoregression have been used routinely in longitudinal data analysis where accurate modeling of the mean demands accounting of the correlations. Such simple matrices play a central role in the popular areas of generalized estimating equations [Liang and Zeger, 1986b] and random-effect models [Pinheiro and Bates, 2000]. In fact, a growing menu of structured matrices (containing at least 18, as of 1992) is available in software packages such as SAS, R and Stata.

3. In view of 1-2, interpretation of α_i's is not clear and does not follow the familiar regression practice that *for every unit increase in a covariate, its slope is the expected change in the response*. It begs the question: What is a unit increase in a matrix?

Fortunately, the interpretation is more clear in the more recent matrix-on-scalar regression setup in Zhang and Li [2023] and Kim and Zhang [2024] where they swap the roles of (α, U), so that genuine subject-level scalar covariates are used with matrix coefficients. Their alternative LCM is

$$\boldsymbol{\Sigma}(\boldsymbol{x}) = \mathbf{B}_0 + x_1\mathbf{B}_1 + \ldots + x_q\mathbf{B}_q, \tag{6.4}$$

where $\boldsymbol{x} = (x_1, \ldots, x_q)'$ is the subject-level covariates, $\mathbf{B}_0$ is a symmetric positive-definite matrix, $\mathbf{B}_i$'s are symmetric matrices. In a sense, the model (6.4) is a more faithful matrix-version of the classical multiple regression, but with a huge number of parameters, namely $(q+1)\frac{p(p+1)}{2}$. As in regression, the "matrix-intercept" term $\mathbf{B}_0$ has the interpretation as the overall or the population level covariance, and the matrix $\mathbf{B}_i$ represents the effect of covariate x_i on the heterogeneous covariance matrix. Kim and Zhang [2024] method of moment estimation is computationally efficient and does not impose much distributional assumptions on the data. It leads to estimating its matrix parameters within the framework of the multivariate linear regression problem. Of course, conditions are needed to ensure positive-definiteness of the estimated covariance matrix, so that still the first drawback of LCM persists. Analog of (6.4), for precision matrices or Gaussian graphical models is developed in Zhang and Li [2023].

6.1.1.2 Are There Matrix Covariates?

The notion of matrix covariates U_k's is relatively new and does not seem natural at the first encounter. In fact, it is seldom the case that one collects matrix-valued covariates directly from experiments. We discuss three directions where considerable effort is spent to come up with reasonable matrix covariates for covariance regressions. In general, the progress to develop meaningful (symmetric) matrix covariates has been rather slow and more research is needed.

Structured Matrices: In the literature of generalized estimation equations (GEE) and random-effects models it is common to use a single structured (parameterized) covariance matrix. Some notable examples of such structures are the autoregression of order one (AR(1)), compound symmetry (CS) and moving average of order one (MA(1)) covariance matrices; see Section 3.4 for more details. Many more and commonly used examples of such matrices are provided by SAS PROC Mixed [SAS Institute Inc., 2009] and other software packages.

Similarity Matrices: One may construct matrix covariates using pertinent scalar covariates from an experiment. A major contribution in this direction due to Zou et al. [2017] writes a covariance matrix as a linear combination of similarity matrices of the vector covariates. More concretely, for a response vector $Y = (Y_1, \ldots, Y_p)' \in \mathbb{R}^p$, with the vector of covariates $z_j = (z_{j,1}, \ldots, z_{j,q})' \in \mathbb{R}^q$ for the component Y_j, $j \in [p]$, one constructs $p \times p$ symmetric covariate matrices $U_k = (u^{(k)}_{j_1,j_2})$, $k \in [q]$; see (6.2). A similarity measure [Johnson and Wichern, 2008] between two responses Y_{j_1} and Y_{j_2} based on the closeness of their continuous covariate values is a decreasing function of their distance such as $u^{(k)}_{j_1,j_2} = \exp(-|z_{j_1,k} - z_{j_2,k}|^2)$. For binary or qualitative covariates, the similarity takes the value 1, if the two covariates belong to the same category, and 0 otherwise. For example, in networks, $u^{(k)}_{j_1,j_2} = 1$, if the nodes j_1 and j_2 are connected, and zero otherwise. For parameter identifiability, the diagonals of similarity matrices are set to zeros. For an application in portfolio selection, Zou et al. [2017] constructs similarity matrices for $q = 5$ covariates measuring a firm's characteristics. Their idea of relating covariance matrices to linear combination of similarity matrices of covariates is a major step forward, but their covariates are variable-specific and not individual-specific (e.g., covariates of genes but not of individuals). As such, the estimated covariance could not account for individual-level heterogeneity due to clinical covariates and genotypes.

Tree models: Trees are undirected graphs without cycles, for more on trees and graphs see Chapter 9. They naturally give rise to LCMs with specific U_i's and meaningful coefficients α_i's. The most prominent example of tree structures are Brownian motion tree models [Felsenstein, 1973] for the evolution of continuous traits in humans or animals. Given a rooted tree T on a set of nodes $V = \{1, \ldots, r\}$ and with p leaves ($p < r$), the corresponding Brownian motion tree model consists of all covari-

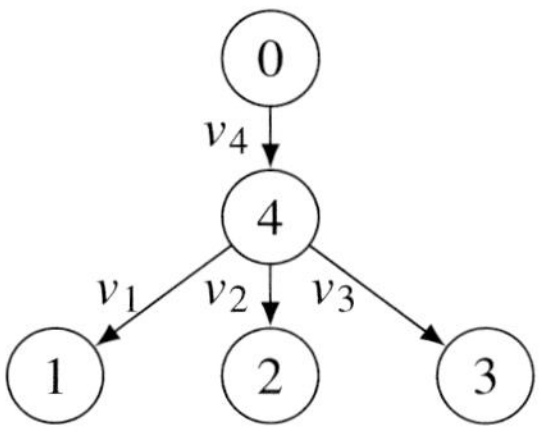

FIGURE 6.1
A star tree corresponding to (6.6).

ance matrices of the form

$$\mathbf{\Sigma}_v = \sum_{i \in V} v_i e_{de(i)} e'_{de(i)}, \tag{6.5}$$

where $e_{de(i)}$'s are p-vectors of 0's and 1's, with entry 1 at position j if leaf j is a descendant of node i and is 0, otherwise. The parameters v_i's describe branch lengths, and the covariance between any two leaves i and j is the amount of shared ancestry between these leaves, i.e., it is the sum of all the branch lengths from the root of the tree to the least common ancestor of i and j. Note that every leaf is a descendant of itself. For example, for T a star tree on three leaves, displayed in Figure 6.1, we have

$$\mathbf{\Sigma}_v = \begin{pmatrix} v_1 + v_4 & v_4 & v_4 \\ v_4 & v_2 + v_4 & v_4 \\ v_4 & v_4 & v_3 + v_4 \end{pmatrix}, \tag{6.6}$$

where v_1, v_2 and v_3 parameterize the branches leading to the leaves and v_4 parameterizes the root branch. Evidently, the linear structure of $\mathbf{\Sigma}_v$ is that of an LCM where (α_i, U_i)'s are prescribed by the structure of the underlying tree; see Problem 6.4.

In network tomography, where it is of interest to better understand the structure of the connections in the Internet [Tsang et al., 2001], a common assumption is that the underlying network has a tree structure. Then a Brownian motion tree corresponds to an intuitively appealing model where the correlation in arrival time is given by the sum of the edge lengths along the path shared by the messages. A problem of great interest is learning the underlying tree structure. The most popular heuristic for searching the tree space is Felsenstein's pruning algorithm [Felsenstein, 1973, 1981].

6.1.1.3 Quadratic Covariance Regression Models

Unlike the LCM, the model for covariance matrix in Hoff and Niu [2011] is quadratic in the covariates with matrix parameters so as to ensure its positive-definiteness. We refer to this class of covariance models as the *quadratic* covariance regression models, reserving the name *covariance regression models* for the broader task of expressing covariance matrices in terms of covariates.

The simplest quadratic covariance regression model is

$$\mathbf{\Sigma}(\boldsymbol{x}) = B_0 + B\boldsymbol{x}\boldsymbol{x}'B', \ \boldsymbol{x} \in \mathbb{R}^q, \tag{6.7}$$

where B_0, a $p \times p$ positive definite matrix, and B a $p \times q$ real matrix, are the new parameters, and $\boldsymbol{x}$ is a q-dimensional vector of covariates. It expresses $\boldsymbol{\Sigma}$ as sum of a "baseline" covariance matrix B_0 and a $p \times p$ rank-1 positive semi-definite matrix which varies with the q-vector of covariates $\boldsymbol{x}$. This covariance regression model is closely related to a dynamic model for time-varying covariance matrices; see (6.25).

The model and ease of interpretation of its parameters can be traced to the simple latent factor or the random-effects model:

$$\boldsymbol{Y}_i = \boldsymbol{\mu}_i + \gamma_i B\boldsymbol{x}_i + \boldsymbol{\epsilon}_i, \quad i = 1, \ldots, n, \tag{6.8}$$

where $\boldsymbol{\epsilon}_i$ and γ_i are mean-zero, uncorrelated random variables with

$$\operatorname{Cov}(\boldsymbol{\epsilon}_i) = B_0, \quad \operatorname{Var}(\gamma_i) = 1.$$

In the context of factor models (see Chapter 7) $B\boldsymbol{x}_i$ is the (subject-dependent) factor loadings and γ_i's are the corresponding factor scores. The latter can also be interpreted as the additional subject-level variability beyond the error $\boldsymbol{\epsilon}_i$, in the context of random-effects models.

In spite of its simplicity and ease of interpretation, these models suffer from two drawbacks:

1. The number of parameters increases quadratically in p and linearly in q.
2. The difference between $\boldsymbol{\Sigma}(\boldsymbol{x})$ and the baseline matrix B_0 is restricted to be a rank-1 matrix.

Higher rank models can be written easily by adding additional quadratic terms like $B_1\boldsymbol{x}\boldsymbol{x}'B_1$, which would further increase the number of parameters and may make it difficult to scale to higher dimensions. The Bayesian nonparametric covariance regression models in Fox and Dunson [2015] is a far-reaching and fascinating extension of (6.8), which scales up readily to higher dimensions; see Section 7.4.3 for a brief review.

As for data analysis, adapting the quadratic covariance regression model, Niu and Hoff [2019] analyze a four-dimensional health outcome and Seiler and Holmes [2017] employ a simpler model with $B_0 = \sigma^2 I_p$. Analysis of a large and multilevel neuroimaging dataset is described in the Case Study III in Section 6.4. The more general Bayesian nonparametric covariance regression models in Fox and Dunson [2015] is applied to the Google Flu Trends data.

In a geometric sense described next, the model (6.7) seems to be a natural generalization of mean regression to the covariance matrices. To see this point, recall that a mean regression with q covariates, $\boldsymbol{\mu} = B\boldsymbol{x}$, is a linear map between two linear vector spaces, namely from $\mathbb{R}^q$ to $\mathbb{R}^p$. Similarly, since covariance matrices lie in the cone of positive-definite matrices, the quadratic covariance regression model can be viewed as a map from a cone of $q \times q$ positive-definite matrices to the cone of $p \times p$ positive-definite matrices. While this generalization is interesting and natural from the geometric perspective, the C-GLM in (6.1) seems to be far more versatile both from mathematical and practical perspectives. For its applications in learning the sparsity patterns of $\boldsymbol{\Sigma}$ from that of $\log \boldsymbol{\Sigma}$, see Battey [2017], and for solving GEE via quadratic inference functions (QIF), see Qu et al. [2000] and Section 6.3.3.

6.1.1.4 Log-Linear Covariance Models

A strong motivation for the important class of log-linear covariance models introduced in this section, is the phenomenal role of its univariate counterpart in handling variance heterogeneity in regression analysis while ensuring nonnegativity of the variance estimates.

The key mathematical fact supporting log-linear models for a general positive-definite covariance matrix with the spectral decomposition $\mathbf{\Sigma} = P\Lambda P'$, is the convenient form of its *matricial logarithm* defined by

$$\log \mathbf{\Sigma} = P \log \Lambda P', \quad \text{where } \log \Lambda = \text{diag}(\log \lambda_1, \ldots, \log \lambda_p). \tag{6.9}$$

Note that $\log \Lambda$ is a diagonal matrix with diagonal entries $\log \lambda_i$.

Evidently, $\log \mathbf{\Sigma}$ is a symmetric matrix with $q = p(p+1)/2$ unconstrained entries taking values in $(-\infty, \infty)$, see Section 3.5, and when $\mathbf{\Sigma}$ is diagonal then $\log \mathbf{\Sigma} = \text{diag}(\log \sigma_{11}, \ldots, \log \sigma_{pp})$, and (6.10) reduces to the familiar log-linear models for heterogeneous variances enjoying a long history in econometrics and other areas, see Carroll and Ruppert [1988] and references therein.

The logarithmic transformation of covariance matrices has been pursued by Leonard and Hsu [1992] and Chiu et al. [1996] who introduced the *log-linear covariance* (LLM) models for $\mathbf{\Sigma}$ as

$$\log \mathbf{\Sigma} = \alpha_1 U_1 + \cdots + \alpha_q U_q, \tag{6.10}$$

where U_i's are known basis matrices as in (6.2), but the α_i's are now unconstrained. This is effectively entry-wise modeling of the logarithm of a covariance matrix and Chiu et al. [1996] expressed each entry of $\log \mathbf{\Sigma}$ as linear combinations of the covariates in a vector $\boldsymbol{x}$ collected from an experiment.
Perhaps, a more interpretable version of log-linear covariance models can be defined using the simple idea of separating the correlations and standard deviations via the factorization $\mathbf{\Sigma} = DRD$, with R the matrix of correlations and D the diagonal matrix of standard deviations of the variables. Then one may apply the log transformation to both matrices. Interestingly, $\log D$ leads to log-linear models for marginal variances with a fairly well-developed methodology, but it seems only recently Archakov and Hansen [2021] have explored the idea of reparameterizing R using its matrix logarithm $\log R$. It turns out that its lower off-diagonal entries are unconstrained. In fact, stacking up these into a vector $\gamma = \gamma(R)$, Archakov and Hansen [2021] show that this transformation defines a one-to-one correspondence between the set of $p \times p$ positive-definite correlation matrices and the vector space $\mathbb{R}^d$ where $d = p(p-1)/2$. They develop a fast algorithm for computing the inverse map. Consequently, a positive-definite $p \times p$ covariance matrix corresponds to a unique vector in $\mathbb{R}^{p+d}$ consisting of the p log-variances and d unconstrained γ's, which is ideal and facilitate modeling of covariance matrices in terms of an unrestricted vector.

Replacing $\mathbf{\Sigma}$ in (6.10) by a correlation matrix R may facilitate the task of interpretation, because at least for $p = 2$ the (1,2) entry of $\log R$ is the Fisher's z-transform of ρ, the correlation coefficient of the two variables; see Problem 3.22. It turns out that aspects of this interpretation carry over when $p > 2$, so that it can be viewed as a generalization of the Fisher's z-transform to higher dimensions. Indeed, it is expected to enjoy the variance-stabilizing and normality of such transformation so far as the off-diagonal entries of $\log R$ are concerned; see Archakov and Hansen [2021] for more details. The unconstrained entries of the log transformation of the correlation matrix makes it possible to use covariates in its parsimonious modeling. Fortunately, this idea has been pursued recently in Hu et al. [2024], where generalized linear models are developed for the non-redundant entries of $(D, \log R)$ along with an MLE methodology for inference. The effectiveness of the procedure is illustrated by applying it to two longitudinal datasets.

In a related, but in a slightly different spirit, Chattopadhyay et al. [2024] develop models for pairwise correlations in terms of covariates by writing linear models for their Fisher z-transforms. In addition, a major novelty of their methodology is the use of a composite pairwise likelihood which avoids the positive-definiteness restriction on the correlation matrix. They rely on a Minorize–Maximize (MM) algorithm which is quite effective for the efficient estimation of a large number of parameters.

Unfortunately, the great ability of the log transformation to remove the positive-definiteness constraint is tempered by the observation that $\log \boldsymbol{\Sigma}$ is a highly nonlinear operation on $\boldsymbol{\Sigma}$ (Section 3.5). Consequently, there is a consensus that the α_i's lack statistical interpretation [Brown et al., 1994], this is further complicated by *the fact that a sub-matrix of* $\boldsymbol{\Sigma}$ *does not necessarily relate to or coincide with a corresponding submatrix of* $\log \boldsymbol{\Sigma}$; see P4 in Remark 3.

6.1.2 Modeling *Aspects* of the Covariance Matrix

The previous approaches, model the *whole* covariance matrix as the outcome or parameter of interest which could make interpretation a challenge. In this section, we focus on modeling aspects or certain parts of a covariance matrix extractable from various matrix factorizations such as Cholesky and spectral decompositions, principal component analysis (PCA) and factor models. These usually lead to more interpretable components of the covariance matrix.

6.1.2.1 GLM via the Cholesky Decomposition

In this section, we show that a neat compromise between the constrained LCM (6.2) and the unconstrained log-linear covariance models (6.10) can be reached by realizing that the two factors T and $\log D$ of the modified Cholesky decomposition of a covariance matrix, $T\boldsymbol{\Sigma}T' = D$, are unconstrained, see Chapter 4. In general, a notable drawback of this idea is that it requires implicitly an order among the columns of the covariance matrix or the variables.

In fact, the map $\boldsymbol{\Sigma} \longrightarrow (T, \log D)$ provides the needed link for the C-GLM. More precisely, the non-redundant entries of the lower triangular matrix T and the diagonal matrix $\log D$ are unconstrained and interpretable as regression coefficients and prediction error variances when a variable is regressed on its predecessors. Consequently, bona fide GLMs for the entries of the two factors of a covariance (precision) matrix in terms of covariates can be written. Of course, an immediate consequence is that for any estimate of the Cholesky factors, the estimated covariance matrix and precision matrix $\widehat{\boldsymbol{\Sigma}}^{-1} = \widehat{T}'\widehat{D}^{-1}\widehat{T}$ are guaranteed to be positive-definite.

Following the GLM's tradition, it is plausible to write parameteric models for the *individual* entries of the Cholesky factors using covariates [Pourahmadi, 1999]. Suitable graphical tools such as regressogram may help suggest parametric models for ϕ_{tj} and $\log \sigma_t^2$, for $t = 1, \ldots, p$; $j = 1, \ldots, t-1$:

$$\log \sigma_t^2 = z_t'\lambda, \ \phi_{tj} = z_{tj}'\gamma, \tag{6.11}$$

where z_t, z_{tj} are $q \times 1$ and $d \times 1$ vectors of known covariates, $\lambda = (\lambda_1, \ldots, \lambda_q)'$ and $\gamma = (\gamma_1, \ldots, \gamma_d)'$ are parameters related to the innovation variances and dependence in Y, respectively. The most common covariates employed in the analysis of several longitudinal datasets are

$$z_t = (1, t, t^2, \ldots, t^{q-1})', z_{tj} = (1, t-j, (t-j)^2, \ldots, (t-j)^{d-1})',$$

simply the powers of times and time-lags. The AIC or BIC can be used [Pan and MacKenzie, 2003] to determine the orders (d, q). For a simple data analysis example, see Pourahmadi [2013, Section 3.11].

The idea of Cholesky-based GLM for covariance matrices has been used widely in areas far beyond the traditional longitudinal data. For example, in a series of papers by Gioia et al. [2024] and presented briefly in the Case Study II in Section 6.4 deals with a timely and practical example of forecasting daily electricity demand in UK using a spatially and temporally correlated dataset. They replace the parametric models in (6.11) for the individual entries of the Cholesky factors by a more elaborate additive (semiparametric) models involving relevant covariates for each entry. Similarly, the Bayesian graphical regression models in Ni et al. [2019] are simply DAG models allowing the graph structure (entries of the Cholesky factors) to vary with a finite number of discrete or continuous covariates.

Remark 3 *A Covariance Link Function Wishlist:*

The known link functions for $\boldsymbol{\Sigma}$ share many desirable properties of a GLM link functions, but not all due to positive-definiteness or perhaps the special (two-way) nature of covariance matrices. A link function g for the covariance matrix of a random vector $Y = (Y_1, \ldots, Y_p)'$ is expected to have at least the following four desirable properties:

P_1: *$g(\cdot)$ is bijective, that is for any covariance matrix $\boldsymbol{\Sigma}$, $g(\boldsymbol{\Sigma})$ maps it to a unique vector v with $q = p(p+1)/2$ unconstrained entries and vice versa;*

P_2: *The elements of v are (statistically) interpretable preferably as certain regression parameters;*

P_3: *The parameterization, $v = g(\boldsymbol{\Sigma})$, is "invariant" to the ordering of the random variables that define $\boldsymbol{\Sigma}$;*

P_4: *The parameters of any sub-matrix of* $\boldsymbol{\Sigma}$ *relates to or coincides with those of a corresponding submatrix of* $g(\boldsymbol{\Sigma})$.

Unfortunately, as seen in previous sections the log-link lacks P_2 and P_4, while the Cholesky link lacks P_3. In summary, after about five decades of research and progress in covariance modeling starting with the linear covariance models [Anderson, 1973], there is no bona fide analog of GLM for covariance matrices. Developing such a link function or covariance GLM (C-GLM), though, challenging is much needed and remains an open problem.

6.1.2.2 Covariate Assisted PCA

Arguably, principal component analysis (PCA) is one of the most fundamental tools in the high-dimensional data analysis and the modern machine learning. However, the traditional PCA and its close relative, the factor models, are notorious for not being covariate-friendly mostly because the normalized eigenvectors are severely constrained, see Chapter 7. This section directly introduces covariates into several covariance matrices by employing the conceptual and computational underpinnings of PCA. It searches for those directions of variability that can be explained by the covariates. The key assumption is that all covariance matrices commute or can be diagonalized simultaneously which is a serious restriction when there are too many covariance matrices.

Often for theoretical and practical considerations one may need to deal with several covariance matrices. For example, in fMRI studies, networks and finance one has to deal with multiple covariance matrices which may have some common features. Incorporating such features leads to parsimonious models for covariance matrices. In finance, the heteroscedasticity is captured via either variants of ARCH [Engle, 2002] or stochastic volatility models [Harvey et al., 1994]. The former directly specifies the volatility (covariance) matrix as a linear combination of lagged volatilities and squared returns as in (6.25), which suffers from the curse of dimensionality and is typically limited to datasets with 5 or fewer dimensions. Fortunately, there are variants of principal components models designed for characterizing covariance heterogeneity when the observations can be clustered into a few groups [Flury, 1988, Boik, 2002, Hoff, 2009]. A closely related alternative to PCA or the spectral decomposition is the more abstract idea of reducing subspaces where it is assumed that all differences among several covariance matrices are confined to a common low-dimensional subspace, see Cook and Forzani [2008] for an early version and Franks and Hoff [2019] for an empirical Bayes generalization. When observations cannot be clustered or grouped as above, then more sophisticated methods involving continuous covariates are needed.

Consider a study consisting of n subjects, let the time-varying response for subject $i, i = 1, \ldots, n$ be the p-vectors $Y_{it}, t = 1, \ldots, T_i$. The observations are assumed to be independent and identically distributed random samples from a multivariate normal distribution with mean zero and covariance matrix $\boldsymbol{\Sigma}_i$, which may depend on the explanatory variables $\boldsymbol{x}_i$. For example, the responses could be a sample of brain fMRI measurements of p regions, and $\boldsymbol{x}_i$ a q-vector of covariates related to the fMRI measurements, both collected from subject i, for further details see Zhao et al. [2019] and the Case Study III in Section 6.4.

For the moment, assume that there exists a direction in the data or a p-vector $\boldsymbol{\gamma}$ such that the projection $z_{it} = \boldsymbol{\gamma}' Y_{it}$ has a variance that depends on some covariates. More specifically, it satisfies a multiplicative heteroscedasticity model:

$$\log \mathrm{Var}(z_{it}) = \log(\boldsymbol{\gamma}' \boldsymbol{\Sigma}_i \boldsymbol{\gamma}) = \boldsymbol{\beta}' \boldsymbol{x}_i, \tag{6.12}$$

with the q-vector $\boldsymbol{\beta}$ as the coefficients in the log-linear models for the transformed responses. Note that for $p = 1$ this is simply a log-linear model for the variance discussed earlier. An interesting question here is how does the variances in (6.12) and the vector $\boldsymbol{\gamma}$ relate to the eigenvalues and eigenvectors of the covariance matrices? (See Problem 6.9.)

A key advantage of the idea of looking for certain variability directions in the data is that unlike the models in Anderson [1973], Hoff and Niu [2011] and Zhao et al. [2019], it does not seek to model the whole $\boldsymbol{\Sigma}_i$ in terms of basis matrices. Indeed, (6.12) writes a log-linear model for variances after a rotation of the response vector. As such the linear form allows easy interpretation of the regression coefficient as in (generalized) linear models. Furthermore, the projection idea enables development and computational scalability similar to the popular PCA techniques, see Case Study III.

The parameters $(\boldsymbol{\gamma}, \boldsymbol{\beta})$ are estimated by maximizing the Gaussian likelihood function of the data z_{it} in the projection space subject to the constraint $\boldsymbol{\gamma}' H \boldsymbol{\gamma} = 1$ for a positive definite matrix H, where $\boldsymbol{\Sigma}_i$'s are replaced by $\mathbf{S}_i$'s the sample covariance for subject i, see Zhao et al. [2019] for details. For high-dimensional data when $T_i < p$, for all i, the $\mathbf{S}_i$'s are singular matrices and can be replaced by their Ledoit-Wolf shrinkage estimators [Zhao et al., 2021].

As in PCA there is also the option of projecting in several directions simultaneously [Park, 2023] or replacing the vector $\boldsymbol{\gamma}$ by a $p \times d$ matrix Γ. Then, the right-hand side is replaced by a diagonal matrix, diag($B\boldsymbol{x}_i + u_i$), where B is a $d \times d$ matrix of regression coefficients and u_i's are normally distributed random effects. More precisely, one may consider the model

$$\log \mathrm{var}(\Gamma' \boldsymbol{Y}_i) = \log(\Gamma' \boldsymbol{\Sigma}_i \Gamma) = \mathrm{diag}(B\boldsymbol{x}_i + u_i), \tag{6.13}$$

where the log on the left is acting on a $d \times d$ matrix, or the matricial logarithm of a matrix encountered in the log-linear models (see Section 6.1.1.4) with its interpretation challenge. This technique is applied to the neuroimaging data of the Case Study III, where $d = 1$ is selected, however, for the sake of illustration and practice the results in Park [2023] for $d = 2$ are reported as well.

6.1.3 Linear Entropic Covariance Models

For a long time, it was thought that $\log \boldsymbol{\Sigma}$ is the only reasonable transformation leading to unconstrained symmetric matrices. In this section, a broad class of real-valued differentiable functions $g(\cdot)$ is introduced where their gradients $\nabla g(\mathbf{\Sigma}) = (\frac{\partial g}{\partial \sigma_{ij}})$ are $p \times p$ (not necessarily unconstrained) symmetric matrices which can be considered as a reparameterization of $\mathbf{\Sigma}$.

Recall from Section 3.5 that for a function $g : \mathbb{R} \longrightarrow \mathbb{R}$, the notation $g(\mathbf{\Sigma})$ stands for the matrix obtained from $\mathbf{\Sigma}$ through its spectral decomposition by replacing its eigenvalues λ_i's by $g(\lambda_i)$'s. The most interesting examples of scalar links are the so-called *spectral functions* $g(\mathbf{\Sigma}) = \text{tr}(g(\mathbf{\Sigma})) = \sum_{i=1}^{p} g(\lambda_i)$. Their gradient matrices are of the form $g'(\mathbf{\Sigma})$, so that everything is controlled by the function $g(\cdot)$ and its derivative. Some examples of such functions are given in the Table 6.1, where the first function corresponds to $g(x) = -\log(x)$, $x > 0$, with the derivative $g'(x) = -\frac{1}{x}$, $x > 0$, and its gradient matrix is the negative of the precision matrix; the second function corresponds to $g(x) = \frac{1}{2}x^2$ with derivative $g'(x) = x$, its gradient matrix is $g'(\mathbf{\Sigma}) = \mathbf{\Sigma}$. Thus, writing linear models for these gradient matrices is precisely the same as the LCM, the one corresponding to the fourth row of the table is the class of log-linear models. As such, the framework is broad enough to subsume most of the earlier covariance regression models.

Let $\mathbb{S}$ be the real vector space of symmetric $p \times p$ matrices with the standard inner product $(A, B) = \text{tr}(AB)$ and the induced Frobenius norm $||A||_F$, and $\mathbb{S}_+$ be the manifold of $p \times p$ positive-definite matrices. Consider the class of real-valued, strictly convex and differentiable functions $g : \mathbb{S}_+ \longrightarrow \mathbb{R}$ with the $p \times p$ gradient matrix $\nabla g(\mathbf{\Sigma}) = (\frac{\partial g}{\partial \sigma_{ij}})$. Under mild conditions on the function, it is well-known that the map $\nabla g : \mathbb{S}_+ \longrightarrow \mathbb{S}$ is one-to-one, so the gradient matrix can be viewed as a reparameterization of $\mathbf{\Sigma}$ with the range $\mathbb{L}_+ = \nabla g(\mathbb{S}_+)$. Note that restricting its range to a fixed affine subspace $\mathcal{L}$ of $\mathbb{S}$ allows writing an LCM for the gradient matrix. Thus, for a fixed $\mathcal{L}$, we assume throughout that

$$\mathcal{M}_g(\mathcal{L}) = \{\mathbf{\Sigma} \in \mathbb{S}_+; \nabla g(\mathbf{\Sigma}) \in \mathcal{L}\} = \mathbb{L}_+ \cap \mathcal{L} \neq \emptyset. \tag{6.14}$$

Restricting the range of $L = \nabla g(\mathbf{\Sigma})$ as above is the starting point of Zwiernik [2025] novel idea of *linear entropic covariance models*. It is capable of unifying diverse class of existing covariance models including some of those discussed above by postulating linear models for the gradient of g. Some interesting and common examples of $g(\cdot)$ are listed in the first two columns of Table 6.1; the relevance of the latter two columns will be explained later when the attention is on estimation. The negative of the first and the fourth functions in the table are known as the Gaussian and von Neumann entropies, respectively, and hence the name for the class of linear covariance models.

6.1.4 Multivariate Binary Data: Ising Model with Covariates

For a random vector $Y = (Y_1, \ldots, Y_p)'$ with binary entries, the dependencies across all pairs Y_i and Y_j can be modeled by specifying a probability distribution for Y. The most prominent distribution for this task is Ising models [Ising, 1925] with the probability distribution containing a component of the form $\exp\{\sum_{1 \leq i \leq j \leq p} \theta_{ij} y_i y_j\}$, with $\theta_{i,j}$ denoting the interaction term between the corresponding pair. The $p \times p$ symmetric matrix $\mathbf{\Theta} = (\theta_{ij})$ is commonly referred to as the matrix of pairwise interaction coefficients of this joint probability model.

TABLE 6.1
Link functions, their gradients, conjugate functions, and Bregman divergence.

$g(\mathbf{\Sigma})$	$L = \nabla g(\mathbf{\Sigma})$	$g^*(\mathbf{L})$	$D_g(\mathbf{S}, \mathbf{L})$
$-\log\det(\mathbf{\Sigma})$	$-\mathbf{\Sigma}^{-1}$	$-p - \log\det(-\mathbf{L})$	$-\log\det((-\mathbf{L})\mathbf{S}) + \mathrm{tr}((-\mathbf{L})\mathbf{S} - \mathbf{I}_p)$
$\frac{1}{2}\|\mathbf{\Sigma}\|_F^2$	$\mathbf{\Sigma}$	$\frac{1}{2}\mathrm{tr}(\mathbf{L}^2)$	$\frac{1}{2}\|\mathbf{L} - \mathbf{S}\|_F^2$
$\mathrm{tr}(\mathbf{\Sigma}^{-1})$	$-\mathbf{\Sigma}^{-2}$	$-2\mathrm{tr}(\sqrt{-\mathbf{L}})$	$\mathrm{tr}(\mathbf{S}^{-1}) - 2\mathrm{tr}(\sqrt{-\mathbf{L}}) + \mathrm{tr}(\mathbf{L}S)$
$-\mathrm{tr}(\mathbf{\Sigma} - \mathbf{\Sigma}\log(\mathbf{\Sigma}))$	$\log(\mathbf{\Sigma})$	$\mathrm{tr}(e^{\mathbf{L}})$	$-\mathrm{tr}(\mathbf{S} - \mathbf{S}\log(\mathbf{S})) + \mathrm{tr}(e^{\mathbf{L}}) - \mathrm{tr}(\mathbf{LS})$

Interestingly and somewhat surprisingly, as in the Gaussian models the interaction coefficients are directly tied to the conditional dependencies between Y_i and Y_j given the remaining elements. This is known as the Hammersley-Clifford equivalence [Hammersley and Clifford, 1971]: if θ_{ij} is close to (far) away from zero, the conditional dependence between Y_i and Y_j is weak (strong), with the limiting case of $\theta_{ij} = 0$ implying conditional independence. The similarity of the interpretation of the pairwise interaction coefficients with that of the precision matrix in the Gaussian setup has contributed greatly to developing analogous sparse estimation techniques for Ising models which we review briefly next.

In analogy with the development of GLasso (Chapter 9) there are a number of approaches for regularized estimation of the symmetric matrix $\mathbf{\Theta}$, starting with Banerjee et al. [2008] using a block-coordinate descent algorithm in solving a penalized likelihood function; Ravikumar et al. [2011] considered a neighborhood selection method by fitting a separate logistic regression with the Lasso penalty on the coefficients of each binary response. The pseudo-likelihood function employed in Höfling and Tibshirani [2009] involves the conditional probability function of a component Y_i given the rest which is free of the normalizing constant. An advantage of the shift to pseudo-likelihood is that it obviates the computational challenge of dealing with the normalizing constant of the multivariate binary distribution:

$$C(\mathbf{\Theta}) \exp\{\sum_{i=1}^{p} \theta_{ii} y_i + \sum_{1 \le i < j \le p} \theta_{ij} y_i y_j\}, \tag{6.15}$$

where $C(\mathbf{\Theta})$ is a sum of 2^p terms.

In analogy with covariance regression one may write such models for the entries of the interaction matrix as in Cheng et al. [2014] whose *Ising regression model* expresses the pairwise interaction coefficients θ_{ij} in terms of a covariate vector $\boldsymbol{x}$. More recently, Tho et al. [2024] has proposed a novel LCM-type, matrix-on-matrix regression model which regresses the matrix of interaction coefficients $\mathbf{\Theta}$ on a set of similarity matrices U_k's. For estimation of the LCM parameters, a regularized pseudo-likelihood with an adaptive Lasso penalty as in (2.20) is employed. It should be noted that the interpretation of the regression parameters are different in Cheng et al. [2014] and Tho et al. [2024]. In fact, in the first model the interaction coefficients for different pairs (i, j) depend on the same covariate vector $\boldsymbol{x}$, instead of the pairwise similarity measures $U_{k.ij}$ in the LCM model.

Tho et al. [2024] has analyzed the U.S. Senate roll call voting data with a binary response representing a senator's voting (Yea or Nay) to a particular piece of legislation or bill. The pairwise interaction coefficient θ_{ij} captures the conditional dependence relationship between the i-th and j-th senators' voting patterns for $p = 100$ senators. The similarity matrices U_k's are constructed from senators' covariates such as political parties, ages, occupations and social network profiles. Here the regression coefficients α_k's in LCM offer quantification of how the k-th similarity matrix covariate of the senators affects their conditional dependence when it comes to voting patterns.

6.2 Model Estimation

We touch-base with estimation of a few covariance regression models starting with the maximum likelihood estimation of the LCM parameters for multivariate normal data. For discrete and nonGaussian data, the increasing roles of LCM in the generalized estimating equations (GEE) and quadratic inference functions (QIF) are reviewed.

Two key advantages of C-GLM for covariance estimation are:

1. The number of parameters in $\boldsymbol{\alpha}$ is fixed and does not depend on p, the data dimension. This is in sharp contrast with more standard methods for covariance estimation such as the sample covariance matrix, Ledoit and Wolf [2004]'s shrinkage estimator, banding and thresholding (Chapter 8), where effectively the number of parameters is quadratic in p.

2. Simply vectorizing both sides of the LCM, see also (6.20), reduces their estimation problems to that of a multivariate linear regression model with $p(p+1)/2$ responses and the vec(U_i)'s as the covariates. Thus, computation and inference based on (asymptotic) properties of the MLE, OLS and GLS estimators of $\boldsymbol{\alpha}$ can be studied using the familiar regression machinery for various (n, p).

Let $\boldsymbol{Y}_1, \ldots, \boldsymbol{Y}_n$ be a sample of size n from a $N_p(\mu, \boldsymbol{\Sigma})$ distribution with the mean vector modeled as $\mu = \boldsymbol{X\beta}$ where $\boldsymbol{X}$ is a $p \times r$ known matrix of rank r, and the covariance matrix $\boldsymbol{\Sigma}$ is modeled as in (6.2) or its various generalizations. The sample mean and sample covariance of the data are $(\bar{\boldsymbol{Y}}, \mathbf{S})$. The focus is on the challenges one faces when computing the MLE of the LCM parameters for a given initial estimate $\hat{\boldsymbol{\Sigma}}$.

6.2.1 MLE of the LCM Parameters

Substituting for $(\mu, \boldsymbol{\Sigma})$ in the normal likelihood function, after some manipulation it can be expressed in terms of the regression parameters $\boldsymbol{\beta} = (\beta_1, \ldots, \beta_r)'$ and covariance parameters $\alpha = (\alpha_1, \ldots, \alpha_q)'$. The score equations for a given initial value $\hat{\boldsymbol{\Sigma}}$ are:

$$(X'\hat{\Sigma}^{-1}X)\hat{\beta} = X'\hat{\Sigma}^{-1}\bar{Y}, \tag{6.16}$$

$$[\text{tr}(\hat{\Sigma}^{-1}U_i\hat{\Sigma}^{-1}U_j)]\hat{\alpha} = [\text{tr}(\hat{\Sigma}^{-1}U_j\hat{\Sigma}^{-1}\mathbf{C})], \tag{6.17}$$

where $[\text{tr}(\hat{\Sigma}^{-1}U_i\hat{\Sigma}^{-1}U_j)]$ is a $q \times q$ matrix and $[\text{tr}(\hat{\Sigma}^{-1}U_j\hat{\Sigma}^{-1}C)]$ is a $q \times 1$ vector with indicated generic entries and

$$\mathbf{C} = \mathbf{S} + (\bar{Y} - \hat{\mu})(\bar{Y} - \hat{\mu})', \text{for} \quad \hat{\mu} = X\hat{\beta}, \tag{6.18}$$

see Problem 6.3. Evidently, the form of the score equations suggests the following iterative scoring method, solving in each step a set of linear equations [Anderson, 1973, 2003]:

1. Given an initial value $\hat{\Sigma}$, solve the first linear equations in β and compute $\mathbf{C}$,
2. Solve the second linear equations in α to arrive at the next (updated) estimate $\hat{\Sigma}$.

Unfortunately, in spite of simplicity of this iterative method, the following two computational problems persist:

- The need to monitor the values of $\hat{\alpha}$ to ensure the positive-definiteness of $\hat{\Sigma} = \hat{\Sigma}(\alpha)$,
- The iterative method is not guaranteed to converge.

However, there is a special class of linear Gaussian covariance models, denoted by Sz in honor of Szatrowski [1980], for which the maximum likelihood estimator has an explicit solution. The class consists of models where the covariance and precision matrices are linear combinations of the same basis matrices, but with different scalar coefficients:

$$\Sigma^{\pm 1} = \sum_{i=1}^{q} \alpha_i^{\pm} U_i, \tag{6.19}$$

Important examples of this class are the compound symmetry matrices and the Brownian motion tree models on the star tree with equal variances. For example, the 4×4 matrix $A = a_0 I + a_1 \mathbf{1}_4 \mathbf{1}_4' + a_2 U_2$, with

$$U_2 = \begin{pmatrix} 1 & 1 & 0 & 0 \\ 1 & 1 & 0 & 0 \\ 0 & 0 & 1 & 1 \\ 0 & 0 & 1 & 1 \end{pmatrix},$$

belongs to the Sz class, see Problems 6.6, 6.7 and 6.8 for more examples. The main result in Szatrowski [1980] shows that the MLE for linear Gaussian LCM has an explicit representation as a known linear combination of entries of the sample covariance matrix, if and only if the covariance belongs to the Sz class. Furthermore, for this restrictive Sz class, the MLE is the arithmetic mean of the corresponding elements of the sample covariance matrix and the scoring method in Anderson [1973] yields the MLE in one iteration when initiated at any positive definite matrix in the Sz class of covariances.

On the more practical side, an excellent review of the MLE procedures for the model (6.2) and their applications to the problem of testing homogeneity of the covariance matrices of several dependent multivariate normals are presented in Jiang et al. [1999]. They derive a likelihood ratio test, and show how to compute the MLE of $\boldsymbol{\Sigma}$, in both the restricted (null) and unrestricted (alternative) parameter spaces using SAS PROC MIXED software with several worked out examples.

6.2.2 MLE of Log-Linear and Cholesky Factors

Maximum likelihood estimation of the parameters in the log-linear covariance models (6.10) and their asymptotic properties are studied in Chiu et al. [1996] along with the analysis of two real data sets. An interesting application of such models to spatial autoregressive (SAR) models and some of its computational advantages are discussed in LeSage and Pace [2007]. A penalized least squares estimation of log-linear covariance models involving expansion in basis matrices is studied in Battey [2017], with an interesting finding that a sparse $\log \boldsymbol{\Sigma}$ potentially gives rise to a $\boldsymbol{\Sigma}$ which is significantly less sparse.

An algorithm for computing the MLE of the parameters (γ, λ) introduced in (6.11) of the Cholesky factors using the iterative Newton-Raphson algorithm with Fisher scoring is given in Pourahmadi [2000] along with the asymptotic properties of the estimators. A pleasant and unexpected result is the asymptotic orthogonality of the MLE of the parameters λ and γ, in the sense that their Fisher information matrix is block-diagonal; see Ye and Pan [2006], Pourahmadi [2004] and references therein.

6.2.3 MLE, OLS and GLS: Large p and $n = 1$

The case of $p \to \infty$ and $n = 1$ is of great interest in areas like time series analysis and finance where the risks or covariance matrices are time-varying, and one often relies on their rolling-window estimators in a very small neighborhood of a time point. This is in contrast to the classical setup where the consistency and asymptotic normality of the estimators of the parameters are established for p fixed and $n \to \infty$ [Anderson, 1973].

It is instructive to consider the LCM as the mean in the regression

$$YY' = \alpha_0 I_p + \alpha_1 U_1 + \ldots + \alpha_q U_q + \varepsilon, \tag{6.20}$$

for a centered column vector Y with the true covariance matrix $E(YY') = \boldsymbol{\Sigma}_0 = \boldsymbol{\Sigma}(\boldsymbol{\alpha}^0)$ corresponding to the true parameter vector $\boldsymbol{\alpha}^0$, and ε a random matrix with $E(\varepsilon) = 0$. Though, estimation of the $q + 1$ parameters based on a single observation is problematic, this issue disappears if additional information accrues in larger dimensions. In fact, Zou et al. [2017] study asymptotic properties of the maximum likelihood, the ordinary least-squares and the generalized least-squares estimators of the LCM coefficients when $n = 1$ and $p \to \infty$. A key point is that in LCM the number of parameters is fixed and does not depend on p. This is not the case in approximate factor models in Section 7.3 where the dimension p goes to infinity and rp, the number of parameters in the loading matrix depends on.

While the MLE is obtained by assuming normality of the data $Y \sim N_p(\boldsymbol{\mu}, \boldsymbol{\Sigma}(\boldsymbol{\alpha}))$, no distributional assumptions are needed for the OLS and GLS, other than existence of the first few moments. Specifically, the OLS estimator is defined as the minimizer of the Frobenius norm:

$$Q_{ols}(\boldsymbol{\alpha}) = ||YY' - \boldsymbol{\Sigma}(\boldsymbol{\alpha})||_F^2,$$

and its efficiency is improved by minimizing a corresponding weighted criterion

$$Q_{gls}(\boldsymbol{\alpha}) = \text{vec}(YY' - \boldsymbol{\Sigma}(\boldsymbol{\alpha}))(\widehat{\boldsymbol{\Sigma}}^{-1} \otimes \widehat{\boldsymbol{\Sigma}}^{-1})\text{vec}(YY' - \boldsymbol{\Sigma}(\boldsymbol{\alpha})),$$

where $\widehat{\boldsymbol{\Sigma}}^{-1} = \boldsymbol{\Sigma}^{-1}(\widehat{\boldsymbol{\alpha}})$ is a consistent estimator of $\boldsymbol{\Sigma}^{-1}$ and $\otimes$ represents the Kronecker product of the two indicated matrices.

For the existence and positive-definiteness of the above covariance matrix estimators, it is prudent to work with a nonempty parameter space $\mathcal{P} = \{\boldsymbol{\alpha} : \boldsymbol{\Sigma}(\boldsymbol{\alpha}) > 0\}$. It turns out that $\mathcal{P}$ is an open and nonempty set by imposing the following familiar conditions from regression and covariance estimation theory [Zou et al., 2017]:

(C1) For symmetric matrices $U_k \in \mathbb{R}^{p\times p}, k = 1, \ldots, q$, there exits $c > 0$ such that

(C2) There exists $c > 0$ such that $\lambda_p(\boldsymbol{\Sigma}_0) > c$, for any $p \geq 1$.

(C3) The matrix $p^{-1}[\text{tr}(\boldsymbol{\Sigma}_0^d U_k \boldsymbol{\Sigma}_0^d U_\ell)]_{(q+1)\times(q+1)}$ converges to a positive definite matrix $Q_d \in \mathbb{R}^{(q+1)\times(q+1)}$, respectively, for the integers $d = -2, -1, 0, 1$ with $\boldsymbol{\Sigma}_0^0 := I_p$.

Under these conditions, Zou et al. [2017] show that the MLE, OLS and GLS estimators are asymptotically normal with covariance matrices $2Q_{-1}^{-1}, 2Q_0^{-1}Q_1Q_0^{-1}$, and $2Q_{-1}^{-1}$, respectively. While the maximum likelihood estimator is computationally complex and the ordinary least squares estimator is inefficient, the generalized least squares estimator mitigates the computational burden and achieves the same asymptotic efficiency as the maximum likelihood estimator. The convergence rates of the covariance matrix estimation errors with respect to both the spectral norm and the Frobenius norm are obtained and they indicate that the resulting covariance matrix estimators are consistent as either p or n (or both) goes to infinity.

If it happens that the sample size $n > 1$, then the estimation can proceed replacing the right hand side of $Q_{ols}(\boldsymbol{\alpha})$ above by the obvious choice:

$$\sum_{i=1}^{n} ||\boldsymbol{Y}_i\boldsymbol{Y}_i' - \boldsymbol{\Sigma}(\boldsymbol{\alpha})||_F^2, \tag{6.21}$$

and earlier asymptotic distributions remain valid.

6.2.4 Estimation of Linear Entropic Models

Next, we review a good blend of topics related to estimation and conceptual developments. Given a sample of size n from a population with mean zero and covariance matrix $\mathbf{\Sigma}_0 \in \mathbb{S}_+$ and $\nabla\mathbf{\Sigma}_0 \in \mathcal{L}$ one may estimate the covariance matrix using the Gaussian log-likelihood, even for non-Gaussian data the normal log-likelihood can be viewed as a loss function which leads to an asymptotically optimal estimation procedure so long as all fourth order cumulants of the underlying distribution vanish [Browne, 1974]. While the Gaussian log-likelihood is a strictly concave function in $\mathbf{\Theta} = \mathbf{\Sigma}^{-1}$, this is not the case for the general entropic covariance models. In fact, optimizing the Gaussian log-likelihood could be quite complicated since the linear constraints in L translate into non-linear constraints in the coefficients. Its solution has led to the topic of estimation of linear covariance models via the dual MLE providing efficient alternatives to the MLE [Christensen, 1989, Kauermann, 1996, Lauritzen and Zwiernik, 2022]. An alternatively idea is to use least-squares/generalized least-squares estimators [Browne, 1974].

Bregman estimation procedure for linear entropic models offers a natural alternative to the MLE. It generalizes the use of Gaussian likelihood for linear models for precision matrix, and the least-squares estimation for linear covariance models. The key ingredient is the Bregman divergence:

$$D_g(\mathbf{S}, \mathbf{\Sigma}) = g(\mathbf{S}) - g(\mathbf{\Sigma}) - \text{tr}(\nabla g(\mathbf{\Sigma})\mathbf{S} - \mathbf{\Sigma}), \tag{6.22}$$

as a function of two covariance matrices. Note that the divergence $D_g(\mathbf{S}, \mathbf{\Sigma})$ is non-negative due to strict convexity of g, with equality if and only if $\mathbf{S} = \mathbf{\Sigma}$, see for example (2.47). Their explicit forms for four examples are given in the last column of Table 6.1. In particular, for the Gaussian entropy $g(\mathbf{\Sigma}) = -\log\det(\mathbf{\Sigma})$, the Bregman divergence reduces to the familiar Kullback-Liebler divergence between two mean zero normal distributions with covariances $\mathbf{S}$ and $\mathbf{\Sigma}$.

With $\mathbf{S}$ as the sample covariance matrix of the data, the Bregman estimator of $\mathbf{\Sigma}_0$ is obtained by minimizing the Bregman divergence over the parameters of the entropic model. More precisely, it is the global minimizer of $D_g(\mathbf{S}, \mathbf{\Sigma})$ subject to $\nabla g(\mathbf{\Sigma}) \in \mathcal{L} \cap \mathbb{L}_+$. A crucial aspect of solving the optimization problem is to show that the Bregman divergence is a strictly convex function of the gradient $\mathbf{L}$, and that when g is essentially smooth, then a unique minimizer exists if $\mathbf{S}$ is positive-definite.

The conjugate function of $g(\mathbf{\Sigma})$ is defined as

$$g^*(\mathbf{L}) = \sup\{\text{tr}(\mathbf{\Sigma L}) - g(\mathbf{\Sigma})\},$$

over $\mathbf{\Sigma} \in \mathbb{S}$. The third column of the table provides its form for the four examples. Since $\text{tr}(\mathbf{\Sigma L}) - g(\mathbf{\Sigma})$ is strictly concave, it is evident that domain of g^* contains $\mathbb{L}_+$. It follows directly from its definition that

$$g(\mathbf{\Sigma}) + g^*(\mathbf{L}) - (\mathbf{\Sigma}, \mathbf{L}) \geq 0,$$

for all $\mathbf{\Sigma}, \mathbf{L} \in \mathbb{S}$. But, for $\mathbf{\Sigma}, \mathbf{L} \in \mathbb{S}_+$, it can be shown that

$$g(\mathbf{\Sigma}) + g^*(\mathbf{L}) - (\mathbf{\Sigma}, \mathbf{L}) = 0 \iff \mathbf{L} = \nabla g(\mathbf{\Sigma}) \iff \mathbf{\Sigma} = \nabla g^*(\mathbf{L}).$$

Consequently, the second equivalence shows that ∇g and ∇g^* are inverses of each other in the sense that $\mathbf{L} = \nabla g(\boldsymbol{\Sigma})$ if and only if $\boldsymbol{\Sigma} = g^*(\mathbf{L})$.

The Bregman divergence can be written as

$$D_g(\mathbf{S}, \nabla g^*(\mathbf{L})) = g(\mathbf{S}) + g^*(\mathbf{L}) - \mathrm{tr}(\mathbf{LS}),$$

By slight abuse of notation replacing $\nabla g^*(\mathbf{L})$ by $\mathbf{L}$, we write $D_g(\mathbf{S}, \mathbf{L})$, as in Table 6.1. In analogy to log-likelihood function, we solve the constrained optimization problem:

$$\begin{aligned} &\max - g^*(\mathbf{L}) + \mathrm{tr}(\mathbf{LS}) \\ &\text{s.t.}\, \mathbf{L} \in \mathcal{L}, \end{aligned}$$

where $\mathcal{L}$ is an affine subspace defining linear constraints on a covariance. For example, if $g(\boldsymbol{\Sigma}) = -\mathrm{tr}(\boldsymbol{\Sigma} - \mathbf{S}\log(\boldsymbol{\Sigma}))$, then $\mathcal{L}$ is defined through the constraint imposed by (6.10).

This optimization framework naturally extends the setups of least-squares and maximum likelihood estimation methods.

6.3 Generalized Estimating Equations (GEE)

Accounting for correlations among responses when estimating the mean or regression parameters has always been a topic of great interest in statistics. Longitudinal studies since the 1980's have been one of the main source of generating correlated data over time where observations closer in time are more correlated than those far apart. Of course, time series, spatial, spatial-temporal and environmental studies are the more common areas of data analysis dealing with correlated data and stand to benefit greatly from the flexibility of GEE in handling data with diverse non-Gaussian distributions, correlation structures and data irregularities such as missing values, unbalanced data and outliers.

For Gaussian distributions, it is natural to use the likelihood function to estimate and carry out inference about the model parameters. Otherwise, the most popular method of estimating the regression parameters in the correlated data is the method of Generalized Estimating Equations (GEE) in Liang and Zeger [1986a]. It is quite popular in many application areas, very flexible and allows glaring departures from the basic assumptions of inference in the standard multiple linear regressions. The starting point of GEE is a regression model for the mean of vector of responses without making any distributional assumptions. It works well for continuous and discrete (binary and count) data. To account for the dependence in the data, GEE relies on the notion of *working correlation matrices* with a few unknown parameters to be estimated. Selection of a candidate working correlation matrix and estimation of its parameters continue to be a source of lively debate and challenge for GEE.

6.3.1 The GEE Setup

In longitudinal studies it is common to denote by Y_{ij} a response variable measured for the subject $i = 1, \ldots, n$ at times $t_{ij},\ j = 1, \ldots, T_i$. The response vector $\boldsymbol{Y}_i = (Y_{i1}, \ldots, Y_{iT_i})'$ collects the measurements over time for the ith subject. For the time being, we work with the balanced data with $T_i = T$ for all i.

The starting point of GEE is writing models for the mean and variance of Y_{ij}:

$$\mu_{ij} = E(Y_{ij}) = h(\boldsymbol{x}_{ij}^{'}\boldsymbol{\beta}),\ \text{Var}(Y_{ij}) = g(\mu_{ij}, \phi_j),$$

where $h(\cdot)$ and $g(\cdot)$ are known functions, $\boldsymbol{x}_{ij}$ is an r-vector of covariates and $\boldsymbol{\beta}$ is an unknown r-dimensional parameter vector, and ϕ_j's are the dispersion parameters. The GEE solves an unbiased estimating equation for $\boldsymbol{\beta}$, namely

$$\sum_{i=1}^{n} \dot{\mu}_i{}' V_i^{-1}(\boldsymbol{Y}_i - \boldsymbol{\mu}_i) = 0, \tag{6.23}$$

where $\boldsymbol{\mu}_i = E(\boldsymbol{Y}_i) = (\mu_{i1}, \ldots, \mu_{iT})'$ is the mean vector, with gradient vector $\dot{\boldsymbol{\mu}}_i = \partial\boldsymbol{\mu}_i/\partial\boldsymbol{\beta}$, and $V_i = \text{Cov}(\boldsymbol{Y}_i)$ is a $T \times T$ unknown and unstructured covariance matrix with too many parameters. Its empirical estimator based on the sample data could be unstable when the sample size is relatively small compared with a large number of covariance parameters. Indeed, the root of the equation (6.23) can be traced to the quasi-likelihood estimating equation of Wedderburn [1974].
The novel and fruitful idea of GEE in Liang and Zeger [1986a] was to replace $\text{Cov}(\boldsymbol{Y}_i)$ by a very structured matrix of the form $V_i = A_i^{1/2} R(\rho) A_i^{1/2}$, where A_i is the diagonal matrix of marginal variances of the Y_{ij}'s and $R(\rho)$ is a *working correlation matrix* such as the AR(1), CS, etc. from Section 3.4 with one or two parameters. These correlation parameters are usually estimated via the method-of-moments. For example, for binary responses, the diagonal entries of A_i are given by $\phi\mu_{ij}(1-\mu_{ij})$ where ϕ is a common dispersion parameter. Thus, the three components that determine and may contribute to the statistical performance of an estimator of $\boldsymbol{\beta}$ are the mean-variance functions (h, g) and the correlation matrix R.

The next example provides further details on the nature of the mean function h, and g the variance function for continuous, count and binary response variables.

Example 17 *1. Continuous Response Variable: The plausible mean-variance functions are the identity link for the mean, and the variance not depending on the mean as in the normal distribution,*

$$\mu_{ij} = E(Y_{ij}) = \boldsymbol{x}_{ij}^{'}\boldsymbol{\beta}, \quad Var(Y_{ij}) = \phi,$$

where $\boldsymbol{\beta}$ and the time-invariant scale parameter ϕ need to be estimated along with the correlation parameter ρ in

$$Corr(Y_{ij}, Y_{ik}) = \rho^{|k-j|},$$

where $\rho \in [0, 1]$.

2. *Count Response: The plausible functions for the count data collected from the Poisson-like family of distributions are the log link for the mean, and variance depending on the mean as in*

$$\log(\mu_{ij}) = \boldsymbol{x}'_{ij}\boldsymbol{\beta}, \quad Var(Y_{ij}) = \phi\mu_{ij},$$

where $\boldsymbol{\beta}$ *and the time-invariant scale parameter* ϕ *need to be estimated. The working correlation matrix can be taken as the AR(1) or CS.*

3. *Binary Response: For a binary or Bernoulli random variable with* $P(Y_{ij} = 1) = \mu_{ij}$*, the mean is the probability of success. Thus, one is in the familiar logistic regression setup with*

$$\log(\frac{\mu_{ij}}{1-\mu_{ij}}) = \boldsymbol{x}'_{ij}\boldsymbol{\beta}, \; Var(Y_{ij}) = \mu_{ij}(1-\mu_{ij}),$$

where the dispersion parameter $\phi = 1$*. The correlation matrix can be taken to be the AR(1) correlation matrix with one parameter. It is also common to use pairwise log odds ratio to introduce measures of association among the binary responses [Fitzmaurice et al., 2011, page, 298].*

It is known that with reasonable choices of the three components, the resulting GEE estimator of $\boldsymbol{\beta}$ is consistent even when the working correlation structure is misspecified. Unfortunately, there are simple cases for which the estimator of the correlation parameter ρ does not exist [Crowder, 1995]. More generally, estimation of ρ could be cumbersome, but since the target of inference is $\boldsymbol{\beta}$ and not ρ, it is not unusual somehow to bypass or downplay the estimation of the correlations. For more details, discussions and examples of popular distributions in the context of GEE, see Fitzmaurice et al. [2011, Chapter 11] .

Since misspecification is common in GEE, various properties of GEE estimators have been studied under misspecification of the working correlation structure [Crowder, 1995, Wang and Carey, 2003], and empirically when the number of covariates is diverging [Xu et al., 2022]. In addition, Wang et al. [2012a], Zheng et al. [2014], and Lv et al. [2015] have considered various penalized GEEs for variable selection and possibly high-dimensional data. Robust GEE estimators of the mean and covariance have been proposed in Zheng et al. [2013].

6.3.2 Hands-on R: GEE

In this section, we use the GEE method in Section 6.3, to illustrate and estimate the effect of ozone pollution on tree growth using the R package glmtoolbox [Vanegas et al., 2023]. The dataset spruces is borrowed from Diggle et al. [2002] and is available in the R package glmtoolbox. The dataset contains four variables on 79 trees. The variable size is the response variable that indicates the tree size. The explanatory variables treat and days that describe the atmospheric treatment (normal

TABLE 6.2
The first 6 observations of spruce dataset.

	tree	days	size	treat
1	N1T01	152.00	92.76	normal
2	N1T01	174.00	156.02	normal
3	N1T01	201.00	177.68	normal
4	N1T01	227.00	223.63	normal
5	N1T01	258.00	225.88	normal
6	N1T01	469.00	301.87	normal

with 25 trees or ozone-enriched with 54 trees) and the number of days after the treatment began, respectively. Finally, the tree variable identifies individual trees. Table 6.2 displays the first 6 observations. Examining box-plot of the data reveals that ozone suppresses tree growth, and that under the two types of atmosphere the mean as well as the dispersion of the tree size are non-decreasing and non-linear functions of the time. This suggests to model the mean to include a polynomial effect of the time and dummy variable to indicate the type of atmosphere in which the trees grew.

Following Vanegas et al. [2023] we use Gamma logarithmic link with variance function $V(\mu) = \mu^2$, and three different working correlation matrices (Independence, Exchangeable, and AR(1)) to fit the data.

```
##Import required packages and data
library(glmtoolbox)
data(spruces)

## Fit different working matrices
m1 <- glmgee(size ~ poly(days,4) + treat , id=tree ,
             family=Gamma(log), data=spruces)
m2 <- update(m1, corstr="Exchangeable")
m3 <- update(m1, corstr="AR-M-dependent(1)")
```

We use Correlation Information Criterion (CIC), Quasi-likelihood under Independence model Criterion, and Gosho-Hamada-Yoshimura's Criterion , QIC and GHYC, respectively, available in glmtoolbox to choose the more suitable structure for the correlation matrix.

```
a <- CIC(m1, m2, m3)
b <- QIC(m1, m2, m3)
c <- GHYC(m1, m2, m3)
cbind(a,QIC=b[,"QIC"],GHYC=c[,"GHYC"])
  Object         Correlation      CIC      QIC       GHYC
1     m1        Independence 23.43487 42067.65 116.41461
2     m2        Exchangeable 23.43487 42067.65  40.95970
3     m3 AR-M-dependent(1) 23.66161 42086.20  11.26152
```

From the output, it can be seen that first-order autoregression (AR-1) is more adequate structure for the correlation matrix. To summarize the estimation results for this model, we use the summary function.

```
summary(m3)
```

6.3.3 Quadratic Inference Functions (QIF)

The QIF method has the goal of improving the efficiency of estimators of the mean parameters by minimizing a quadratic function [Qu et al., 2000]. It employs LCM to approximate the inverse working correlation matrix as a linear combination of a few simple structured matrices.

QIF starts with approximating the inverse of a correlation matrix as an LCM-type linear combination

$$R^{-1} \approx \sum_{i=0}^{q} \alpha_i U_i,$$

where the basis matrices U_i's have 0−1 entries with $U_0 = I$. Note that this approximate working correlation matrix does not involve the mean-variance parameters as in GEE.

Substituting V^{-1} using the LCM form of R^{-1} in (6.23), the GEE is then approximated by the estimating function $\bar{g}_n = \frac{1}{n}\sum_{i=1}^{n} g_i$, which is a linear combination of the so-called extended scores

$$g_i = [\dot{\boldsymbol{\mu}}_i' A_i^{-1}(\boldsymbol{Y}_i - \boldsymbol{\mu}_i), \dot{\boldsymbol{\mu}}_i' A_i^{-1/2} U_1 A_i^{-1/2}(\boldsymbol{Y}_i - \boldsymbol{\mu}_i), \ldots, \dot{\boldsymbol{\mu}}_i' A_i^{-1/2} U_q A_i^{-1/2}(\boldsymbol{Y}_i - \boldsymbol{\mu}_i)]'.$$

Note that the $(q+1)r$ dimension of the vector $\bar{g}_n$ is greater than the number of unknown parameters r, so that $\boldsymbol{\beta}$ cannot be found uniquely by solving the approximate GEE, $\bar{g}_n = 0$. The QIF resolves this challenge and estimates $\boldsymbol{\beta}$ by minimizing a quadratic function

$$Q(\beta) = n\bar{g}_n' \Omega^{-1} \bar{g}_n,$$

where $\Omega = \text{Cov}(\bar{g}_n)$, and it is usually estimated by $W_n = n^{-1}\sum_{i=1}^{n} g_i g_i'$.

The quadratic function provides an objective function for the regression parameter estimation without requiring estimation of the basis matrix coefficients α_i's. Nevertheless, it does improve the efficiency of regression parameter estimation. It turns out that the QIF estimator of $\boldsymbol{\beta}$ is closely related to the generalized method of moments estimator [Hansen, 1982] and enjoys its asymptotic properties.

If within the QIF setup there is interest in estimating the correlation matrix itself or the coefficients α_i's, one can easily shift attention to selection of the correlation structure. For this, again the inverse of the correlation matrix is approximated by $\sum_{i=0}^{q} \mathbf{U}_i \boldsymbol{\alpha}_i$, where now it is a linear combination of groups of basis matrices $\mathbf{U}_i = \{U_{i,1}, \ldots, U_{i,q_i}\}$, each group representing a number of correlation structures and $\boldsymbol{\alpha}_i = \{\alpha_{i,1}, \ldots, \alpha_{i,q_i}\}$. In the absence of prior information about the correlation structure, one may use a linear representation of a simple and complete set of basis matrix with $0-1$ entries. The selection of the correlation structure R is performed through identifying nonzero entries of the vectors $\boldsymbol{\alpha}_i$, by minimizing the discrepancy between the estimating function using the sample correlation matrix $\tilde{R}^{-1}$ and the estimating function using its basis matrices representation. The discrepancy $\mathbf{D}_i$, between the two estimating functions for the ith cluster is measured by

$$\mathbf{D}_i = \dot{\boldsymbol{\mu}}'(\tilde{\boldsymbol{\beta}}) A_i^{-1/2}\{\tilde{R}^{-1/2} - \mathbf{U}_1\boldsymbol{\alpha}_1 - \ldots - \mathbf{U}_q\boldsymbol{\alpha}_q\} A_i^{-1/2}(\boldsymbol{Y}_i - \boldsymbol{\mu}_i(\tilde{\boldsymbol{\beta}})),$$

where $\tilde{\boldsymbol{\beta}}$ is the estimator of $\boldsymbol{\beta}$ via the GEE with $R = I$, and $\tilde{R}$ is the sample correlation estimator of the residuals based on $\tilde{\boldsymbol{\beta}}$. The Euclidean norm of $\mathbf{D} = (\mathbf{D}_1', \ldots, \mathbf{D}_n')'$ is expected to be sufficiently small if R^{-1} is approximated well by the selected groups of basis matrices which is achievable for q large. However, to balance model sufficiency and model complexity, one may need to select or estimate the correlation structure by minimizing the Euclidean norm of $\mathbf{D}$ and penalize models with too many basis matrices. Consequently, the coefficient vectors α_i's are estimated by minimizing the so-called group Lasso objective function

$$\sum_{i=1}^{n} \mathbf{D}_i'\mathbf{D}_i + np \sum_{j=2}^{q} p_\lambda(||\boldsymbol{\alpha}_j||_1),$$

where $p_\lambda(\cdot)$ is the smoothly clipped absolute deviation (SCAD) penalty; see Example 2 in Chapter 2 and Zhou and Qu [2012] for more details. A similar idea for log-linear covariance models involving basis matrix expansion is presented in Battey [2018].

Since the introduction of QIF in [Qu et al., 2000], the idea has attracted increasing attention and been studied in Song et al. [2009], Han and Song [2011] and Westgate and Braun [2013]. Qu and Li [2006] proposed an extension of QIF to varying coefficient models. More recently, Lai et al. [2020] considered feature screening of QIFs for high-dimensional longitudinal data, while Dumitrescu et al. [2021] proposed a survey-weighted QIF approach for longitudinal survey data. Luo and Song [2019], Luo et al. [2023] have extended QIF estimation to the streaming data setting.

A recent, detailed and profound study of the theoretical properties of QIF estimators and comparison with GEE estimators is presented in Muller et al. [2024]. Their sobering conclusion is that the choice of working correlation can have surprisingly large effects on the performance of the GEE estimators.

6.3.4 Joint Estimation of Mean-Correlation Parameters

The classical GEE literature often treats the working correlation matrix either as known or as a nuisance parameter [Liang and Zeger, 1986a, Qu et al., 2000]. In this section as a departure from the classical setup, both the mean and correlation parameters are treated as equally important in the modeling process.

Separate estimation of $\boldsymbol{\alpha}$ and $\boldsymbol{\beta}$ has been abandoned in Tho et al. [2024] by joining their two estimating equations. The correlation matrix of the response vector is an LCM-type model in terms of similarity matrices of the covariates:

$$R(\boldsymbol{\rho}) = I_p + \rho_1 U_1 + \ldots + \rho_q U_q,$$

where I_p is the $p \times p$ identity matrix, U_i's are some similarity matrices and $\boldsymbol{\rho} = (\rho_1, \ldots, \rho_q)'$ is the vector of correlation parameters. These are restricted so that the resulting matrix is a valid correlation matrix, namely being positive-definite with 1's on the diagonal and off-diagonal entries in $[-1, 1]$.

The correlation parameters have simple, direct interpretation as assessing the impact of the similarities on the correlation between responses. For example, in ecology where similarity matrices are constructed from trait information of different species, a higher, positive value of ρ_k implies that, conditional on the other trait values, two species with more similar values in their kth trait variable are expected to have a stronger positive correlation after accounting for differences in their mean response.

The vector of correlation parameters can be estimated by minimizing (6.21) or its weighted version. It is often desired to do inference on both the mean and the correlation parameters. Of course, this requires an asymptotic theory for their joint estimators. When the number of responses p grows with the sample size n, Tho et al. [2024] seems to be the first to establish estimation consistency and asymptotic normality of the mean-correlation parameter estimators with differing rates of convergence for each group of parameters. As expected, simultaneous estimation of both sets of parameters substantially increases the difficulty in deriving the theoretical properties of their estimators.

6.4 Three Case Studies

In this section through three case studies from three distinct application areas we illustrate the role of LCM, GLM and their extensions in modeling covariance matrices using covariates. Even though they are collected in this chapter, we refer to them throughout the book.

6.4.1 Case Study I: Finance and Portfolio Selection

Covariances of asset returns are important for risk management and portfolio selection in financial markets. The rate of returns of stocks are usually studied for their dependence and volatility and put to use in portfolio optimization and risk forecasting. Their covariance matrix is known [Fama and French, 1992] to depend on the relative amounts of firm and market level information capitalized into stock prices. Therefore, it is desirable to model or explain such a covariance matrix in terms of the firm's auxiliary information or fundamentals.

The data for the first example from Zou et al. [2017] is the quarterly returns of $p = 660$ stocks in Shanghai Stock Exchange and Shenzhen Stock Exchange from 2007 to 2011, with $T = 20$ quarters in total. For each stock (firm) the auxiliary information are:

- IND (industry classified into 14 categories);
- LEV (leverage computed by liability-to asset ratio)
- CF (cash flow of the firm);
- SIZE (measured by the logarithm of market value);
- BM (book-to-market ratio)

The authors [Zou et al., 2017] employ the LCM in (6.2) to model the covariance matrix through the similarity matrices of the five covariates. They assess and compare the performance of the portfolio selection via the Markowitz (mean-variance) optimization [Markowitz, 1952] relative to other covariance estimation methods like the Ledoit and Wolf [2004]'s shrinkage estimator.

The stock data for the tth quarter is denoted by $(Y_t, \boldsymbol{x}_t)$, where $Y_t \in \mathbb{R}^p$ and $\boldsymbol{x}_t = (\boldsymbol{x}_t^{(1)}, \ldots, \boldsymbol{x}_t^{(5)})' \in \mathbb{R}^{p\times 5}$. The returns (in percentages) of the 660 stocks is standardized by subtracting their sample means. The corresponding similarity matrices U_k, $k = 1, \ldots, 5$ are constructed as follows: For the categorical variable IND, the off-diagonal element of its similarity matrix U_1 is set to be 1 if two stocks belong to the same industry, and 0 otherwise. This is slightly different from that of using similarity matrices for continuous covariates. Before constructing the similarity matrices for the other four variables, they were standardized to have a zero mean and unit variance for each given quarter. For the other four covariates corresponding to $k = 2, \ldots, 5$, the off-diagonal elements of the similarity matrices U_k's are set to $\exp\{-(x_{j_1k} - x_{j_2k})^2\}$, where $j_1 \neq j_2$ represent different stocks, and diagonal elements are set to zero.

The covariance matrix of returns often exhibits time-varying characteristics so that $\boldsymbol{\Sigma}_t = \mathrm{Cov}(Y_t)$ is not fixed. Thus, essentially, there is only a single data point available at time t to fit the LCM and estimate the $\boldsymbol{\Sigma}_t$ or the case of the sample size $n = 1$. Fortunately, with a fixed number of parameters in the LCM and a large $p = 660$, one can effectively rely on information accumulating for large p, instead of large n, to estimate efficiently the α_k's in (6.2). In fact, Zou et al. [2017] rely on this observation and show that the (feasible) generalized least-squares estimator improves the efficiency of the ordinary least-squares estimator and avoids the computational challenges of the maximum likelihood estimation; see Section 6.2.

Any estimated covariance matrix can be used to determine optimal portfolio weights in the portfolio selection method. The problem of portfolio selection aims to construct a portfolio or find the possibly nonnegative weights $w = (w_1, \ldots, w_p)'$ so that the risk of the portfolio defined as $w'\boldsymbol{\Sigma}_t w = \mathrm{Cov}(w'Y_t)$ is as small as possible. For a fixed estimated covariance matrix, this amounts to the following optimization problem [Markowitz, 1952]:

$$\begin{aligned} &\min_w w'\hat{\Sigma}w, \\ &\text{s.t. } w'\mathbf{1} = 1, \end{aligned} \tag{6.24}$$

where $\mathbf{1} = (1, \ldots, 1)'$, and the weight w represents individual investments to ensure full capital allocation, i.e., $w'\mathbf{1} = 1$. This optimization problem does not impose any short-sale constraints on w and has a closed form solution. If the no-shortsale constraint $w \geq 0$ is enforced, then numerical optimization becomes necessary.

In an elaborate study of the data of this example, Zou et al. [2017] demonstrate, among other important findings, that the out-of-sample portfolio returns, calculated using a covariance matrix estimator obtained from the LCM, outperform the returns from a portfolio constructed using the Ledoit and Wolf [2004]'s shrinkage estimator of the covariance matrix.

There has been an ongoing debate on the choice between (mean-variance) optimal weights and the naive diversification (equal weights) portfolio. Recently, Heide et al. [2024] investigates the role of various covariance estimators on the mean-variance portfolio selection. They construct optimal commodity futures portfolios based on 12 promising covariance matrix estimators (sample, shrinkage, sparse,...) and compare their out-of-sample investment performance to the simple, equally weighted investment strategy using bootstrap testing. Somewhat surprisingly, their conclusion is that neither the naive allocation approach nor the advanced covariance estimators can outperform the traditional sample covariance matrix. Their empirical result seems to be fairly robust to modifications of the research design (including alternative investigation periods, data frequencies, estimation window sizes, holding period lengths, weight constraint specifications, and transaction cost levels), but their conclusion contradicts the recurrent suggestion of the literature of this area that the sample covariance matrix should not be used for the purpose of portfolio optimization. If nothing else, this calls for further well-designed research in the area.

In the context of multivariate time series data from finance, realizing that the covariance matrix of returns is often heterogeneous or time-varying, a plausible alternative is developing dynamic models for $\mathbf{\Sigma}_t = \mathrm{Cov}(Y_t)$ where the time t is effectively treated as a covariate. An early version of this idea, with positive-definiteness in mind, is the simplest BEKK model [Engle and Kroner, 1995] for the sequence $\{\mathbf{\Sigma}_t\}$ defined by

$$\mathbf{\Sigma}_t = \mathbf{B}_0 + \mathbf{B}_1\mathbf{\Sigma}_{t-1}\mathbf{B}_1' + \mathbf{B}_2(Y_{t-1}Y_{t-1}')\mathbf{B}_2', \tag{6.25}$$

where $\mathbf{B}_0$ is a positive-definite (intercept) matrix, $\mathbf{B}_i, i = 1, 2$ are two arbitrary $p \times p$ matrices, see also (6.7). It was quickly realized that such models, cannot handle even moderate-size portfolios. A simpler version of it for correlation matrices with scalar coefficients, known as the dynamic conditional correlation (DCC) models [Engle, 2002] and [Engle et al., 2019] can deal with larger portfolios. A simple DCC for a quasi-correlation matrix Q_t is defined by

$$Q_t = (1 - \alpha - \beta)C + \alpha\mathbf{s}_t\mathbf{s}_t' + \beta Q_{t-1}, \tag{6.26}$$

where the matrix C is the (unconditional) correlation matrix of the returns, $\mathbf{s}_t$ is the vector of standardized returns and (α, β) are the scalar DCC parameters. The matrix Q_t can be interpreted as a conditional pseudo-correlation matrix, or a conditional covariance matrix of standardized residuals. In general, it is not a correlation matrix since its diagonal elements are not exactly one. Fortunately, a solution of such a model can be transformed to the conditional correlation and the conditional covariance matrices via

$$R_t = \mathrm{diag}Q_t^{-1/2}Q_t\mathrm{diag}Q_t^{-1/2}, \; \mathbf{\Sigma}_t = D_tR_tD_t, \tag{6.27}$$

where D_t is the diagonal matrix of standard deviations of the returns.

Note that the correlation matrix C serves as the target or intercept of the ARCH/-GARCH dynamics in (6.26). Its estimation from data is important so far as performance of a portfolio is concerned. While the sample correlation matrix is a natural choice, it is well-known that it is a notoriously bad estimator when the sample size is small relative to the dimension p. Improved estimators of C based on random matrix theory (RMT) as in Heide et al. [2024] are ideal replacement. Once C is determined, the DCC parameters (α, β) are estimated using composite likelihood as in Engle et al. [2019]. Note that the DCC model (6.26) is a dynamic version of the LCM studied earlier.

6.4.2 Case Study II: Forecasting Electricity Net-Demand in UK

Forecasting regional electricity net-demand is essential for reliable and economic power system operation, and energy trading. It is well-known that electricity net-demand depend on many environmental covariates like region, temperature, wind speed, time of year (day, weak, months, ...). For a large country, instead of forecasting region by region, it is more prudent to focus on the more general task of forecasting the joint distribution of the net-demands which requires accounting for the cross-regional (spatial) and temporal dependencies.

For example, the Great Britain (GB) is divided into $p = 14$ regions from North to South which amounts to a sensible partial ordering (direction) for this problem. Figure 6.2 displays the selected regions and boundaries using a publicly available data from http://data.nationalgrideso.com. In the figure, the numbers are the labels for the regions in the partial ordering, and the positions in the response vector $Y_t = (Y_{t,1}, \ldots, Y_{t,14})'$ at time t. Each entry in the response vector is the standardized net-demand of each region viewed as a time series measured at 30-minute intervals from the 1st of January 2014 to the 31st of December 2016, a total of $n = 83024$ observations for each region. The following weather-related and socio-economic covariates (for more details, see Gioia et al. [2024]) are also available over the period 2014-2018.

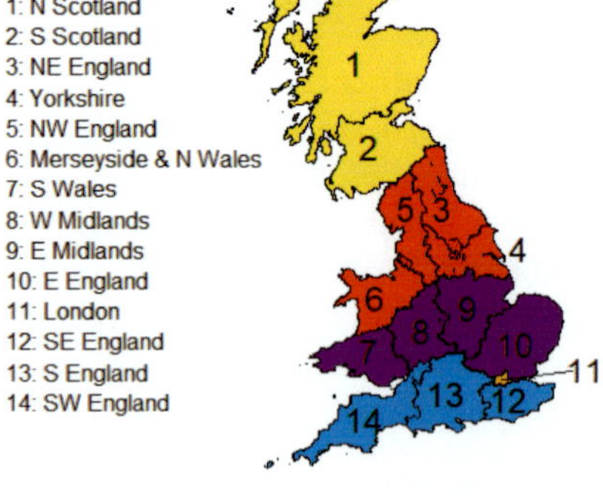

FIGURE 6.2
A map of the 14 regions of Great Britain electricity grid. The numbers correspond to a partial ordering and the position in the response vector $\boldsymbol{y} \in \mathbb{R}^{14}$.

- Day of the week (dow_i),
- As above but with additional levels for public holidays(dow^+_{ij}),
- Time since the 1st January 2014 (t_i),
- School holidays with three levels to distinguish Christmas (shol_{ij}),
- Wind generator capacity in GB,
- Time of day (tod),
- Day of the year (doy),
- Day-ahead electricity price (n2ex),
- Net-demand at a 24 hours lag,
- Daily mean precipitation,
- Temperature,
- 48 hours rolling mean of the temperature,
- Product of mean solar irradiance and solar generation capacity,
- Mean wind speed at 10 meters,
- Mean wind speed at 100 meters.

The response vector $\boldsymbol{Y}_i, i = 1, \ldots, n$ is spatially and temporally correlated, and it seems plausible to start by assuming that it has a multivariate normal distribution: $\boldsymbol{Y}_i \sim N_{14}(\mu_i, \boldsymbol{\Sigma}_i)$. Its time-varying mean and covariance matrix could be modeled using the data on regional net-demand in GB and the available covariates. It is well-known that incorporating covariates in modeling a covariance matrix and ensuring its positive-definiteness are challenging tasks. However, the modified Cholesky decomposition of a covariance matrix readily resolves both issues, particularly when there is a sensible partial ordering (North to South) of the 14 regions of GB as in Figure 1.

Following the notational convention in Section 6.1.2.1 and Gioia et al. [2024], the MCD of $\boldsymbol{\Sigma}_i$ is written as $\boldsymbol{\Sigma}_i^{-1} = T_i' D_i^{-2} T_i$, where D_i^2 is a diagonal matrix and T_i a unit lower triangular matrix. We collect the 14 mean, and 105 unconstrained parameters of (T_i, D_i) in the 119-vector $\eta_i, i = 1, \ldots, n$, where the first 14 entries $\eta_{ij}, j = 1, \ldots, 14$ stand for the means of the regions, and the remaining $\eta_{ij}, j = 15, \ldots, 119$ will stand for the unconstrained parameters of D_i with $D_{jj} = \exp(\eta_{j+14}), j = 1, \ldots, 14$ and T_i with T_{ij}, $1 \leq j < i \leq 14$ correspond to η_{49} to η_{119}, respectively. It is important to keep in mind that the effects acting on D_i are not directly modeling the regional variances (i.e., the diagonal elements of $\boldsymbol{\Sigma}_i$), but the residual variance of the net-demand in each region, after having conditioned on the preceding regions. Similarly, the effects acting on T_i modulate the dependence between regions but they do not directly control correlations, which depend on the elements of both Cholesky factors.

The parameters of the Cholesky factors can be modeled using generalized additive models [Hastie and Tibshirani, 1990] or generalized additive model for location, scale and shape [Rigby and Stasinopoulos, 2005]. They are popular in the electricity demand forecasting [Fan and Hyndman, 2012], partly because they strike a balance between predictive performance and interpretability. Evidently, model selection for the means are easier and more intuitive, but more challenging for the entries of the Cholesky factors when there are a large number of covariate effects. For example, using net-demand data, one of the postulated models for the MCD parameters is:

$$\eta_{ij} = g_{j1}(t_i) + g_{j2}(t_i^2) + g_{j3}(\text{dow}_i) + f_{j1}^{10}(doy_i) + f_{j2}^{10}(tod_i) +$$
$$\text{wcap}_i \times f_{j3}^{5}(\text{wsp}_{i\ell_j}^{100}) + f_{j4}^{5}(\text{irr}_{i\ell_j}) + f_{j5}^{5}(\text{temp}_{i\ell_j}) + f_{j6}^{5}(\text{rain}_{i\ell_j}^{1/2}) + f_{j7}^{5}(n2ex_i),$$

where g_{jk}'s are the parametric (linear) effects and f_{jk}'s are the smooth effects captured by splines; the spline bases dimensions are indicated by the superscripts. The subscript ℓ_j indicates the row of D_i or T_i on which the j-th linear predictor appears, with $\text{wsp}_{i\ell_j}^{100}$, $\text{irr}_{i\ell_j}$ and so on the weather forecasts corresponding to the ℓ_j-th region.

The regression interpretation of the MCD parameterization is helpful in specifying the candidate weather forecasts and modeling the non-redundant entries of the rows of MCD. Indeed, the forecasts for the North of Scotland which corresponds to the first region in Figure 6.2, are used to model only $(D_i)_{11}$, those for the South of Scotland the second region in Figure 6.2, are used to model $(D_i)_{22}$ and second row of T_i, and so on.

The data from 2014 to 2016 is used to generate a large list of candidate covariate effects, ordered in terms of decreasing importance in cumulative log-likelihood. We choose the number of effects to add to the final multivariate Gaussian model, that is where to stop along the ordered effect list, by maximizing the out-of-sample predictive performance on 2017 net-demand.

Some noteworthy model selection results captured in Figure 6.3 are summarized next:

1. Confirmation that most of the effects act on the diagonal elements of D_i and that the time of day, tod_i, affects all such elements. It is not surprising to see that the cumulative log-likelihood gain of the tod_i effect is particularly large in more urbanized areas, such as the Midlands (Regions 8 and 9) and London (Region 11). The time of day is used to model also several elements of T_i. The strongest such effect in terms of cumulative log-likelihood gain, acts on the element corresponding to London and the West Midlands (Regions 11 and 8). The effect of the day of the year, doy_i, is used to model many elements of D_i and T_i. Thus, it should not be surprising to see this effect appearing on the 8th and 11th row of Figure 3, which correspond again to the highly urbanized West Midlands (Region 8) and London (Region 11).

2. In accordance with intuition, wind speed, wsp_{ij}^{100}, is selected to model the elements of D_i corresponding to windy regions, such as the South East of England (Region 12), and the North (Region 1) and South (Region 2) of Scotland.

3. The time of day and solar irradiance, irr_{ij} , are both strongly related to solar energy production, hence the effects of both variables are selected to model the elements of D_i corresponding to several Southern regions, which have high solar generation capacity.

4. The set of effects, selected by the semi-automated procedure proposed in Gioia et al. [2024], matches intuition in many respects. However, more than half of the

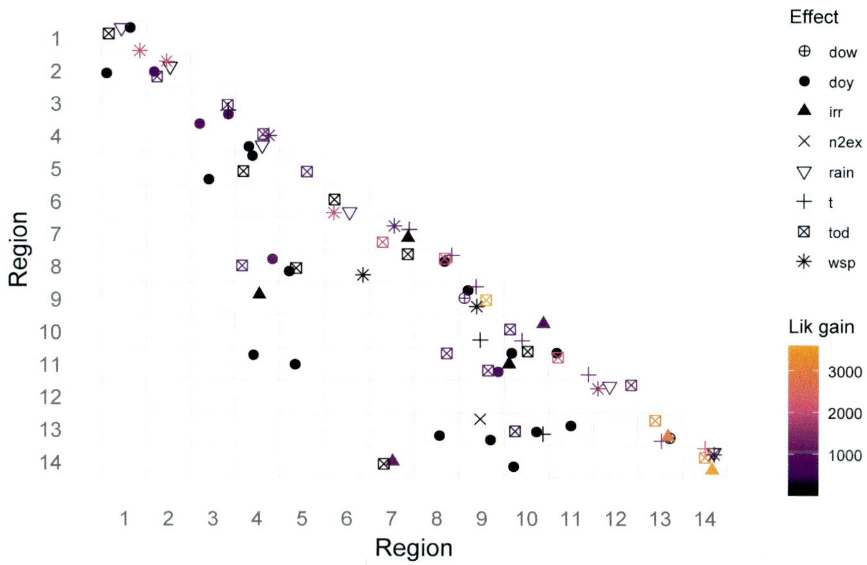

FIGURE 6.3
Model selection results. The diagonal corresponds to the elements of $\boldsymbol{D}_i$, the rest to those of $\boldsymbol{T}_i$. The symbols represent different effects and their color is determined by the cumulative log-likelihood gain. Regions 1 - 14 correspond to the regions listed in Figure 6.2.

selected effects are related to calendars variables, namely progressive time, time of day, day of year and day of the week. Further, external temperature has not been selected to model any element of T_i or D_i. Hence, it is interesting to assess how the predictive performance of the model depends on the set of candidate covariates discussed above.

Having chosen the model structure, then one may explore the model output and evaluate the accuracy of the resulting forecasts on 2018 data. Empirical results in Gioia et al. [2024] show that the time-varying additive covariance matrix models significantly outperform, in terms of forecast performance on the test set, a model where the covariance matrix is static.

6.4.3 Case Study III: Brain Functional Connectivity

Analysis of dependency among multiple time series of brain signals from distinct regions of the brain is called the functional connectivity (FC) of the brain [Friston, 2011]. The task of developing a brain network and displaying its functional connectivity starts with defining a set of nodes corresponding to various brain regions of interest (ROIs), and attaching to each node a time series of brain imaging data. Once the nodes are labelled, the network connections or edges between them are estimated by relying on various measures of statistical dependencies from multivariate time series analysis. These measures could be either in the time-domain (autoregressive models, Granger's causality) or the spectral-domain (spectral coherence, time series graphical models), see Chapter 11.

Some of the statistical techniques used are usually time series analogs of network construction for independent multivariate data which may use Pearson's correlation coefficients [Hutchison et al., 2013], partial correlations or Gaussian graphical models [Whittaker, 1990]. Historically, they have not employed the subject level covariates or model the heteroscedasticity in the construction of the FC networks. Though it is recognized that the subject-level covariates are playing an important role in modern FC analysis.

PCA-based methods are also popular in FC analysis and modeling covariance matrices because of their simplicity, desirable interpretability, and their potential ability to incorporate covariates. Decomposing covariance matrices into separate components can enable the identification of coherently active brain subnetworks [Poldrack et al., 2011], and usually a few principal components are needed to explain the variation in neuroimaging data. Interestingly, Zhao et al. [2019]'s covariate-assisted PCA method (see Section 6.1.2.2) is designed to model covariance matrices using covariates so that changes in functional connectivity are explained in terms of demographic and clinical factors such as age, gender, and cognitive behavioral functions. Fortunately, this also makes it possible to compare FCs among different subgroups.

Here we apply covariate assisted PCA (see Section 6.1.2.2 Smith et al. [2009] and Zhao et al. [2019]), to compare the connectivity of short sleepers ($\leq$ 6 hours) with conventional sleepers (7 to 9 hours) and male with female participants from the Human Connectome Project (HCP) [Van Essen et al., 2013]. We use the functional-resting fMRI (rfMRI) data of 730 subjects (489 conventional and 241 short sleepers) where each subject underwent 4 complete 15-minute sessions (with repetition time TR = 750ms, which corresponds to 1200 time points per session for each subject). Consequently, for each subject we observe 4800×15 time series. The final rfMRI time-series data has been preprocessed through temporal and spatial preprocessing using independent component analysis [Van Essen et al., 2013]. For illustration, we focus on 15 components or parcels displayed in Figure 6.4.

These 15 parcels correspond to visual areas (R1, R3, R4, and R8), sensorimotor (R7 and R11), cognition-language (R2, R5, R10, and R14), perception-somesthesis-pain (R2, R6, R10, and R14), cerebellum (R9), executive control (R12), auditory (R12 and R13), and default network (R15). For this experiment, the design matrix has three columns corresponding to the intercept, sleep duration, and gender. The off-diagonal elements in covariance matrix $\mathbf{\Sigma}_i$ represent the pairwise functional connectivity between node networks and $\mathbf{\Sigma}_i$ represents the brain functional architecture of subject i.

The number of projection directions, which turns out to be 1 in our case, is arrived at using average deviation from diagonality criterion [Zhao et al., 2019], where for a positive definite symmetric matrix A, the deviation from diagonality is defined as $\det\{\text{diag}(A)\}/\det A$.

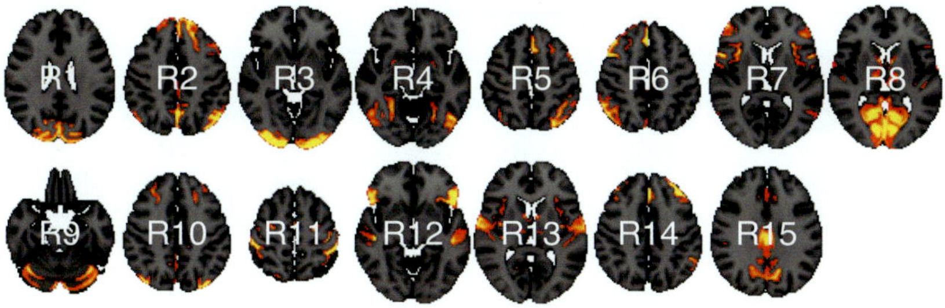

FIGURE 6.4
Parcels derived from spatial group-ICA with 15 components. These parcels corresponds to visual areas (R1, R3, R4, and R8), sensorimotor (R7 and R11), cognition-language (R2, R5, R10, and R14), perception-somesthesis-pain (R2, R6, R10, and R14), cerebellum (R9), executive control (R12), auditory (R12 and R13), and default network (R15) [Smith et al., 2009].

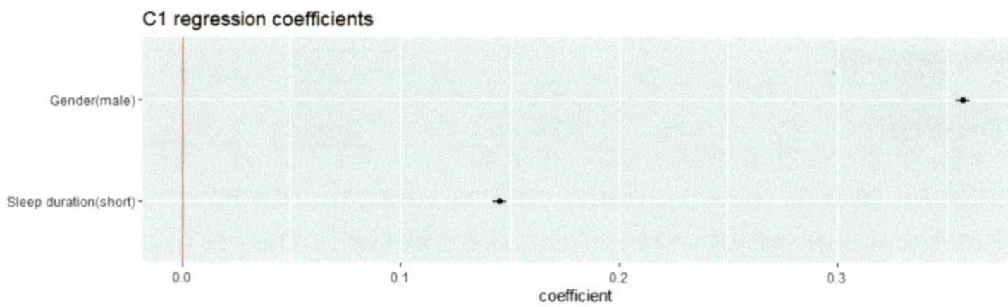

FIGURE 6.5
Regression coefficients associated with the component.

Figure 6.5 presents the model coefficients β with 95% confidence intervals, obtained from bootstrapping [Zhao et al., 2019]. Both covariates associated with gender and sleep duration are significant. The loading coefficients γ are summarized in Figure 6.6 with corresponding brain parcels on the negative and positive sides.

From (6.13), for subject i with subject-specific pairwise correlation matrix $\mathbf{\Sigma}_i$, the log-variance of the kth component can be written as

$$\log((\boldsymbol{\gamma}^{(k)})'\mathbf{\Sigma}_i\boldsymbol{\gamma}^{(k)}) = \log(1 + \sum_{j \neq j'}^{p} \gamma_j^{(k)}\gamma_{j'}^{(k)}\sigma_{jj',i}) = \beta_0^{(k)} + x_i\beta_1^k + u_i^{(k)} \quad (6.28)$$

Thus, for any pair (j, j') the interpretation of the results depends on the signs of the regression coefficient β and corresponding product of loadings coefficients $\gamma_j^{(k)}\gamma_{j'}^{(k)}$. In particular, if $\beta_i > 0$, then two parcels with the same loading sign have a higher FC and a lower FC if the loading signs are different. Therefore, both short sleepers and males have a higher FC than conventional sleepers and females. For example, FC between cerebellum R9 and visual R1 areas is weaker for short sleepers, compared to conventional sleepers. A similar finding was also reported by [Park, 2023].

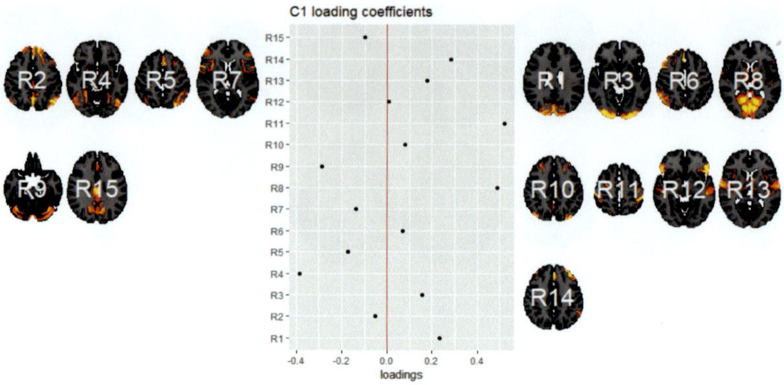

FIGURE 6.6
Loading coefficients associated with the components

6.5 Problems

6.1 Use the definition of the covariance matrix of a random vector in (3.2) and show that when the mean vector of responses is expressed in terms of covariates, namely $\mu = X\beta$, then the resulting covariance matrix could be a function of covariates. Discuss the connections with quadratic covariance regression models in Hoff and Niu [2011].

6.2 Show that in the covariance regression models, the restriction that the difference between $\mathbf{\Sigma}(x)$ and the baseline matrix A to be a rank-1 matrix amounts to requiring that the extreme residuals of one component of the response vector Y to occur with extreme residuals of the other components [Hoff and Niu, 2011].

6.3 Derive the score equations (6.16) for $(\alpha, \boldsymbol{\beta})$.

6.4 Write the covariance matrix in (6.6) in the LCM form by identifying the matrices U_i's.

6.5 Show that the compound symmetry matrices and the Brownian motion tree models on the star tree with equal variances belong to the Sz class of LCMs.

6.6 Let $\mathbf{\Sigma}$ have a linear representation of the form

$$a_0 I + a_1 P_1 + \ldots + a_m P_m,$$

for some scalars a_j's where the P_j's are projection matrices closed under multiplication in the sense that, for every (i, j), there exists k such that $P_i P_j = P_k$; that is, the pairs of subspaces corresponding to the projections are either nested or orthogonal.

(a) Show that

$$\boldsymbol{\Sigma}^{-1} = b_o I + b_1 P_1 + \ldots + b_m P_m,$$

where the coefficients b_j's are found by solving the equations generated by the relationship

$$\boldsymbol{\Sigma}\boldsymbol{\Sigma}^{-1} = I.$$

(b) Show that if the P_j's are orthogonal and idempotent, then

$$a_0 b_0 = 1, \ a_0 b_j + a_j b_0 + a_j b_j = 0, \ \text{for every } j > 1.$$

(c) If $a_0 = 1$, then $b_0 = 1$ and $b_j = -a_j/(1 + a_j)$ for all $j \geq 1$.

6.7 Consider the 4×4 matrix $A = a_0 I + a_1 U_1 + a_2 U_2$, where I is the identity matrix, all entries of U_1 is 1 and

$$U_2 = \begin{pmatrix} 1 & 1 & 0 & 0 \\ 1 & 1 & 0 & 0 \\ 0 & 0 & 1 & 1 \\ 0 & 0 & 1 & 1 \end{pmatrix}.$$

Show that it belongs to the Sz class by checking that

(a) $P_1 = 0.25 U_1$ and $P_2 = 0.5 U_2$ are projection matrices onto the subspaces generated by $(1, 1, 1, 1)'$ and $\{(1, 1, 0, 0), (0, 0, 1, 1)\}$, respectively. Furthermore, the subspaces are nested, with $P_1 P_2 = P_2 P_1 = P_1$.

(b) Find the inverse A using the result from the previous problem.

6.8 For $a > b$, consider the matrix

$$A = \begin{pmatrix} a+b & 0 & a-b \\ 0 & 2a & 0 \\ a-b & 0 & a+b \end{pmatrix} = 2a \begin{pmatrix} \frac{1}{2} & 0 & \frac{1}{2} \\ 0 & 1 & 0 \\ \frac{1}{2} & 0 & \frac{1}{2} \end{pmatrix} + 2b \begin{pmatrix} \frac{1}{2} & 0 & \frac{-1}{2} \\ 0 & 0 & 0 \\ \frac{-1}{2} & 0 & \frac{1}{2} \end{pmatrix}.$$

(a) Show that the expression in the right hand side is its spectral decomposition corresponding to the eigenvalues $2a$ and $2b$ with multiplicities 2 and 1, respectively.

(b) Show that its inverse is given by

$$A^{-1} = \frac{1}{4ab} \begin{pmatrix} a+b & 0 & b-a \\ 0 & 2b & 0 \\ -b-a & 0 & a+b \end{pmatrix} = \frac{1}{2a} \begin{pmatrix} \frac{1}{2} & 0 & \frac{1}{2} \\ 0 & 1 & 0 \\ \frac{1}{2} & 0 & \frac{1}{2} \end{pmatrix} + \frac{1}{2b} \begin{pmatrix} \frac{1}{2} & 0 & \frac{-1}{2} \\ 0 & 0 & 0 \\ \frac{-1}{2} & 0 & \frac{1}{2} \end{pmatrix},$$

which has the same form as A.

(c) Find another example of such 3×3 matrices. Let $\mathcal{A}$ be the set of all such matrices. Show that $\mathcal{A}$ is an algebra in that if $A \in \mathcal{A}$ then $A^2 \in \mathcal{A}$.

6.9 Investigate and discuss how does the variances in (6.12) relate to the eigenvalues of the covariance matrices? What about the connection between γ and the eigenvectors?

7

PCA and Factor Models

When it comes to dimension reduction, principal component analysis (PCA) is a simple and yet the most powerful method using the eigendecomposition of covariance matrices. As a fundamentally important statistical tool for high-dimensional data analysis, it searches for those directions (linear combinations of the *observed* variables) carrying the most information or having the largest variance. The closely related factor models seek to explain the high correlations in the multivariate data through linear combinations of a few common latent (*unobserved*) factors. In spite of this distinction, PCA-based estimation methods are the backbone of computational algorithms for estimating the factor loadings and scores in the classical and modern approximate factor models. The exact or classical orthogonal factor models decompose covariance matrices as low rank plus diagonal matrices, by contrast the modern approximate factor models, originally inspired by high-dimensional data arising in business and economics, lead to the important class of low-rank plus sparse covariance matrices. These decompositions are of great interest in the modern high-dimensional statistics and machine learning. We present a simple method, known as the Principal Orthogonal ComplEment Thresholding (POET) estimator, which applies the additional step of adaptive thresholding to the PCA residuals to arrive at a sparse estimate of the error covariance matrix. The POET estimator of the loading matrix is not usually sparse, but its large number of entries can be reduced either via regularized likelihood estimation or modeling them using covariates. Fortunately, a general class of covariance regressions [Fox and Dunson, 2015] relies on factor models where the entries of the loading matrix and the factors are expressed in terms of covariates.

7.1 Principal Component Analysis (PCA)

In statistics, signal processing and machine learning given a random vector $Y = (Y_1, \dots, Y_p)'$ with the covariance matrix $\mathbf{\Sigma}$, it is desirable to find directions $\mathbf{c} \in \mathbb{R}^p$ such that $\text{Var}(\mathbf{c}'Y) = \mathbf{c}'\mathbf{\Sigma}\mathbf{c}$ is the largest possible subject to $||\mathbf{c}||_2 = 1$. Since the variance can be increased by multiplying any $\mathbf{c}$ by a constant, this indeterminacy is eliminated by restricting $\mathbf{c}$ to have unit length, that is, $\mathbf{c}'\mathbf{c} = 1$. Furthermore, the constraint on the length of $\mathbf{c}$ makes the results insensitive to the units of measurements in Y. In spite of this, $\mathbf{c}$ is unique up to a sign, to fix this sign issue in most computer packages the first entry of $\mathbf{c}$ is fixed to be positive.

DOI: 10.1201/9781003638896-7

7.1.1 Definition and Basic Properties

The *principal component analysis* (PCA) is concerned with explaining the covariance matrix $\mathbf{\Sigma}$ of Y through a few, say k, linear combinations of its components. The hope is that some new variables $Z_1, \ldots, Z_k$ with larger variances will help to

1. provide a simpler and more parsimonious approximation to $\mathbf{\Sigma}$,

2. reduce the dimension of the data from p to k, where these linear combinations are possibly easier to interpret and visualize for $k = 2, 3$.

Let us start with p arbitrary linear combinations

$$Z_i = \mathbf{c_i}'Y = c_{i1}Y_1 + c_{i2}Y_2 \quad + \cdots + c_{ip}Y_p, \quad i = 1, \ldots, p \tag{7.1}$$

where c_{ij}'s are some scalars, $\mathbf{c}_i = (c_{i1}, \ldots, c_{ip})'$ with

$$\text{Var}(Z_i) = \mathbf{c}_i'\mathbf{\Sigma}\mathbf{c}_i, \quad \text{Cov}(Z_i, Z_j) = \mathbf{c}_i'\mathbf{\Sigma}\mathbf{c}_j, \ i, j = 1, \ldots, p.$$

Then, the *principal components* (PCs) of Y are those *uncorrelated* linear combinations $Z_1, \ldots, Z_p$ whose variances are as large as possible subject to the length and orthogonality constraints stated below: The *first principal component* Z_1 is obtained by finding $\mathbf{c}_1$ so that $\text{Var}(Z_1) = \mathbf{c}_1'\mathbf{\Sigma}\mathbf{c}_1$ is maximized subject to

$$\mathbf{c}_1'\mathbf{c}_1 = 1.$$

The *second principal component* Z_2 is obtained by finding $\mathbf{c}_2$ so that $\text{Var}(Z_2) = \mathbf{c}_2'\mathbf{\Sigma}\mathbf{c}_2$ is maximized subject to

$$\mathbf{c}_2'\mathbf{c}_2 = 1, \quad \text{Cov}(Z_1, Z_2) = \mathbf{c}_1'\mathbf{\Sigma}\mathbf{c}_2 = 0.$$

Finally, the *i-th principal component* is obtained by finding $\mathbf{c}_i$ so that $\text{Var}(Z_i) = \mathbf{c}_i'\mathbf{\Sigma}\mathbf{c}_i$ is maximized subject to

$$\mathbf{c}_i'\mathbf{c}_i = 1, \quad \mathbf{c}_j'\mathbf{\Sigma}\mathbf{c}_i = 0, \text{ for } j = 1, 2, \ldots, i-1.$$

The spectral decomposition of the covariance matrix is the key to solving the PCA problem; see (3.19), (3.21) and Problem 3.18. We assume that the eigenvalues λ_i's are distinct and arranged in descending order with their corresponding eigenvectors denoted by $\mathbf{e}_i$. Then, from the spectral decomposition of $\mathbf{\Sigma}$, it follows that $\mathbf{e}_i$ is the i-th normalized eigenvector of $\mathbf{\Sigma}$ and $\text{Var}(\mathbf{e}_i'Y) = \lambda_i$ is its i-th largest eigenvalue. The ith *principal component (PC)* of Y is the linear combination $Z_i = \mathbf{e}_i'Y$, and its variance is given by the eigenvalue λ_i. The hope is that the first few larger eigenvalues or the corresponding PCs explain a big chunk of the total variance of Y defined by $\text{tr}(\mathbf{\Sigma}) = \sum_{i=1}^{p} \lambda_i$. Although so far we have described the population PCs based on $\mathbf{\Sigma}$, but given data the sample PCs are found simply replacing $\mathbf{\Sigma}$ by the sample covariance matrix $\mathbf{S}$ of the data and the corresponding sample eigenvalues and sample eigenvectors.

The next example highlights two fundamentally important properties of the eigenvalues and eigenvectors of compound symmetric matrices so far as PCs and spiked covariance matrices are concerned; see Section 7.1.3.

Example 18 (The First PC of the CS matrix) *Recall that the eigenvector of the compound symmetry (CS) covariance matrix*

$$\mathbf{\Sigma} = \left[(1-\rho)I + \rho\mathbf{1}_n\mathbf{1}_n'\right],$$

corresponding to the largest eigenvalue $\lambda_1 = 1 + (p-1)\rho$, *for a* ρ *positive, is*

$$\mathbf{e}_1 = \frac{1}{\sqrt{p}}(1,\ldots,1)' = \frac{1}{\sqrt{p}}\mathbf{1}_p.$$

Consequently, its first PC given by

$$Z_1 = \mathbf{e}_1'Y = \frac{1}{\sqrt{p}}\sum_{j=1}^{p} Y_j = \sqrt{p}\bar{Y},$$

is simply a multiple of the average of all the p variables.

As such it can be interpreted as a sort of overall index with equal weights. It explains a proportion of

$$\frac{\lambda}{p} = \frac{1+(p-1)\rho}{p},$$

the total population variance. For example, for $p = 10$ *and* $\rho = 0.90$, *the first PC explains* 91% *of the population variation; for* $\rho = 0.50$, *it explains* 55% *of the variation.*

The first few PCs coming out of PCA are usually dense with mostly nonzero coefficients in the linear combinations and hard to interpret. The case of ease of interpretation of Z_1 in the example above in spite of nonzero coefficients is an exception due to equality of the weights. Finding sparse PCs or linear combinations with many zero coefficients is of great interest so far as interpretability is concerned [Jolliffe et al., 2003, Witten et al., 2009] and Pourahmadi [2013, Section 4.4]. They tend to simplify the relationship between the PCs and the original variables. The method of Principal Variables [McCabe, 1984] which aims to select a subset of the original variables pushes the interpretability idea as much as possible and is PCA's sparsest and most interpretable alternative.

7.1.2 Asymptotic Distribution of Sample Eigenvalues: p fixed

Given the sample data or the sample covariance matrix $\mathbf{S}$, to compute the sample PCs it suffices to replace the population covariance matrix by $\mathbf{S}$ and proceed as in the previous section. It is well-known [Anderson, 2003, page 473] that for p fixed and n large the sample covariance matrix, sample eigenvalues and eigenvectors converge to their population counterparts as stated next.

Theorem 9 *For a sample of size n from a multivariate normal distribution where the eigenvalues of the covariance matrix are distinct, let* $\hat{\boldsymbol{\lambda}} = (\hat{\lambda}_1,\ldots,\hat{\lambda}_p)$ *and* $\Lambda = diag(\lambda_1,\ldots,\lambda_p)$. *Then, we have as* $n \to \infty$,

(a)

$$\sqrt{n}(\hat{\boldsymbol{\lambda}} - \boldsymbol{\lambda}) \rightarrow N_p(\mathbf{0}, 2\Lambda^2).$$

(b)

$$\sqrt{n}(\hat{\boldsymbol{e}}_i - \boldsymbol{e}_i) \rightarrow N_p(\mathbf{0}, \boldsymbol{V}_i),$$

where

$$\boldsymbol{V}_i = \lambda_i \sum_{k=1, k \neq i}^{p} \frac{\lambda_k}{(\lambda_k - \lambda_i)^2} \boldsymbol{e}_k \boldsymbol{e}'_k.$$

(c) Each $\hat{\lambda}_i$ *is distributed independently of the corresponding sample eigenvector* $\hat{\mathbf{e}}_i$.

The presence of the difference $\lambda_k - \lambda_i$ in the denominator of $\boldsymbol{V}_i$ reveals the importance of the eigengap or separation of the eigenvalues in the asymptotic properties of the eigenstructure of the sample covariance matrix. An encyclopedic coverage of PCA can be found in Jolliffe [2002], the textbook by Johnson and Wichern [2008] provides a readable account of the PCA methodologies along with a comprehensive analysis of a good number of genuine data examples.

Not surprisingly, the asymptotic behavior of the sample eigenstructure is quite different in the high-dimensional data situations where $p = p_n$ grows with n in such a way that $\frac{p}{n} \rightarrow c > 0$. Consequently, much care is needed in dealing with a host of issues involving the sample eigenvalues and eigenvectors which may impact the important task of selecting the number of leading PCs in practice. In this connection, it is notable that in Example 18, its first eigenvalue converges to infinity for p large while all the other eigenvalues are equal and remain bounded. Perhaps, this is one of the simplest, most concrete and prevalent example related to *spiked covariance matrices* which are of central importance in studying the asymptotic behavior of PCA and factor models for high-dimensional data.

7.1.3 Spiked Covariance Models

The eigenvalues of the important class of spiked covariance matrices have the following simple and canonical form:

$$\lambda_1 \geq \ldots \geq \lambda_r > 1, \quad \lambda_{r+1} = \ldots = \lambda_p = 1, \tag{7.2}$$

where the number of "spikes" r and their magnitudes are fixed and independent of p. Note that the first r eigenvalues of $\mathbf{\Sigma}$ are strictly larger than 1, the remaining eigenvalues are equal to one, so that the spectrum of $\mathbf{\Sigma}$ has r spikes and hence the name, coined by Johnstone [2001].

For p fixed and n large the sample covariance matrix, sample eigenvalues and eigenvectors converge to their population counterparts, see Theorem 9. However, this is not the case for the high-dimensional data situations where $p = p_n$ grows with n in such a way that $\frac{p}{n} \rightarrow c > 0$. In fact, in high-dimensional regimes and for the single spiked covariance model ($r = 1$) it is known that the leading eigenvector of the sample covariance matrix can actually be orthogonal to the leading eigenvector of the population covariance matrix if its corresponding eigenvalue is not sufficiently large [Paul, 2007]. A more precise statement of known results is given next.

Theorem 10 *Under the above conditions on p and n [Donoho et al., 2016] ,*

(a) *When* $\lambda_1 = \ldots = \lambda_r = 1$, *the empirical distribution of the sample eigenvalues converges as* $n \to \infty$, *to a nondegenerate absolutely continuous distribution with the probability density function*

$$f(x) = \frac{1}{2\pi c x}\sqrt{(c_+ - x)(x - c_-)}, \tag{7.3}$$

the so called Marchenko–Pastur distribution, supported on the interval $[c_-, c_+]$ *where* $c_\pm = (1 \pm \sqrt{c})^2$.

(b) *Each of the largest r sample eigenvalues* $\hat{\lambda}_i$ *converges in probability to the deterministic limit* $b(\lambda_i)$, *where for* $i = 1, \ldots, r$

$$b(\lambda_i) = \lambda_i(1 + \frac{c}{\lambda_i - 1}), \quad \textit{for } \lambda_i > c_+,$$

and

$$b(\lambda_i) = 1, \quad \textit{for } \lambda_i \leq c_+.$$

(c) *The corresponding top r sample eigenvectors* $\widehat{\boldsymbol{e}}_i$ *make asymptotically deterministic angles with the top r population eigenvectors* $\boldsymbol{e}_i$. *Specifically,* $|(\widehat{\boldsymbol{e}}_i, \boldsymbol{e}_j)| \to 0$, *in probability if* $i \neq j$, *and*

$$|(\widehat{\boldsymbol{e}}_i, \boldsymbol{e}_i)| \to c(\lambda_i),$$

where

$$c(\lambda) = \sqrt{\frac{1 - c/(\lambda - 1)^2}{1 + c/(\lambda - 1)}}, \quad \textit{for } \lambda > c_+,$$

and $c(\lambda) = 0$, *for* $\lambda \leq c_+$.

In other words, the top sample eigenvalues are *shifted* upwards from their population counterparts by a displacement of asymptotically predictable size. The quantity $c(\lambda)$ can be viewed as the cosine of the angle between the two normalized vectors or their correlation coefficient. Thus (c) reveals that the sample eigenvectors are *rotated* away from their population counterparts through an angle of asymptotically predictable size controlled by the corresponding eigenvalue. In particular, $\widehat{\boldsymbol{e}}_i$ and $\boldsymbol{e}_j$ are asymptotically orthogonal if $\lambda_i \leq c_+$. For more details on regularization and consistency of PCA for high-dimensional data situations, see Pourahmadi [2013, Chapter 7] and the references therein.

7.1.4 Sample PCs: Scree Plot vs ScreeNOT

The question of how many principal components to keep in various applications of PCA is central to its success in data analysis and dimension reduction. Fortunately, spiked covariance matrices provide the ideal mathematical structure and heuristic behind many methods including the classic graphical tool of *scree plot* for selecting the number of spikes r.

The scree plot is the most popular ad hoc graphical guide [Cattell, 1966] for selecting the number of leading PCs. It simply plots the ordered sample eigenvalues $\widehat{\lambda}_i$ versus $i = 1, \ldots, p$ and looks for an elbow or drop in the plot. The are many approaches in the literature for determining the number of principal components which can be divided into three groups: subjective methods like the scree plot, penalized methods like AIC and BIC, and computational procedures like cross-validation.

A recent method for choosing r which enjoys many optimality properties relative to scree plot, is based on hard-thresholding the singular-values of the data matrix $\boldsymbol{Y}$. It has the goal of minimizing the reconstruction error in the Frobenius norm [Donoho et al., 2023]. More precisely, starting with the singular value decomposition of the data matrix

$$\boldsymbol{Y} = \sum_{i=1}^{\min(n,p)} d_i \boldsymbol{u}_i \mathbf{v}_i',$$

it searches for a threshold τ to separate the "signal" components in the SVD of $\boldsymbol{Y}$ from the "noise":

$$\widehat{\boldsymbol{Y}}_\tau = \sum_{i=1}^{\min(n,p)} d_i I_{\{i:d_i>\tau\}} \boldsymbol{u}_i \mathbf{v}_i', \tag{7.4}$$

so that the reconstruction error $E||\boldsymbol{Y} - \widehat{\boldsymbol{Y}}_\tau||_F^2$ is minimized. Under rather broad conditions inspired by random matrix theory, Donoho et al. [2023] showed that the optimal choice of the threshold is

$$\tau_{\text{opt}} \approx \frac{4}{\sqrt{3}} \frac{s_{\text{med}}}{\sqrt{0.6528 \times n}}, \tag{7.5}$$

where s_{med} is the median of the sample singular values d_i's of the data matrix. They show the superior theoretical and empirical performance of this new technique of *singular-value thresholding* which can be viewed as a principled competitor to the traditional scree plot. Even though it is applied directly to the data matrix $\boldsymbol{Y}$, it seems the same thresholding idea should work for the sample covariance matrix used in PCA and factor models.

Another modern alternative to scree plot for selecting r is a ratio-based method [Lam and Yao, 2012] involving the ratio of successive ordered sample eigenvalues. More precisely, let $\hat{\lambda}_1 \geq \hat{\lambda}_2 \geq, \ldots, \geq \hat{\lambda}_p \geq 0$ be the eigenvalues of the sample covariance matrix. Then, the ratio-based estimator for r is defined as

$$\hat{r} = \underset{1 \leq j \leq r_0}{\text{argmax}} \frac{\hat{\lambda}_j}{\hat{\lambda}_{j+1}}, \tag{7.6}$$

where r_0 is an upper bound for r such as the integer part of $p/2$. The circle of ideas and techniques related to the ratio-based method is very popular in statistics and econometrics literature where there is great interest in finding suitable conditions for the consistency of the selection method. The paper Zhang et al. [2025] provides an excellent review of the literature and a new criteria based on the proportion of variance explained.

In addition to critical role of r in dimension-reduction (see Section 7.1.5), the number of PCs selected is important in the popular *Principal Components Regression* (PCR). PCR is an approach to linear regression when there are too many covariates or multicollinearity is present. It combines PCA with least-squares regression. Given the $n \times p$ design matrix of covariates and scalar responses, PCR first computes PCA of the covariates and selects the r leading PCs, then performs an ordinary least-squares regression of responses on these selected PCs or the derived features. In modern datasets where p is often comparable to or larger than n, the general idea of dimensionality reduction before regression is attractive. Intuitively, PCR can be viewed as the search for a *latent and low-dimensional structure* where there are a few high-variability directions that mostly explain the response, hidden in many low-variability noise directions. In this sense, PCR is estimating a *latent subspace* which fits the framework of latent factor models presented in Section 7.2.

7.1.5 Hands-On R: PCA

In the analysis of marketing datasets, especially consumer survey data, the focus is often on identifying smaller sets of variables that reflect a reduced number of underlying concepts, such as customer satisfaction or product perception. In this section, we use PCA to plot and interpret perceptual maps [Chapman and Feit, 2019, Chapter 8]. Perceptual maps visualize how consumers perceive and evaluate products or brands in relation to one another. We use data from a well-known case study in the marketing literature, where the goal is to understand how consumers perceive the Infiniti G20 relative to its competitors. The data, available in the complementary R package `recode`, is based on a customer survey of 100 respondents from Nissan's target segments, described as individuals between the ages of 25 and 35 with annual household incomes between 50,000 and 100,000. Participants were asked to rate their perceptions (on a scale from 1 to 7) of 10 different cars (Infiniti G20, Ford T-bird, Audi 90, Toyota Supra, Eagle Talon, Honda Prelude, Saab 900, Pontiac Firebird, BMW 318i, and Mercury Capri) across 15 attributes. This results in a dataset of size 1000×15. The attributes are coded as follows in the data:

- Attractive - `Attract`
- Quiet - `Quiet`
- Unreliable - `Unreliab`
- Poorly built - `Pbuilt`
- Interesting - `Interest`
- Sporty - `Sporty`
- Uncomfortable - `Uncomfort`
- Roomy - `Roomy`

TABLE 7.1
Example of consumer survey data

	Brand	Attract	Quiet	Unreliab	Pbuilt	Interest	Sporty	Uncomfort	Roomy	Esyserv	Prestige	Common	Econ	Success	Avantg	Pvalue
1	Infinity G20	6	6	5	2	4	4	3	4	5	6	3	3	6	4	4
2	Ford T-Bird	5	4	6	4	5	5	4	4	5	4	3	4	5	3	4
3	Audi 90	6	5	6	3	4	4	2	5	4	7	3	4	6	4	4
4	Toyota Supra	7	4	3	2	4	6	4	3	5	6	3	3	6	5	3
5	Eagle Talon	5	4	5	4	4	5	4	3	5	4	4	5	5	4	4
6	Honda Prelude	6	6	5	3	3	5	3	4	5	6	4	5	7	4	3

- Easy Service - Esyserv
- High prestige - Prestige
- Common - Common
- Economical - Econ
- Successful - Success
- Avant-grade - Avantg
- Poor value - Pvalue

Table 7.1 reports a small segment of the dataset. We start by reading the data and then standardizing the attributes, which helps the interpretability of the results.

```
library(recode)
data(brand)

head(data)
# Standardize data
brand.sc <- brand
brand.sc[, 2:16] <- scale(brand[, 2:16])
```

We perform PCA on the survey data using prcomp() function.

```
# Run PCA
brand.pca <- prcomp(brand.sc[, 2:16], scale=TRUE)

# Summarize the result
summary(brand.pca)
```

Table 7.2 reports the partial results of applying PCA to the data. It shows that the first two components account for around 65% of the variance in the ratings. Our focus is on plotting the perceptual mapping with regard to the first two components.

In Figure 7.1 we display the scree plot. The latter plots the number of components against the proportion of the variance explained. The scree plot is used for selecting the number of components using, so called "elbow" criterion, i.e., selecting all

TABLE 7.2
Importance of components

	PC1	PC2	PC3	PC4	PC5	PC6
Standard deviation	2.6519	1.6556	1.2659	1.1512	0.7569	0.6675
Proportion of Variance	0.4688	0.1827	0.1068	0.0883	0.0382	0.0297
Cumulative Proportion	0.4688	0.6516	0.7584	0.8468	0.8849	0.9146

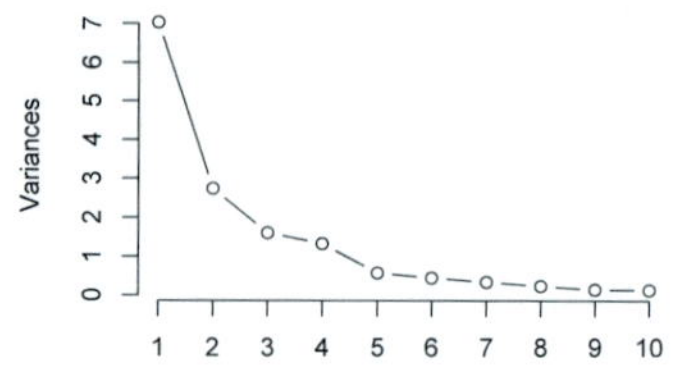

FIGURE 7.1
Scree plot

components just before the line flattens out. Here, we rely on ScreeNOT algorithm, discussed in Section 7.1.4 and available in R package ScreeNOT, to interpret results using the number of components equal to 2. Based on the scree plot, in Problem 7.4, it is suggested that an avid reader repeat our analysis using the first three PCs.

```
brand.pca <- prcomp(brand.sc[, 2:16], scale=TRUE)

require(ScreeNOT)
print(adaptiveHardThresholding(brand.sc[, 2:16], k = 5)$r)
[1] 2
```

Next, we use the R function biplot() to display perceptual mapping for cars positioning in Figure 7.2.

```
biplot(brand.pca, main="Car positioning", cex=c(1.1, 1),
    expand = 1)
```

The first two principal components reveal how consumers perceive different cars

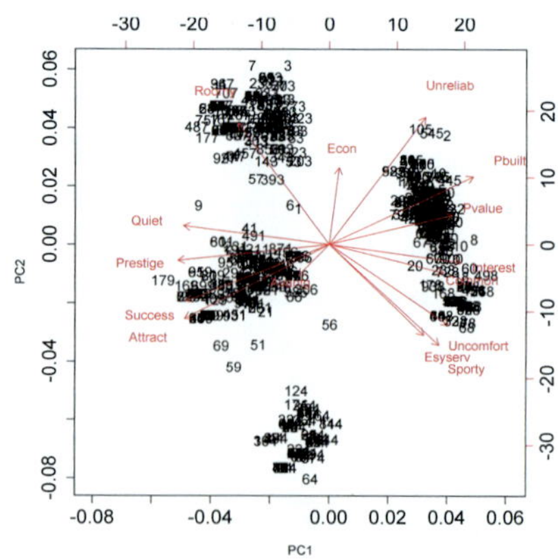

FIGURE 7.2
Car positioning

TABLE 7.3
Importance of components for aggregated data

	PC1	PC2	PC3	PC4	PC5	PC6
Standard deviation	2.7893	1.7151	1.3131	1.1841	0.7063	0.6175
Proportion of Variance	0.5187	0.1961	0.1149	0.0935	0.0333	0.0254
Cumulative Proportion	0.5187	0.7148	0.8297	0.9232	0.9565	0.9819

based on attributes. We can see that there are four different adjective or attribute clusters based on consumer's perception rating. However, visualizing and interpreting individual respondents is challenging due to the density of the clusters. It is recommended to perform PCA using aggregated ratings by car names, where the average respondent rating for each car is reported. This approach reduces the data to a size of 10×15.

```
brand.mean <- aggregate(.~Car_Name, data=brand.sc , mean)

# use brand for the row names
rownames(brand.mean) <- brand.mean[, 1]
# remove brand name column
brand.mean <- brand.mean[, -1]

# Run PCA
pca.mean <- prcomp(brand.mean, scale=TRUE)

# Summarize the result
summary(pca.mean)
```

Table 7.3 reports the partial results of applying PCA to the aggregated data. It shows that the first two components account for around 71% of the variance in the aggregated ratings. As before, our focus will be on plotting the perceptual mapping with regard to the first two components.

Next, we use R function biplot() to display perceptual mapping for the car positioning in Figure 7.3.

```
biplot(pca.mean, main="Car positioning for aggregated data", cex=c
    (1.1, 1),
    expand = 1)
```

We interpret the adjective clusters and relationships and see four areas with well-differentiated sets of adjectives and brands that are positioned in proximity:

- Audi 90 and Saab 900 were perceived as roomy, however, they were opposite to being perceived as an easy serve.
- BMW 318i, Honda Prelude, and Infinity G20 were considered as avand-grade, successful and attractive.
- On the other hand, Eagle Talon, Mercury Capri, Ford T-bird (the latter two are overlapped in the figure) were perceived as poorly built and poor in value.

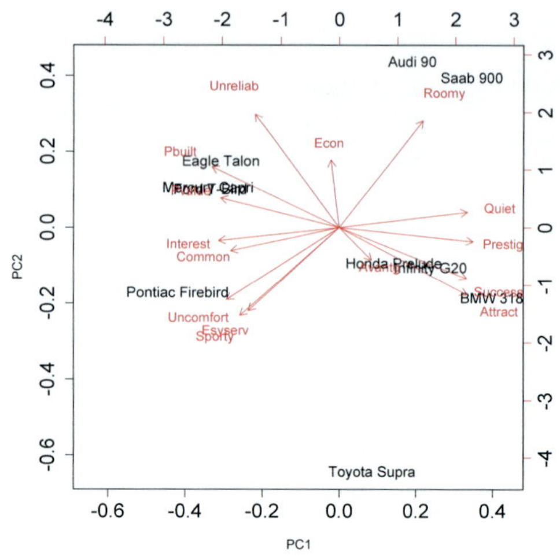

FIGURE 7.3
Car positioning for aggregated data

- Finally, Pontiac Firebird were considered close to being common and uncomfortable.

From the perceptual map, we can conclude that the main competitors of the Infinity G20, based on market perception, are the Honda Prelude and the BMW 318i. If Nissan stakeholders want to differentiate their product from these competitors, one possibility is to improve the perception of their product as prestigious, where there are no competitors, or to shift the perception of the Infinity G20 to be perceived as a sporty car.

7.1.6 Regression Models for PCA

A number of models for PCA have been considered in the literature. The popular *probabilistic PCA* approach in Tipping and Bishop [1999] assumes that each row of the $n \times p$ data matrix $\boldsymbol{Y}$ is independently drawn from a multivariate normal distribution with mean zero and a spiked covariance matrix.

By contrast, the recent PCA model in Perry et al. [2025] postulates a regression-like model or a low-rank structure for the mean, rather than for the covariance matrix of the data matrix. Their data-generating model is of the form

$$\boldsymbol{Y} = \Theta + E,$$

where $\Theta \in \mathbb{R}^{n \times p}$ is a fixed (non-random) mean matrix, and $E \in \mathbb{R}^{n \times p}$ is a noise matrix with independent and identically distributed entries with $N(0, \sigma^2)$. In terms of *matrix-normal* distributions one may write a model for such data matrix as

$$\boldsymbol{Y} \sim \mathrm{MN}(\Theta, I_n, \sigma^2 I_p) \quad \text{or} \quad \mathrm{vec}(\boldsymbol{Y}) \sim N_{np}(\mathrm{vec}(\boldsymbol{\Theta}), \sigma^2 I_p \otimes I_n), \tag{7.7}$$

with the mean Θ, independent rows and columns with covariance matrices $\sigma^2 I_p$ and I_n, respectively.

The singular value decompositions of $\boldsymbol{Y}$ and Θ play a central role in introducing a viable population parameter for the *proportion of variance explained* which makes it possible to have point estimate for the number of PCs retained and conduct inference, namely find confidence intervals and p-values as in the traditional statistical models. This setup seems to be the most promising and nuanced regression model formulation for the PCA thus far, we skip the details and refer the readers to Perry et al. [2025] for more detailed coverage, review of the literature and illustrations using data.

7.2 Factor Analysis

Factor analysis, developed by psychologists, had the goal of identifying a few meaningful common (random, latent) *factors* to explain the exceedingly high correlations seen among test scores of students on school subjects like algebra, geometry, biology,..... Nowadays, such high correlations are routinely encountered in economics and financial data among others, because many measured variables are impacted by the same factors stemming from market sectors and regional weather conditions, etc. In short, the goal of factor models is: *use a low-dimensional latent vector to describe the dependence in high-dimensional data.*

7.2.1 Factor Models

A factor model expresses a high-dimensional, observable p-vector $Y = (Y_1, \dots, Y_p)'$, with covariance matrix $\boldsymbol{\Sigma}$ in terms of a low-dimensional r-vector of *latent factors* $F = (f_1, \dots, f_r)'$ by

$$Y_i = \sum_{j=1}^{r} \ell_{ij} f_j + \varepsilon_i, \quad i = 1, \dots, p,$$

where ε_i's are the errors. Its vector form is

$$Y = LF + \boldsymbol{\epsilon}, \tag{7.8}$$

where $L = (\ell_{ij})$ is a $p \times r$ matrix of *factor loadings*, and $\boldsymbol{\epsilon}$ is the vector of uncorrelated errors. It is assumed that Y, F and $\boldsymbol{\epsilon}$ have zero means with

$$\mathrm{Cov}(\boldsymbol{\epsilon}) = \Psi = \mathrm{diag}(\psi_1, \dots, \psi_p), \quad \mathrm{Cov}(F, \boldsymbol{\epsilon}) = 0, \quad \mathrm{Cov}(F) = I_r, \tag{7.9}$$

so that the latent factors F and errors $\boldsymbol{\epsilon}$ are uncorrelated or orthogonal.

From the simple assumptions in (7.9) and computing $\mathrm{Cov}(Y) = \mathrm{Cov}(LF + \boldsymbol{\epsilon})$, one obtains a decomposition of the covariance matrix

$$\boldsymbol{\Sigma} = L\mathrm{Cov}(F)L' + \Psi = LL' + \Psi, \tag{7.10}$$

as a sum of *low-rank and diagonal* matrices. Of course, such a model with r much smaller than p is highly desirable in practice so far as dimension reduction and interpretation of the common factors F are concerned. A factor model with assumptions in (7.9) is usually referred to as the classical *orthogonal (strict, exact) factor model*, as opposed to the *approximate factor models* where Ψ is a sparse matrix; see Section 7.3.

The orthogonal factor model or the decomposition (7.10) is a reparameterization of the covariance matrix. In particular, the diagonal entries of $\boldsymbol{\Sigma}$ relate to the entries of L and Ψ via

$$\sigma_{ii} = \text{Var}(Y_i) = \ell_{i1}^2 + \cdots + \ell_{ir}^2 + \psi_i, \tag{7.11}$$

The quantity $\ell_{i1}^2 + \cdots + \ell_{ir}^2 = c_i^2$, or the portion of variance of Y_i explained by the r common factors, is called its *communality* and the unexplained portion ψ_i in (7.11) is its *idiosyncratic, uniqueness* or *specific variance*.

Is there any difference between factor and regression models? At first glance, the factor model (7.8) appears just like the multivariate linear regression model in (5.6) with L as the regression coefficient matrix and F as the covariates. Except that both L and F are unknown, since the "covariates" F are latent or random and not *observed* as in the regression models. This leads to the sticky issue of identifiability of the pair (L, F) discussed in the next section. Nevertheless, the loading matrix L has the interpretation as the regression coefficient in the sense that if $\ell_{ij} = 0$, then the factor f_j disappears or does not *load* on the linear model for Y_i. In this sense, a sparse L with many zeros in each row is highly desirable. In fact, starting from the early days of factor analysis various forms of sparsity of L has been suggested to partially resolve the notorious identifiability problem.

7.2.2 The Identifiability Problem

The root of the problem is to note that for any invertible $r \times r$ matrix H, the two distinct pairs (L, F) and $(LH, H^{-1}F)$ satisfy (7.8) or (7.10), and hence L and F are not identifiable.

There are many classical and modern approaches to deal with this problem. They amount to fixing at zero certain entries of the loading matrix L. For example, a popular classical approach to identification is to set the upper $r \times r$ submatrix of L to be the identity or a unit lower triangular matrix, see Anderson [2003, page 574] for more discussions and its history in psychological testing dating back to Thurstone [1947]. Another approach to identifiability of L is to impose a diagonality constraint like

$$L'\Psi^{-1}L = D, \tag{7.12}$$

where D is a diagonal matrix with diagonal entries arranged in decreasing order. The ensuing $r(r+1)/2$ constraints are helpful in identifying L up to possible sign changes in its columns.

Finally, the Kaiser [1958]'s *Varimax* criterion searches for a rotation or an orthogonal matrix P so that LP has many zeros while the new rotated factors $P'F$ have meaningful substantive interpretation. This criterion is widely used by psychologists, and to a lesser extent by statisticians. The freedom in choosing P has been the source of many challenges in applying factor analysis to social sciences and related areas. Next, we introduce the *Varimax* criterion. For a given loading matrix L, let $\tilde{\ell}^*_{ij} = \frac{\ell_{ij}}{c_i}$ be its entries scaled by the (positive) square root of communalities. The Varimax procedure selects an orthogonal matrix P that maximizes the quantity

$$V(P) = \frac{1}{p}\sum_{j=1}^{r}\left[\sum_{i=1}^{p}(\tilde{\ell}^*_{ij})^4 - \frac{1}{p}\left(\sum_{i=1}^{p}\tilde{\ell}^{*2}_{ij}\right)^2\right],$$

which is a non-convex function in P. In words, it searches for an orthogonal matrix to spread out the squares of the loadings on each factor as much as possible. Alternatively, it searches for a sparse L or to group together large and negligible coefficients in any column of the rotated matrix of factor loadings. For a more recent discussion, detailed illustration of sparsity property of Varimax, and connection between factor analysis and independent component analysis (ICA), see Rohe and Zeng [2023].

Remark 4 *There are more recent application areas where the factors F are observable and hence the identification problem of (L, F) disappears. For example, a trend in finance and economics that emerged in the 1960's employs factors that are fixed and not latent. An important early example is the popular capital asset pricing model (CAPM), which is a single factor model and the proxy for the market factor is a value-weighted market index like the S&P 500 [Sharpe, 1970]. In multifactor models, the macroeconomic variables such as the inflation rate, interest rate, risk premia are used as the factors. More specifically, the factors in the Fama and French [1992] popular three-factor model are portfolio-based firm characteristics such as the firm's size, market value and the industry type.*

7.2.3 Estimation: The PCA Method

Accurate estimation of (L, F, Ψ), the factor loadings, latent factors and the specific variances are important in statistics and many applications. For example, in financial applications it is desirable to understand accurately how each individual stock depends on the unobserved common factors in order to appreciate its relative performance and risks. While good estimation of the factor loading L is of great interest in the traditional applications of factor analysis, that of the latent factors F is becoming of central importance in the newer area of independent component analysis (ICA), see Hyvärinen et al. [2004]. We note that in the classical applications, usually the dimension p of Y is fixed and much smaller than the sample size n, but in modern high-dimensional data both p and n are allowed to grow together.

Given data $\boldsymbol{Y}_1, \ldots, \boldsymbol{Y}_n$ with the sample covariance matrix $\mathbf{S}$, from the orthogonal factor model

$$\boldsymbol{Y}_i = LF_i + \epsilon_i, \quad i = 1, \ldots, n, \tag{7.13}$$

its parameters can be estimated using either the PCA method or the maximum likelihood method. While the former does not require the normality assumption of the data nor the knowledge of r (the number of common factors), they are required for the latter method to write the likelihood function.

The number of factors r is usually unknown and needs to be estimated using the available sample data. There are several methods available to determine r for the standard factor model, such as the information criterion in Bai and Ng [2002], the random matrix theory method in Onatski [2010]. Here, we use the ratio-based method in Lam and Yao [2012], Ahn and Horenstein [2013] involving the sample eigenvalues defined in (7.6).

We focus here on the PCA estimation method which is popular and rather simple to implement. It starts with computing the sample eigendecomposition or the ordered eigenvalue-eigenvector pairs of the sample covariance matrix.

$$(\widehat{\lambda}_1, \widehat{e}_1), (\widehat{\lambda}_2, \widehat{e}_2), \ldots, (\widehat{\lambda}_p, \widehat{e}_p),$$

For any $r < p$, the matrix of factor loadings is estimated by

$$\widehat{L} = (\widehat{\ell}_{ij}) = \left[\sqrt{\widehat{\lambda}_1}\widehat{e}_1 \vdots \sqrt{\widehat{\lambda}_2}\widehat{e}_2 \vdots \cdots \vdots \sqrt{\widehat{\lambda}_r}\widehat{e}_r\right], \tag{7.14}$$

and the estimated specific variances are the diagonal elements of the matrix $\mathbf{S} - \widehat{L}\widehat{L}'$:

$$(\widehat{\psi}_1, \ldots, \widehat{\psi}_p) = \text{diag}(\mathbf{S} - \widehat{L}\widehat{L}').$$

For the moment, using $\hat{r}$ from (7.6) and treating the estimated values of $\widehat{\theta} = (\widehat{L}, \widehat{\Psi})$ as their true values, for each i the model (7.8) then appears like a linear regression model with F_i as the unknown regression coefficients. Using the weighted least-squares, the estimated latent factors or the *factor scores* for the ith case is given by

$$\widehat{F}_i = (\widehat{L}'\widehat{\Psi}^{-1}\widehat{L})^{-1}\widehat{L}'\boldsymbol{Y}_i, \ i = 1, \ldots, n. \tag{7.15}$$

For the likelihood-based method of estimating the parameters, the PCA estimates are commonly used as the initial values for the iterative method of maximizing the likelihood function. In particular, when the observations constitute a random sample from a multivariate normal distribution with a nonsingular covariance matrix structured as in (7.10), the negative loglikelihood function of $\boldsymbol{\theta} = (L, \Psi)$ is of the form

$$L(\boldsymbol{\theta}|\boldsymbol{Y}_1, \ldots, \boldsymbol{Y}_n) = \log|LL' + \Psi| + \text{tr}\{(LL' + \Psi)^{-1}\mathbf{S}\}, \tag{7.16}$$

where $\mathbf{S}$ is the sample covariance matrix. The matrices $\widehat{L}$ and $\widehat{\Psi}$ resulting from maximizing $L(\cdot|\boldsymbol{Y}_1, \ldots, \boldsymbol{Y}_n)$ with respect to the entries of L and Ψ subject to the constraint (7.12) are the maximum likelihood estimates (MLE) of $\boldsymbol{\theta} = (L, \Psi)$; see Anderson [2003, Section 14.3] for more details and review of the literature on MLE of factor models.

Maximum likelihood estimation of the parameters of approximate factor models for time series and panel data is discussed in Section 7.3 where sparse estimates of L is introduced by adding a multiple of the ℓ_1-norm of L to the loglikelihood function.

7.2.4 Hands-on R: Factor Analysis

We apply factor analysis to stock return volatility data from Demirer et al. [2018], where the goal is to estimate global bank network connectedness. The raw dataset includes 96 banks from 29 developed and emerging economies, covering the period from September 12, 2003, to February 7, 2014. For illustration, we focus on countries where the number of banks exceeds 4, resulting in a subset of 54 banks. Table 7.4 lists a sample (10 out of 54) of variable names, bank names, and their respective countries. The dataset, consisting of $n = 2676$ observations and $p = 54$ variables, is available in the recode package and consists of daily volatility series for each bank stock. For the complete list, refer to help(globalbank) in R.

Our objective is twofold: to provide a low-dimensional summary of the data using factor models and to ensure that each factor has a meaningful interpretation. Following Rohe and Zeng [2023], we achieve the latter through a Varimax rotation. For the implementation, we use the R package vsp, available on CRAN.
We start by importing the required packages and the dataset.

```
library(recode)
library(vsp)

data(globalbank)
# Display data
head(globalbank)
# Convert to matrix
data_mat = as.matrix(globalbank)
colnames(data_mat) = colnames(globalbank)
```

The last two lines of the code convert globalbank into a matrix and assign column names to it. This step is necessary because the main function, vsp(), in the vsp pack-

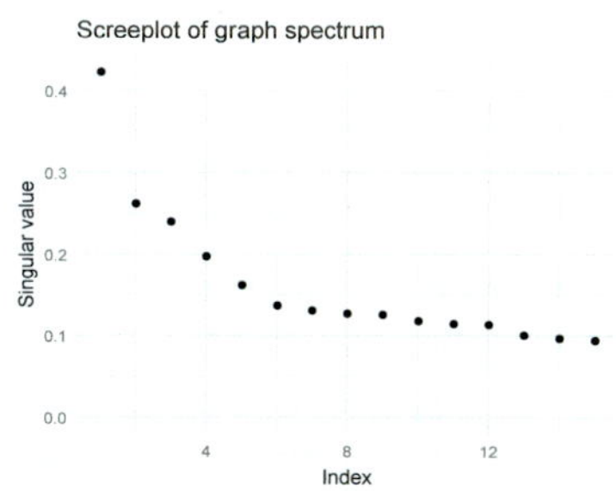

FIGURE 7.4
Screeplot for bank connectivity dataset

TABLE 7.4
Bank connectivity dataset

	Variable name	Bank	Country
1	hsba.ln	HSBC	Great Britain
2	X8306.to	MTBH	Japan
3	jpm	JPM	USA
4	barc.ln	BARC	Great Britain
5	bac	BAC	USA
6	c	Citi	USA
7	X8411.to	MZH	Japan
8	rbs.ln	RBS	Great Britain
9	X8316.to	SMTM	Japan
10	wfc	WFC	USA

age requires a matrix as input. Next, we use vsp() function to implement factor analysis. In this case, we selected 15 factors using scree plot described next. Similar to the PCA analysis discussed in Section 7.1.5, a scree plot can be useful for determining the appropriate number of factors. Figure 7.4 shows the scree plot.

```
# Run factor analysis
x = vsp(data_mat, rank = 15)
```

To obtain the Varimax-rotated loadings L and factors F, we use the functions get_varimax_y() and get_varimax_z(), respectively.

```
# Obtain Varimax rotated loadings and factors
L = get_varimax_y(x)
F = get_varimax_z(x)
```

Another useful indicator for the number of factors, as suggested by Rohe and Zeng [2023], is to look for the so-called radial streaks in the pairwise scatter plot of Varimax-rotated factors. Figure 7.5 presents scatter plots for the first 10 Varimax-rotated factors. The upper triangular portion of the plot shows the pairwise correlations between factors. The ideal outcome is for the axes to align with the streaks, indicating nearly zero correlation between factors, which results in sparse loadings. In Problem 7.5, the reader is encouraged to replicate the above steps and interpret the results using the number of factors equal to 7, which corresponds to the number of countries.

Figure 7.6 displays the estimated loading matrix L, where the rows are ordered by the country to which each bank belongs, and the columns represent the factors.

We observe that the first factor captures cross-country correlations between banks, while the remaining factors reflect country-specific correlations. For example, factors 11-14 represent correlations specific to USA.

In PCA and factor analysis one often has the option of using the sample covariance matrix or the correlation matrix of the standardized variables. While there are no clear guidelines which one should be preferred in practice, if feasible it is advisable to use

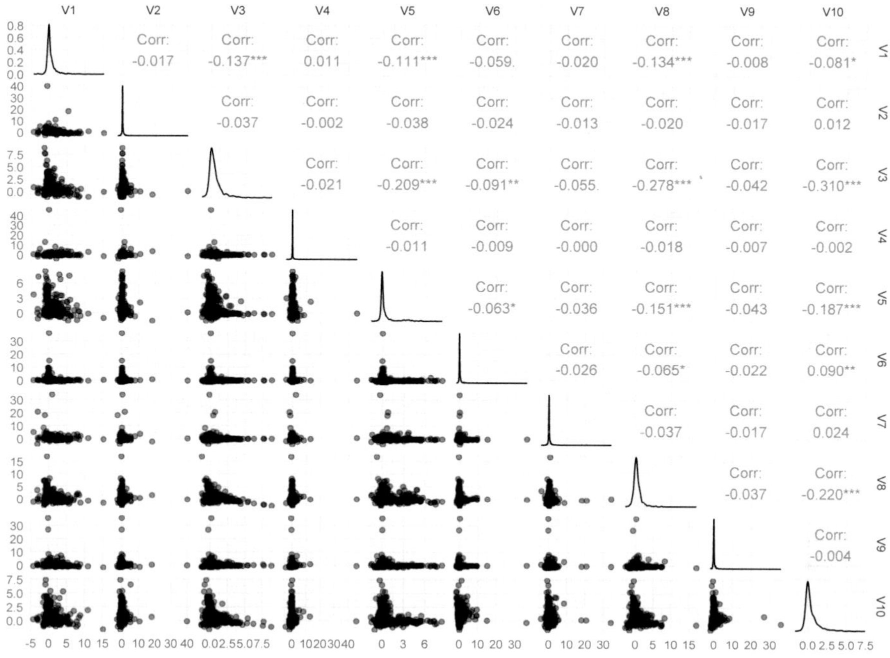

FIGURE 7.5
Scatter plot of Varimax rotated factors

both, compare the results and decide based on the interpretability and the relevance of the results; see Problem 7.5.

7.3 Approximate Factor Models: Large p PCA

It was noted rather early by Chamberlain and Rothschild [1983] in analyzing high-dimensional time series data arising from business and economics that the assumption of Ψ being diagonal is no longer viable in the orthogonal (strict) factor models. Their *approximate factor model*

$$Y_i = \chi_i + \epsilon_i = LF_i + \epsilon_i, \; i = 1, \ldots, n, \tag{7.17}$$

at the first glance appears just like the standard factor model with χ_i and ϵ_i as the common and idiosyncratic components. It satisfies all the assumptions in (7.9) of the standard factor models except that the error covariance matrix $\mathrm{Cov}(\epsilon) = \boldsymbol{\Sigma}_\epsilon$ is sparse rather than being diagonal. The vast literature in this area tends to downplay

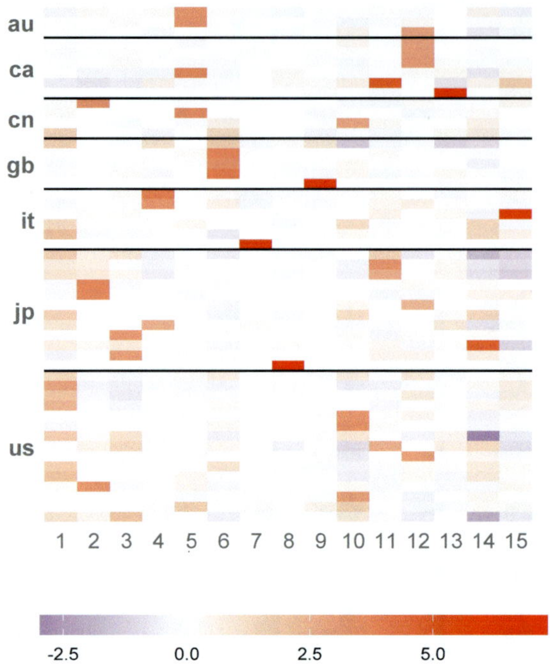

FIGURE 7.6
Factor loading matrix

the temporal correlations and treat the data as if they were i.i.d. It is common to apply the PCA estimation method (Section 7.2.3) to the sample covariance matrix or the lag-0 autocovariance matrix of the multivariate time series data while ignoring all the higher-lags.

7.3.1 Low-Rank Plus Sparse Covariance Matrices

Sparsity is the most common assumption in estimating high-dimensional covariance matrices. However, it may not hold in applications where the variables of interest are affected by the same common factors. For example, it is known that in financial markets stocks may share common market risks or belong to the same market sector so that their returns are highly correlated. In microarray, genes from the same pathway may be co-regulated by a small amount of regulatory factors, which makes the gene expression data highly correlated.

The model (7.17) naturally leads to the important class of low-rank plus sparse covariance matrices of the form

$$\mathbf{\Sigma} = \mathbf{\Sigma}_{\chi} + \mathbf{\Sigma}_{\epsilon} = LL' + \mathbf{\Sigma}_{\epsilon}, \tag{7.18}$$

where the loading matrix L is a $p \times r$ matrix with r smaller than p and $\mathbf{\Sigma}_{\epsilon}$ is a sparse covariance matrix instead of being diagonal as in the classical factor model (7.10).

As before, estimating the triplet $(L, F_i, \mathbf{\Sigma}_\epsilon)$ of factor loadings, factor scores and the error covariance matrix using the available data is of central importance. However, this is a more challenging problem upon realizing that the number of parameters in $\mathbf{\Sigma}_\epsilon$ is potentially of order p^2 rather than p as in the (strict) orthogonal factor models.

Moreover, relaxing the diagonality of $\mathbf{\Psi}$ gives rise to a new (additive) identification problem by noting that for any orthonormal matrix $P \neq \boldsymbol{I}_p$, one can rewrite the model as

$$\boldsymbol{Y}_i = L^* F_i + \boldsymbol{\epsilon}_i^*, \; i = 1, \ldots, n, \tag{7.19}$$

where

$$L^* = PL, \quad \boldsymbol{\epsilon}_i^* = (\boldsymbol{I}_p - P) L F_i + \boldsymbol{\epsilon}_i,$$

which can be shown to be an approximate factor model (Problem 7.10). It turns out that one can disentangle the common and idiosyncratic components when $p \longrightarrow \infty$ [Chamberlain and Rothschild, 1983]. Intuitively, such separation is possible, asymptotically for p going to infinity and when the covariance matrix of the common component $\mathbf{\Sigma}_\chi$ or LL' is spiked (big) with a few large eigenvalues converging to infinity at a rate like $p^\beta, \beta \leq 1$, while the eigenvalues of $\mathbf{\Sigma}_\epsilon$ remain bounded (small) or grow slowly relative to p. The factors or the common components are said to be *pervasive* if $\beta = 1$ and *weakly pervasive* for $\beta < 1$. Factors being pervasive means that they have strong impact on most of the variables Y_j's. This is similar to the accumulation of information for large p in the regression setup in Section 6.2.3. Thus, under the pervasiveness as p increases, the information about the common factors accumulates fast, but not so for the idiosyncratic components. This effectively makes it possible to separate the two matrices given their sum. We note that in sparse estimation of the loading matrix there is interest in less pervasive factors [Daniele et al., 2024].

7.3.2 A Rational for PCA Estimation Method

Given the popularity of PCA method of estimation of high-dimensional approximate factor models, we show that for p large the columns of the factor loading matrix $L = (\boldsymbol{\ell}_1, \ldots, \boldsymbol{\ell}_r)$ are close to the leading eigenvectors of $\mathbf{\Sigma}$ under mild conditions.

First, without loss of generality it is plausible to assume that the columns of L are orthogonal. For this Fan et al. [2020c, Section 10.2] notes that the free choice of H, in Section 7.2.2, can be made in such a way that the columns of LH are orthogonal; indeed, for any given L one may choose the matrix H to do the task of implementing the Gram-Schmidt orthogonalization process on the columns of L. Thus, from here on it is assumed that the columns of L are orthogonal with $\ell_j = ||\boldsymbol{\ell}_j||_2^2$. Next, we show that under the pervasiveness condition and for p large, the leading eigenvectors of $\mathbf{\Sigma}$ are nearly the same as the columns of the factor loading matrix L. Since the columns of L are orthogonal and can be arranged so that ℓ_j's are distinct and sorted in descending order. Then, it follows that

$$(LL')\boldsymbol{\ell}_j = \ell_j \boldsymbol{\ell}_j, \; j \leq r, \tag{7.20}$$

so that $\boldsymbol{\ell}_j / \sqrt{\ell_j}$ are the normalized eigenvectors of LL'. Realizing that $\mathbf{\Sigma}$ is a perturbation of LL' by the matrix $\mathbf{\Sigma}_\epsilon$, using Weyl and Davis-Kahan inequalities on eigenvalues and eigenvectors of perturbed matrices in Theorem 2, it follows that under the pervassiveness condition the normalized columns of L are close to e_j's, the normalized eigenvectors of $\mathbf{\Sigma}$, in the sense that

$$||e_j - \boldsymbol{\ell}_j/\sqrt{\ell_j}||_2 = O(p^{-1}||\mathbf{\Sigma}_\epsilon||_2),\ j \le r, \tag{7.21}$$

and furthermore

$$|\lambda_j - \ell_j| \le ||\mathbf{\Sigma}_\epsilon||_2,\ j \le r; \quad |\lambda_j| \le ||\mathbf{\Sigma}_\epsilon||_2,\ j > r.$$

These inequalities and asymptotic results provide strong justification for relying on the eigendecomposition of the sample covariance matrix of the data for estimation of the triplet parameters of an approximate factor model, and eventually estimation of large covariance matrices via low-rank plus sparse decompositions as presented in the next section.

7.3.3 Principal Orthogonal ComplEment Thresholding (POET)

In this section, we consider estimation of the approximate factor models or low-rank plus sparse covariance matrices using the PCA method followed by adaptive thresholding of the error matrix. Given data $\boldsymbol{Y}_1, \ldots, \boldsymbol{Y}_n$, it is of interest to find r, the number of common factors, and to consistently estimate the triplet parameters $(L, F_i, \mathbf{\Sigma}_\epsilon)$ of the factor model, and finally the covariance matrix itself. Unlike the classical factor models where p is fixed, here the dimension $p = p_n$ depends on n and could grow to infinity at a rate faster than n.

In the classical or p fixed case, one usually relies on the eigendecomposition of the sample covariance matrix to estimate the factor loadings and the principal component scores as the estimate of the common factors F_i's; see Section 7.2.3 and Anderson [2003, Chapter 10]. For high-dimensional data, a novel method in Fan et al. [2013], known as the Principal Orthogonal ComplEment Thresholding (POET) estimator, applies the additional step of adaptive thresholding (8.23) to the PCA residuals to arrive at a sparse estimate of $\mathbf{\Sigma}_\epsilon$. More precisely, for given r it proceeds as follows:

1. Start with the spectral decomposition of the sample covariance matrix:

$$\mathbf{S} = \sum_{i=1}^{r} \widehat{\lambda}_i \widehat{\mathbf{e}}_i \widehat{\mathbf{e}}_i' + \widehat{R},$$

   where $\widehat{R} = (r_{ij})$ is the matrix of residuals.

2. Estimate the pair (L, F_i) in (7.17) by

$$\widehat{L} = (\widehat{\lambda}_1^{1/2} \widehat{\mathbf{e}}_1, \ldots, \widehat{\lambda}_r^{1/2} \widehat{\mathbf{e}}_r), \tag{7.22}$$

   and

$$\widehat{F}_i = \text{diag}(\widehat{\lambda}_1, \ldots, \widehat{\lambda}_r)^{-1} \widehat{L}' \boldsymbol{Y}_i, \quad i = 1, \ldots, n. \tag{7.23}$$

3. Apply the adaptive thresholding (8.23) to $\widehat{R}$, the matrix of residuals, to obtain

$$\widehat{R}^{\delta} = (\widehat{r}_{ij}^{\delta}),$$

where

$$\widehat{r}_{ij}^{\delta} = \widehat{r}_{ij} I(|\widehat{r}_{ij}| \geq \delta_{ij}),$$

and $\delta_{ij} > 0$ is an entry-dependent adaptive threshold. Recall that the universal (standard) thresholding corresponds to $\delta_{ij} = \delta$ and an important example of adaptive thresholding is

$$\delta_{ij} = \delta(\widehat{r}_{ii}\widehat{r}_{jj})^{1/2},$$

for a given $\delta > 0$.

4. Finally, a POET estimator of the population covariance matrix $\mathbf{\Sigma}$ is given by

$$\widehat{\mathbf{\Sigma}}^{\delta} = \sum_{i=1}^{r} \widehat{\lambda}_i \widehat{\mathbf{e}}_i \widehat{\mathbf{e}}_i' + \widehat{R}^{\delta}. \tag{7.24}$$

It does not involve optimization and is very easy to compute. The consistency and rate of convergence of $\widehat{R}^{\delta}$ and $\widehat{\mathbf{\Sigma}}^{\delta}$ and their inverses are established in Fan et al. [2013] under some standard conditions.

Surprisingly, this simple and transparent procedure reduces to some earlier covariance estimators for specific choices of (δ, r) in (7.24):

1. For $\delta = 0$, it reduces to the sample covariance matrix $\mathbf{S}$.

2. For $\delta = 1$, it reduces to the covariance estimator using a standard factor model.

3. For $r = 0$, it reduces to the thresholded estimator of Bickel and Levina [2008a] or the adaptive thresholded estimator of Cai and Liu [2011].

A POET estimator of components of a factor model and the covariance matrix can also be derived as the least-squares estimator, which relates neatly to the approach in Stock and Watson [2002b] and Stock and Watson [2002a], see Fan et al. [2020c, page 479]. The close connection between factor models and PCA has led to the popularity of PCA-based estimation methods in finance and econometrics in the last three decades. Further discussions and applications of factor models and PCA can be found in the Chapter 11 on time series analysis. The review paper Fan et al. [2016a] provides a comprehensive coverage of topics related to the role of factor models.

7.3.4 Factor-Adjusted Regression Models

It was pointed out in Chapter 2 that Lasso regression performs poorly in terms of variable selection when the covariates are highly correlated. A way to resolve this issue is to reduce the correlations by developing a factor model for the covariates. Once the factor loadings and common factors (scores) are consistently estimated, they can be added as additional covariates to the regression model.

More precisely, we consider the standard linear regression and factor models

$$\boldsymbol{Y}_i = X_i'\boldsymbol{\beta} + \epsilon_i, \quad X_i = LF_i + \tilde{X}_i,\ i = 1, \ldots, n,$$

where F_i is an r-vector of latent factors. Substituting the factor model for X_i in the regression model, one arrives at

$$\boldsymbol{Y}_i = \tilde{X}_i'\boldsymbol{\beta} + F_i'\gamma + \epsilon_i,\ i = 1, \ldots, n, \tag{7.25}$$

where $\gamma = L'\boldsymbol{\beta}$ is the new additional regression parameters. If $\tilde{X}_i, F_i$ were observable, the above would be a regression model with $p+r$ parameters, but with the added advantage that the new "covariates" ($\tilde{X}_i$ and F_i) are much less correlated due to the factor model construction. One may apply a regularized regression method to it without worrying about the negative impact of correlations. A minor disadvantage to note is that $\gamma = L'\boldsymbol{\beta}$ is constrained and not a free parameter.

However, in reality $\tilde{X}_i$ and F_i are not observable, but they can be consistently estimated using the PCA or the POET method. Treating them as the covariates in the regression model (7.25), Fan et al. [2020a] developed a regularized model selection method called *factor-adjusted regularized model (FarmSelect)*, with an accompanying R package. Model selection consistency is established under mild conditions, with numerical studies supporting finite-sample performance in terms of both model selection and out-of-sample prediction. FARMSelect is flexible and applicable to variety of correlated high-dimensional data from finance and economics.

7.3.5 Diffusion Index Forecasting Models

A *diffusion index model* [Stock and Watson, 2002b] for forecasting a scalar time series $\{x_t\}$ is an approximate factor model for a high-dimensional time series $\{\boldsymbol{Y}_t\}$ in (7.17) amended by the following h-step ahead forecast equation:

$$x_{t+h} = \boldsymbol{\beta}' F_t + e_{t+h}, \tag{7.26}$$

where h is the forecast horizon. It involves the latent factors F_t, where $\boldsymbol{\beta} = (\beta_1, \ldots, \beta_r)'$ is the vector of coefficients for the r latent factors, and the forecast errors e_t is a sequence of zero-mean uncorrelated random variables with finite variance.

Model (7.26) provides a forecast for x_{t+h} based on the common factors F_t, which are termed *diffusion indices*. Since these factors are unobserved, Stock and Watson [2002a] developed a simple two-step estimation procedure. In the first step, PCA of the sample covariance matrix estimates F_t via the sample principal components, denoted $\hat{F}_t$. In the second step, $\boldsymbol{\beta}$ in (7.26) is estimated by ordinary least squares with $\hat{F}_t$ replacing F_t.

The empirical results in Stock and Watson [2002a] and thereafter show that this approach to forecasting offers noticeable improvements over the more standard univariate or low-dimensional (small-d) forecasting methods. Due to its simplicity and improved performance, it continues to remain the most popular approach in the modern high-dimensional time series data analysis.

The hard work in Stock and Watson [2002b] consists of showing that, under general conditions, as d and n go to infinity together at certain rate:

1. $\hat{F}_t$ converges in probability to F_t,
2. the forecast derived from (7.26) using the sample $\hat{F}_t$ converges in probability to the mean-square efficient (oracle) forecast that would be obtained if F_t were known.

The rather surprising and deep finding here is that the standard PCA for fixed d can be extended to the high-dimensional case, where both d and n go to infinity.

7.4 Advanced Topics

This section presents several advanced topics related to factor models. First, an overview of estimation of factor models via penalized likelihood is given where the POET method plays a key role to reduce the computational burden. Then, the focus is shifted to the structure of the loading matrix L through parsimony and use of covariates in a semi-parametric setup. Finally, a rather broad class of covariance regression models is presented where the covariance matrix flexibly depends on covariates and scales easily to high-dimensional datasets that can handle irregular data features like missing values, outliers, etc.

7.4.1 Estimation Via Penalized Likelihood

There is a large literature on estimating the approximate factor model using PCA methods, starting with Stock and Watson [2002a] and Bai and Ng [2002], where the observations are treated as a homogeneous independent sample. It is known that the PCA estimation method is inefficient when cross-sectional heterogeneity and temporal dependence are present, as in time series and panel data. In this situation, the maximum likelihood method provides a better alternative and has been pursued by Bai and Liao [2016] and Daniele et al. [2024], among many others. A brief review of these methods is presented in this section, where it is assumed that r, the number of factors, is known and does not depend on n and p.

A two-step procedure in Bai and Liao [2016] estimates $(L, \mathbf{\Sigma}_\epsilon)$ separately. In the first step, it applies the POET estimator of the previous section to arrive at a sparse estimate of $\mathbf{\Sigma}_\epsilon$, denoted by $\widehat{\mathbf{\Sigma}}_\epsilon^{(1)}$. In the second step, the factor loading L is estimated by minimizing the negative (quasi) log-likelihood function of $\boldsymbol{\theta} = (L, \widehat{\mathbf{\Sigma}}_\epsilon^{(1)})$:

$$L(\boldsymbol{\theta}|\boldsymbol{Y}_1, \ldots, \boldsymbol{Y}_n) = \log|LL' + \widehat{\mathbf{\Sigma}}_\epsilon^{(1)}| + \mathrm{tr}\{(LL' + \widehat{\mathbf{\Sigma}}_\epsilon^{(1)})^{-1}\mathbf{S}\}, \tag{7.27}$$

where $\mathbf{S}$ is the sample covariance matrix of the data. The matrix $\widehat{L}$, obtained from minimizing $L(\cdot|\boldsymbol{Y}_1, \ldots, \boldsymbol{Y}_n)$ with respect to the entries of L for a given $\widehat{\mathbf{\Sigma}}_\epsilon^{(1)}$, is called its maximum quasi-likelihood estimator. The indeterminacy of $\widehat{L}$ up to a rotation, discussed in Section 7.2.2, is resolved by constraining $\widehat{L}, F_i$ so that

$$\widehat{L}'\left(\widehat{\mathbf{\Sigma}}_{\epsilon}^{(1)}\right)^{-1}\widehat{L} = \widehat{\Delta}, \quad \frac{1}{n}\sum_{i=1}^{n} F_i F_i' = I_r, \tag{7.28}$$

where $\widehat{\Delta}$ is a diagonal matrix with distinct diagonal entries arranged in a decreasing order; see (7.12). This restriction guarantees a unique solution . up to a column sign change for L. Thus, we define

$$\widehat{L}^{(1)} = \underset{L\in\Theta}{\operatorname{argmin}}(\log|LL' + \widehat{\mathbf{\Sigma}}_{\epsilon}^{(1)}| + \operatorname{tr}\{(LL' + \widehat{\mathbf{\Sigma}}_{\epsilon}^{(1)})^{-1}\mathbf{S}\}), \tag{7.29}$$

where Θ is a set of feasible loading matrices. Once $\widehat{L}^{(1)}$ is found, the estimated factor scores $\widehat{F}_i$ is computed using weighted LS; see (7.15). The two steps are carried out iteratively using the updates

$$\widehat{\boldsymbol{\epsilon}}_i = \boldsymbol{Y}_i - \widehat{L}^{(1)}\widehat{F}_i,$$

as the new data in the next iteration, and the procedure is continued until a convergence criterion is met.

The Joint Method in Bai and Liao [2016] estimates the $(L, \mathbf{\Sigma}_{\epsilon})$ by minimizing the weighted ℓ_1-penalized likelihood function:

$$(\widehat{L}^{(2)}, \widehat{\mathbf{\Sigma}}_{\epsilon}^{(2)}) = \underset{L,\mathbf{\Sigma}_{\epsilon}}{\operatorname{argmin}}(\log|LL' + \mathbf{\Sigma}_{\epsilon}| + \operatorname{tr}\{(LL' + \mathbf{\Sigma}_{\epsilon})^{-1}\mathbf{S}\}) + \lambda p(\mathbf{\Sigma}_{\epsilon})), \tag{7.30}$$

where only the off-diagonal entries of $\mathbf{\Sigma}_{\epsilon}$ are penalized using a weighted sum of their absolute values

$$p(\mathbf{\Sigma}_{\epsilon}) = \sum_{i\neq j} w_{ij}|\sigma_{\epsilon,i,j}|,$$

where the weights $w_{ij} \geq 0$, and λ is a tuning parameter.

Asymptotic analysis for both methods is developed in Bai and Liao [2016], establishing consistency, uniform rate of convergence and the limiting distribution of the estimators for the two-step procedure. As expected, the computation and asymptotic analysis of the joint estimation method are more challenging as it involves penalizing a large covariance matrix of the errors. We note that in a related work, Bai and Li [2016] also estimated the factor loadings by maximizing the Gaussian quasi-likelihood, which addresses the heterogeneity but within the setup of orthogonal factor models in which the idiosyncratic errors are uncorrelated.

Sparse estimation of the factor loading matrix L and $\mathbf{\Sigma}_{\epsilon}$ within the context of approximate factor models is pursued in Daniele et al. [2024] via a two-step procedure. In the first step, ℓ_1-penalization is applied to L while restricting $\mathbf{\Sigma}_{\epsilon}$ to be diagonal. In the second step, $\mathbf{\Sigma}_{\epsilon}$ is estimated using the POET method, which amounts to soft-thresholding the sample covariance matrix of the residuals of the estimated approximate factor model. This two-step procedure handles the sparsity of the two matrices separately, and consequently avoids some expected computational challenges.

7.4.2 Covariate-Dependent Loading Matrix

It is known that accurate estimation of the eigenvectors and hence the columns of the loading matrix L require a relatively large sample. The idea of using covariates in modeling the loading matrix and hence reducing its large number of parameters was pursued in Connor and Linton [2007] and Connor et al. [2012] using semi-parametric factor models.

More specifically, let $\boldsymbol{x}_i = (x_{i1}, \ldots, x_{iq})'$ be the q-vector of observed covariates associated with the ith response at time t. Connor and Linton [2007]'s semi-parametric factor model is

$$Y_{it} = \sum_{k=1}^{q} g_k(\boldsymbol{x}_i) f_{kt} + \epsilon_{it}, \quad i = 1, \ldots, p, t = 1, \ldots, n, \tag{7.31}$$

where $g_k(\boldsymbol{x}_i)$ is an unknown function of the q covariates. A popular idea in dealing with the curse of dimensionality in the nonparametric curve estimation is to use an additive model for $g_k(\cdot)$'s:

$$g_k(\boldsymbol{x}_i) = g_{k1}(x_{i1}) + \ldots + g_{kq}(x_{iq}).$$

In this direction, Fan et al. [2016b] go a step further presenting evidence that such semi-parametric factor models might be restrictive for some applications as the (fixed) covariates may not capture completely the structure of factor loadings. Instead, they proposed the following more flexible semi-parametric random-effect factor model:

$$Y_{it} = \sum_{k=1}^{q} [g_k(\boldsymbol{x}_i) + \gamma_{ik}] f_{kt} + \epsilon_{it}, \quad i = 1, \ldots, p, t = 1, \ldots, n, \tag{7.32}$$

where γ_{ik}'s are zero-mean random variables. Their analysis and empirical results of the returns of the stocks in the S&P500 index and four exogenous covariates (size, value, momentum, and volatility), following Connor et al. [2012], provide convincing evidence that the simpler model (7.31) do not perform well relative to that of the more elaborate model (7.32).

7.4.3 Bayesian Nonparametric Covariance Regressions

A rather general class of covariance regressions where the covariance matrix depends flexibly on covariates and easily scales to high-dimensional datasets is provided in this section. Fox and Dunson [2015] rely on factor models and reduce the task of modeling large covariance matrices to that of flexible Bayesian nonparametric regression models where the entries of the loading matrix and the factors both are dependent on covariates. In contrast to Hoff and Niu [2011] quadratic regression framework (6.7), entries of the loading matrix and latent factor are expressed in terms of a collection of basis functions that vary nonparametrically over the predictor space.

More precisely, for a collection of covariance matrices $\boldsymbol{\Sigma}_{\mathcal{X}} = \{\boldsymbol{\Sigma}(x), x \in \mathcal{X}\}$, where $\mathcal{X}$ is subset of a Euclidean space, consider the multivariate Gaussian nonparametric mean-covariance regression model

$$\boldsymbol{Y}_i = \mu(\boldsymbol{x}_i) + \boldsymbol{\epsilon}_i, \; \boldsymbol{\epsilon}_i \sim N_p(0, \boldsymbol{\Sigma}(\boldsymbol{x}_i)), \; i = 1, \ldots, n,$$

where $\boldsymbol{Y}_i$ is a response vector, $\boldsymbol{x}_i \in \mathbb{R}^q$ is the vector of covariates and $\boldsymbol{\epsilon}_i$'s are independent.

For p large, reparameterizing the mean vector or writing a factor model as in Fox and Dunson [2015] which allows the factors to depend on the covariates, one arrives at

$$\boldsymbol{Y}_i = L(\boldsymbol{x}_i)F_i + \boldsymbol{\epsilon}_i,\ F_i \sim N_r(\mathbf{B}(\boldsymbol{x}_i), I_r),\ \boldsymbol{\epsilon}_i \sim N_p(0, \boldsymbol{\Sigma}_0),\ i = 1, \ldots, n,$$

where $L(\boldsymbol{x}_i) \in \mathbb{R}^{p \times r}$ with $r < p$, and $F_i \sim N_r(\mathbf{B}(\boldsymbol{x}_i), I_r)$ are the loading matrix and latent factors, respectively, and $\boldsymbol{\Sigma}_0$ is a diagonal matrix. It is instructive to note that the vector $\mathbf{B}(\boldsymbol{x})$ has the important role of capturing the evolution of the latent factors over $\boldsymbol{x}$.

Now, integrating out the F_i's leads to the following nonparametric mean-covariance model

$$\mu(\boldsymbol{x}_i) = L(\boldsymbol{x}_i)\mathbf{B}(\boldsymbol{x}_i), \quad \boldsymbol{\Sigma}(\boldsymbol{x}_i) = L(\boldsymbol{x}_i)L(\boldsymbol{x}_i)' + \boldsymbol{\Sigma}_0, \tag{7.33}$$

which is a parsimonious model for larger p values; see Problem 8.12. However, for p large, still estimating the high-dimensional factor loading matrix $L(\boldsymbol{x}_i)$ can be computationally challenging. As a way to resolve this issue, one may take the factor loadings to be a linear combination of a much smaller set of basis elements, or factorize it as

$$L(\boldsymbol{x}_i) = \Theta\xi(\boldsymbol{x}_i),$$

where $\Theta \in \mathbb{R}^{p \times b}$ is a fixed matrix of coefficients mapping the matrix of basis functions $\xi(\boldsymbol{x}_i) \in \mathbb{R}^{b \times r}$, with $b, r < p$, to the covariate-dependent loadings matrix $L(\boldsymbol{x}_i)$. Substituting this in (7.33) leads to

$$\boldsymbol{\Sigma}(\boldsymbol{x}_i) = \Theta\xi(\boldsymbol{x}_i)\xi(\boldsymbol{x}_i)'\Theta' + \boldsymbol{\Sigma}_0, \tag{7.34}$$

which is a quadratic covariance regression model in Θ; see (6.7).

The smoothness of the response mean and covariance matrix among different experiment conditions $\boldsymbol{x}_i$ is modeled in Fox and Dunson [2015] using Bayesian setup with Gaussian processes as priors on both latent factor $\mathbf{B}(\boldsymbol{x}_i)$ and loading matrix basis $\xi(\boldsymbol{x}_i)$. A key advantage of this general Bayesian methodology over the previous approaches is its fundamental property of combining the use of a shrinkage prior with a latent factor model. This enables one to handle high-dimensional datasets in the presence of irregularities in the data such as missing values, outliers, etc.

In particular, for $q = 1$ and $\boldsymbol{x}$ representing the time, the evolution of F_i could be described by various popular time series models such a autoregressions, etc. The Bayesian nonparametric covariance regression methodology outlined has been used to fully analyze the well-known Google flu data in Fox and Dunson [2015] for $p = 183$ regions of the USA, and $q = 1$ where the predictor $x_i = i$ is a discrete time index indicating the current week for $i = 1, \ldots, 370$.

7.5 Problems

7.1 For the moment think of (7.1) as a noise-free regression of Z_i on Y where the regression coefficients are the e_{ij}'s. Show that

$$\mathrm{Cov}(Y_j, Z_i) = \lambda_i e_{ij},$$

and the correlation coefficient between Z_i and Y_j is given by

$$\rho_{Z_i,Y_j} = \sqrt{\frac{\lambda_i}{\sigma_{jj}}} \cdot e_{ij},$$

which is proportional to e_{ij} and its sign is determined by that of the e_{ij}, the entries of the corresponding eigenvector.

7.2 Consider the 3×3 covariance matrix

$$\boldsymbol{\Sigma} = \begin{pmatrix} 4 & -2 & 3 \\ -2 & 6 & -6 \\ 3 & -6 & 10 \end{pmatrix}.$$

(a) Show that for $r = 1$ it can be decomposed as (7.10) with

$$L = (1, -2, 3)' \text{ and } \boldsymbol{\Psi} = \mathrm{diag}(3, 2, 1).$$

(b) Show that the communality of Y_2 is $c_2^2 = 4$ and its specific variance is $\psi_2 = 2$. Repeat the calculations for the other two variables. Compute the communality and specific variance for the variables Y_1 and Y_3.

(c) Verify that the loading matrix $L^* = -L = (-1, 2, -3)'$, also satisfies the above decomposition.

7.3 For the 3×3 covariance matrix

$$\boldsymbol{\Sigma} = \begin{pmatrix} 1 & 0.9 & 0.7 \\ 0.9 & 1 & 0.4 \\ 0.7 & 0.4 & 1 \end{pmatrix},$$

compute the nonredundant entries of L and $\boldsymbol{\Psi}$ for an orthogonal factor model with $r = 1, 2$. Discuss whether the solutions are feasible.

7.4 Repeat the PCA analysis in Section 7.1.5 to implement and interpret perceptual map using the number of components equal to 3.

7.5 Consider data analysis in Section 7.2.4:

(a) Repeat the analysis in Section 7.2.4, but use the function scale() to standardize the dataset. Are there any differences in the results when factor analysis is implemented using the covariance matrix versus the correlation matrix?

(b) Now, repeat the factor analysis in Section 7.2.4 and try to interpret the factors when the number of selected factors are equal to 7 (the number of countries). Use the function headmap.2 from the R package gplots to produce a similar plot, as in Figure 7.6.

7.6 Consider the SW2002 dataset available from recode package. The SW2002 object is a list that contains two elements: data- the dataset as a matrix and desc - description of the variables. The dataset, from [Stock and Watson, 2002a], contains $p = 215$ monthly US macroeconomic variables spanning the period 1959-1998 with $n = 492$ observations. Using this dataset, repeat the analysis in Section 7.2.4, and summarize clearly your finding and conclusions.

7.7 Varimax encourages sparsity:

(a) Consider a point $x = (x_1, x_2) \in \mathbb{R}^2$ with $x_1^2 + x_2^2 = 1$. Show that the function $f(x_1, x_2) = x_1^4 + x_2^4 - 1$ subject to x being on the circle, is maximized when either x_1 or x_2 is zero.

(b) Repeat (a) for a function defined on $\mathbb{R}^p$.

7.8 The CS$(1, \rho)$ model in (3.12) can be viewed as a single-factor model where the loading vector $\mathbf{1}_p$ is known and x is the single latent factor and unknown. Show that X satisfies the equation

$$\frac{1}{p}\sum_{i=1}^{p} Y_i = pX + \frac{1}{p}\sum_{i=1}^{p} \varepsilon_i,$$

and hence it can be estimated by

$$\widehat{X} = \frac{1}{p}\sum_{i=1}^{p} Y_i = \widehat{Y}.$$

Discuss why the quality of this estimator improves with increasing dimension p.

7.9 From the parsimony perspective, compare the mean model in (7.33) with the model

$$\boldsymbol{Y}_i = L(\boldsymbol{x}_i)\boldsymbol{\beta}_i + \boldsymbol{\mu}(\boldsymbol{x}_i) + \boldsymbol{\epsilon}_i,$$

where $\boldsymbol{\beta}_i \sim N_r(\mathbf{0}, \boldsymbol{I}_r)$ and $\boldsymbol{\mu}(\boldsymbol{x}_i)$ is a standard p-dimensional mean regression.

7.10 (a) Show that the model (7.19) is an approximate factor model.

(b) Compute $\mathrm{Cov}(\boldsymbol{\epsilon}_i^*, \boldsymbol{\epsilon}_j^*)$ for $i \neq j$. Are they uncorrelated?

(c) Compute $\mathrm{Cov}(F_i, \boldsymbol{\epsilon}_j^*)$ and discuss the nature of their dependence.

7.11 For a zero-mean random p-vector Y with the covariance matrix $\boldsymbol{\Sigma}$ with the spectral decomposition (3.16), for a given integer $p > r > 0$, consider the optimization (regression) problem [Baldi and Hornik, 1989]:

$$\min_{(C,L)\in\mathbb{R}^{p\times r}\times\mathbb{R}^{r\times p}} \sum_{k=1}^{p} E(Y_k - C_k LY)^2,$$

where C_k is the kth row of the matrix C.

(a) Argue that the optimal solution L projects Y to an r-dimensional subspace and C reconstructs Y from this linear subspace.

(b) Show that the optimal C, L in terms of the normalized eigenvectors of $\boldsymbol{\Sigma}$ are:

$$C = (\boldsymbol{e}_1, \ldots, \boldsymbol{e}_r), \quad L = C'.$$

7.12 Consider the approximate factor model

$$\boldsymbol{Y}_t = LF_t + \boldsymbol{\epsilon}_i t \ t = 1, \ldots, n.$$

(a) Show that it can be written in matrix form as

$$\boldsymbol{Y}' = LF + \boldsymbol{\mathcal{E}},$$

where $\boldsymbol{Y}$ is the $n \times p$ data matrix, F and $\boldsymbol{\mathcal{E}}$ are the $r \times n$ and $p \times n$ matrices of the factors and the idiosyncratic errors.

(b) Use the above matrix form to obtain

$$\boldsymbol{YY}' = F'(L'L)F + \boldsymbol{\mathcal{E}}'\boldsymbol{\mathcal{E}} + (G + G'), \quad G = \boldsymbol{\mathcal{E}}'LF.$$

(c) Divide both sides of (b) by p, and take the limit as $p \to \infty$, to obtain

$$\Gamma_0 = F'F + \sigma^2 I_n,$$

assuming the convergence of the triplet

$$\frac{1}{p}(L'L, G, \sum_{i=1}^{p} \varepsilon_{it}^2) \to (I_r, \boldsymbol{0}, \sigma^2).$$

(d) One may estimate the $n \times n$ matrix Γ_0 by

$$\widehat{\Gamma}_0 = \frac{1}{p}\boldsymbol{YY}',$$

which is noticeably different from the traditional $p \times p$ sample covariance matrix $\mathbf{S}$, used when p is fixed.

(e) From part (c), it is evident that the eigenvalues and eigenvectors of $F'F$ can be obtained from those of $\widehat{\Gamma}_0$. Thus, in principle, the eigenvectors corresponding to its r largest eigenvalues can be used to form consistent estimate of the latent factors F_t.

(This problem presents a step-by-step overview of an early attempt [Connor and Korajczyk, 1986, G. and R., 1988] to link PCA for p large to consistent estimation of an approximate factor model and possible the negative role of dependence between the factors and noise processes.) .

7.13 Discuss the validity of following potential consequences of the convergence of the three components in part (c) of the previous problem:

(a) Convergence of the first component involving L, implies that its columns are asymptotically orthogonal,

(b) Convergence of the second component implies that the latent factors F_t and the idiosyncratic error are asymptotically uncorrelated,

(c) convergence of the third component for any fixed t implies a sort of cross-sectional homogeneity in $\boldsymbol{\epsilon}_t$.

8

Shrinkage and Thresholding

Structure and sparsity are quite helpful in reducing the large number of parameters of a large covariance matrix with shrinkage and thresholding as two popular methods for estimating them. The first shrinks the sample covariance matrix toward a simple structure like the identity matrix using a convex combination $(1 - \delta)\mathbf{S} + \delta \boldsymbol{I}$, where $0 \leq \delta \leq 1$ is the *shrinkage intensity*. The second replaces small entries of $\mathbf{S}$ with magnitudes less than a threshold λ by zero. We review diverse methods of motivating the shrinkage toward this simple convex combination. While shrinkage estimators have a longer history, they shrink only the eigenvalues and leave the eigenvectors intact; in contrast, thresholding methods affect both the eigenvalues and eigenvectors of the sample covariance matrix which makes them more attractive for PCA and factor analysis. Both are computationally simple. For shrinkage, a target, choice of a loss function and shrinkage intensity are needed, but for thresholding, selection of the shrinkage parameter λ is the key in implementing it. There are situations in finance and microarray analysis where certain groups of variables are highly correlated due to common latent variables and their covariance matrices are far from being sparse, but can be decomposed as a sum of low-rank and sparse covariance matrices of the type seen in factor models. More general and elaborate methods like ℓ_1-based regularization and Graphical Lasso for covariance/precision matrix estimation are discussed in Chapter 9.

8.1 Shrinking the Sample Eigenvalues

It should come as no surprise that a general covariance matrix has far too many parameters compared to a mean vector. A way to reduce the number of entries to be estimated is to make sparsity structural assumptions so that the parameters can be estimated consistently. Popular families of sparsity-structured (S-structured) covariance matrices include bandable, sparse and spiked covariance matrices, and covariances with a tensor product structure. Heuristically, very large covariance matrices could be *sparse* or have many small entries, so that componentwise regularization or estimating small entries by zeros seems natural. At one extreme, experience in the context of classification in machine learning [Domingos and Pazzani, 1997] and microarray data [Dudoit et al., 2002] suggests that using a diagonal covariance matrix or ignoring

DOI: 10.1201/9781003638896-8

most of the off-diagonal entries (naive Bayes) can lead to better classification rules than those based on estimating the whole covariance matrix.

It was noted by Stein [1975], and supported by later researchers, that the sample covariance matrix $\mathbf{S}$ is a notoriously bad estimator of $\boldsymbol{\Sigma}$ in that the largest (smallest) eigenvalue of $\mathbf{S}$ tends to be larger (smaller) than its population counterpart. Moreover, for $n < p$ the sample covariance matrix is singular and cannot be inverted, an operation which is needed in many multivariate statistical methods. In the absence of any knowledge about the structure of $\boldsymbol{\Sigma}$, such as sparsity, bandedness or factor structures, a good strategy is to correct for the bias of the eigenvalues by focusing on the class of *orthogonally invariant or equivariant* estimators, which only modifies the sample eigenvalues and leaves the eigenvectors intact.

Let $\lambda = (\lambda_1, \ldots, \lambda_p)', \lambda_1 \geq \ldots \geq \lambda_p > 0$, be the ordered eigenvalues of the population covariance matrix. We denote the ordered sample eigenvalues by $\widehat{\lambda} = (\hat{\lambda}_1 \ldots, \hat{\lambda}_p)'$ and let $\hat{P} = (\hat{\mathbf{e}}_1, \ldots, \hat{\mathbf{e}}_p)$ be the orthogonal matrix of the corresponding normalized eigenvectors of $\mathbf{S}$. The generic form of an orthogonally invariant proto-estimator of $\boldsymbol{\Sigma}$ is

$$\widehat{\boldsymbol{\Sigma}} = \hat{P} D \hat{P}' = \sum_{i=1}^{p} d_i \hat{\mathbf{e}}_i \hat{\mathbf{e}}_i', \quad d_i = g^{-1}[\hat{\mathbf{e}}_i' g(\boldsymbol{\Sigma}) \hat{\mathbf{e}}_i], \ i = 1, \ldots, p, \tag{8.1}$$

where $g : \mathbb{R}^+ \longrightarrow \mathbb{R}$, is a smooth invertible function, which modifies the eigenvalues, but leaves the eigenvectors intact and D is a diagonal matrix with diagonal entries d_i. The proposal in (8.1) is a proto-estimator, not a genuine estimator, since it depends on the unknown $\boldsymbol{\Sigma}$.

The amount or type of shrinkage of the sample eigenvalues is determined by the choice of a risk function and its ensuing minimizer. For example, Stein [1975] employed the entropy loss or the Kullback-Liebler divergence of two multivariate normal distributions with different covariance matrices in (4.38) and derived its approximate minimizer. For details, see Stein [1975], Ledoit and Wolf [2004] and Pourahmadi [2013, Section 2.2]. A more timely and detailed discussion with closed form formulas for the optimal eigenvalue shrinkers are worked out in Donoho et al. [2016] for a menagerie of 26 loss functions for covariance estimation including the Stein, Entropy, Frobenius Norm, Operator Norm, Nuclear Norm and Condition Number losses. The functions g corresponding to some of these loss functions are $x, \frac{1}{x}, \sqrt{x}, \log(x), x^2, \frac{1}{x^2}$. The paper by Ledoit and Wolf [2022] presents a comprehensive review of the power of non-linear shrinkage and its importance for analysis of high-dimensional data in business and economics.

8.1.1 Shrinking the Sample Covariance Matrix

In this section, we provide some basic results on improving the sample covariance matrix by shrinking the eigenvalues of the sample covariance matrix toward a central value.

For simplicity we set the mean vector to be $\boldsymbol{\mu} = \mathbf{0}$, and let $\boldsymbol{Y}$ be the $n \times p$ data matrix so that $\mathbf{S} = \frac{1}{n}\boldsymbol{Y}'\boldsymbol{Y}$, is its sample covariance matrix. From the success of the ridge estimators in the regression setup, it seems a simple way to address the singularity of $\mathbf{S}$ when $p > n$ is to replace it by $\mathbf{S} + \kappa \mathbf{I}_p$, which is then invertible for any $\kappa > 0$. There is a host of estimators of this form known as shrinkage estimators where though motivated from various perspectives, at the end they amount to shrinking the sample eigenvalues of $\mathbf{S}$ toward a central value. Their generic form is

$$\widehat{\mathbf{\Sigma}} = \alpha_1 I + \alpha_2 \mathbf{S}, \tag{8.2}$$

where α_1 and α_2 are scalars chosen to minimize certain loss/risk functions. Heuristically, they are motivated by recalling that the sample covariance matrix is unbiased for $\mathbf{\Sigma}$, but unstable when $n \ll p$. At the other extreme, a structured covariance estimator like the identity matrix, has very little variability, but can be severely biased. Thus, a natural compromise between these two extremes is perhaps a convex combination of the form (8.2) [Ledoit and Wolf, 2004]. In fact, a motivation of Ledoit and Wolf [2004] was to carry Stein [1975]'s shrinkage program to the next level: shrink a covariance like a mean, see also Pourahmadi [2013, Section 2.4].

Next, we take a slight detour and present as another motivation a non-traditional approach to arriving at (8.2) due to Maurya [2016] which penalizes the larger spread of the eigenvalues.

8.1.2 Penalizing the Spread of the Eigenvalues

The intuitive idea of shrinking the distorted sample eigenvalues toward a central value is the key in this approach to shrinking. It relies on a natural measure of spread of the eigenvalues, namely their variance:

$$v^2 = \frac{\sum_{i=1}^{p} (\lambda_i - \bar{\lambda})^2}{p}, \tag{8.3}$$

with $\bar{\lambda}$ the average of the p population eigenvalues.

A way to improve the sample covariance matrix is to penalize the larger spread of the eigenvalues as quantified by v^2. More precisely, an improved estimator is defined as the solution of the optimization problem

$$\widehat{\mathbf{\Sigma}} = \min(||\mathbf{\Sigma} - \mathbf{S}||^2 + \gamma p v^2), \tag{8.4}$$

subject to $\mathrm{tr}(\mathbf{\Sigma}) = \mathrm{tr}(\mathbf{S})$ and $\mathbf{\Sigma}$ be a positive-definite matrix, where $\gamma > 0$ is a penalty on the spread of the population eigenvalues. Evidently, it shrinks more the extreme eigenvalues than the middle ones and hence reduces the bias nonlinearly.

Since v^2 is a quadratic function of the eigenvalues, standard calculations and using the condition $\mathrm{tr}(\mathbf{\Sigma})$=$\mathrm{tr}(\mathbf{S})$ reduce the objective function to

$$\begin{aligned} ||\mathbf{\Sigma} - \mathbf{S}||^2 + p\gamma v^2 &= (1+\gamma)\{\mathrm{tr}\mathbf{\Sigma}^2 - \frac{2}{1+\gamma}\mathrm{tr}\mathbf{\Sigma}(\mathbf{S} + \gamma\tau I)\} + C \\ &= (1+\gamma)\{||\mathbf{\Sigma} - \frac{1}{1+\gamma}(\mathbf{S} + \gamma\tau I)||_F^2\} + C, \end{aligned}$$

where C is a scalar not depending on $\mathbf{\Sigma}$ and is different in each appearance. From which the following closed-form solution is immediate:

$$\widehat{\mathbf{\Sigma}} = \frac{1}{1+\gamma}\mathbf{S} + \frac{\gamma}{1+\gamma}\tau I, \quad \tau = \frac{\operatorname{tr}(\mathbf{S})}{p}. \tag{8.5}$$

Evidently, the solution is a convex combination of the sample covariance matrix $\mathbf{S}$ and a diagonal (highly structured) matrix τI, as in (8.2); see Problem 8.1. Note that while its computation does not require the eigenvalues of $\mathbf{S}$, the eigenvalues of the improved covariance estimator $\widehat{\mathbf{\Sigma}}$ are given by

$$\frac{\hat{\lambda}_i + \gamma\tau}{1+\gamma}, \quad i = 1, \ldots, p,$$

which does depend on those of $\mathbf{S}$ in a transparent and simpler manner than that in (8.1).

We note that though the above covariance estimator is well-conditioned, it is not sparse. For future references, sparsity can be introduced by adding the penalty term $\lambda||\mathbf{\Sigma}^-||_1$ to (8.5), where $||\boldsymbol{A}^-||_1$ stands for the sum of absolute values of the off-diagonal entries of a symmetric matrix $\boldsymbol{A}$. Interestingly, the new optimization problem still has a closed-form solution involving the soft-thresholding operator:

$$\widehat{\mathbf{\Sigma}}_\nu = \frac{1}{1+\gamma}\operatorname{sign}(\mathbf{S})\{\mathbf{S} + (\tau\gamma - \lambda/2)I\}_+. \tag{8.6}$$

However, this covariance estimator is not necessarily positive-definite. Maurya [2016] restricts the tuning parameters (λ, γ) to ensure positive-definiteness. The optimal tuning parameter (λ, γ) is usually found by minimizing a 5-fold cross-validation error. The R package JPEN, developed by Maurya [2016], implements the above computations.

8.1.3 Bayes and Empirical Bayes Estimators

This section presents a Bayesian motivation for the class of estimators of the form (8.2). For a quick review of general Bayesian covariance estimation, see Ledoit and Wolf [2004] and Pourahmadi [2013, Section 3.13] and references therein.

Recall that in the univariate normal data setup, Bayesian estimation of σ^2 usually employs the *inverse Gamma* prior, its analog in the multivariate setup is the *inverse Wishart* distribution; see (4.5)–(4.6) and Anderson [2003, Chapter 7] for definitions and details on the derivations of the corresponding posterior distribution and mode. A reader satisfied with similarity of the forms of (8.2), (8.9) and (8.10), may skip this section in the first reading.

A typical formulation of Bayesian framework for estimation of covariance matrices [Evans, 1965] is to assume normality of the data and the inverse Wishart distribution as a prior for the precision matrix:

$$Y|\mathbf{\Sigma} \sim N_p(0, \mathbf{\Sigma}), \quad \mathbf{\Sigma}^{-1} \sim W_p(F^{-1}/\nu, \nu), \tag{8.7}$$

where the matrix F and the scalar ν are the hyperparameters, assumed known for the time being. It follows that

$$E(\mathbf{\Sigma}^{-1}) = F^{-1},$$

so that the degrees of freedom ν reflects the degree of belief one has in the prior mean, namely a larger ν corresponds to a stronger belief about the prior mean F^{-1}.

Given the $n \times p$ data matrix $\boldsymbol{Y}$ with the sample covariance matrix $\mathbf{S}$, and computing the posterior distribution of $\mathbf{\Sigma}$ with the above prior one obtains

$$\mathbf{\Sigma}|\mathbf{S} \sim W_p^{-1}\left([(n-1)\mathbf{S} + \nu F]^{-1}, n + p + \nu\right), \tag{8.8}$$

which is also an inverse Wishart distribution. Its mode or the Bayes estimate is given by

$$\hat{\mathbf{\Sigma}}_B = \frac{n-1}{n-1+\nu}\mathbf{S} + \frac{\nu}{n-1+\nu}F, \tag{8.9}$$

which turns out to be a convex combinations of the sample covariance matrix $\mathbf{S}$ and F. It is instructive to note that when $\nu \longrightarrow \infty$ or when the prior belief is very strong, the Bayes estimator reduces to the prior mean F and the contribution of the data or $\mathbf{S}$ washes out.

Thus far the hyperparameters F and ν were assumed known. Treating (F, ν) as fixed unknowns, one may estimate them from the marginal distribution of the data using the method of maximum likelihood which is also the method of *empirical Bayes* [Haff, 1980, Efron, 2010]. Once (F, ν) is estimated, the empirical Bayes estimator of the population covariance matrix is obtained by plugging-in for the parameters in (8.9) leading to

$$\widehat{\mathbf{\Sigma}}_{EB} = \frac{np - 2n - 2}{n^2 p}\alpha_{EB} I_p + \frac{n}{n+1}\mathbf{S}, \tag{8.10}$$

where $\alpha_{EB} = [\det(\mathbf{S})]^{1/p}$ and the subscript EB stands for empirical Bayes. Note that the EB estimators work so long as $n > p$, but the class of estimators in the next section do away with this requirement.

8.1.4 Ledoit-Wolf Shrinkage Estimators

Instead of the convex combination of $\mathbf{S}$ and the highly structured identity matrix in (8.2), motivated by (8.7) one may consider replacing the identity matrix by a more general target matrix $F = F(\theta)$ with a few unknown parameters denoted by θ. Two key properties of a good target matrix pointed by Ledoit and Wolf [2004] are:

1. it should have a small number of parameters to be estimated,

2. it should capture important structural characteristics of the true covariance matrix.

The popular identity matrix as a target, is a parameter-free target that satisfies the first requirement, but may fail to capture the true covariance structure. In addition to the identity matrix, other popular examples of F are the *compound symmetry* in Example 9, *single-index factor model* in Chapter 7, AR(1), MA(1) with two parameters in Chapter 11. Once a target matrix is selected, computing the shrinkage intensity is an important problem and has been studied extensively in Ledoit and Wolf [2004], see also Schafer and Strimmer [2005].

Once θ is estimated by $\hat{\theta}$ using the available data, setting $\hat{F} = (\hat{f}_{ij}) = F(\hat{\theta})$, the estimator

$$\widehat{\mathbf{\Sigma}}_{LW} = (1-\delta)\mathbf{S} + \delta\hat{F}, \tag{8.11}$$

where $0 \leq \delta \leq 1$ is the *shrinkage intensity* is called a Ledoit-Wolf shrinkage estimator.

The loss function for estimating the shrinkage intensity based on the Frobenius norm between the true and estimated covariance matrices is

$$||\mathbf{\Sigma} - \hat{\mathbf{\Sigma}}_{LW}||^2 = ||\mathbf{\Sigma} - \delta\hat{F} - (1-\delta)\mathbf{S}||_F^2, \tag{8.12}$$

which is quadratic in δ and does not involve inversion of the matrix $\mathbf{S}$. The corresponding risk

$$R(\delta) = E(||\mathbf{\Sigma} - \delta\hat{F} - (1-\delta)\mathbf{S}||_F^2),$$

rewritten as

$$\begin{aligned} R(\delta) &= \sum_{i=1}^{p}\sum_{j=1}^{p} E[\delta\hat{f}_{ij} + (1-\delta)s_{ij} - \sigma_{ij}]^2 \qquad (8.13) \\ &= \sum_{i=1}^{p}\sum_{j=1}^{p}\{\mathrm{Var}[\delta\hat{f}_{ij} + (1-\delta)s_{ij}] + [E(\delta\hat{f}_{ij} + (1-\delta)s_{ij} - \sigma_{ij})]^2\} \\ &= \sum\sum\{\delta^2\mathrm{Var}(\hat{f}_{ij}) + (1-\delta)^2\mathrm{Var}(s_{ij}) + 2\delta(1-\delta)\mathrm{Cov}(\hat{f}_{ij}, s_{ij}) \\ &\quad + \delta^2(f_{ij} - \sigma_{ij})^2\}, \end{aligned}$$

is quadratic in δ, where we are using $E(\hat{f}_{ij}) = f_{ij}$ and $E(s_{ij}) = \sigma_{ij}$. Now, minimizing it with respect to δ is a gratifying calculus exercise which leads to the closed-form solution

$$\tilde{\delta} = \frac{\sum\sum \mathrm{Var}(s_{ij}) - \mathrm{Cov}(\hat{f}_{ij}, s_{ij})}{\sum\sum \mathrm{Var}(\hat{f}_{ij} - s_{ij}) + (f_{ij} - \sigma_{ij})^2}, \tag{8.14}$$

see Problem 8.4. It highlights the relevance and impact of $\mathrm{Var}(s_{ij})$, $\mathrm{Cov}(\hat{f}_{ij}, s_{ij})$ and the degree of misspecification $f_{ij} - \sigma_{ij}$ on the magnitude of the shrinkage intensity. For example, all other terms kept fixed, larger values of $\mathrm{Var}(s_{ij})$ lead to a larger $\tilde{\delta}$ and hence to a smaller weight for the sample covariance matrix $\mathbf{S}$.

This role becomes even more clear when n is large, the asymptotic behavior of the shrinkage intensity is of the form

$$\tilde{\delta} = \frac{C}{n} + O(\frac{1}{n^2}), \quad C = \frac{\pi - \rho}{\gamma}, \tag{8.15}$$

where

$$\pi = \sum\sum \text{AsyVar}(\sqrt{n}s_{ij}),$$
$$\rho = \sum\sum \text{AsyCov}(\sqrt{n}\hat{f}_{ij}, \sqrt{n}s_{ij}),$$
$$\gamma = \sum\sum (f_{ij} - \sigma_{ij})^2,$$

are, respectively, sum of asymptotic variances, covariances and the misspecification measures. As expected, the shrinkage intensity increases for larger π, and decreases for larger γ and ρ where the latter measures the covariance between $\mathbf{S}$ and $\hat{F}$.

Note that $\widehat{\mathbf{\Sigma}}$ as a convex combination of F and $\mathbf{S}$ is not a bona fide *estimator*, because it involves scalars π, ρ, γ which depend on the unknown $\mathbf{\Sigma}$. Of course, these can be estimated by replacing the population covariance matrix $\mathbf{\Sigma}$ by its sample counterpart $\mathbf{S}$. It turns out that establishing consistency of these three parameters, the shrinkage intensity and estimators of the type (8.11) bring to fore an interesting interplay among the sample size, dimension, moments and the dependence structure of the data matrix.

More precisely, writing $\pi = \sum\sum \pi_{ij}$ where $\pi_{ij} = \text{AsyVar}(\sqrt{n}s_{ij})$, and similarly for ρ and γ, using standard sample estimators of these quantities (see Section 8.3.2 and Ledoit and Wolf [2004]), and plugging-in for π, ρ and γ one arrives at $\hat{C}$, and hence a shrinkage covariance estimator

$$\widehat{\mathbf{S}} = \frac{\hat{C}}{n}\hat{F} + \left(1 - \frac{\hat{C}}{n}\right)\mathbf{S}. \tag{8.16}$$

Extensive simulation study in Ledoit and Wolf [2004] provides ample evidence in support of the superiority of this class of shrinkage estimators compared to the sample covariance matrix, the class of empirical Bayes estimators [Haff, 1980], and a few others reviewed in Lin and Perlman [1985]. A matlab code for implementing this estimator is available at www.ledoit.net.

8.1.5 The R Package "corpcor"

The choice of the target F is important in the performance of shrinkage estimators and it should be guided by the true (lower-dimensional) structure of the unknown covariance matrix. A requirement which is hard to observe in practice. In this section, we present six different target covariance matrices along with analytical formulae for their shrinkage intensities given in the popular R package corpcor [Schafer and Strimmer, 2005].

The corpcor implements a slightly different and perhaps more practical version of Ledoit-Wolf estimator of a covariance matrix by separately shrinking the variances and correlations toward their respective targets. More precisely, for a random vector $Y = (Y_1, \ldots, Y_p)'$ with the covariance matrix $\mathbf{\Sigma} = (\sigma_{ij})$, it relies on the variance-correlation separation of the form:

$$\mathbf{\Sigma} = DRD, \quad D = \text{diag}(\sqrt{\sigma}_{11}, \ldots, \sqrt{\sigma}_{pp}), \quad R = (\text{Corr}(Y_i, Y_j)). \tag{8.17}$$

Then, suitable shrinkage estimates of D^2 and R are computed and plugged-in in (8.17). The shrinkage estimate of the variances is obtained by shrinking the sample variances of the p variables toward their median; its shrinkage intensity is computed using the analytic formula in Opgen-Rhein and Strimmer [2007]. The correlation matrix is estimated by shrinking the sample correlation matrix toward the identity target matrix; its shrinkage intensity is computed using the analytic formulae in Schafer and Strimmer [2005, Table 2]. Their six target matrices (A-F) can be described precisely in terms of the specific values of parameters of the compound symmetry matrix $\mathrm{CS}(\rho, \sigma^2)$ of Example 9 and its matrix version as follows:

1. Targets A, B and C correspond, respectively, to CS(0,1), CS(0,σ^2) and $\mathrm{CS}(\rho, \sigma^2)$, with 0, 1, 2 parameters to be estimated.

2. With a slight abuse of notation, one may view (8.17) as a matricial $\mathrm{CS}(R, D^2)$. Then, with $\boldsymbol{J}_p$ a $p \times p$ matrix of all ones representing a perfect positive correlation, the Targets D, E and F correspond, respectively, to matricial CS with R and D^2 given by
$$\boldsymbol{I}_p, \quad \mathrm{diag}(\sigma_1^2, \ldots, \sigma_p^2),$$
$$\boldsymbol{J}_p, \quad \mathrm{diag}(\sigma_1^2, \ldots, \sigma_p^2),$$
$$\rho \boldsymbol{I}_p + (1-\rho)\boldsymbol{J}_p, \quad \mathrm{diag}(\sigma_1^2, \ldots, \sigma_p^2), \tag{8.18}$$
with p, p, and $p+1$ parameters to be estimated.

The parameters σ_i^2, the common variance σ^2 and correlation ρ of the target matrices are estimated by the corresponding diagonal entries, average of the diagonal entries and off-diagonal entries of the sample covariance matrix **S** and the sample correlation matrix, respectively. Estimators of the shrinkage intensity invariably depend on the estimators of $\mathrm{Var}(s_{ij})$ where one may rely on the estimators of the relevant entries of the matrix in (4.7) as a natural starting point. The empirical results in Schafer and Strimmer [2005] show that the method is both computationally and statistically efficient, works well for "small n, large p" and always return a positive-definite and well-conditioned covariance matrix estimator.

As noted earlier the choice of a target matrix is difficult in practice. A nonparametric shrinkage method based on repeated sample splitting in Abadir et al. [2014] is target-free in the sense that it does not rely on choosing a target F and requires minimal distributional assumptions on the data. It is easy to implement, though it takes a lot of computational time and cannot handle large dimensions.

8.2 Thresholding the Sample Covariance Matrix

The idea of thresholding from nonparametric function estimation has proved useful in the estimation of large covariance matrices where it is known that the sample covariance matrix $\mathbf{S} = \frac{1}{n}\boldsymbol{Y}'\boldsymbol{Y}$ performs poorly in high dimensions. Thus, when the population covariance matrix is sparse, one may regularize $\mathbf{S}$ by applying the hard- or soft-thresholding operator (2.13) to it elementwise, effectively replacing its smaller entries by zeros.

8.2.1 Hard- and Soft-Thresholding

Using the hard-thresholding operator T_λ for a $\lambda \geq 0$, given $\mathbf{S} = (s_{ij})$ we define a covariance estimator by $T_\lambda(\mathbf{S}) = \left(s_{ij}\mathbf{1}(|s_{ij}| \geq \lambda)\right)$. This amounts to replacing by zero all small entries of $\mathbf{S}$ with absolute value less than the tuning parameters λ. Note that the transformation T_λ preserves symmetry and is invariant under permutations of variable indices, but does not necessarily preserve positive definiteness. Its biggest advantage is its simplicity as it carries no major computational burden compared to its competitors like the covariance graphical Lasso in Chapter 9.

Similarly, a soft-thresholded covariance estimator is defined by

$$\widehat{\mathbf{\Sigma}}_\lambda = S(s_{ij}, \lambda), \tag{8.19}$$

where $\lambda \geq 0$ is a tuning parameter and $S(\cdot, \lambda)$ the soft-thresholding operator defined in Chapter 2. It turns out that such a simple covariance estimator is the solution of the following convex optimization problem:

$$\begin{aligned}
\widehat{\mathbf{\Sigma}}_\lambda &= \underset{\mathbf{\Sigma}}{\operatorname{argmin}} \left\{ \frac{1}{2} ||\mathbf{\Sigma} - \mathbf{S}||_F^2 + \lambda ||\mathbf{\Sigma}||_1^- \right\} \\
&= \underset{\mathbf{\Sigma}}{\operatorname{argmin}} \left\{ \frac{1}{2} \sum_{i=1}^{p} \sum_{j=1}^{p} (\sigma_{ij} - s_{ij})^2 + \lambda \sum_{i=1}^{p} \sum_{j \neq i}^{p} |\sigma_{ij}| \right\} \\
&= \underset{\mathbf{\Sigma}}{\operatorname{argmin}} \sum_{i=1}^{p} \sum_{j \neq i}^{p} \left\{ \frac{1}{2} (\sigma_{ij} - s_{ij})^2 + \lambda |\sigma_{ij}| \right\},
\end{aligned} \tag{8.20}$$

where $||\cdot||_F$ stands for the Frobenius norm and $||\cdot||_1^-$ stands for the ℓ_1-norm of the off-diagonal elements of its matrix argument. Note that the optimization problem above is the same as minimizing for each $i \neq j$, the expression inside the last curly bracket which is of the form (2.10) and hence amounts to soft-thresholding the individual entry s_{ij}.

The choice of the threshold parameter $\lambda = \lambda_n$ is important in implementing this procedure in practice and a cross-validation method is usually employed for its selection. When the same tuning parameters λ_n is used for all the off-diagonal entries, the process is referred to as a *universal thresholding*. However, the entries of the sample covariance matrix are known to be heteroscedastic in the sense that s_{ij}'s have different variances which depend on the distribution of the pairs (Y_i, Y_j). Setting

$$\theta_{ij} = \text{Var}[(Y_i - \mu_i)(Y_j - \mu_j)], \tag{8.21}$$

then it is known that when $\boldsymbol{Y}$ is Gaussian than

$$\sigma_{ii}\sigma_{jj} \leq \theta_{ij} \leq 2\sigma_{ii}\sigma_{jj}, \tag{8.22}$$

so that the condition $\max_i \sigma_{ii} \leq C$ in (8.31) ensures that the variances of the entries of the sample covariance are uniformly bounded. However, outside this class it makes sense to threshold differentially to account for the possible differences in the variability of the different entries. To this end, an *adaptive* thresholded estimator of the form

$$\widehat{T}^*_{\lambda_{ij}}(\mathbf{S}) = (T_{\lambda_{ij}}(s_{ij})), \tag{8.23}$$

where each entry has a different threshold

$$\lambda_{ij} = \delta\sqrt{\frac{\widehat{\theta}_{ij}\log p}{n}}, \tag{8.24}$$

was introduced by Cai and Liu [2011], where δ is a tuning parameter that can be fixed at 2 or chosen empirically through cross-validation with

$$\widehat{\theta}_{ij} = \frac{1}{n}\sum_{k=1}^{n}[(Y_{ki} - \bar{Y}_i)(Y_{kj} - \bar{Y}_j) - s_{ij}]^2,$$

where for an $n \times p$ data matrix $\boldsymbol{Y} = (Y_{ki})$, $\bar{Y}_i$ stands for the mean of the entries of the i-th column.

It is known [Bickel and Levina, 2008b] that under some regularity conditions such sparse covariance estimator is consistent in high dimensions. A consequence of the consistency of the thresholded estimator is that its positive-definiteness is guaranteed in the asymptotic setting with high probability. However, the finite-sample estimators are not necessarily positive-definite matrices as illustrated by Xue et al. [2012] using the Michigan lung cancer data with $n = 86$ tumor samples from patients with lung adenocarcinomas and 5217 gene expression values for each sample. Randomly choosing p genes ($p = 200, 500$) they obtained the $p \times p$ soft-thresholding sample correlation matrix for these genes, and repeated the process ten times for each p, each time the thresholding parameter λ was selected via a 5-fold cross validation. It was found that none of the soft-thresholding estimators were positive-definite. On the average, there were 22 and 124 negative eigenvalues for the soft-thresholding estimator for $p = 200$ and 500, respectively, see also Section 8.3.1 for more examples and discussions.

A popular way to deal with the lack of positive-definiteness is to utilize the eigen-decomposition

$$\mathbf{S} = \sum_{i=1}^{p}\widehat{\lambda}_i\widehat{e}_i\widehat{e}'_i, \tag{8.25}$$

and project it into the convex cone of positive-definite matrices to obtain a positive semi-definite estimator by setting

$$\tilde{\mathbf{S}}^+ = \sum_{i=1}^{p}\max(\widehat{\lambda}_i, 0)\widehat{e}_i\widehat{e}'_i.$$

Unfortunately, this strategy does not bode well with the sparsity, because the projection destroys the sparsity patterns. In particular, for the Michigan lung cancer data, after the semi-definite projection the soft-thresholding estimator had no zero entry. However, the alternative

$$\mathbf{S}_{+} = \mathbf{S} + |\widehat{\lambda}_p| \cdot I\{\widehat{\lambda}_p < 0\}\boldsymbol{I}_p, \tag{8.26}$$

in Cai et al. [2016] preserves the structure and sparsity with interesting additional properties; see Problem 8.10.

8.2.2 Generalized Thresholding

There is a more general class of *generalized thresholding operators* which combines thresholding with shrinkage [Rothman et al., 2009]. It includes as special cases the hard and soft thresholding, smoothly clipped absolute deviation (SCAD), and adaptive Lasso. For a $\lambda \geq 0$, the generalized thresholding operator is a function $\mathcal{T}_\lambda : \mathbb{R} \to \mathbb{R}$ satisfying the following conditions for all $x \in \mathbb{R}$:

(i) $|\mathcal{T}_\lambda(x)| \leq |x|$,

(ii) $\mathcal{T}_\lambda(x) = 0 \quad \text{for} \quad |x| \leq \lambda$,

(iii) $|\mathcal{T}_\lambda(x) - x| \leq \lambda$.

We note that the conditions (i)–(ii) enforce shrinkage and thresholding, and finally the condition (iii) restricts the amount of shrinkage to no more than λ.

The generalized thresholded covariance estimator is computed by applying the operator $\mathcal{T}_\lambda(\cdot)$ to each entry of $\mathbf{S}$:

$$\widehat{T}_\lambda(\mathbf{S}) = (\mathcal{T}_\lambda(s_{ij})).$$

In high dimensions it has good theoretical properties [Rothman et al., 2009] and carries almost no computational burden other than the selection of the threshold parameter.

Unlike banding and tapering introduced next, thresholding does not require the variables to be ordered so that the estimator is invariant to permutation of the variables. Banding is a simple way of estimating large covariance matrices by setting to zero a contiguous group of entries away from the main diagonal. It starts with estimating the main diagonal, then successively the first, second, . . . , ℓ-th subdiagonals. Choosing the *banding* parameter ℓ is the key challenge. Given a $p \times p$ sample covariance matrix $\mathbf{S} = (s_{ij})$ and an integer ℓ, $0 \leq \ell \leq p$, its ℓ-banded version defined by

$$B_\ell(\mathbf{S}) = [s_{ij}\mathbf{1}(|i - j| \leq \ell)], \tag{8.27}$$

serves as an estimator for Σ [Bickel and Levina, 2008a] where $\mathbf{1}(\cdot)$ is the indicator function. The idea of banding covariance matrices has its root in time series data analysis where the observations have a natural time order, and class of moving average models of order ℓ implicitly does the job. Thus, the ideal setup for banding occurs when the variables have a natural order or their indices can be arranged so that entries of the covariance matrix farther away from the main diagonal are negligible, namely $|i-j| > k$ implies that $\sigma_{ij} = 0$. This occurs, for example, for some time series and longitudinal data. The performance of a banding estimator depends critically on the optimal choice of the banding parameter which is usually done via cross-validation. However, there is no guarantee that a banded covariance estimator is positive-definite.

A smoother version of banding called *tapering* gradually shrinks the off-diagonal entries toward zero using a so-called *taper function*. Its suitable choice guarantees the positive-definiteness of the tapered covariance estimator. Like banding, tapering a covariance matrix has its roots in time series analysis and signal processing. It can be used to improve the performance of linear discriminant analysis [Bickel and Levina, 2004] and Kalman filter ensembles [Furrer and Bengtsson, 2007]. A tapered estimator of $\mathbf{\Sigma}$ for a given tapering matrix $W = (w_{ij})$, replaces $\mathbf{S}$ by $\mathbf{S}_W = \mathbf{S} * W = (s_{ij}w_{ij})$ where $(*)$ denotes the Schur (element-wise) matrix multiplication. When $W = (w_{ij})$ is a positive-definite symmetric matrix, then $\mathbf{S}_W$, as the Schur product of two positive-definite matrices, is guaranteed to be positive-definite (Problem 3.7). We note that banding corresponds to the *rectangular* weight $W = (w_{ij}) = (\mathbf{1}(|i-j| \leq \ell)$, which is not necessarily a positive definite matrix. The choice of a smoother positive-definite tapering matrix W with off-diagonal entries gradually decaying to zero will ensure the positive-definiteness as well as optimal rate of convergence of the tapered estimator.

The consistency in operator norm of these and other covariance estimators is important in applications as it implies the consistency of the eigenvalues and eigenvectors, and hence the consistency of the principal component analysis [Johnstone and Lu, 2009a], and related eigenstructure-based methods in multivariate statistics when n is small and p is large; see Section 8.4.

Historically, the notion of bandedness is tied to an ordering of the variables. However, recently Bien [2019] has defined bandedness with respect to a known graph of the variables. For example, such graphs are available in diverse situations like:

1. Image data: For any $p_1 \times p_2$ pixeled image, the $p = p_1 \times p_2$ pixels lie on a two-dimensional lattice graph where it is expected that nearby pixels would be more correlated than those far apart.

2. fMRI data: Related to the previous example, but instead one observes three-dimensional images of the brain over time. where the voxels (three-dimensional pixels) lie on a three-dimensional lattice.

3. Gene expression data: Genes encoding proteins that are known to interact may potentially be more likely to be co-expressed than genes whose proteins do not interact. Constructed protein interaction networks [Das and You, 2012] can be taken as seed graphs, leveraging protein interaction data for better estimation of the covariance matrix of gene expression data.

For the analysis of data in the above situations it is common to vectorize the data and ignore the graph structure when estimating the covariance matrix. However, ignoring the banded structure with respect to a known graph would lead to undully complications when in fact a much simpler approach would work.

8.3 Hands on R: Comparing Thresholding Estimators

In this section, we use the dataset from a cardiovascular microarray study, available from recode package to implement covariance matrix estimation using hard-, soft-, and adaptive- thresholding estimators with 5-fold cross-validation. The dataset has 63 subjects with 44 healthy controls and 19 cardiovascular patients. For each subject, 20,426 genes have been measured.

We start by importing the required packages and the dataset.

```
## Import required packages and dataset
require(recode)
data(cardiodata)
```

Similar to Xue et al. [2012], we use F-statistic to rank the genes by how much discriminative information they provide. Then pick the top 50 and bottom 100 genes so we have both informative and noninformative genes using the generate_cardio_fstat() function in the package recode. For more details on selection, see Rothman et al. [2009].

```
top = 50
last = 100
## Generate F-stat
df = generate_cardio_fstat(cardiodata, top = top, last = last)
```

By doing so, we expect a weak dependence between top and the bottom genes. For all three thresholding estimators, we use cv.covEstimation() function, available in recode package. In addition to thresholding estimators, this function also provides options method = "tapering", and method = "banding", which we do not explore here. The estimated covariance matrix is displayed using plot_mat() function.

```
## Implement hard thresholding
est_hard = cv.covEstimation(df, k.min = 0.01, method =
    "thresholding", thr.method = "hard")
## Plot the result
plot_mat(est_hard$Sigma_fit, main = "Hard thresholding")
## Print selected threshold
print(est_hard$cv.k)
[1] 0.04090747
```

As can be seen from Figure 8.1, the estimated sparsity pattern well matches the expected sparsity pattern. The above statement is true for the other two methods, which are implemented below.

```
## Implement soft thresholding
est_soft = cv.covEstimation(df, k.min = 0.01,
    method = "thresholding", thr.method = "soft")
plot_mat(est_soft$Sigma_fit, main = "Soft thresholding")
```

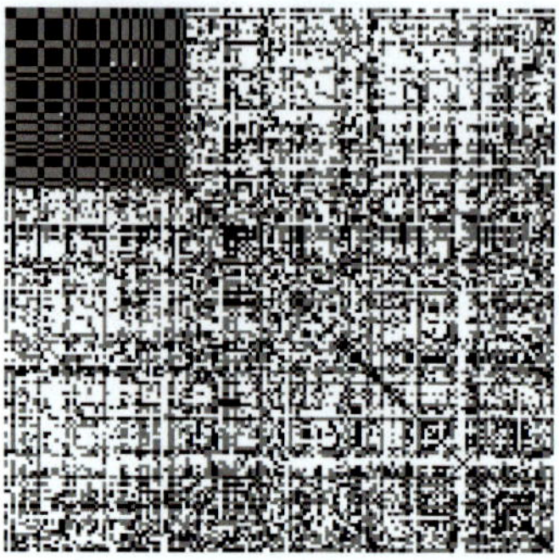

FIGURE 8.1
Covariance estimation using hard thresholding.

```
## Implement adaptive soft thresholding
est_ad_soft = cv.covEstimation(df, k.min = 0.01,
    method = "thresholding", thr.method = "adaptive")
plot_mat(est_ad_soft$Sigma_fit , main = "Adaptive Soft thresholding")
```

As we mentioned above, all three methods do not guarantee positive definiteness of the estimated covariance matrix. This can be easily observed in Figure 8.2, where the eigenvalues of all three estimated covariance matrices are plotted.

```
par(mfrow = c(1,3))
plot(eigen(est_hard$Sigma_fit)$values , type = "l",
       ylab = "Eigenvalue", col = "red",
       main = "Eignevalues of Hard thresholding")
abline(h = 0)
plot(eigen(est_soft$Sigma_fit)$values ,type = "l",
    ylab = "Eigenvalue", col = "red",
    main = "Eignevalues of Soft thresholding")
abline(h = 0)
plot(eigen(est_ad_soft$Sigma_fit)$values , type = "l",
    ylab = "Eigenvalue", col = "red",
    main = "Eignevalues of Adaptive Soft thresholding")
abline(h = 0)
par(mfrow = c(1,1))
```

8.3.1 The Positive-Definiteness Dilemma

Ensuring positive-definiteness of an estimated covariance matrix is a well-known and sticky problem in statistics. Most of the asymptotic results in the next section, and in general, imply positive-definiteness of an estimated covariance matrix when n and

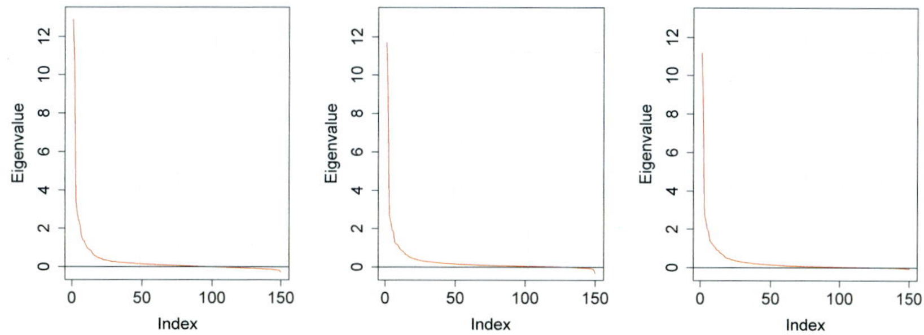

(a) Hard (b) Soft (c) Adaptive soft

FIGURE 8.2
Plot of estimated eigenvalues for different thresholding methods.

p are large. However, this sort of results, as important as they are, do not guarantee positive-definiteness of covariance estimators for the finite sample situations encountered in practice. For example, appearance of negative eigenvalues in the first two panels of Figure 8.2 indicates that some of the estimated covariance matrices are not positive-definite.

There are two solutions to this dilemma that we are aware of due to Fan et al. [2011] and Khare et al. [2024]. The first proposal is a clever threshold parameter selection presented next, and the second and more general one based on the idea of hidden Cholesky factor is presented in Section 9.8.

The threshold parameter in Fan et al. [2011] is selected in such a way to guarantee the finite sample positive-definiteness of the thresholded covariance matrix. We use the notation $\hat{\Sigma}(\delta)$ to emphasize the dependence of a thresholded covariance matrix on the threshold δ, the idea is to choose δ in a range where the minimum eigenvalue $\lambda_{\min}[\Sigma(\delta)] > 0$. That is, define

$$\delta_{\min} = \inf\{\delta > 0 : \lambda_{\min}[\Sigma(M)] > 0, \forall M > \delta\}, \tag{8.28}$$

and note that $\delta_{\min}$ always exists, since for larger δ, the estimated covariance matrix is diagonal with positive minimum eigenvalue. Then, for all $\delta > \delta_{\min}$ the estimated covariance matrix $\Sigma(\delta)$ is positive definite.

We illustrate this technique using adaptive soft thresholding technique (8.23) and the data set considered in the last section. The find.thr.min() function in recode package implements the estimation of $\delta_{\min}$.

```
## define the grid for threshold
delta.grid = seq(0.5, 1.5, length.out = 30)

## Find the minimum threshold
find.delta = find.thr.min(df, k.grid = delta.grid, thr.method = "
    adaptive")
```

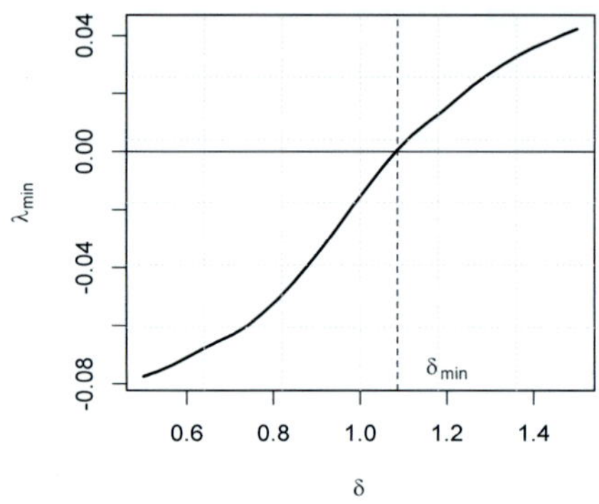

FIGURE 8.3
Minimum eigenvalue as a function of δ for adaptive soft thresholding.

```
delta.min = find.delta$k.min
eigen.store = find.delta$eigen.store

## plot the result
plot(delta.grid, eigen.store,type = "l", lwd = 2,
     xlab = expression(delta), ylab = expression(lambda[min]))
text(1.2, -0.075, expression(delta[min]))
grid(nx = 6, ny = 6)
abline(h = 0)
abline(v = delta.min, lty = 2)
```

Figure 8.3 plots the estimated minimum eigenvalues over the specified grid of thresholds. For this example, $\delta_{min} \approx 1.08$. Then, we recommend to run cross-validation using cv.covEstimation() function with the argument k.min = 1.08. By doing this, the estimated covariance matrix is guaranteed to be positive definite.

```
est_ad_soft = cv.covEstimation(df, k.min = delta.min,
     method = "thresholding", thr.method = "adaptive")
```

8.3.2 Comparing Shrinkage Estimators and the Sample Covariance

In this section, we use the R package corpcor to compare shrinkage and sample covariance estimators in terms of their condition numbers and positive definiteness when $n < p$.

We analyze the gene protein dataset (oslo2rppa), available in the recode package, which was measured using Reverse Phase Protein Array (RPPA) [Aure et al., 2015]. The dataset includes $p = 100$ genes and $n = 280$ breast cancer samples collected from hospitals in Oslo. To create a $n < p$ scenario, we employ the first 50 observations.

We start by importing the required packages and the dataset.

```
require(recode)
require(corpcor)

# Use only the first 50 observations
data(oslo2rppa)
oslo_trim = oslo2rppa[1:50,]
```

Next, the functions cov(), cov.shrink() (R package corpcor), and cov_shrink_F() (R package recode) are used to obtain the sample, shrinkage covariance matrices, respectively.

```
# Sample covariance matrix
emp_cov = cov(oslo_trim)

# Shrinkage covariance matrix
shrink_cov = cov.shrink(oslo_trim)

# Shrinkage covariance matrix Case F
shrink_cov_F = cov_shrink_F(oslo_trim)
```

The cov.shrink() function, relies on the decomposition in (8.17) to combine shrinkage estimators of the variance and correlation matrix. The former can be obtained using the var.shrink() function, which computes the empirical variance of each random variable and shrinks them toward their median. Its shrinkage intensity lambda.var, derived analytically [Schafer and Strimmer, 2005], is given by

$$\lambda_{\text{var}} = \frac{\sum_{i=1}^{p} \text{Var}(s_{ii})}{\sum_{i=1}^{p} (s_{ii} - \text{median}(s))^2}. \tag{8.29}$$

The cor.shrink() function computes the shrinkage estimate of the correlation matrix by shrinking the sample correlations toward the identity target matrix. Its shrinkage intensity is given by Schafer and Strimmer [2005]

$$\lambda_{\text{cor}} = \frac{\sum_{i \neq j} \text{Var}(r_{ij})}{\sum_{i \neq j} r_{ij}^2}.$$

The cov_shrink_F() function computes the shrinkage estimate of the sample covariance matrix of the data using the target matrix F in Schafer and Strimmer [2005] or in (8.18). The shrinkage intensity for the variance is the same as (8.29) and the shrinkage intensity for the correlation matrix is given by

$$\lambda_{\text{cor}} = \frac{\sum_{i \neq j} \text{Var}(s_{ij}) - \bar{r} f_{ij})}{\sum_{i \neq j} (s_{ij} - \bar{r}\sqrt{s_{ii}s_{jj}})^2},$$

where $\bar{r}$ is the average of the off-diagonals of sample correlation matrix and

$$f_{ij} = \frac{1}{2}\{\sqrt{\frac{s_{jj}}{s_{ii}}}\text{Cov}(s_{ii}, s_{ij}) + \sqrt{\frac{s_{ii}}{s_{jj}}}\text{Cov}(s_{jj}, s_{ij})\}.$$

The computed shrinkage intensities are reported below

TABLE 8.1
Comparing rank and condition number of sample and shrinkage covariance matrices.

Covariance	Rank	Condition
Sample	49	∞
Shrinkage	100	33.16
Shrinkage Case F	100	18819.37

```
# shrinkage intensity for correlation
print(attr(shrink_cov,"lambda"))
[1] 0.387
# shrinkage intensity for variance
print(attr(shrink_cov,"lambda.var"))
[1] 0.865

# shrinkage intensity for correlation case F
print(attr(shrink_cov,"lambda"))
[1] 0.431
```

Compared to the sample covariance, the shrinkage estimate offers several advantages. In particular, they are computationally efficient, achieve lower mean squared error, remain positive-definite and well-conditioned, and only assume the existence of the first two moments without additional distributional assumptions.

To compare the positive-definiteness of the above two matrices, use the commands

```
# compare positive definiteness
is.positive.definite(emp_cov)
[1] FALSE
is.positive.definite(shrink_cov)
[1] TRUE
is.positive.definite(shrink_cov_F)
[1] TRUE
```

As expected, the sample covariance matrix is not positive-definite, but the shrinkage covariance matrix is. The next code extract and table 8.1 compare ranks and condition numbers by using the rank.condition() function.

```
# compare ranks and condition number
rank.condition(emp_cov)
rank.condition(shrink_cov)
```

The rank of the sample covariance matrix is 49 compared to the full rank 100 of the shrinkage covariance matrices. However, the condition number of the Case F is significantly larger than for the case where target matrix is identity. Figure 8.4 plots the eigenvalues of the corresponding matrices.

Another salient advantage of this type of shrinkage estimator is its flexible form, which allows for efficient matrix inversion. Thus, if the goal is to estimate the precision matrix, it can be computed efficiently using the invcov.shrink() function.

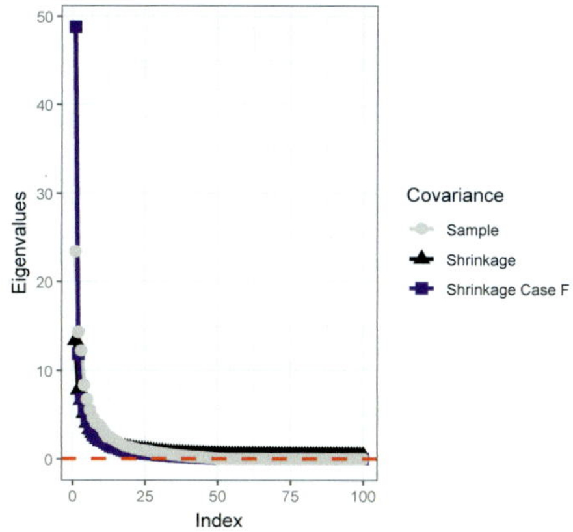

FIGURE 8.4
Eigenvalues of the sample and shrinkage covariance matrices.

8.4 Some Asymptotic Properties

Asymptotic properties of some covariance estimators are briefly reviewed in this section. The focus is on thresholding methods and their high-dimensional consistency when both p and n are large, the convergence of $\frac{\log p}{n}$ to zero plays a central role. To emphasize the dependence of the dimension p on the sample size n one should denote it by p_n, but this could lead to cumbersome notation in $\mathbf{\Sigma}_p$, etc, so we suppress dependence on n when there is no confusion.

8.4.1 Consistency and Rates

We begin by highlighting the special role of variances in the asymptotic analysis and state the following generic result which gives a bound on the error of estimating a single entry of a covariance matrix [Ravikumar et al., 2011].

Lemma 2 *For a sample of size n with sample covariance matrix* $\mathbf{S} = (s_{ij})$ *from a normal population with mean zero and covariance matrix* $\mathbf{\Sigma} = (\sigma_{ij})$*, it holds that*

$$P(|s_{ij} - \sigma_{ij}| > \delta) \leq 4\, exp\left\{-\frac{n\delta^2}{3200\, \max\limits_i \sigma_{ii}^2}\right\}, \tag{8.30}$$

for each (i, j) *and all* $\delta \in (0, 40 \max\limits_i\ \sigma_{ii})$.

Thus, so long as the population variances are bounded, the probability of the estimation error exceeding a fixed level goes to zero exponentially fast in n. What is the situation when the variances are unbounded or when the interest lies in finding error bounds for the whole covariance estimator?

It turns out that the notion of sparsity is a major player in establishing the consistency of the whole covariance matrix estimator and finding error bounds under various norms. For the consistency of thresholded estimators in the operator norm we consider two classes of matrices enjoying a notion of "approximate sparsity":

$$\mathcal{U}_\tau(q, s_0(p)) = \left\{\boldsymbol{\Sigma} : \sigma_{ii} \leq C, \textstyle\sum_{j=1}^{p} |\sigma_{ij}|^q \leq s_0(p), \text{for all}\, i\right\}, \tag{8.31}$$

for $0 \leq q \leq 1$. Evidently, for $q = 0$ the inequality in (8.31) reduces to

$$\sum_{j=1}^{p} \mathbf{1}(\sigma_{ij} \neq 0) \leq s_0(p),$$

so that the matrices in the class are more sparse for smaller values of $s_0(p)$. In this sense, and because of dependence of $s_0(p)$ on p the above condition could be viewed as an implicit definition of *sparsity*. A similar, but smaller class of sparse matrices used in the study of norm consistency of the thresholded precision matrices is

$$\mathcal{U}_\tau(q, s_0(p), \epsilon) = \{\boldsymbol{\Sigma} : \boldsymbol{\Sigma} \in \mathcal{U}_\tau(q, s_0(p)) \quad \text{and} \quad \lambda_{min}(\boldsymbol{\Sigma}) \geq \epsilon > 0\}. \tag{8.32}$$

Consistency of the thresholded covariance and precision matrix estimators under some conditions on the on p_n, n, λ_n is given next [Bickel and Levina, 2008b].

Theorem 11 *For a sample of size n from Y, a zero mean Gaussian random vector with the covariance matrix* $\boldsymbol{\Sigma}$*, if*

$$\lambda_n = C\sqrt{\frac{\log p}{n}} \quad \text{and} \quad \frac{\log p}{n} = o(1), \tag{8.33}$$

for a sufficiently large C, then uniformly on $\mathcal{U}_\tau(q, s_0(p))$*, we have,*

$$||T_{\lambda_n}(\mathbf{S}) - \boldsymbol{\Sigma}|| = O_P\left(s_0(p)\left(\frac{\log p}{n}\right)^{(1-q)/2}\right). \tag{8.34}$$

Moreover, for the precision matrix, uniformly on $\mathcal{U}_\tau(q, s_0(p), \epsilon)$

$$||T_{\lambda_n}(\mathbf{S})^{-1} - \boldsymbol{\Sigma}^{-1}|| = O_P\left(s_0(p)\left(\frac{\log p}{n}\right)^{(1-q)/2}\right). \tag{8.35}$$

It is instructive to note how the rate of convergence depends explicitly on the dimension p and $s_0(p)$, the number of nonzero elements of the population covariance matrix $\boldsymbol{\Sigma}$. Furthermore, the rate here is similar to rates of form $(\frac{s \log p}{n})^{1/2}$ obtained for sparse covariance (precision) matrix estimators in several other situations where s is the number of nonzero off-diagonal entries of $\boldsymbol{\Sigma}$, its inverse or Cholesky factor [Lam and Fan, 2009]. Rothman et al. [2008] shows that the rate of convergence of the ℓ_1-penalized normal likelihood estimator of the precision matrix in the Frobenius norm is of the order $(\frac{s \log p}{n})^{1/2}$, where $s = s_n$ is the total number of nonzero elements in the precision matrix. This rate reveals that the price one pays for high-dimensionality of the data is merely a logarithmic factor of the dimension p.

Analogs of Theorem 11 can be found in Cai and Liu [2011] for *adaptive thresholding* where it is shown to attain the optimal rate of convergence under the spectral norm for the following class of covariance matrices which includes $\mathcal{U}_\tau(q, s_0(p))$. The new class defined by

$$\mathcal{U}_\tau{}^*(q, s_0(p)) = \left\{ \boldsymbol{\Sigma} : \max_i \sum_{j=1}^{p} (\sigma_{ii}\sigma_{jj})^{(1-q)/2} |\sigma_{ij}|^q \leq s_0(p) \right\}, \tag{8.36}$$

no longer requires that the variances σ_{ii} to be uniformly bounded and allows $\max_i \sigma_{ii} \longrightarrow \infty$. In fact, it can be shown that the choice of $\delta = 2$ in (8.24) is optimal for support recovery in the sense that an adaptive thresholding estimator with any smaller δ would fail to recover the support of $\boldsymbol{\Sigma}$ exactly with probability converging to 1.

In Cai et al. [2016] data-driven adaptive procedures are reviewed for estimation of various S-structured matrices such as the bandable, Toeplitz, sparse, and sparse spiked covariance matrices. They provide a thorough coverage and discussion of the minimax rate of convergence and related optimality properties of the thresholded covariance and precision matrix estimators.

8.5 Problems

8.1 Complete the derivation of the solution in (8.5).

8.2 Repeat the analysis in Section 8.3, but this time use top 30 and bottom 150 genes. Estimate the covariance matrix using hard, soft and adaptive soft thresholding. Display the estimated covariance matrices.

8.3 Repeat analysis in Section 8.3.2 but this time use p = 150 genes to create a $n > p$ scenario. Estimate the sample and shrinkage covariance matrices and report the rank and condition number. Discuss the results and compare it with the results based on $n < p$ scenario.

8.4 Establish the identity (8.13), find its minimizer and the minimum value of the risk function.

8.5 A common loss function for choosing α_i's in (8.2) is

$$L(\widehat{\boldsymbol{\Sigma}}, \boldsymbol{\Sigma}) = ||\widehat{\boldsymbol{\Sigma}} - \boldsymbol{\Sigma}||_F^2 = p^{-1}\text{tr}(\widehat{\boldsymbol{\Sigma}} - \boldsymbol{\Sigma})^2.$$

This is simply the Frobenius norm except that it is divided by the dimension p, so that norm pf the identity matrix is one and does not depend on the dimension p.

(a) Show that the optimal choice of α_1 and α_2 minimizing the corresponding risk $\mathrm{E}L(\widehat{\mathbf{\Sigma}}, \mathbf{\Sigma})$ depends only on the following four scalar functions of the true (but unknown) covariance matrix $\mathbf{\Sigma}$:

$$\mu = \mathrm{tr}(\mathbf{\Sigma})/p, \alpha^2 = ||\mathbf{\Sigma} - \mu I||_F^2, \beta^2 = E||\mathbf{S} - \mathbf{\Sigma}||_F^2, \delta^2 = E||\mathbf{S} - \mu I||_F^2. \quad (8.37)$$

(b) Show that

$$\alpha^2 + \beta^2 = \delta^2,$$

for the three terms from (8.37).

(c) For the shrinkage "estimator"

$$\widehat{\mathbf{\Sigma}} = \frac{\beta^2}{\delta^2}\mu I + \frac{\alpha^2}{\delta^2}\mathbf{S},$$

show that the corresponding minimized risk is

$$E||\widehat{\mathbf{\Sigma}} - \mathbf{\Sigma}||_F^2 = \frac{\alpha^2\beta^2}{\delta^2}.$$

(d) Note that $\widehat{\mathbf{\Sigma}}$ is not really an estimator because all the four coefficients $\mu, \alpha, \beta, \gamma$ depend on the unknown $\mathbf{\Sigma}$. Given a sample data of size n or its sample covariance matrix $\mathbf{S}$ propose a method of estimating these four parameters.

8.6 Show the inequalities in (8.22) hold for a multivariate normal distribution.

8.7 Simulate $\boldsymbol{X}_1, \ldots, \boldsymbol{X}_n \sim N_p(\mathbf{0}, \mathbf{\Sigma})$ for each of the following cases. Let $\boldsymbol{S}$ be the sample covariance matrix and p_i be the ith percentile of its diagonal entries.

(a) $\mathbf{\Sigma}$ is the compound symmetry matrix with $\sigma^2 = 1, \rho = 0.5$ and $(n, p) = (50, 25)$. Hard-threshold the sample covariance matrix with the threshold parameter $\lambda = p_1$, compute and count the number of negative eigenvalues of the soft-thresholded covariance matrix. Examine the sparsity of

$$\mathbf{S}^+ = \sum_{i=1}^{p} \max(\hat{\lambda}_i, 0)\hat{e}_i\hat{e}_i',$$

obtained from the spectral decomposition of $\boldsymbol{S}$ replacing the negative eigenvalues by zero.

(b) Repeat part (a) with λ selected via 5-fold cross-validation.

(c) Repeat (a) and (b) with $(n, p) = (25, 50)$.

8.8 Repeat (a)–(c) above when $\mathbf{\Sigma}$ is the AR(1) covariance matrix with $\sigma^2 = 1, \rho = 0.5$.

8.9 Repeat Problems 2 and 3 when the hard-thresholding function is replaced by soft-thresholding, SCAD and adaptive thresholding functions, respectively.

8.10 Show that the matrix $\mathbf{S}_+$ in (8.26) is positive semi-definite, and satisfies

$$||\mathbf{S}_+ - \mathbf{\Sigma}|| \leq 2||\mathbf{S} - \mathbf{\Sigma}||.$$

9

Undirected Graphical Models

Graphical models are means to depict dependencies or associations among characteristics or random variables of interest in a study. The hope is that a graph with its display of nodes and edges in the plane will illuminate better the inherent interrelationships among the variables than the mathematical expression of their joint distribution. Nowadays, diverse forms of graphs are used extensively in neurosciences, genetics, business and economics, social and computer network analysis, see Whittaker [1990], Hastie et al. [2009]. Given data $\boldsymbol{Y}_1, \ldots, \boldsymbol{Y}_n$ or a sample of size n from $N_p(\mathbf{0}, \boldsymbol{\Sigma})$ our focus is on the construction of Gaussian graphical models using the graphical Lasso (GLasso) algorithm by detecting zeros in the *precision matrix* $\boldsymbol{\Theta} = \boldsymbol{\Sigma}^{-1}$. A quick overview of graphical models and a few examples of graphs and their corresponding precision matrices are provided in Section 9.1. A penalized likelihood method and the GLasso algorithm for sparse estimation of the precision matrix are developed in Section 9.2. The level of coverage in the first two sections is designed to prepare a novice to apply GLasso and related algorithms to genuine high-dimensional data. Generalizations and various improvements of the GLasso such as thresholded GLasso, latent variable GLasso, and simple-GLasso are reviewed in subsequent sections. A nearly parallel development for *Covariance graphical models* where the interest is in the zeros of the covariance matrix itself and the marginal independence of the corresponding random variables is discussed in Section 9.6.2. The more specialized directed acyclic graphs (DAGs), their connections with Cholesky factors of precision matrices and Veechia's likelihood approximation in spatial statistics are the focus of the next few chapters.

9.1 Gaussian Graphical Models

The graph of a random vector $Y = (Y_1, \ldots, Y_p)'$, denoted by $G = (V, E)$, where $V = \{1, 2, \ldots, p\}$ is the set of p nodes (vertices) corresponding to the p variables and E is the set of edges. An edge between Y_i and Y_j is present if and only if Y_i and Y_j are conditionally dependent given the other variables. The simplest and most popular case is when Y follows a multivariate normal distribution $N_p(\boldsymbol{\mu}, \boldsymbol{\Sigma})$. Then, the zero entries of its precision matrix $\boldsymbol{\Theta} = \boldsymbol{\Sigma}^{-1}$ hold the key to the conditional independence of pairs of variables given the rest. Indeed, the (i, j)-th entry of $\boldsymbol{\Theta}$ is zero if and only if Y_i and

DOI: 10.1201/9781003638896-9

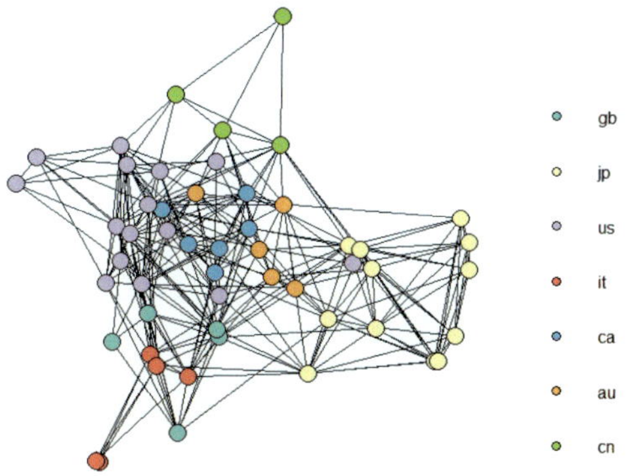

FIGURE 9.1
The GLasso solution. Colors indicate the country. The labels gb corresponds to Great Britain, us - USA, ca - Canada, cn - China, jp - Japan, it - Italy, and au - Australia.

Y_j are conditionally independent given the rest. In this case, a graph displaying the conditional (in)dependence of the coordinates of a Gaussian random vector is called its *Gaussian graphical model* (GGM).

As introductory and illustrative examples, we consider two real-world datasets from the complementary R package recode that accompanies this book.

Example 19 Global Banks Network: *We utilize stock return volatility data (*globalbank*) from Demirer et al. [2018], where the goal was to estimate the global bank network connectedness. Here, we illustrate the role of graphical lasso (GLasso) algorithm in discovering connectedness of the banks. Details on* R *implementation is relegated to Section 9.7. Figure 9.1 illustrates the result. Colors in the figure indicate the corresponding country of the bank. Adding colors, it can be seen more clearly that banks from the same country tend to group together, while also they are connected to banks from the other countries.*

Example 20 Gene Protein: *We consider gene protein dataset (*oslo2rppa*) measured by Reverse Phase Protein Array [Aure et al., 2015]. The dataset consists of* $p = 100$ *genes and* $n = 280$ *breast cancer samples collected from hospitals in Oslo.*

The goal of Gene protein network analysis is to examine, validate and interpret protein-protein interactions. Figure 9.2 displays the GLasso output, where the tuning parameter is selected using BIC; see Section 9.2.4. As shown in Section 9.7.1, the selected tuning parameter is $\lambda \approx 0.03$*; see Problem 9.7. It is expected that key breast* *cancer genes will be represented as hubs in the resulting networks. For example, Estrogen Receptor 1 (ESR1) protein-coding gene in the Figure 9.2. A common approach to validate edges is to compare the estimated interactions with validated interactions present in the STRING database (*https://string-db.org/*).*

Example 21 EEG Data: *We apply GLasso to an EEG data set from the alcoholism study, available in Begleiter [1999]. Data were initially obtained from 64 electrodes placed on different locations of subjects' scalps that captured EEG signals at 256 Hz during a one-second period (256 time points). The measurements were taken from 122 subjects, of which 77 belonged to the alcoholic group and 45 to the control group. The stimuli were pictures from a picture set. Each subject completed 120 trials under the following three conditions: single stimulus (a single picture), two matched stimuli and two unmatched stimuli. We only consider the average of all trials for each subject under the single stimulus condition. Research has shown that, relative to control subjects, alcoholic subjects demonstrate unstable or poor rhythm and lower signal power in the α-band signal. Thus we band-pass the EEG signals to extract the α frequency band in the range of 8–12.5 Hz. The goal of the study is to examine regional asymmetric patterns between alcoholic and control subjects. Figure 9.3 shows the GLasso output for subjects in the alcoholic and the control group, respectively. To*

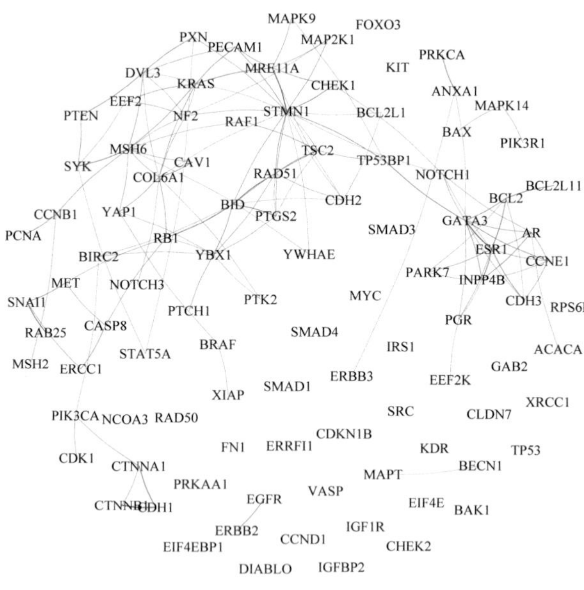

FIGURE 9.2
The GLasso solution for the gene dataset

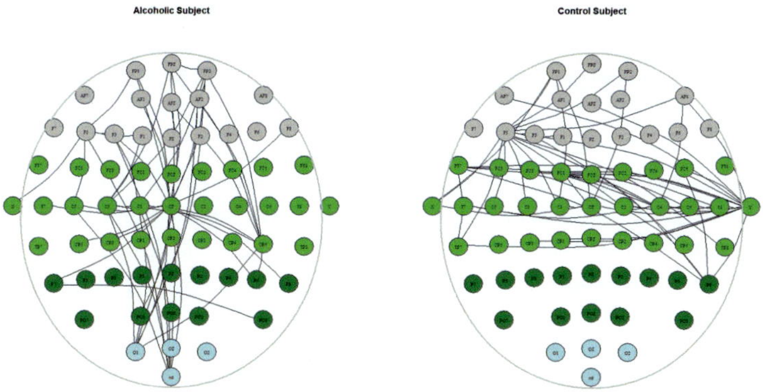

FIGURE 9.3
Left graph plots the GLasso estimated network for the alcoholic group and right graph plots the estimated network for the control group

differentiate regions of the brain, we use gray to highlight nodes in the frontal region, green and dark green for the central and parietal regions, and light blue indicates the occipital region. We can see that the control subject contains more edges connecting the left frontal and central regions to the right central region. The alcoholic subject, on the other hand, contains more edges connecting the middle frontal regions to the middle central, parietal and occipital regions than the control subject.

9.2 The Graphical Lasso Problem

In practice, given sample data from a multivariate normal distribution, graph construction requires selection of possible edges in the graph or estimation of location of zero entries of the precision matrix. There are a number of methods suitable for the task, but the gold standard is penalized likelihood estimation of precision matrices. In this section, we focus on developing the background concepts and the role of Lasso regression in sparse estimation of precision matrices. In high-dimensional situations, Lasso regression and regularization techniques in general have revolutionized machine learning and the related areas. It turns out that such regression-based successes will carry over to the more difficult task of covariance estimation and graph selection if one could reformulate such tasks as regression problems. Fortunately, the entries of $\mathbf{\Sigma}^{-1}$ have regression-based interpretations so that sparsity in the precision matrix can be attained by imposing variety of Lasso penalties on its entries, factors, etc.

Given the sample data $\boldsymbol{Y}_1, \ldots, \boldsymbol{Y}_n \sim N_p(\mathbf{0}, \boldsymbol{\Sigma})$, the statistical problem is to use the data and propose a method for selecting the correct graph with high probability. Let $\boldsymbol{\Theta} = \boldsymbol{\Sigma}^{-1}$ be the precision matrix and $\boldsymbol{S} = \frac{1}{n}\sum_{i=1}^{n} \boldsymbol{Y}_i \boldsymbol{Y}_i'$ the sample covariance matrix of the data. For $p < n$, $\mathbf{S}$, the maximum likelihood estimator (MLE) of $\boldsymbol{\Sigma}$, is known to perform poorly when p is relatively large [Stein, 1975, Ledoit and Wolf, 2004]. Indeed, the MLE of the precision matrix may not exist when $p > n$, as in this case there are more parameters than observations. In such situations it is natural to assume sparsity and regularization is the natural recipe to follow.

The ℓ_1-*penalized negative log-likelihood* function for sparse estimation of the precision matrix is

$$\ell(\boldsymbol{\Theta}) = -\log|\boldsymbol{\Theta}| + \operatorname{tr}(\mathbf{S}\boldsymbol{\Theta}) + \lambda||\boldsymbol{\Theta}||_1, \tag{9.1}$$

where λ is a tuning parameter and $||\boldsymbol{\Theta}||_1$ is the ℓ_1-norm of the matrix. In some formulations the diagonal entries of the precision matrix are not penalized; see Problem 9.5. Though the penalized likelihood is convex in $\boldsymbol{\Theta}$, there is no closed-form solution for its unique minimum for a given $\lambda > 0$. An iterative block coordinate descent algorithm for computing the solution, popularly known as graphical Lasso (GLasso) algorithm, is fast and widely used in applications. Many alternatives and improvements of the GLasso algorithm are reviewed and presented in this chapter. Two natural and relevant conceptual/theoretical questions of interest are: How does the penalized MLE (GLasso solution) of the precision matrix depend on $\mathbf{S}$? How does it relate to the simpler and explicitly computed thresholded $\mathbf{S}$? These questions are pursued next while developing and assessing various properties of the GLasso algorithm.

9.2.1 The GLasso Algorithm

The central role of GLasso algorithm in sparse estimation of precision matrices is similar to that of the least square estimation in linear models. The power of the GLasso algorithm lies in its simplicity, reliance on the iterative block coordinate descent and fast Lasso regressions at the core of its computational engine. There are many extensions and improvements of the GLasso algorithm going far beyond the i.i.d. multivariate normal setup with enhanced algorithmic convergence, computational speed and statistical properties. In particular, Chapter 11 develops time series analogs of GLasso for temporally and cross-sectionally dependent data, where instead of sparsity of $\boldsymbol{\Sigma}^{-1}$ one is seeking sparsity of the inverse matrix spectral density function.

The connections between the GLasso solution, the sample covariance matrix $\mathbf{S}$ and possibly a Lasso regression is usually revealed through the subgradient of the penalized log-likelihood function (9.1):

$$\boldsymbol{\Theta}^{-1} - \mathbf{S} - \lambda \cdot \Gamma = \mathbf{0}. \tag{9.2}$$

Here $\Gamma = (\gamma_{ij}) = \operatorname{sgn}(\boldsymbol{\Theta})$ is a symmetric matrix with zero diagonal entries, $\gamma_{ij} = \operatorname{sign}(\theta_{ij})$, if $\theta_{ij} \neq 0$, otherwise $\gamma_{ij} \in [-1, 1]$. Looking at (9.2), the role of (Lasso) regression is not immediately obvious, it takes a bit of ingenuity to recast (9.2) into the framework of regularized linear regression. To explain the connection, define the auxiliary matrix $W = \mathbf{\Theta}^{-1}$, and for the moment pretend that it is known and think of it as a proxy for the covariance matrix of Y. From here on, we only focus on computing the last row/column of $\mathbf{\Theta}$. The plan is to show that computing its first $p-1$ entries can be reduced to the Lasso solution or finding the regression coefficients $\boldsymbol{\beta}$ when Y_p is regressed on the rest, i.e., $Y_1, \ldots, Y_{p-1}$. This requires partitioning $\mathbf{\Theta}$, W and $\mathbf{S}$ as:

$$\mathbf{\Theta} = \begin{pmatrix} \mathbf{\Theta}_{11} & \boldsymbol{\theta}_{12} \\ \boldsymbol{\theta}_{21} & \theta_{22} \end{pmatrix}, \; W = \begin{pmatrix} W_{11} & w_{12} \\ w_{21} & w_{22} \end{pmatrix}, \; \mathbf{S} = \begin{pmatrix} S_{11} & s_{12} \\ s_{21} & s_{22} \end{pmatrix}, \tag{9.3}$$

where $\mathbf{\Theta}_{11}$ and others are the $(p-1) \times (p-1)$ principal minors of their respective matrices and θ_{22}, w_{22}, s_{22} are scalars, etc.

The appearance of W in (9.2) and in subsequent calculations is the source of some mysteries and confusions in understanding various aspects of GLasso. These are discussed and resolved in Mazumder and Hastie [2011] within the framework involving W. In Section 9.3 we show how to bypass W, and present a much simpler and more direct derivation of an algorithm for minimizing the objective function (9.1), where the connection with Lasso regression appears rather naturally.

Now, we present the derivation of the GLasso algorithm and the role of the auxiliary matrix W. Using $W\mathbf{\Theta} = I$ and the compatible partition of $\mathbf{\Theta}$, it follows that

$$\begin{aligned} W_{11}\boldsymbol{\theta}_{12} + w_{12}\theta_{22} &= \mathbf{0}, \\ w'_{12}\boldsymbol{\theta}_{12} + w_{22}\theta_{22} &= 1. \end{aligned} \tag{9.4}$$

From the first identity in (9.4) one obtains

$$w_{12} = -W_{11}\boldsymbol{\theta}_{12}/\theta_{22} = W_{11}\boldsymbol{\beta}, \tag{9.5}$$

where $\boldsymbol{\beta} = -\boldsymbol{\theta}_{12}/\theta_{22}$ is the desired vector of regression coefficients, see Section 4.1.2, and

$$\text{sgn}(\boldsymbol{\beta}) = -\text{sgn}(\boldsymbol{\theta}_{12}) = -\gamma_{12},$$

using the fact that $\boldsymbol{\beta}$ and $\boldsymbol{\theta}_{12}$ have opposite signs.

Now, the equations arising from the upper-right block of (9.2) are

$$\mathbf{w}_{12} - \mathbf{s}_{12} - \lambda\boldsymbol{\gamma}_{12} = 0. \tag{9.6}$$

Substituting (9.5) into (9.6) we arrive at

$$W_{11}\boldsymbol{\beta} - s_{12} + \lambda \cdot \text{sgn}(\boldsymbol{\beta}) = 0, \tag{9.7}$$

which is recognized as the estimating equations for the following Lasso regression for a given $W_{11} \succ 0$:

$$\min_{\beta} \frac{1}{2}\boldsymbol{\beta}' W_{11}\boldsymbol{\beta} - \boldsymbol{\beta}' s_{12} + \lambda\|\boldsymbol{\beta}\|_1, \tag{9.8}$$

except that $\boldsymbol{X}'\boldsymbol{X}$ is replaced by W_{11}, see Lemma 1. Thus, in principle, $\boldsymbol{\beta}$ can be computed via a Lasso regression algorithm. Then, since $\boldsymbol{\theta}_{12} = -\boldsymbol{\beta}\theta_{22}$ it follows that $\boldsymbol{\theta}_{12}$ is known up to the scale factor θ_{22}, and the latter is computed from the second equation in (9.4) as

$$\theta_{22}^{-1} = w_{22} - w_{12}'\boldsymbol{\beta}. \tag{9.9}$$

Stacking up these two components as $(\boldsymbol{\theta}_{12}', \theta_{22})'$ gives the last row/column of the matrix $\boldsymbol{\Theta}$.

Of course, the Lasso computation in (9.8) is done efficiently using the coordinate descent algorithm in Section 2.1.3. The procedure above can be repeated for any other column of $\boldsymbol{\Theta}$, by first permuting its rows and columns to make the target column the last one.

These developments for computing sparse precision matrices is the crux of the *graphical Lasso (GLasso) algorithm.* It usually starts with $W = \mathbf{S} + \lambda I$ where λ is a tuning parameter whose selection along with other implementation issues are discussed next.

Algorithm 1 GLasso

```
Initialize W = S + λI, for a given λ
for i = 1 to p until convergence do
    Partition matrix W into two parts, j^th row and column, and the rest denoted
    symbolically by W_11
    Solve the Lasso estimating equations (9.7) using the coordinate-descent al-
    gorithm to obtain β̂
    Update w_12 = W_11 β̂
end for
In the final cycle (for each j) solve for θ̂_12 = −β̂ θ̂_22, with θ̂_22^{-1} = w_22 − w'_12 β̂
Output W and Θ
```

Effective implementation of the GLasso algorithm requires selection of the starting value, tuning parameter, Lasso algorithm, etc., which are discussed later. Some technical issues are discussed next by relying on deeper concepts from convex analysis. A reader less interested in these theoretical aspects, can skip to Section 9.3 for a simpler and more direct derivation of the GLasso algorithm. For a data example, see Section 9.7.1.

9.2.2 Limitations and Technical Details of GLasso

There are a number of questions surrounding the GLasso algorithm—what exactly it estimates, whether it guarantees convergence, if it maintains positive-definiteness of the matrices involved, and how well does it perform overall? These concerns were highlighted in Mazumder and Hastie [2011], and here we discuss them in detail. This section can be skipped in the first reading and returned to later.

I. One of the central questions is: *What does GLasso algorithm estimate?* At first glance, the answer seems obvious. Our goal appears to be to minimize the following penalized likelihood with respect to the precision matrix $\mathbf{\Theta}$:

$$\min_{\mathbf{\Theta} \succ 0} \; -\log|\mathbf{\Theta}| + \operatorname{tr}(\mathbf{S}\mathbf{\Theta}) + \lambda\|\mathbf{\Theta}\|_1, \tag{9.10}$$

However, it turns out that GLasso does *not* solve this primal problem directly. Instead, it solves the *dual* problem with respect to a different matrix, $\mathbf{W}$. For details on primal and dual problems, see Section 2.6.

In fact, GLasso is a *dual ascent algorithm*, meaning it performs optimization steps on the dual of the original problem. The dual of the penalized likelihood can be written as [Mazumder and Hastie, 2011, Section 4.1]:

$$\max_{\mathbf{W}:\|\mathbf{W}-\mathbf{S}\|_\infty \leq \lambda} \; \log|\mathbf{W}|, \tag{9.11}$$

which suggests that GLasso is seeking a matrix $\mathbf{W}$ that is both close to the sample covariance matrix $\mathbf{S}$ (within an ℓ_∞ distance of λ), and has a large determinant. Maximizing the log-determinant encourages $\mathbf{W}$ to be far from singular, i.e., well-conditioned.

To better understand this dual problem, we can look at how it behaves when we partition the matrix $\mathbf{W}$. Using Schur's complement, the objective function $\log|\mathbf{W}|$ can be decomposed as:

$$\log|\mathbf{W}| = \log|\mathbf{W}_{11}| + \log(\mathbf{s}_{22} + \lambda - \mathbf{w}_{12}'\mathbf{W}_{11}^{-1}\mathbf{w}_{12}),$$

which allows us to rewrite the problem as a quadratic program in the off-diagonal block $\mathbf{w}_{12}$:

$$\begin{aligned} &\min_{\mathbf{w}_{12}} \; \mathbf{w}_{12}'\mathbf{W}_{11}^{-1}\mathbf{w}_{12}, \\ &\text{s.t. } \|\mathbf{w}_{12} - \mathbf{s}_{12}\|_\infty \leq \lambda. \end{aligned} \tag{9.12}$$

So, rather than solving directly for $\mathbf{\Theta}$, the precision matrix, the GLasso algorithm works on $\mathbf{W}$, which is a proxy for the covariance matrix.

II. Now, solving (9.12) directly is computationally expensive, because it involves matrix inversions. The GLasso algorithm cleverly avoids this by solving the *dual of this dual*, which brings us back to a lasso-type problem (9.8) where the primal $\boldsymbol{\beta}$ and dual $\mathbf{w}_{12}$ variables are linked by $\mathbf{w}_{12} = \mathbf{W}_{11}\boldsymbol{\beta}$. This approach dramatically reduces computational cost, but it also introduces subtle mysteries. One key issue lies in the update step for $\boldsymbol{\beta}$, which is tied directly to the entries of the precision matrix $\mathbf{\Theta}$. In the lasso-style update:

$$\mathbf{W}_{11}\boldsymbol{\beta} - s_{12} + \lambda \cdot \operatorname{sgn}(\boldsymbol{\beta}) = 0, \tag{9.13}$$

the sub-matrix $\mathbf{W}_{11}$ remains fixed and exactly one column of $\mathbf{W}$ changes. However, the update of $\boldsymbol{\theta}_{12}$ of the precision matrix requires the update of the full matrix

W. Only updating a slice like $\mathbf{w}_{12}$ while holding the rest of **W** constant leads to an inconsistency: the updates to $\boldsymbol{\beta}$ (and thus $\boldsymbol{\theta}_{12}$) are made under the assumption that **W** is not changing, when in fact it is. This causes GLasso to behave *non-monotonically*. In other words, the optimization objective might not decrease at every step, even though we're supposed to be minimizing.

III. Another natural question is about *positive-definiteness*. Does GLasso maintain it throughout the iterations?

For the working matrix **W**, the answer is yes, as long as the initialization satisfies the constraint $||\mathbf{W}-\mathbf{S}||_\infty \leq \lambda$. Under this condition, one can show that **W** remains positive definite at each step (Problem 9.3).

However, things are trickier for $\boldsymbol{\Theta}$. Since GLasso never updates $\boldsymbol{\Theta}$ directly, instead, it infers it from the updated **W**, the computed $\boldsymbol{\Theta}$ at each iteration may not be a valid inverse of **W**, and hence may not be positive definite. Only once the algorithm has fully converged do we have $\boldsymbol{\Theta} = \mathbf{W}^{-1}$, and only then is positive-definiteness restored.

In summary, while GLasso is a computationally efficient method, it suffers from several drawbacks as pointed out earlier. These caveats are crucial to understand when interpreting the results of the algorithm or proposing alternatives as in the next two subsections.

9.2.3 The P- and DP-GLasso Alternatives

This alternative algorithm in Mazumder and Hastie [2011] was designed to address several issues we discussed earlier. In particular, the non-monotonic updates and the indirect optimization over the precision matrix $\boldsymbol{\Theta}$.

The key innovation lies in reformulating the optimization step, where instead of working directly with $\mathbf{w}_{12}$ an alternative expression for this quantity (derived in Problem 9.2) allows the core update equation (9.6) to be rewritten more naturally as:

$$\boldsymbol{\Theta}_{11}^{-1}\boldsymbol{\theta}_{12}w_{22} + \mathbf{s}_{12} + \lambda\boldsymbol{\gamma}_{12} = \mathbf{0}, \tag{9.14}$$

where $\boldsymbol{\Theta}_{11}^{-1}$ is expressed in terms of blocks of **W** as $\boldsymbol{\Theta}_{11}^{-1} = \mathbf{W}_{11} - \mathbf{w}_{12}\mathbf{w}_{12}'/w_{22}$, and the scalar w_{22} is held fixed and defined as $w_{22} = 1/(\theta_{22} - \boldsymbol{\theta}_{12}'\boldsymbol{\Theta}_{11}^{-1}\boldsymbol{\theta}_{12})$.

To make the computation more tractable, note that the substitution $\boldsymbol{\alpha} = \boldsymbol{\theta}_{12}w_{22}$ in equation (9.14) transforms it to the normal equations for the standard Lasso problem:

$$\min_{\alpha} \frac{1}{2}\alpha'\boldsymbol{\Theta}_{11}\alpha + \alpha'\mathbf{s}_{12} + \lambda||\alpha||_1.$$

This reformulation is more stable and interpretable, and allows us to optimize directly in terms of the precision matrix $\boldsymbol{\Theta}$, the actual quantity of interest, rather than indirectly through **W**. To further improve computational efficiency, a faster variant called DP-GLasso was developed. It incorporates numerical optimization techniques discussed in Section 9.3.1, allowing for quicker convergence without sacrificing accuracy.

9.2.4 Tuning Parameter Selection

In the regularized estimation of the precision matrix in general, and in the GLasso and its improvements, the value of the tuning parameter λ is directly connected to the sparsity of $\mathbf{\Theta}$. In the sense that the higher the λ, the sparser is the estimated inverse covariance matrix. For tuning parameter selection in GLasso, Friedman et al. [2008] suggested K-fold cross-validation. The cross-validation procedure for GLasso is similar to that for Lasso regression discussed in Section 2.4. For GLasso

$$\mathrm{CV}(\lambda) = \frac{1}{K}\sum_{k=1}^{K} L(\mathbf{S}_k, \hat{\mathbf{\Theta}}(\lambda), \lambda),$$

where $L(S_k, \hat{\mathbf{\Theta}}(\lambda), \lambda) = -\log|\hat{\mathbf{\Theta}}| + \mathrm{tr}(\mathbf{S}_k\hat{\mathbf{\Theta}})$, $\mathbf{S}_k$ is the sample covariance matrix computed using data in the kth fold, and $\mathbf{\Theta}(\lambda)$ is the estimated precision matrix using data in the other $K-1$ folds.

However, cross-validation is known to be fairly liberal for model selection, allowing many false positives and overfitting the data [Wasserman and Roeder, 2009]. To address this issue, Foygel and Drton [2010] proposed extended Bayesian information criterion (eBIC) and demonstrated its strong improvement in false discovery rate over ordinary BIC and the cross-validation.

We review popular information criteria for selection of λ such as the Akaike information criterion (AIC), Bayesian information criterion (BIC) that rely on penalizing model complexity in terms of the number of parameters. Ideally, the selected tuning parameter should produce an undirected graph that is sufficiently complex to be interesting, sufficiently sparse to be interpretable, and, more importantly, should be supported by data.

From the likelihood function in (9.10), the AIC approximation for the graphical model is:

$$\mathrm{AIC}(\lambda) = -n\log|\mathbf{\Theta}| + n\mathrm{tr}(\mathbf{S\Theta}) + E, \tag{9.15}$$

where E is the number of non-zero elements in $\mathbf{\Theta}$ for a given λ. Thus, given a list of candidate tuning parameters, we choose λ which minimizes (9.15). Similarly, an approximation of the BIC is:

$$\mathrm{BIC}(\lambda) = -n\log|\mathbf{\Theta}| + n\mathrm{tr}(\mathbf{S}\Theta) + E\log(p). \tag{9.16}$$

The GLasso and other sparse precision matrix estimation techniques using cross-validation to choose the tuning parameter λ, could be expensive computationally as they require computing solutions over a full regularization path. Fortunately, there are methods with the goal of reducing computation and the impact of the selected tuning parameter on the end result. For example, Liu and Wang [2012]'s TIGER (Tuning-Insensitive Graph Estimation and Regression) is *tuning-insensitive* in the sense that it automatically adapts to the unknown sparsity pattern and is asymptotically tuning-free. In the finite sample settings, it requires minimal efforts to select the tuning or the regularization parameter. Like GLasso, it estimates the precision matrix $\mathbf{\Theta}$, in a column-by-column fashion where each column is computed by solving a Lasso regression problem using the SQRT-Lasso in Belloni et al. [2011], which is also tuning-insensitive. The R package bigmatrix implementing TIGER, under a parallel computing framework, reveals that TIGER is faster and more scalable than the glasso [Friedman et al., 2008] and huge [Zhao et al., 2012] packages in R. GRankLasso, a tuning-free alternative to GLasso, in Tran and Yu [2022], estimates each column of the precision matrix using Rank Lasso regression [Wang et al., 2020]. It replaces the squared error loss by a rank-based loss—related to Wilcoxon scores, which achieves robustness to outliers and non-Gaussian data. More importantly, its tuning parameter λ_j is determined via simulation from a known distribution of rank-based subgradients, removing the need for manual tuning.

9.3 A Simple and Direct Derivation (S-GLasso)

Although DP-GLasso successfully overcomes the drawbacks of the GLasso algorithm, it continues to rely on (9.2)–(9.3) to derive the updates. This seems to complicate the derivation and distracts from the overall understanding of the algorithm. Thus, it is natural to ask: Is it possible to bypass (9.2)–(9.3) and derive the updates more directly from (9.1)?

In this section, a simpler and more direct derivation of the updates of $\mathbf{\Theta}$ is presented based on first decoupling the penalized likelihood function through a novel reparameterization, then ting the differentiating it. Interestingly, this idea was used first by Wang [2014] in regularized likelihood estimation of covariance matrices.

We start with the reparameterization

$$\boldsymbol{\beta} = \boldsymbol{\theta}_{12},\ \gamma = \theta_{22} - \boldsymbol{\beta}' \mathbf{\Theta}_{11}^{-1} \boldsymbol{\beta}.$$

Inserting the partitioned $\mathbf{\Theta}$ in the objective function (9.10) and after some straightforward algebra, it follows that:

$$\begin{aligned} &1)\ \log\det(\mathbf{\Theta}) = \log(\gamma) + c, \\ &2)\ \mathrm{tr}(\mathbf{S}\mathbf{\Theta}) = 2\mathbf{s}_{12}'\boldsymbol{\beta} + s_{22}(\gamma + \boldsymbol{\beta}'\mathbf{\Theta}_{11}^{-1}\boldsymbol{\beta}) + c, \\ &3)\ \|\mathbf{\Theta}\|_1 = 2\|\boldsymbol{\beta}\|_1 + \gamma + \boldsymbol{\beta}'\mathbf{\Theta}_{11}^{-1}\boldsymbol{\beta} + c, \end{aligned} \tag{9.17}$$

where c is a universal constant different in each appearance. Substituting these in the penalized (negative) normal log-likelihood function, it magically decouples as a sum of two convex functions of γ and $\boldsymbol{\beta}$:

$$\ell(\gamma, \boldsymbol{\beta}) = \{-\log\gamma + (s_{22} + \lambda)\gamma\} + \{(s_{22} + \lambda)\boldsymbol{\beta}'\mathbf{\Theta}_{11}^{-1}\boldsymbol{\beta} + 2\mathbf{s}_{12}'\boldsymbol{\beta} + 2\lambda\|\boldsymbol{\beta}\|_1\}. \tag{9.18}$$

Its key features are that it depends only on $\mathbf{\Theta}_{11}$ and completely bypasses $\mathbf{W}$ and the second expression is readily recognized as a Lasso regression problem.

As for developing an algorithm, we note that for a given tuning parameter λ and fixed $(\boldsymbol{\beta}, \boldsymbol{\Theta}_{11})$, minimizing (9.18) with respect to γ gives the simple closed-form solution

$$\gamma = \frac{1}{s_{22} + \lambda}. \tag{9.19}$$

Then, for a fixed γ, minimizing (9.18) over the $\boldsymbol{\beta}$ with $V = \boldsymbol{\Theta}_{11}^{-1}$, is the same as solving a Lasso regression problem

$$\min_{\boldsymbol{\beta}} \ (s_{22} + \lambda)\boldsymbol{\beta}' V \boldsymbol{\beta} + 2\mathbf{s}_{12}'\boldsymbol{\beta} + 2\lambda\|\boldsymbol{\beta}\|_1, \tag{9.20}$$

where the usual cross-product matrix S_{11} is replaced by $\boldsymbol{V} = \boldsymbol{\Theta}_{11}^{-1}$. The solution for its j-th element β_j is given by

$$\beta_j = \frac{S(-[(\mathbf{s}_{12})_j + \sum_{k \neq j} v_{jk}\beta_k s_{22}])}{v_{jj}(s_{22} + \lambda)}, \tag{9.21}$$

where $S(\cdot, \lambda)$ is a soft-thresholding operator. The update (9.21) is iterated for $j = 1, 2 \ldots, p-1, 1, 2 \ldots$, till convergence. These are the key iterative steps of the S-Glasso algorithm summarized in Algorithm 2.

Algorithm 2 S-GLasso

1. Initialize $\boldsymbol{\Theta}^{(0)} = (\text{diag}(\mathbf{S}) + \lambda \boldsymbol{I})^{-1}$, for a given λ,
2. Repeat for $i = 1$ **to** p until convergence,
   a. Partition $\boldsymbol{\Theta}^{(k)}$ and $\mathbf{S}$ as in (9.3),
   b. Compute γ as in (9.19).
   c. Solve the Lasso regression (9.20) by repeating (9.21) until convergence,
   d. Update $\boldsymbol{\theta}_{12}^{(k+1)} = \boldsymbol{\beta}$, $\theta_{22}^{(k+1)} = \gamma - \boldsymbol{\beta}' \boldsymbol{\Theta}_{11}^{-1} \boldsymbol{\beta}$,
3. Output $\boldsymbol{\Theta}$.

The iterations in (9.21) requires the computationally expensive inversion of the matrix $\boldsymbol{\Theta}_{11}$. One may deal with this computational burden by solving the dual of (9.20) as a box-constrained quadratic programming solved in Mazumder and Hastie [2011].

9.3.1 Speeding up the S-GLasso

The iterations in (9.21) requires inverting the matrix $\boldsymbol{\Theta}_{11}$ which is computationally expensive when p is large. To overcome this computational burden one can solve the dual of (9.20), as a box-constrained quadratic programming with constraints $u_i \in [-\lambda, \lambda]$ for $i = 1, \ldots, p-1$:

$$\begin{aligned} &\min_{\boldsymbol{u}} \frac{1}{2}(\mathbf{s}_{12} + \boldsymbol{u})' \boldsymbol{\Theta}_{11} (\mathbf{s}_{12} + \boldsymbol{u}), \\ &\text{s.t. } u_i \in [-\lambda, \lambda]; \ i = 1, \ldots, p-1. \end{aligned} \tag{9.22}$$

This is solved using the coordinate descent algorithm by minimizing for each coordinate

$$u_i = \Pi_{[-\lambda,\lambda]}\left(\frac{-b_i - \sum_{j\neq i}(\boldsymbol{\Theta}_{11})_{ij}u_j}{(\boldsymbol{\Theta}_{11})_{ii}}\right), \tag{9.23}$$

where $\boldsymbol{b} = \boldsymbol{\Theta}_{11}\mathbf{s}_{12}$ and the projection operator $\Pi_{[-\lambda,\lambda]}(x) = -\lambda$, if $x < -\lambda$, $\Pi_{[-\lambda,\lambda]}(x) = \lambda$, if $x > \lambda$ and $\Pi_{[-\lambda,\lambda]}(x) = x$, otherwise. The update (9.23) is iterated for $j = 1, \ldots, p-1, 1, 2, \ldots,$ until convergence.

These modifications can be incorporated by substituting the substeps (c) and (d) in the Algorithm 2 by

c′. Solve the dual (9.22) by repeating (9.23) until convergence and compute $\boldsymbol{\beta} = \boldsymbol{\Theta}_{11}^{(k+1)}(\mathbf{s}_{12} + \hat{\boldsymbol{u}})/(s_{22} + \lambda)$,
d′. Update $\boldsymbol{\theta}_{12}^{(k+1)} = \boldsymbol{\beta}$, $\theta_{22}^{(k+1)} = \gamma - (\mathbf{s}_{12} + \hat{\boldsymbol{u}})'\boldsymbol{\beta}/s_{22}$.

It can be shown that $\boldsymbol{\beta}$ and γ are computable from the optimal $\hat{\boldsymbol{u}}$ using the formula $\boldsymbol{\beta} = \boldsymbol{\Theta}_{11}(\mathbf{s}_{12} + \hat{\boldsymbol{u}})/(s_{22} + \lambda)$. Then the row/column of $\boldsymbol{\Theta}$ can be updated as $\boldsymbol{\theta}_{12} = \boldsymbol{\beta}$ and $\theta_{22} = \gamma - (\mathbf{s}_{12} + \boldsymbol{u})'\boldsymbol{\beta}/s_{22}$.

9.3.2 Key Differences and Similarities

Here, we highlight some similarities and differences among GLasso, DP-GLasso, and S-GLasso algorithms.

The key difference between S-GLasso and DP-GLasso is in the derivation of the updates. While DP-GLasso employs (9.2)–(9.3), S-GLasso utilizes the decoupled likelihood (9.18) to compute $\boldsymbol{\Theta}$ directly. This obviates the need for working with $\mathbf{W}$. Despite this difference, S-GLasso and DP-GLasso algorithms both employ block coordinate descent optimization to minimize the primal objective function, while GLasso minimizes the dual of (9.1). Another beneficial property shared with DP-GLasso is the maintenance of positive definiteness of $\boldsymbol{\Theta}^{(k+1)}$ in each iteration, as long as the initial matrix $\boldsymbol{\Theta}^{(0)}$ is positive-definite, as the matrix $\boldsymbol{\Theta}_{11}^{(k)}$ remains fixed during a row/column update, and γ is positive.

The next two sections highlight the important roles of the idea of thresholding in improving the performance of and expanding the scope of GLasso.

9.4 Thresholded and Latent Variable GLasso

This section provides some generalizations and improvements of the GLasso such as the thresholded GLasso and latent variable GLasso.

9.4.1 Thresholded Graphical Lasso

Thresholding after regularization has become popular in the last decade or so in the high-dimensional data analysis. It is inspired mostly by realizing that Lasso regression produces many small, nonzero estimates, and that thresholding Lasso regression is a way of producing more zeros or debiasing the Lasso estimators; see Section 2.2.

The graphical Lasso is also known to select many false positive edges [Lafit et al., 2019]. The thresholded GLasso estimator proposed in Wang and Allen [2022] aims to zero out small entries or removing the weak edges. Given a GLasso estimator $\widehat{\boldsymbol{\Theta}}^{\lambda}$ with tuning parameter λ, the ijth entry of its (hard) thresholded estimator is

$$\widehat{\boldsymbol{\Theta}}_{ij}^{\lambda,\tau} = \widehat{\boldsymbol{\Theta}}_{ij}^{\lambda} I(\widehat{\boldsymbol{\Theta}}_{ij}^{\lambda} > \tau),$$

where τ is the threshold parameter and $I(\cdot)$ is the indicator function of a set. Similar to the tuning parameter selection in Lasso, one may use information criteria or cross-validation for selecting the tuning and threshold parameters. For more details the reader is referred to Wang and Allen [2022], where they show that the thresholded GLasso estimator is consistent and sparsistent under weaker conditions than the irrepresentability condition needed in establishing ℓ_∞ consistency of the graphical Lasso, see Ravikumar et al. [2011].

Interestingly, one may apply thresholding to other existing graph selection methods such as CLIME, see Section 4.4.3 and Cai and Liu [2011], and node-wise regression, see Section 4.4.1 and Meinshausen and Bühlmann [2006], leading to estimators with improved theoretical and empirical properties.

9.4.2 Latent Variable Graphical Lasso

This section deals with the problem of graph selection in the presence of unobserved (latent) variables. Its message is that simply *thresholding* the GLasso, neighborhood selection, or CLIME estimators can have superior properties in terms of graph selection consistency and empirical performance than the latent variable GLasso estimator in Chandrasekaran et al. [2012].

In many applications, like neuroscience, genome and finance, often there is a set of observed variables Y_O while knowing that there is lurking around another unobserved set of correlated latent (hidden, missing) variables denoted by Y_H. In this situation, the precision matrix of Y_O cannot be sparse, but is more likely to be sparse when Y_O is conditioned on Y_H. For example, in neuroscience, it is of great interest to uncover the connectivity of neurons from large scale neural recordings using various graphical model selection and estimation techniques. Since the number of neurons in the brain is in the order of billions, and even modern technologies can only record from a small subset of them, then one is faced with the challenge of graph selection in the presence of latent (unobserved) variables.

For such situations, Chandrasekaran et al. [2012] proposed a regularized maximum-likelihood framework which approximates the precision matrix as a sum of sparse and low-rank matrices. More precisely, let the random vector $Y = (Y_O', Y_H')'$ have mean zero and a covariance matrix $\boldsymbol{\Sigma}$, partitioned precision matrix $\boldsymbol{\Theta}$ with blocks $\boldsymbol{\Theta}_O, \boldsymbol{\Theta}_H, \boldsymbol{\Theta}_{O,H}$ capturing the dependencies in (observed, unobserved) random variables. It follows from the inversion formula for a partitioned matrix, see Theorem 7 in Chapter 4, that

$$\tilde{\mathbf{\Theta}}_O = \tilde{\mathbf{\Sigma}}^{-1} = \mathbf{\Theta}_O - \mathbf{\Theta}_{O,H}\mathbf{\Theta}_H^{-1}\mathbf{\Theta}_{H,O} = \mathbf{\Theta}_O - \mathbf{L},$$

where $\tilde{\mathbf{\Theta}}_O$ is the precision matrix of observed variables given the hidden and $\mathbf{L} = \mathbf{\Theta}_{O,H}\mathbf{\Theta}_H^{-1}\mathbf{\Theta}_{H,O}$ is a low-rank matrix with its rank not exceeding $\dim(Y_H)$. The matrix $\tilde{\mathbf{\Theta}}_O$ is not necessarily sparse. Let $\mathbf{S}$ be the sample covariance of the observed data from Y_O, then the solution $(\widehat{\mathbf{\Theta}}_O, \widehat{\mathbf{L}})$ of minimizing

$$\mathrm{tr}[\mathbf{S}(\mathbf{\Theta}_O - \mathbf{L}) - \log[\det(\mathbf{\Theta}_O - \mathbf{L})] + \lambda[\gamma||\mathbf{\Theta}_O||_1 + \mathrm{tr}(\mathbf{L})],$$

over the set $\{\mathbf{\Theta}_O - \mathbf{L} \succ 0, \mathbf{L} \succeq 0\}$ provides possibly sparse estimators for them where λ, γ are the penalty parameters.

Evidently, this convex formulation [Chandrasekaran et al., 2012] addresses the problem, but faces a number of challenges from both computational and statistical perspectives, and requiring the restrictive irrepresentability condition. By contrast, Wang and Allen [2022] shows the superior performance of the simple and attractive idea of applying a hard thresholding operator to a GLasso or any reasonable existing graph selection estimators. For an example of implementing this idea with real data, see Section 9.7.

9.5 Asymptotic Normal Thresholding (ANT) Methods

This section provides a central limit theorem for the estimators of an entry of the precision matrix which is much needed for inference regarding edge selection in the Gaussian graphical models and constructing confidence intervals for the underlying parameters.

A fundamental problem for statistical inference in the Gaussian graphical model is that of finding conditions under which there exists an asymptotically efficient estimator of each entry θ_{ij} of the precision matrix, in the sense that it is asymptotically normal with its variance matching that of the maximum likelihood estimator (Fisher information) in the classical setting of p fixed. Minimal conditions under which such results hold in the high-dimensional data situations are presented in Ren et al. [2015]. A bivariate regression setup involving the parameter θ_{ij} and solving it within the framework of scaled Lasso regression in Section 2.3 plays a key role.

9.5.1 A Bivariate Scaled Lasso Regression

We rely on a bivariate regression setup to study the asymptotic distribution of estimators of entries of a precision matrix. This goes beyond the univariate regression formulation in Chapter 4 for estimating θ_{ij}.

Let $\boldsymbol{Y}$ be the $n \times p$ data matrix where its ith row stands for the ith sample, and $\boldsymbol{Y}_A$ be the sub-matrix of $\boldsymbol{Y}$ with its columns indexed by a subset A of $\{1, 2, \ldots, p\}$.

Consider the following data version of a multivariate regression model

$$\boldsymbol{Y}_A = \boldsymbol{Y}_{A^c}\boldsymbol{\beta} + \boldsymbol{\epsilon}_A,$$

where $\boldsymbol{\beta} = \boldsymbol{\beta}_{A^c,A}$, is a $(p - |A|) \times |A|$ coefficient matrix and $|A|$ is the cardinality of the set A with $\text{Cov}(\boldsymbol{\epsilon}_A) = \boldsymbol{\Sigma}_{2\cdot1}^{-1}$, for the notation see Chapter 4.

In what follows, we fix $A = \{i, j\}$ and focus on the data version of regression of two variables on the rest (see Theorem 8 (b)) for estimating the $(p - 2) \times 2$ matrix of regression coefficients $\boldsymbol{\beta}$ and the 2×2 covariance matrix of the regression residuals:

$$\Theta_{A,A} = \boldsymbol{\Sigma}_{2\cdot1}^{-1} = \begin{pmatrix} \theta_{i,i} & \theta_{i,j} \\ \theta_{j,i} & \theta_{j,j} \end{pmatrix}.$$

It is plausible that inverting an efficient estimator of $\Theta_{A,A}$ should yield an efficient estimator of the entries of $\boldsymbol{\Sigma}_{2\cdot1}$. This idea will work if $\boldsymbol{\beta}$ was known, because then $\boldsymbol{\epsilon}_A'\boldsymbol{\epsilon}_A/n$ would be a natural estimator of $\Theta_{A,A}$. However, since $\boldsymbol{\beta}$ is unknown, one needs to estimate it first, and then plug in its estimator to arrive at the regression residuals $\widehat{\boldsymbol{\epsilon}}_A$ leading to the estimator

$$\widehat{\Theta}_{A,A} = \widehat{\boldsymbol{\epsilon}}_A'\widehat{\boldsymbol{\epsilon}}_A/n. \tag{9.24}$$

At this point one may ask the natural question: What is the suitable estimator of $\boldsymbol{\beta}_{A^c,A}$ for the plan? We implement the plan outlined above by estimating the two columns of $\boldsymbol{\beta}$ separately as follows: For each $m \in A = \{i, j\}$, apply a scaled Lasso estimator, see (2.29), to the univariate linear regression of $\boldsymbol{y}_m$ against $\boldsymbol{Y}_{A^c}$ to obtain $(\widehat{\boldsymbol{\beta}}_m, \widehat{\theta}_{m,m}^{1/2})$. Then, using $(\widehat{\boldsymbol{\beta}}_i, \widehat{\boldsymbol{\beta}}_j)$ we construct $\widehat{\boldsymbol{\epsilon}}_A$, where the desired estimator of $\widehat{\Theta}_{A,A}$ is given by $\widehat{\boldsymbol{\epsilon}}_A'\widehat{\boldsymbol{\epsilon}}_A/n$ leading to an estimator $\widehat{\theta}_{i,j}$ for $\theta_{i,j}$. Then, repeat this procedure for all other entries. Note that the variance of this estimator plays important roles in what follows, so we define

$$F_{ij} = (\theta_{ii}\theta_{jj} + \theta_{ij})^{-1}, \; \hat{F}_{ij} = (\hat{\theta}_{ii}\hat{\theta}_{jj} + \hat{\theta}_{ij})^{-1}, \tag{9.25}$$

as the Fisher information for estimating θ_{ij} and its estimator, respectively.

9.5.2 Asymptotic Normality and Selection Consistency

In this section we focus on the class of sparse precision matrices with bounded eigenvalues (Section 8.4) with at most $k_{n,p}$ nonzero entries in each column.

The following theorem, a small part of the result in Ren et al. [2015, Theorem 1], shows that a sufficient condition for obtaining a $n^{-1/2}$-consistent estimator of θ_{ij} is that $k_{n,p} = O(\frac{\sqrt{n}}{\log p})$, and when $k_{n,p} = o(\frac{\sqrt{n}}{\log p})$, then the proposed estimator is asymptotically efficient.

Theorem 12 *For a sample of size n from* $Y \sim N_p(\boldsymbol{\mu}, \boldsymbol{\Sigma})$, *assume that* $3 \leq k_{n,p} \leq c_0 n/\log p$, *for a sufficiently small constant* $c_0 > 0$ *and* $p > k_{n,p}^{\nu}$ *for some* $\nu > 2$.

Then, the estimator $\widehat{\theta}_{ij}$ in (9.24) is asymptotically efficient with a consistent variance estimate and

$$\sqrt{n\hat{F}_{ij}}(\hat{\theta}_{ij} - \theta_{ij}) \to^{D} N(0,1), \quad \frac{\hat{F}_{ij}}{F_{ij}} \to 1. \tag{9.26}$$

uniformly for all i, j, and $\mathbf{\Theta}$, provided that $s = o(\sqrt{n}/\log p)$ where s is the maximum degree of the graph of the precision matrix, with F_{ij} and $\hat{F}_{ij}$ are given in (9.25).

This fundamental asymptotic normality result for individual entries is obtained without requiring the irrepresentability condition or the ℓ_1 constraint on the precision matrix which are commonly required [Ravikumar et al., 2011]. It is known that the estimation of each θ_{ij} at the parametric $n^{-\frac{1}{2}}$ convergence rate requires a sparsity condition equivalent to the sample size being of order $(s \log p)^2$. To this end, an adaptive estimator of individual θ_{ij} can be used which enjoys asymptotic normality and efficiency when $n \geq (s \log p)^2$.

For a given (i, j), two immediate consequences of (9.26) are efficient interval estimation of θ_{ij}, and efficient test for the existence of an edge between nodes i and j in a Gaussian graphical model, i.e., the hypotheses $\theta_{ij} = 0$. More importantly, based on the asymptotic normality of $\widehat{\theta}_{ij}$, one may devise an adaptive thresholding method to correctly distinguish zero and nonzero entries of the precision matrix. For this, note that the variance term $\theta_{ii}\theta_{jj} + \theta_{ij}$, needs to be estimated first and the threshold parameter varies with (i, j):

$$\widehat{\tau}_{ij} = \sqrt{\frac{2\eta(\widehat{\theta}_{ii}\widehat{\theta}_{jj} + \widehat{\theta}_{ij}) \log p}{n}}, \tag{9.27}$$

for a fixed tuning parameter $\eta > 2$. Then, an adaptive support recovery procedure based on ANT is given by

$$\widehat{\mathbf{\Theta}}^{\text{ANT}} = (\hat{\theta}_{ij}^{\text{ANT}})_{p \times p}, \quad \hat{\theta}_{ij}^{\text{ANT}} = \widehat{\theta}_{ij} I(\widehat{\theta}_{ij} \geq \widehat{\tau}_{ij}), \ i \neq j, \tag{9.28}$$

and $\hat{\theta}_{ii}^{\text{ANT}} = \widehat{\theta}_{ii}$. This procedure is adaptive in the sense that the threshold parameter is entry-dependent.

Let the sets $E = E(\mathbf{\Theta}) = \{(i, j); \theta_{ij} \neq 0\}$ and $\boldsymbol{S}(\mathbf{\Theta}) = \{\text{sgn}(\theta_{ij}), 1 \leq i, j \leq p\}$ be the support and signs of nonzero entries of a precision matrix, respectively. Then, assuming the uniform signal strength condition,

$$\theta_{ij} \geq 2\sqrt{\frac{2\eta(\widehat{\theta}_{ii}\widehat{\theta}_{jj} + \widehat{\theta}_{ij}) \log p}{n}}, \quad \forall \theta_{ij} \in E(\mathbf{\Theta}), \tag{9.29}$$

the following holds.

Theorem 13 *For a sample of size n from $Y \sim N_p(\boldsymbol{\mu}, \mathbf{\Sigma})$, assume that $3 \leq k_{n,p} \leq c_0 n/\log p$ with a sufficiently small constant c_0 and $p > k_{n,p}^{\nu}$ for some $\nu > 2$. Then,* *the ANT recovers the support of* $\mathbf{\Theta}$, *and also the signs of its nonzero entries exactly in the sense that*

$$\mathbb{P}[\mathcal{S}(\widehat{\mathbf{\Theta}}^{ANT}) = \mathcal{S}(\mathbf{\Theta})] \to 1,$$

as $n \to \infty$.

An extensive simulation study was conducted in Ren et al. [2015, Section 6] to compare the performance of various graph selection procedures in the literature. The findings demonstrate the selection consistency property of the ANT method and substantial false positive for the GLasso and CLIME. Perhaps, this is due to the ANT taking the advantage of an additional thresholding step which the GLasso and CLIME do not.

Remark 5 *An astute reader might wonder about the connection between the univariate regression formulation in Chapter 4 for estimating* θ_{ij} *and the bivariate formulation for the same task in Section 9.5.1. It turns out that they are asymptotically equivalent [Ren et al., 2015, Section 5.3], but the latter method enjoys a simpler form and provides a better understanding of the fundamental limit of the Gaussian graphical model in making statistical inference. The two theorems stated above and the discussions around them are only a small portion of some deep theoretical contributions of the influential work in Ren et al. [2015]. The reader may refer to it for more discussions and insights into the problem of when and under what conditions one can estimate the low-dimensional parameters at the parametric square-root rate in a large Gaussian graphical model.*

9.6 Covariance Graphs

In a Gaussian covariance graph, absence of an edge is decided based on marginal independence rather than conditional independence between variables. While less well-known than the Gaussian graphical models, covariance graphs are becoming more popular [Cox and Wermuth, 1996, Drton and Richardson, 2008]. The notion of a relevance network in Butte et al. [2000] is closely related to covariance graphs, where genes with a pairwise correlation exceeding a threshold are connected by an edge.

For a given pattern of zeros, the problem of maximum likelihood estimation (MLE) of the nonzero entries of the covariance matrix is solved iteratively through a sequence of linear regressions [Chaudhuri et al., 2007]. For unknown zero patterns solving regularized normal likelihood requires an analog of the GLasso algorithm using a coordinate descent algorithm.

9.6.1 MLE of Covariance Matrices with Zero Entries

Covariance matrices with prespecified zero entries often appear in variety of applications. In this section, we present an important and interesting regression-based iterative conditional fitting (ICF) algorithm due to Chaudhuri et al. [2007] for computing the maximum likelihood estimate of a covariance matrix with a known zero pattern. The algorithm is efficient and solves a sequence of least-squares optimization problems iteratively. Its convergence has been proved only when the sample covariance matrix has full rank, but Cibinel et al. [2024] establishes its convergence in general, particularly when $p > n$. Another regression formulation using a penalized normal likelihood is given in Hastie et al. [2009, Section 17.3] and solved via an iterative Glasso-type algorithm.

For an $n \times p$ data matrix $\boldsymbol{Y}$ from a zero-mean multivariate normal distribution with covariance matrix $\boldsymbol{\Sigma}$, and a fixed i let $\boldsymbol{y}_i$ be its ith column and $\boldsymbol{Y}_{-i}$ be the rest. We focus on the ith row/column of the population covariance matrix and partition $\boldsymbol{\Sigma}$ and $\mathbf{S}$ as

$$\boldsymbol{\Sigma} = \begin{pmatrix} \boldsymbol{\Sigma}_{-i,-i} & \boldsymbol{\sigma}_{-i,i} \\ \boldsymbol{\sigma}'_{-i,i} & \sigma_{ii} \end{pmatrix}, \ \mathbf{S} = \begin{pmatrix} \mathbf{S}_{-i,-i} & \mathbf{s}_{-i,i} \\ \mathbf{s}'_{-i,i} & s_{ii} \end{pmatrix}$$

where $-i$ stands for the set of integers $\{1, \ldots, p\}$ with i removed. The (negative) loglikelihood conveniently decomposes as

$$\ell_{\boldsymbol{Y}}(\boldsymbol{\Sigma}) = \ell_{\boldsymbol{y}_i|\boldsymbol{Y}_{-i}}(\gamma_i, \boldsymbol{\sigma}_{-i,i}) + \ell_{\boldsymbol{Y}_{-i}}(\boldsymbol{\Sigma}_{-i,-i}), \tag{9.30}$$

namely, as a sum of the loglikelihoods of marginal distribution of $\boldsymbol{Y}_{-i}$ and the conditional distribution of $\boldsymbol{y}_i$ given $\boldsymbol{Y}_{-i}$. Then, the first term or the conditional likelihood reduces to

$$\ell_{\boldsymbol{y}_i|\boldsymbol{Y}_{-i}}(\gamma_i, \boldsymbol{\sigma}_{-i,i}) = \log \gamma_i + \frac{1}{\gamma_i}||\boldsymbol{y}_i - \boldsymbol{Z}_{-i}\boldsymbol{\sigma}_{-i,i})||_2^2, \tag{9.31}$$

where the matrix

$$\boldsymbol{Z}_{-i} = \boldsymbol{Y}_{-i}\boldsymbol{\Sigma}_{-i,-i}^{-1}, \tag{9.32}$$

plays the role of auxiliary covariates for the regression of $\boldsymbol{y}_i$ on the other variables, and

$$\gamma_i = \sigma_{ii} - \boldsymbol{\sigma}'_{-i,i}\boldsymbol{\Sigma}_{-i,-i}^{-1}\boldsymbol{\sigma}_{-i,i}. \tag{9.33}$$

This likelihood is readily recognized as the normal loglikelihood function for the linear regression of $\boldsymbol{y}_i$ on $\boldsymbol{Z}_{-i}$ with the regression coefficients $\boldsymbol{\sigma}_{-i,i}$ and the error variance γ_i, see Problem 3.23 for a population version.

Exploiting this regression idea, the ICF algorithm in Chaudhuri et al. [2007] iterates between estimating the blocks of $(\boldsymbol{\sigma}_{-i,i}, \gamma_i)$ and $\boldsymbol{\Sigma}_{-i,-i}$. When some off-diagonal entries $\boldsymbol{\Sigma}$ are equal to zero, it is helpful to represent its sparsity pattern through a graph $G = (V, E)$ with vertex set $V = \{1, \ldots, p\}$ and the edge set E. Knowledge of the graph G is the same as knowing which off-diagonal entries of $\boldsymbol{\Sigma}$ are zero. In particular, if we denote by $\text{bd}(i)$ the boundary of $i \in V$, that is the set of vertices connected to i in G, then the structure imposed by it implies that the entries of $\boldsymbol{\sigma}_{-i,i}$ are such that $\sigma_{ij} = 0$ for every $j \notin \text{bd}(i)$.

In fact, for a given initial value of $\boldsymbol{\Sigma}_{i,i}$, ICF simply applies the least-squares formula to estimate the non-zero elements of the regression coefficients $\boldsymbol{\sigma}_{-i,i}$ in (9.31) and then computes γ_i using the residual sum of squares.

9.6.2 Covariance Graphical Lasso (C-GLasso)

For unknown pattern of zeros, as in the search for sparsity via GLasso, a more systematic search for sparse covariance matrix is to use the penalized log-likelihood

$$\ell(\boldsymbol{\Sigma}) = \log|\boldsymbol{\Sigma}| + \mathrm{tr}(S\boldsymbol{\Sigma}^{-1}) + \lambda\|\boldsymbol{\Sigma}\|_1. \tag{9.34}$$

Though closely related to (9.10), it is not convex in $\boldsymbol{\Sigma}$ and hence more challenging computationally.

Some early works are Lam and Fan [2009] which studied the theoretical properties of penalized covariance estimator without providing an algorithm for computing it; Bien and Tibshirani [2011] has proposed a majorization-minimization algorithm making use of generalized gradient descent. A likelihood-based method that regularizes the distance from the covariance estimate to a symmetric sparsity set is given in Xu and Lange [2022]. Here, we present an analog of GLasso due to Wang [2014] which is a coordinate descent algorithm which updates one column of the covariance matrix at a time while holding the other part in $\boldsymbol{\Sigma}$ fixed. It effectively exploits the regression idea for sparse estimation of covariance matrices and hence discovers marginal independence structures among the random variables.

For simplicity, as usual, we focus on the last row/column of the covariance matrix and partition $\boldsymbol{\Sigma}$ and $\mathbf{S}$ as

$$\boldsymbol{\Sigma} = \begin{pmatrix} \boldsymbol{\Sigma}_{11} & \boldsymbol{\sigma}_{12} \\ \boldsymbol{\sigma}_{21} & \sigma_{22} \end{pmatrix}, \quad \mathbf{S} = \begin{pmatrix} \mathbf{S}_{11} & \mathbf{s}_{12} \\ \mathbf{s}_{21} & s_{22} \end{pmatrix},$$

where $\boldsymbol{\Sigma}_{11}$ is the $(p-1)\times(p-1)$ principal minor, $\boldsymbol{\sigma}_{12}$ and σ_{22} are vectors and scalars, respectively. Let

$$\boldsymbol{\beta} = \boldsymbol{\sigma}_{12}, \gamma = \sigma_{22} - \boldsymbol{\sigma}'_{12}\boldsymbol{\Sigma}_{11}^{-1}\boldsymbol{\sigma}_{12},$$

be a reparameterization of the last row/column of the covariance matrix. Then, applying the block matrix inversion formula to $\boldsymbol{\Sigma}$ using the blocks $(\boldsymbol{\Sigma}_{11}, \boldsymbol{\beta}, \gamma)$ one obtains

$$\boldsymbol{\Sigma}^{-1} = \begin{pmatrix} \boldsymbol{\Sigma}_{11}^{-1} + \boldsymbol{\Sigma}_{11}^{-1}\boldsymbol{\beta}\boldsymbol{\beta}'\boldsymbol{\Sigma}_{11}^{-1}\gamma^{-1} & -\boldsymbol{\Sigma}_{11}^{-1}\boldsymbol{\beta}\gamma^{-1} \\ -\boldsymbol{\beta}'\boldsymbol{\Sigma}_{11}^{-1}\gamma^{-1} & \gamma^{-1} \end{pmatrix},$$

see Problem 3.23(b). Ignoring the constants not involving $(\boldsymbol{\beta}, \gamma)$, the three terms of the penalized Gaussian likelihood function (9.34) can be expressed in terms of the new parameters as

$$\begin{aligned} &1)\ \log|\boldsymbol{\Sigma}| = \log\gamma, \\ &2)\ \|\boldsymbol{\Sigma}\|_1 = 2\|\boldsymbol{\beta}\|_1 + (\boldsymbol{\beta}'\boldsymbol{\Sigma}_{11}^{-1}\boldsymbol{\beta} + \gamma), \\ &3)\ \mathrm{tr}(\mathbf{S}\boldsymbol{\Sigma}^{-1}) = a\gamma^{-1}, \end{aligned} \tag{9.35}$$

where

$$a = \boldsymbol{\beta}'\boldsymbol{\Sigma}_{11}^{-1}\mathbf{S}_{11}\boldsymbol{\Sigma}_{11}^{-1}\boldsymbol{\beta} - 2\mathbf{s}'_{12}\boldsymbol{\Sigma}_{11}^{-1}\boldsymbol{\beta} + s_{22},$$

depends only on $\boldsymbol{\beta}$, but not γ. A key advantage of this reparameterization is that substituting for the above three quantities in the objective function decouples it into sum of two simpler functions, and reduces the problem to minimizing

$$\ell(\boldsymbol{\beta}, \gamma|\boldsymbol{\Sigma}_{11}) = (\log\gamma + a\gamma^{-1} + \lambda\gamma) + \lambda(2\|\boldsymbol{\beta}\|_1 + \boldsymbol{\beta}'\boldsymbol{\Sigma}_{11}^{-1}\boldsymbol{\beta}), \tag{9.36}$$

over the new parameters $(\boldsymbol{\beta}, \gamma)$. Evidently, this is a bi-convex function and for a fixed $\boldsymbol{\beta}$, minimizing (9.36) over γ leads to the closed-form solution

$$\hat{\gamma} = \begin{cases} a & \text{if } \lambda = 0, \\ \frac{-1+\sqrt{1+4a\lambda}}{2\lambda} & \text{if } \lambda > 0, \end{cases} \tag{9.37}$$

see Problem 9.4. Then, keeping γ fixed and setting

$$\mathbf{V} = \boldsymbol{\Sigma}_{11}^{-1}\mathbf{S}_{11}\boldsymbol{\Sigma}_{11}^{-1}\gamma + \lambda\boldsymbol{\Sigma}_{11}^{-1}, \quad \mathbf{u} = s_{12}'\boldsymbol{\Sigma}_{11}^{-1}\gamma^{-1},$$

the objective function involving $\boldsymbol{\beta}$ reduces to

$$\boldsymbol{\beta}'\mathbf{V}\boldsymbol{\beta} - 2\mathbf{u}'\boldsymbol{\beta} + 2\lambda||\boldsymbol{\beta}||_1,$$

which is readily recognized as a Lasso regression problem. It can be solved efficiently using a coordinate descent algorithm for the jth component of $\boldsymbol{\beta}$, for $j = 1, \ldots, p-1$; see Section 2.1.3. Thus, in principle $\boldsymbol{\beta}$ is computed via a Lasso regression written symbolically as

$$\text{Lasso}(\mathbf{V}, \mathbf{u}, \lambda). \tag{9.38}$$

Combining these two almost explicit solutions, one may update the last row/column as $(\sigma_{12} = \boldsymbol{\beta}, \sigma_{22} = \gamma + \boldsymbol{\beta}'\boldsymbol{\Sigma}_{11}^{-1}\boldsymbol{\beta})$, followed by cycling through all rows/columns until convergence. This algorithm can be viewed as a block coordinate descent method with p blocks of $\boldsymbol{\beta}$'s and another p blocks of γ's, and summarized similar to the Algorithm 1 for GLasso, see Algorithm 3.

Algorithm 3 C-GLasso

Initialize $\boldsymbol{\Sigma}^{(0)} = \text{diag}(\mathbf{S}) + \lambda\boldsymbol{I}$, set $\boldsymbol{\Sigma}^{(k+1)} = \boldsymbol{\Sigma}^{(k)}$, $k = 0, 1, 2 \ldots$
for $i = 1$ to p until convergence **do**
    Partition $\boldsymbol{\Sigma}^{(k)}$ and $\mathbf{S}$
    Compute γ as in (9.37)
    Solve the Lasso regression (9.38)
    Update $\boldsymbol{\sigma}_{12}^{(k+1)} = \boldsymbol{\beta}$, $\sigma_{22}^{(k+1)} = \gamma + \boldsymbol{\beta}'\boldsymbol{\Sigma}_{11}^{-1}\boldsymbol{\beta}$
end for
Output $\boldsymbol{\Sigma}$.

Simulation results in Wang [2014] show that the C-GLasso algorithm is more attractive computationally than the algorithm of Bien and Tibshirani [2011] in terms of simplicity, speed and stability. The R package covglasso implements the C-GLasso algorithm.

9.6.3 Penalizing Trace of Precision Matrices

The C-GLasso and ICF algorithms work for nonsingular covariance matrices, but for high-dimensional data or when $p > n$ the sample covariance matrix $\mathbf{S}$ is singular. In this case, one may rely on the ridge estimator $\mathbf{S} + \kappa\mathbf{I}_p$ which is invertible for any $\kappa > 0$ and corresponds to the shrinkage estimator (8.2) with $\alpha_2 = 1$. It turns out that such estimators are, indeed, the solution of a penalized normal likelihood where the penalty function is $\text{tr}\boldsymbol{\Sigma}^{-1}$ or the sum of inverse eigenvalues of the covariance matrix [Warton, 2008]. Evidently, this penalty function forces the eigenvalues of $\boldsymbol{\Sigma}$ to stay away from zero so that the estimated covariance matrix is a nonsingular matrix.

In analogy with the idea of augmented data in ridge regression, see Problem 2.5, for $\boldsymbol{Y}$, the centered $n \times p$ data matrix let $\boldsymbol{Y}^{(\kappa)} = (\boldsymbol{Y}', \sqrt{n\kappa}\mathbf{I}_p)'$ be the augmented data matrix. Then, it is easy to verify that

$$\text{tr}\boldsymbol{\Sigma}^{-1}(\boldsymbol{Y}^{(\kappa)'}\boldsymbol{Y}^{(\kappa)}) = \text{tr}\boldsymbol{\Sigma}^{-1}(\boldsymbol{Y}'\boldsymbol{Y} + n\kappa\mathbf{I}_p) = \text{tr}(\boldsymbol{\Sigma}^{-1}\boldsymbol{Y}'\boldsymbol{Y}) + n\kappa\text{tr}(\boldsymbol{\Sigma}^{-1}),$$

so that the (negative) loglikelihoods of $\boldsymbol{Y}$ and $\boldsymbol{Y}^{\kappa}$ are closely related. In fact, since

$$\ell_{\boldsymbol{Y}^{(\kappa)}}(\boldsymbol{\Sigma}) = \ell_{\boldsymbol{Y}}(\boldsymbol{\Sigma}) + \kappa\text{tr}(\boldsymbol{\Sigma}^{-1}), \tag{9.39}$$

it can be viewed as a penalized version of $\ell_{\boldsymbol{Y}}$ with trace of the precision matrix serving as the penalty function with κ as its tuning parameter. We show that this simple identity can be used to unify the ICF and C-GLasso algorithms. For this, consider the more general penalty

$$p_{\lambda,\kappa}(\boldsymbol{\Sigma}) = \lambda \sum_{i \neq j} |\sigma_{i,j}| + \kappa \sum_{i=1}^{p} \sigma^{i,i}, \tag{9.40}$$

where the diagonal entry $\sigma^{i,i}$ is the inverse of the error variance in the regression of $\boldsymbol{y}_i$ on $\boldsymbol{Y}_{-i}$; see (4.20). Thus, for a fixed i, the conditional likelihood of $\boldsymbol{y}_i$ given the rest or $\boldsymbol{Y}_{-i}$ is

$$\ell_{\boldsymbol{y}_i|\boldsymbol{Y}_{-i}}(\gamma_i, \boldsymbol{\sigma}_{-i,i}) = \log\gamma_i + \frac{1}{n\gamma_i}||\boldsymbol{y}_i - Z_{-i}\boldsymbol{\sigma}_{-i,i}||^2 + 2\lambda||\boldsymbol{\sigma}_{-i,i}||_1 + \kappa\frac{1}{\gamma_i}, \tag{9.41}$$

with γ_i given in (9.33). Note that for known γ_i, this optimization problem is simply a Lasso regression problem. Otherwise, one is faced with the problem of joint estimation of the regression coefficients and the error variance γ_i as discussed in Chapter 2 following Yu and Bien [2019].

Solution of penalized normal likelihood with the penalty (9.40) seems challenging at first. However, in view of (9.39) it can be solved by simply applying the C-Glasso algorithm in Section 9.6.2 to the augmented data $\boldsymbol{Y}^{(\kappa)}$. As expected, the decomposition (9.30) of the likelihood into marginal and conditional components plays the key role. Here we focus on its analog: for a fixed i the condition likelihood of $\boldsymbol{y}_i$ given the rest or $\boldsymbol{Y}_{-i}$ is

$$\ell_{\boldsymbol{y}_i|\boldsymbol{Y}_{-i}}(\gamma_i, \boldsymbol{\sigma}_{-i,i}) = \log\gamma_i + \frac{1}{n\gamma_i}||\boldsymbol{y}_i^{(\kappa)} - Z_{bd(i)}^{(\kappa)}\boldsymbol{\sigma}_{bd(i),i}||^2 + 2\lambda||\boldsymbol{\sigma}_{bd(i),i}||_1, \tag{9.42}$$

where $\boldsymbol{Y}^{(\kappa)}$ is the augmented data matrix, $Z_{bd(i)}^{(\kappa)}$ is the submatrix of $\boldsymbol{Z}_{-i}^{(\kappa)} = \boldsymbol{Y}_{-i}^{(\kappa)}\boldsymbol{\Sigma}_{-i,-i}^{-1}$, whose column index is in $\text{bd}(i)$ and similarly for $\boldsymbol{\sigma}_{bd(i),i}$.

The application of this penalized likelihood method usually requires an initial grid of penalty parameter values to obtain a path of solutions. Maximum theoretical values of the penalty parameters (λ, κ) and details on the cross-validation approach for their selection and their interplay are given in Cibinel et al. [2024]. They identify the values of the pair of penalty parameters that return a diagonal solution.

9.7 Hands-on R: Inverse Covariance Estimation

9.7.1 Graphical Lasso

We analyze stock return volatility data from Demirer et al. [2018], where the goal was to estimate the global bank network connectedness. The data contains 96 banks from 29 developed and emerging economies (countries) from September 12, 2003, to February 7, 2014. For illustration purposes, we select only economies where the number of banks in each economy is greater than 4, total of 54 banks. (The data is available in recode package.)

Here, our goal is to illustrate the role of graphical lasso algorithm in discovering connectedness of the banks. We use CVglasso package in R to fit graphical lasso algorithm and select the tuning parameter using BIC from a grid of values in the interval [0, 1].

```
## Import required packages
require(recode)
require(CVglasso)

## import and scale data
data(globalbank)
data = scale(globalbank, center = TRUE, scale = TRUE)

## use cvglasso to select threshold
set.seed(1234)
cvg = CVglasso(X = data, nlam = 50, crit.cv = "BIC")
Omega = cvg$Omega
print(cvg$Tuning)
      log10(lam)         lam
[1,]    -0.9410014 0.1145509
```

To visualize the result, we exploit a multidimensional scaling algorithm to calculate proximities between variables. Figure 9.4 illustrates the result. The colors in the figure indicate the corresponding country of the bank. It can be seen, using colors along with the estimated sparse inverse covariance matrix as an input for the network, banks from the same country tend to group together, while connected to banks from other countries.

9.7.2 Thresholded Graphical Lasso

In this section, we compare the performance of the graphical lasso and thresholded graphical lasso in recovering regular and latent graphical models using four metrics: F1 score, true positive rate (TPR), false positive rate (FPR), and true discovery rate (TDR). A higher F1, TPR and TDR values and a smaller FPR imply a better performance. To report comparison results, we use comparegraph() function available in recode package. For both cases, a simulated data is generated using genlatentdata() (for details, type help(genlatentdata)) function and tuning parameters are selected

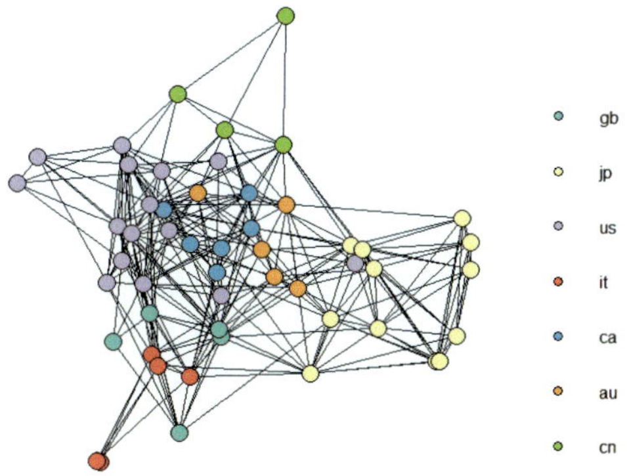

FIGURE 9.4
The GLasso solution. Colors indicate the country. The labels gb corresponds to Great Britain, us - USA, ca - Canada, cn - China, jp - Japan, it - Italy, and au - Australia.

such that methods yield the true number of edges. For details, see Wang and Allen [2022, Section 4] .

We start by considering data generated from regular graphical model, i.e., no latent variables. We specify the number of observations $n = 300$, the number of observed variables $p_o = 30$, and the number of latent variables $p_l = 0$. The probability that there is an edge among only observed, only latent, and observed and latent is 0.2, 0.3 and 0.4, respectively.

```
### Regular graphical model
p_o = 30 # Observed number of  variables
p_l = 0 # Latent number of  variables
N = 300  # Number of observations

# Generate data
lat = genlatentdata(p_o = p_o, p_l = p_l, N = N,
                    prob_o = 0.2, prob_l = 0.3,
                    prob_ol = 0.4, seed = 123)
X = lat$X

true_theta = lat$theta_o ## true theta
S = cov(scale(X))

# Implement glasso
theta_glasso = glasso::glasso(S, rho = 0.131,
            penalize.diagonal = FALSE)$wi
comparegraph(theta_glasso,true_theta )
        F1   TPR   FPR   TDR
[1,] 0.678 0.690 0.217 0.667
```

Next, we implement thresholded GLasso with smaller values of a penalty parameter.

```
# Implement thresholded glasso
theta_thrglasso = glasso::glasso(S, rho = 0.0001,
             penalize.diagonal = FALSE)$wi
theta_thrglasso[abs(theta_thrglasso) < 0.2285] = 0
comparegraph(theta_thrglasso,true_theta )
        F1    TPR   FPR   TDR
[1,] 0.860 0.857 0.086 0.862
```

As can be seen, for all four metrics, the thresholded GLasso performs significantly better than GLasso.

9.7.3 Latent Variable Graphical Lasso

Next, we compare those two methods with the latent graphical model and select the number of latent variables to be $p_l = 20$.

```
### Latent graphical model
p_l = 20 # number of  latent variables

# Generate data
lat = genlatentdata(p_o = p_o, p_l = p_l, N = N,
                    prob_o = 0.2, prob_l = 0.3,
                    prob_ol = 0.4, seed = 123)
X = lat$X

true_theta = lat$theta_o ## true theta
S = cov(scale(X))

# Implement glasso
theta_glasso = glasso::glasso(S, rho = 0.063,
             penalize.diagonal = FALSE)$wi
comparegraph(theta_glasso,true_theta)
       F1    TPR   FPR   TDR
[1,] 0.746 0.893 0.315 0.641

# Implement thresholded glasso
theta_thrglasso = glasso::glasso(S, rho = 0.0001,
                    penalize.diagonal = FALSE)$wi
theta_thrglasso[abs(theta_thrglasso) < 0.259] = 0
comparegraph(theta_thrglasso,true_theta )
        F1    TPR   FPR   TDR
[1,] 0.858 0.845 0.079 0.871
```

Even though TPR of GLasso is higher than for thresholded GLasso, it results in a higher FPR and smaller TDR and F1.

9.7.4 Covariance Graphical Lasso

In this section, we use covglasso and gicf packages in R to implement covariance graphical lasso and GICF algorithm, introduced in Sections 9.6.2 and 9.6.3, respectively. To generate data from clique models, described in Bien and Tibshirani [2011, Section 5] , we use gendataclique() function, available in the recode package. For details, see help(gendataclique). Results are reported using the comparegraph() function. For both models, the tuning parameter λ is selected such that the number of non-zero values corresponds to the oracle 100.

```
require(recode)
require(covglasso)
require(gicf)

n = 100
p = 20
# generate data from clique covariance matrix:
gendata = gendataclique(p = p, n = n, ncliques=4, seed = 123)
X = gendata$X
Sigma = gendata$Sigma

# fit covaraince graphical lasso
fit <- covglasso(X, lambda = 0.079)

# report results
comparegraph(fit$sigma, Sigma)
        F1   TPR    FPR   TDR
[1,] 0.84 0.84 0.057 0.84

# fit GICF
fit_gicf <- gicf(X, lambda = 0.0725, kappa = 0.1)

# report results
comparegraph(fit_gicf$sigma, Sigma)
        F1   TPR  FPR  TDR
[1,] 0.86 0.86 0.05 0.86
```

It can be seen that GICF performs slightly better than covariance glasso. Figure 9.5 displays the true and estimated covariance matrices.

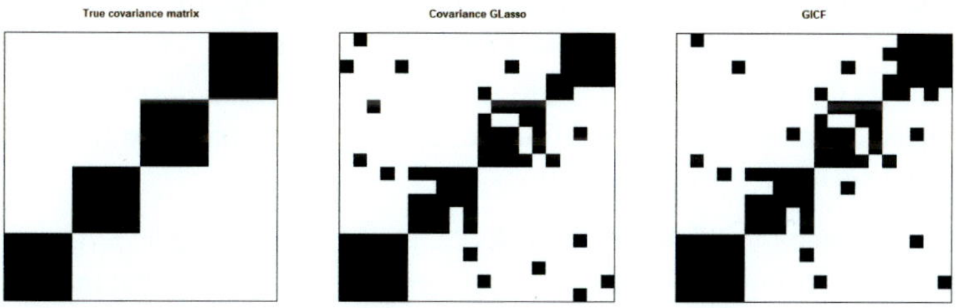

FIGURE 9.5
(Left) True covariance matrix. (Middle) Covariance GLasso. (Right) GICF

9.8 Non-Iterative Algorithms: Hidden Cholesky

Most of the iterative algorithms in this chapter are computationally demanding and do not scale up well to high-dimensional data, and may not lead to positive-definite covariance estimates. They can benefit greatly from a novel and fast *non-iterative* algorithm in Khare et al. [2024] which directly addresses many of these shortcomings of the iterative methods. We present only a brief outline of the ideas and ingredients here, and refer the readers to the original paper for more details and concrete examples.

The *constrained Cholesky approach* [Khare et al., 2024] for sparse estimation of a precision matrix $\boldsymbol{\Theta}$ with the corresponding graph $G = (V, E)$ is based on identifying and working with a decomposable graph cover for G, denoted by $G^D = (V, E^D)$, with more edges than the original graph corresponding to a denser precision matrix $\boldsymbol{\Theta}^D$ with the Cholesky factor L^D. Recall that an undirected graph G is *decomposable* if it is connected and does not contain a cycle of length greater than or equal to four as an induced subgraph. Since the methodology critically depends on a search to match the zeros of the Cholesky factor and that of the precision matrix, we refer to it as the *hidden Cholesky approach* (HCA) in analogy with hidden regression used earlier. We note that the number of nonzero entries of a positive-definite matrix is usually more that of the lower triangular part of its Cholesky factor. The extra non-zero entries in the Cholesky factor are called its *fill-in* entries corresponding to the original matrix; for example in Problem 3.14 the $(4, 2)$ entry of the Cholesky factor is its only fill-in entry.

The HCA consists of the following two steps:

1. First, an estimate $\widehat{\boldsymbol{\Theta}}^D$ of $\boldsymbol{\Theta}^D$ is obtained. For example, if the underlying data distribution is Gaussian, the MLE of $\boldsymbol{\Theta}^D$ is well-defined and has a closed form as long as the sample size n is greater than or equal to the maximal clique size of the decomposable graph G^D. Let $\widehat{L}^D$ be its Cholesky factor where the non-zero entries of its jth column have closed form similar to those in (4.26) or (12.9). The $\widehat{\boldsymbol{\Theta}}^D$ and its Cholesky factor $\widehat{L}^D$ are referred to as *intermediate* estimators. Note that these estimators correspond to the larger number of edges in E^D and their computation do not involve iterations at all.

2. The second step modifies the Cholesky factor $\widehat{L}^D$ by zeroing its *fill-in* entries that arise from working with a decomposable cover; this ensures that the zeros of the precision matrix estimate are aligned with those prescribed by the graph G. In particular, we construct an $\hat{L}$ by sequentially changing each fill-in entry of $\hat{L}^D$ such that the corresponding entry of the estimated precision matrix $\hat{\boldsymbol{\Theta}} = \hat{L}\hat{L}'$ is zero.

The key advantage of this method is that it obviates the need for computationally expensive iterative algorithms even when working with non-decomposable undirected graphical models. More importantly, it enjoys the following additional and attractive properties [Khare et al., 2024]:

1. It has high-dimensional statistical guarantees. Extensive numerical experiments [Khare et al., 2024] show that it achieves scalability without compromising the statistical precision.

2. It can be readily combined with scalable non-iterative thresholding-based methods for sparsity selection to ensure positive-definiteness of the final estimated matrix.

3. It is more robust, in the sense that it does not need to rely on the Gaussian likelihood and hence can handle outliers.

4. Finally, as the dimension p grows, the proportion of graphs for which it will outperform the state-of-the-art method in terms of computational complexity and performance tends to 1.

Positive-definiteness of an estimator of the form $\hat{\boldsymbol{\Theta}} = \hat{L}\hat{L}'$ is rather immediate. Thus, the technique of HCA can used to produce a positive-definite matrix estimator consistent with a sparsity pattern induced by the threshold estimators in Chapter 8 or any ℓ_1-penalized likelihood precision matrix estimators. For example, the CONCORD method [Khare et al., 2014] leads to sparse precision matrix estimators which are not guaranteed to be positive definite. Suitable coupling of these two powerful approaches could lead to many improvements of the iterative algorithms while highlighting the intrinsic role of the *hidden* Cholesky factor in ensuring the positive-definiteness of covariance estimators; see Section 8.3.1.

9.9 Problems

9.1 (a) Sketch the graph of a random vector with the precision matrix of the form

$$\boldsymbol{\Theta} = \begin{pmatrix} \theta_{11} & \theta_{12} & \theta_{13} & \cdots & \theta_{1p} \\ \theta_{12} & \theta_{22} & 0 & \cdots & 0 \\ \theta_{13} & 0 & \theta_{33} & \cdots & 0 \\ \vdots & \vdots & \vdots & \ddots & \vdots \\ \theta_{1p} & 0 & 0 & \cdots & \theta_{pp} \end{pmatrix}.$$

Find the inverse of the matrix Θ.

(b) For $\boldsymbol{\Theta}$ as in (a), find the range of values of $\theta_{1j}, j = 2, \ldots, p$ so that $\boldsymbol{\Theta}$ is a positive-definite matrix.

(c) Let $\boldsymbol{\Sigma} = \boldsymbol{\Theta}^{-1}$ and $\boldsymbol{u}$ be a fixed p-vector. Compute the precision matrix of $\boldsymbol{\Sigma} + \boldsymbol{u}\boldsymbol{u}'$ when it exists, and discuss the nature of its sparsity as it relates to the sparsity of the vector $\boldsymbol{u}$.

9.2 Partition $\boldsymbol{W} = \boldsymbol{\Theta}^{-1}$ and use properties of inverse of a block-partitioned matrix to show that

$$\mathbf{W} = \begin{pmatrix} \mathbf{W}_{11} & \mathbf{w}_{12} \\ \mathbf{w}_{12}' & w_{22} \end{pmatrix} = \begin{pmatrix} \left(\boldsymbol{\Theta}_{11} - \frac{\boldsymbol{\theta}_{12}\boldsymbol{\theta}_{12}'}{\theta_{22}}\right)^{-1} & -\mathbf{W}_{11}\frac{\boldsymbol{\theta}_{12}}{\theta_{22}} \\ -\frac{\boldsymbol{\theta}_{12}'}{\theta_{22}}\mathbf{W}_{11} & \frac{1}{\theta_{22}} - \frac{\boldsymbol{\theta}_{12}'\mathbf{W}_{11}\boldsymbol{\theta}_{12}}{\theta_{22}^2} \end{pmatrix}, \tag{9.43}$$

providing an alternative representation for w_{12} in (9.5).

9.3 In the GLasso algorithm, after updating the w_{12} by $\widehat{w}_{12}$ at each iteration, one obtains the updated

$$\tilde{W} = \begin{pmatrix} W_{11} & \widehat{w}_{12} \\ \widehat{w}_{12}' & w_{22} \end{pmatrix}.$$

Show that, after starting with a positive definite matrix W, the updated $\tilde{W}$ remains positive-definite.
Hint: Use the Schur formula for determinant of partitioned matrices.

9.4 Prove the three statements (equalities) in (9.17). Show that the function in (9.18) is bi-convex in $(\boldsymbol{\beta}, \gamma)$, and the minimizer of the first part has the closed-form given in (9.19).

9.5 (Joint Penalty) Penalizing the diagonal entries of a covariance matrix $\boldsymbol{\Sigma}$ may force some of them to become zero or small which could conflict with its positive-definiteness. As a compromise consider a joint penalty function

$$p_{\lambda,\kappa}(\boldsymbol{\Sigma}) = \lambda||\boldsymbol{\Sigma} - \text{diag}(\boldsymbol{\Sigma})||_1 + \kappa||\text{diag}(\boldsymbol{\Sigma})||_1,$$

with λ and κ as two different tuning parameters for the off-diagonal and diagonal entries, respectively.

(a) Discuss plausible relations between the two penalty parameters λ, κ in $p_{\lambda,\kappa}(\boldsymbol{\Sigma})$. Do you prefer $\lambda < \kappa$?

(b) Discuss plausible relations between λ, κ in the C-GLasso with the mixed penalty function

$$\tilde{p}_{\lambda,\kappa}(\boldsymbol{\Sigma}) = \lambda||\boldsymbol{\Sigma} - \text{diag}(\boldsymbol{\Sigma})||_1 + \kappa||\text{diag}(\boldsymbol{\Sigma}^{-1})||_1.$$

(c) When $\lambda = 0$ in (b), show that the penalized likelihood estimator of $\boldsymbol{\Sigma}$ is given by $\mathbf{S} + \kappa\mathbf{I} =: \mathbf{S}^{(\kappa)}$, the so-called ridge estimator of the covariance matrix [Warton, 2008].

(d) For the augmented data matrix $\boldsymbol{Y}^{(\kappa)} = (\boldsymbol{Y}', \sqrt{n\kappa}\mathbf{I}_p)'$, show that [Cibinel et al., 2024]

$$\ell_{\boldsymbol{Y}}(\boldsymbol{\Sigma}) + \tilde{p}_{\lambda,\kappa}(\boldsymbol{\Sigma}) = \ell_{\boldsymbol{Y}^{(\kappa)}}(\boldsymbol{\Sigma}) + \lambda||\boldsymbol{\Sigma} - \text{diag}(\boldsymbol{\Sigma})||_1 \tag{9.44}$$

$$= \log|\boldsymbol{\Sigma}| + \text{tr}(\boldsymbol{\Sigma}^{-1}\mathbf{S}^{(\kappa)}) + \lambda||\boldsymbol{\Sigma} - \text{diag}(\boldsymbol{\Sigma})||_1. \tag{9.45}$$

Consequently, the penalized likelihood estimator of $\boldsymbol{\Sigma}$ using the data $\boldsymbol{Y}$ and the joint penalty $\tilde{p}_{\lambda,\kappa}(\boldsymbol{\Sigma})$ is the same as the penalized likelihood estimator based on the augmented data and the single penalty $\lambda||\boldsymbol{\Sigma} - \text{diag}(\boldsymbol{\Sigma})||_1$.

9.6 Consider riboflavin dataset from the R package hdi. After importing the package and data, run X <- riboflavin$x to extract gene expression data for Bacillus subtilis with $p = 4088$ genes and $n = 71$ observations.

(a) Standardize the data by using the scale() function..

(b) Use 3-fold cross-validation to select tuning parameter for graphical lasso and estimate the inverse covariance matrix by using the CVglasso() function.

9.7 In Section 9.7.1, using the BIC criterion led to the tuning parameter $\lambda \approx 0.03$.

(a) Use the glasso() function from the R package glasso to estimate the bank connectedness by doubling the tuning parameter to $\lambda = 0.06$.

(b) Use the graph_from_adjacency_matrix() function from the R package igraph and plot the precision matrix from (a).

(c) Compared to the Figure 9.1, is the result sparser or denser?

(d) Repeat (a)–(c), but this time reduce the tuning parameter to $\lambda = 0.01$. Compare and discuss your observations regarding the sparsity of the graphs corresponding to the three tuning parameters $0.01, 0.03$ and 0.06.

9.8 The SPACE algorithm minimizes the objective function:

$$Q_0(\Theta) = \frac{1}{2}\sum_{i=1}^{p}\left(w_i||\boldsymbol{y}_i - \sum_{j\neq i}\rho^{ij}\sqrt{\frac{\theta_{jj}}{\theta_{ii}}}\boldsymbol{y}_j||^2\right) + \lambda\sum_{i<j}|\rho^{ij}|, \tag{9.46}$$

over the partial correlations ρ^{ij}'s where w_i's are some known nonnegative weights.

(a) show that the regression part of the objective function can be rewritten as

$$\begin{aligned} w_i||\boldsymbol{y}_i - \sum_{j\neq i}\rho^{ij}\sqrt{\frac{\theta_{jj}}{\theta_{ii}}}\boldsymbol{y}_j||_2^2 &= \tfrac{w_i}{\theta_{ii}^2}||\textstyle\sum_{j=1}^{p}\theta_{ij}\boldsymbol{y}_j||_2^2 \\ &= \tfrac{w_i}{\theta_{ii}^2}(\boldsymbol{\theta}_i'\boldsymbol{Y}'\boldsymbol{Y}\boldsymbol{\theta}_i), \end{aligned}$$

where $\boldsymbol{\theta}_i$ is the ith column of $\boldsymbol{\Theta}$.

(b) Show that the choice of $w_i = \theta_{ii}^2$, the above identity reduces to

$$w_i||\boldsymbol{y}_i - \sum_{j\neq i}\rho^{ij}\sqrt{\frac{\theta_{jj}}{\theta_{ii}}}\boldsymbol{y}_j||_2^2 = \boldsymbol{\theta}_i'\boldsymbol{Y}'\boldsymbol{Y}\boldsymbol{\theta}_i,$$

which is a quadratic form in $\boldsymbol{\theta}_i$, and when summed over i, it is jointly convex in the elements of $\boldsymbol{\Theta}$. Are there other choices of w_i's that can make the above a quadratic form in the entries of $\boldsymbol{\Theta}$?

(c) Note that the second source of non-convexity in the SPACE objective function is putting the Lasso penalty on the partial correlations ρ^{ij}. Show that this can be resolved by putting the penalty on partial covariances θ_{ij}. Combining these two modifications leads to the new objective function

$$Q_1(\boldsymbol{\Theta}) = -n\sum_{i=1}^{p}\log\theta_{ii} + \frac{n}{2}||\sum_{j=1}^{p}\theta_{ij}\boldsymbol{y}_j||_2^2 + \lambda\sum_{i<j}^{p}|\theta_{ij}|. \tag{9.47}$$

Use the facts that $|x|$ and $-\log x$ are convex functions, and $||\sum_{j=1}^{p}\theta_{ij}\boldsymbol{y}_j||_2^2$ is a positive semi-definite form in $\boldsymbol{\Theta}$, to conclude that the function $Q_1(\cdot)$ is a jointly convex function of its argument, but not necessarily strictly convex when $n < p$.

The regression-based CONvex CORrelation selection methoD (CONCORD) method in Khare et al. [2014] unifies about five penalized precision matrix estimation proposals in the literature. Its key ideas appear in this and the next problem, they lead to a powerful coordinate descent algorithm for minimizing the objective function similar to that for Lasso regression in Section 2.1.3.

9.9 (a) For a fixed (i, j), show that $Q_{1,ij}(\boldsymbol{\Theta})$, the value of the objective function (9.47), is a "quadratic" function of θ_{ij} when all other entries are fixed. More precisely, for $1 \le i \le p$,

$$Q_{1,ii}(\boldsymbol{\Theta}) = -n\log\theta_{ii} + \frac{n}{2}\left(\theta_{ii}^2 s_{ii} + 2\theta_{ii}\sum_{j\neq i}\theta_{ij}s_{ij}\right) + C, \tag{9.48}$$

where the constant C does not depend on θ_{ii} and s_{ij} is the (i, j)th entry of the sample covariance matrix $\boldsymbol{Y}'\boldsymbol{Y}/n$. However, for $1 \le i < j \le p$

$$\begin{aligned} Q_{1,ij}(\boldsymbol{\Theta}) = n\frac{s_{ii}+s_{jj}}{2}\theta_{ij}^2 + n\left(\sum_{j'\neq j}\theta_{ij'}s_{ij'} + \sum_{i'\neq i}\theta_{i'j}s_{i'j}\right)\theta_{ij} \\ + \lambda|\theta_{ij}| + C, \end{aligned} \tag{9.49}$$

is not differentiable with respect to θ_{IJ}.

(b) Due to the presence of "quadratic" forms, the minimizers of the above univariate functions have closed forms possibly involving the soft-thresholding operator. To write the solutions, let $T_{ij}(\boldsymbol{\Theta})$ be the matrix with all the entries of $\boldsymbol{\Theta}$ kept intact except its (i, j)th entry replaced by the value that minimizes $Q_{1,ij}(\boldsymbol{\Theta})$ with respect to θ_{ij} while keeping all other entries fixed. Show that, for $1 \le i \le p$:

$$(T_{ii}(\boldsymbol{\Theta}))_{ii} = \frac{-\sum_{j\neq i}\theta_{ij}s_{ij} + \sqrt{(\sum_{j\neq i}\theta_{ij}s_{ij})^2 + 4s_{ii}}}{2s_{ii}}, \tag{9.50}$$

and for $1 \le i < j \le p$,

$$(T_{ij}(\mathbf{\Theta}))_{ij} = \frac{S_{\frac{\lambda}{n}}\left(-\left(\sum_{j' \neq j} \theta_{ij'} s_{ij'} + \sum_{i' \neq i} \theta_{i'j} s_{i'i}\right)\right)}{s_{ii} + s_{jj}}, \tag{9.51}$$

where $S_\lambda(\cdot)$ is the soft-thresholding operator.

10

Directed Graphs

Most problems in Statistics and Social Sciences are aimed at identifying underlying relationships among variables of observational data. The Gaussian Graphical Models, introduced in Chapter 9, belong to the class of undirected graphs because the edges connecting the nodes have no directions. On the contrary, Directed Graphical Models (DGMs) are powerful tools for representing directions of the flow of effects or the causal structure of such relations. In a DGM, each node i of the directed graph is associated with a variable Y_i, and an edge $Y_i \longrightarrow Y_j$ indicates that the relation has a direction. The implied directions and suggestive terms like "parents, children, descendants" are substitutes for or reminders of the "arrow of time" and the cognizant notions of "past, present, future" from time series or stochastic processes. Such graphs are widely used in epidemiology, econometrics, biology, genomics, medicine, etc. For example, in gene regulatory networks, biologists seek to understand how genes interact and affect their activation levels. For small networks, expert knowledge and reasoning might be used to construct a DGM. However, in most cases, the DGM is unknown and needs to be learned from the observational and/or interventional data. The problem of learning a DGM from data is known as structure learning. This chapter establishes the main properties of DGMs, providing the groundwork for understanding the well-known structure learning algorithms, discussed later in the chapter.

10.1 Directed Graphs: Structural Learning

In a directed graph $G = (V, E)$, with a vertex set V and edge set E, the edge is an ordered pair (Y_i, Y_j) that points from Y_i to Y_j; $Y_i \longrightarrow Y_j$. We denote the set of all parents of node Y_i by $\text{pa}(Y_i)$ or pa_i. A *directed path* in a graph is a sequence of adjacent nodes such that all directed edges have the same direction. For example, in Figure 10.1 $\text{pa}(Y_1) = \emptyset$, $\text{pa}(Y_2) = \{Y_1, Y_3, Y_4\}$, and $Y_1 \longrightarrow Y_3 \longrightarrow Y_2$ is a directed path (the flow of information is active or connected), but $Y_1 \longrightarrow Y_2 \longleftarrow Y_4$ and $Y_4 \longrightarrow Y_2 \longleftarrow Y_3$ are not (the flow is closed or blocked by Y_2). Given a directed path $Y_k \longrightarrow \cdots \longrightarrow Y_j$, then Y_k is an *ancestor* of its *descendant* Y_j. A directed path from a node Y_j back to itself is called a *cycle*. A directed graph without a cycle is known as a Directed Acyclic Graph (DAG). Figure 10.1 is an example of a DAG. A DAG G admits a topological ordering of its nodes, that is there exists a permutation order π such that $\pi(j) < \pi(k)$ only if k is not an ancestor for j.

DOI: 10.1201/9781003638896-10

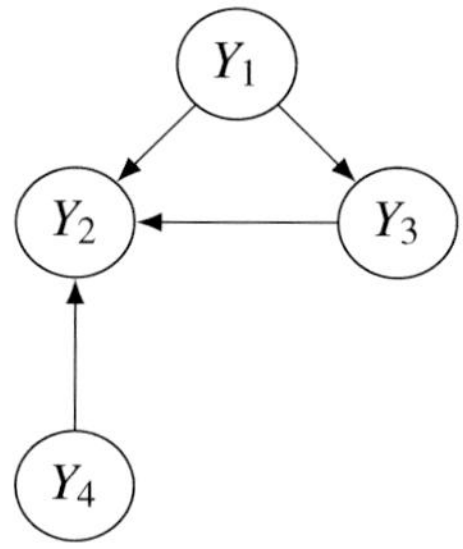

FIGURE 10.1
Example of a DAG

Example 22 (Chain Graphs) *A simple example of directed graphs (as well as of a DAG) is a chain graph, also known as Markov chains (see Figure 10.2(a)), which portrays the AR(1) models in time series analysis. Here the direction of the edges resembles that of time, and the past and future are independent given the present.*

Given a DAG G, careful examination of the orientations of its edges provides a wealth of information about the nature of the flow in a path based on purely graphical notions of d-*connectedness* and d-*separation*. Regardless of the size of a directed graph, these deeper graphical aspects can be understood through simple building blocks of directed graphs (chain, fork, v-structure), which involve only three nodes (see Figure 10.2). We say that a path between nodes Y_i and Y_j is active or d-connected given a set of nodes $\boldsymbol{Z}$ if it contains

1. a chain $\cdots \rightarrow Y_k \rightarrow \cdots$ or fork $\cdots \leftarrow Y_k \rightarrow \cdots$ such that $Y_k \notin \boldsymbol{Z}$.
2. a collider $\cdots \rightarrow Y_k \leftarrow \cdots$ and the collider Y_k or one of its descendants belong to $\boldsymbol{Z}$.

If for at least one pair of nodes $Y_i \in \boldsymbol{Y}_i$ and $Y_j \in \boldsymbol{Y}_j$, the path is not blocked by $\boldsymbol{Z}$, then $\boldsymbol{Y}_i$ and $\boldsymbol{Y}_j$ are d-connected, otherwise we say $\boldsymbol{Y}_i$ and $\boldsymbol{Y}_j$ are d-separated. A d-separation statement of a graph G is denoted by the symbol $\perp\!\!\!\perp_G$.

Example 23 (d-separation) *The set of d-separations implied by the DAG in Figure 10.1 is $Y_4 \perp\!\!\!\perp_G Y_1$ and $Y_4 \perp\!\!\!\perp_G Y_3$. On the other hand, $Y_4 \rightarrow Y_2 \leftarrow Y_1 \rightarrow Y_3$ contains a v-structure $Y_4 \rightarrow Y_2 \leftarrow Y_1$; thus, if $\boldsymbol{Z} = \{Y_2\}$, then Y_4 and Y_3 are d-connected, consequently $Y_4 \not\perp\!\!\!\perp_G Y_3 | \boldsymbol{Z}$.*

Evidently, at this stage, d-separation has absolutely no relation to the probabilistic notion of (in)dependence since as yet no probability distribution is attached to G. When a probability P is attached to G, then the symbol $\perp\!\!\!\perp$ is used to denote independence implied by it. Connecting these apparently different notions of separation and independence is the focus of the next section.

10.1.1 DAGs and Probabilistic Independence

Let Y be a random vector with entries corresponding to the nodes of a DAG, also known as a Bayesian Network. Once a probability distribution is introduced on the DAG, it becomes possible, under some conditions, to relate graphical d-separation statements about nodes to probabilistic conditional independence statements for the random variables in Y given suitable conditioning sets. The size and nature of these conditioning sets are the key players in graphical models.

A random vector Y with distribution P is said to satisfy the *local Markov property* with respect to a DAG G, if any node $Y_i \in V$ is independent of all its non-descendants, given its parents. In Figure 10.1, Y_4 is a non-descendant of Y_3 and $\text{pa}_3 = \{Y_1\}$, then local Markov property requires that $Y_3 \perp\!\!\!\perp Y_4 | Y_1$. Similarly, one can write such conditional statements for the other nodes.

A somewhat stronger notion of Markov property for more general subsets of nodes involving d-separation is the first major step in connecting graphical and probabilistic independence: A random vector Y is said to satisfy the *global Markov property* with respect to a DAG G, if for all pairwise disjoint subsets $\boldsymbol{Y}_i, \boldsymbol{Y}_j, \boldsymbol{Z} \subset V$,

$$\boldsymbol{Y}_i \perp\!\!\!\perp_G \boldsymbol{Y}_j | \boldsymbol{Z} \implies \boldsymbol{Y}_i \perp\!\!\!\perp \boldsymbol{Y}_j | \boldsymbol{Z}. \tag{10.1}$$

In words, under *global Markov property* all graphical d-separation statements in the DAG G hold as conditional independence statements of the probability distribution P. If Y satisfies the global Markov property with respect to a DAG G, then G is called an *independence map* of Y. A DAG G is called a *perfect map* of Y if all the reverse implications in (10.1) hold. Thus, being a perfect map requires the global Markov property and its reverse implication, known as *faithfulness*, to hold. The assumption that G is a perfect map or faithfulness is central to the structural learning methods discussed in the next section. An example of a violation of the faithfulness assumption for discrete variables is given next.

Example 24 (Faithfulness violation) *Consider the data-generating process from the DAG in Figure 10.2(c). Suppose* $Y_1, Y_3 \sim \text{Bernoulli}(\theta)$, $Y_2 = I(Y_1 = Y_3)$, *and* $Y_1 \perp\!\!\!\perp Y_3$. *It can be shown that*

$$P(Y_1 = 1, Y_2 = 0) = \theta(1-\theta) \neq$$
$$P(Y_1 = 1)P(Y_2 = 0) = 2\theta^2(1-\theta).$$

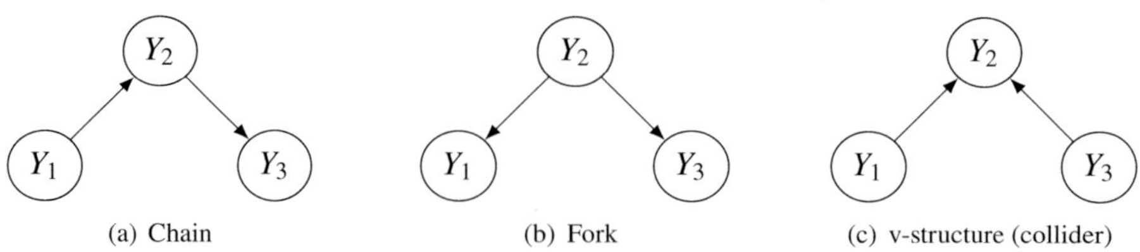

FIGURE 10.2
Three building blocks of DAGs

Thus, $Y_1 \not\perp\!\!\!\perp Y_2$*, i.e.,* $P(Y_1 = 1, Y_2 = 0) \neq P(Y_1 = 1)P(Y_2 = 0)$*. However, when* $\theta = 0.5$*, the equality*

$$P(Y_1 = 1, Y_2 = 0) = P(Y_1 = 1)P(Y_2 = 0),$$

holds and $Y_1 \perp\!\!\!\perp Y_2$*, even though* $Y_1 \not\perp\!\!\!\perp_G Y_2$ *by construction.*

As above examples show, for parametric models the violation of faithfulness is rare since it requires an exact "tuning" of parameter values to violate faithfulness. However, there exist sequences of faithful distributions that converge arbitrarily close to unfaithful distributions Uhler et al. [2013]. Another useful property equivalent to the Markov property, which is helpful in simplifying the joint distribution, is the factorization property.

Definition 1 (Factorization property) *Let G be a DAG over the variables* $Y_1, \ldots, Y_p$*. A probability distribution P has the factorization property according to DAG G if*

$$P(Y_1, \ldots, Y_p) = \prod_{i=1}^{p} P(Y_i | pa_i) \tag{10.2}$$

From here on we assume that the distribution of Y has a positive density with respect to a product measure. Then, it is known that the global and local Markov properties are equivalent and equivalent to the factorization property, see Koller and Friedman [2009]. Moreover, the global Markov property states all the conditional independencies appearing in the factorization above, i.e., it is complete.

Example 23 cont. *The DAG in Figure 10.1 factorizes as*

$$P(Y_1, Y_2, Y_3, Y_4) = P(Y_1)P(Y_4)P(Y_3|Y_1)P(Y_2|Y_1, Y_3, Y_4). \tag{10.3}$$

An advantage of such factorization is that it requires less memory to store and is easier to calculate on the right-hand side. [1]

So far, we've explained how DAGs represent statistical conditional independence relationships through the factorization of probability distributions for observed data. However, interpreting a DAG causally requires moving beyond purely statistical considerations. Causal DAGs make explicit assumptions regarding the system's response to external interventions. This topic is out of the scope of this book. For details, the reader is referred to Pearl [2009] and Peters et al. [2017].

10.1.2 Markov Equivalence Class and CPDAGs

It is easy to see that the DAGs in Figure 10.2(a) and (b) have the same set of d-separations; $Y_1 \perp\!\!\!\perp_G Y_3 | Y_2$. For a given DAG G, the class of all such DAGs is called its *Markov equivalence class* . That is, two DAGs G and G' are Markov equivalent

[1] For example, if in (10.3) $Y_1, \ldots, Y_4$ are binary, then the $2^4 - 1$ dimensional distribution is determined by approximately $2 * 4 - 1$ parameters.

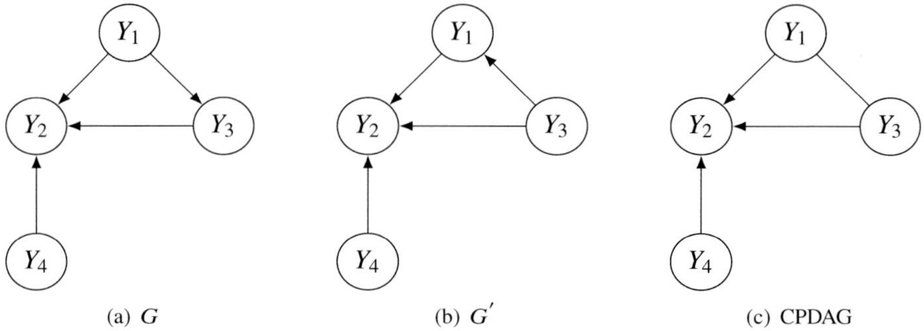

FIGURE 10.3
Graphs G and G' in (a) and (b) composite a Markov Equivalence Class. (c) corresponds to their CPDAG.

if $Y_i \not\perp\!\!\!\perp_G Y_j|Y_m$ is equivalent to $Y_i \not\perp\!\!\!\perp_{G'} Y_j|Y_m$. A simple characteristic of DAGs in a Markov equivalence class is that they have the same skeleton and the same v-structures. The skeleton of a directed graph is obtained by replacing directed edges with undirected ones, as in Figure 10.4(b).

Markov equivalence class is represented by a complete partial DAG (CPDAG) or essential graph, it contains a (directed) edge $Y_i \longrightarrow Y_j$ if and only if all DAGs in the Markov equivalence class contain $Y_i \longrightarrow Y_j$. If at least one member of the Markov equivalence class has $Y_i \longleftarrow Y_j$, then the CPDAG has the undirected edge $Y_i - Y_j$. Most structural learning algorithms, discussed in the next section, recover the underlying DAG up to its Markov equivalence class.

Example 23 cont. (Markov Equivalence Class) *Figure 10.3(a) and (b) comprise a Markov equivalence class, which contains only two DAGs, since the only edge change that does not generate a new v-structure is from $Y_1 \longrightarrow Y_3$ in G to $Y_3 \longrightarrow Y_1$ in G'. Both DAGs imply that $Y_1 \perp\!\!\!\perp_G Y_4|Y_3$. Consequently, any distribution P that can be factorized by one of these graphs can be factorized by the other. In other words, even though the joint distribution $P(Y_1, Y_3)$ can determine whether Y_1 and Y_3 are correlated, but it cannot tell whether the correct orientation is $Y_1 \longrightarrow Y_3$ or $Y_1 \longleftarrow Y_3$. The CPDAG of the Markov equivalence class is shown in Figure 10.3(c).*

Another related graph that connects the conditional independence properties of undirected graphs, discussed in Chapter 9, to directed graphs is a *moral graph* . The moral graph G^m of a DAG G is constructed by first shielding or covering all v-structures and then taking the skeleton of the resulting graph. Shielding a v-structure $Y_1 \longrightarrow Y_2 \longleftarrow Y_3$ means adding an edge between nodes Y_1 and Y_3. If G is a perfect map of Y, then the moral graph G^m is the conditional independence graph of Y. Consequently, the skeleton of a DAG or CPDAG represents a subgraph of its conditional independence graph. An example of a moral graph G^m is displayed in Figure 10.4(c).

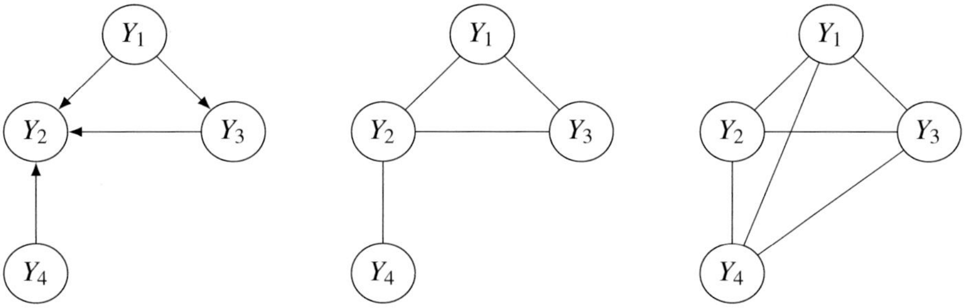

(a) G (b) Skeleton of G (c) Moral graph G^m

FIGURE 10.4
Skeleton and moral graph of a DAG

10.1.3 Structural Equation Models and DAGs

A graphical model associated with a DAG can be equivalently represented through a structural equation model (SEM) [Bollen, 1989], also known as structural causal models (SCM). Specifically, if ε is a vector of independent random noise variables and f_i's are measurable functions, for $i \in V$, then the random vector Y given as a function of its parents and the random noise

$$Y_i := f_i(\boldsymbol{Y}_{\mathrm{pa}_i}, \varepsilon_i),\ i \in V \tag{10.4}$$

is Markov with respect to G. The converse of the latter statements also holds [Wang et al., 2025, Section 3.2]. That is, if the vector Y is Markov with respect to G then there always exist independent random noise ε and function f, such that (10.4) holds. It is important to note that (10.4) should be viewed as a non-symmetric assignment mechanism from parents to the random variable, which is clarified by using := notation. This connection between DAGs and SEMs provides a foundation for modeling causal relationships, where each variable is determined by its direct influences and an independent noise component.

A linear SEM is a special case of (10.4), in which f_i's are linear and the error term is additive, i.e., $Y_i := \sum_{j \in \mathrm{pa}_i} \beta_{ij} Y_j + \varepsilon_i$. In vector formulation, linear SEM can be written as

$$Y := BY + \varepsilon, \tag{10.5}$$

where $B \in \mathbb{R}^{p \times p}$ is a weighted adjacency matrix such that $\beta_{ij} \neq 0$ indicates $Y_j \to Y_i$. From (10.5), the covariance matrix of Y is given by

$$\boldsymbol{\Sigma} = (I - B)^{-1} \Lambda [(I - B)^{-1}]', \tag{10.6}$$

where Λ is a diagonal matrix that contains error variances.

Example 23 cont. *The DAG in Figure 10.1 can be represented as generated from the following linear SEM.*

$$\begin{bmatrix} Y_1 \\ Y_2 \\ Y_3 \\ Y_4 \end{bmatrix} = \begin{bmatrix} 0 & 0 & 0 & 0 \\ \beta_{21} & 0 & \beta_{23} & \beta_{24} \\ \beta_{31} & 0 & 0 & 0 \\ 0 & 0 & 0 & 0 \end{bmatrix} \begin{bmatrix} Y_1 \\ Y_2 \\ Y_3 \\ Y_4 \end{bmatrix} + \begin{bmatrix} \varepsilon_1 \\ \varepsilon_2 \\ \varepsilon_3 \\ \varepsilon_4 \end{bmatrix} \tag{10.7}$$

where $\varepsilon_1, \varepsilon_2, \varepsilon_3, \varepsilon_4 \sim N(0, 1)$ *and* $(Y_1, Y_2, Y_3, Y_4)' \sim N_4(0, \Sigma)$.

10.2 Learning DAG Structure

For the factorization of joint distribution in (10.3) or for identifying the set of d-separations in a DAG, it is typically assumed that its structure is known. However, in practice the structure is often unknown, and the objective is to learn it from the data or a sample of size n from $\boldsymbol{Y}$. A common approach to structure learning involves interventions or randomized control trials. However, such experiments can be unethical, time-consuming, or expensive. Consequently, our goal is to recover the underlying structure of variables Y_i, $i \in V$, using only observational samples from the distribution $\mathbf{P}$.

Example 25 (Flow cytometry dataset) *A well-known dataset in structure learning literature is the flow cytometry dataset, introduced in Sachs et al. [2005]. The goal is to learn the graph of 11 proteins from a combination of observational and experimental data. The* $n = 7466$ *simultaneous measurements of* $p = 11$ *phosphorylated proteins and phospholipids, listed below are obtained from individual primary immune system cells, that were subjected to both general and specific molecular interventions.*

- ***RAF*** *- RAF proto-oncogene serine/threonine-protein kinase*
- ***Mek*** *- Dual specificity mitogen-activated protein kinase*
- ***Plcg*** *- Phospholipase C gamma*
- ***PIP2*** *- Phosphatidylinositol 4,5-bisphosphate*
- ***PIP3*** *- Phosphatidylinositol (3,4,5)-trisphosphate*
- ***Akt*** *- Protein kinase B*
- ***PKA*** *- Protein kinase A*
- ***PKC*** *- Protein kinase C*
- ***P38*** *- p38 MAP kinase*

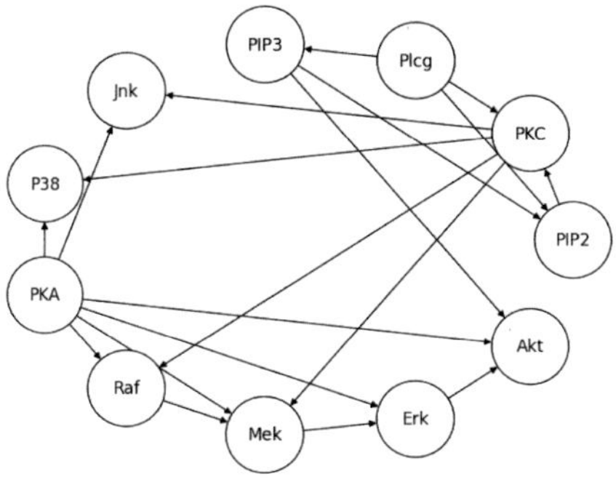

FIGURE 10.5
Causal graph for flow cytometry dataset.

- ***Erk*** *- extracellular signal-regulated kinase*
- ***Jnk*** *- c-Jun N-terminal kinase*

The interesting characteristic of this dataset is that its true causal graph is known from the literature and is given in Figure 10.5. For example, Raf→Mek *indicates that* Raf *activates* Mek, Plcg→PIP2 *indicates that* Plcg *hydrolyzes* PIP2, *etc. In biochemistry, Plcg (phospholipase C gamma) hydrolyzes or breaks PIP2 into two parts by adding water.*

We start by introducing traditional structure learning approaches: constraint-based and score-based, which recover a DAG up to its Markov equivalence class under the faithfulness assumption. Then, we discuss the so-called functional-based approaches which rely on SEM to achieve the *identifiability* of recovering a DAG. We say the structure of a DAG is not identifiable, if there are two models with DAGs G and $G^{'}$ respectively, such that both induce the same joint distribution $\mathbf{P}$, but $G \neq G^{'}$.

For all methods, described in this section, we assume that there are no unobserved or latent common causes (unobserved confounders). This assumption is also known as the *causal sufficiency*.

10.2.1 Constraint-Based Methods

Constraint based methods such as inductive causation (IC) [Pearl, 2009] and PC algorithm [Spirtes and Glymour, 1991] utilize conditional independence tests to detect the existence of edges between each pair of variables and recover a DAG up to its Markov equivalence class. These type of methods assume that the distribution is faithful to the underlying DAG.

Recall that in Chapter 9, when learning an undirected graph, there was no edge between two variables Y_i and Y_j if they were conditionally independent given *all* the other variables. In the case of learning DAGs, however, due to the special nature of d-separations, simply testing for pairwise conditional independence given the rest of the variables, is not enough. One needs to condition on all the smaller subsets of the rest.

For example, in the DAG in Figure 10.4(a), $Y_1 \not\perp\!\!\!\perp Y_4|Y_2$, but $Y_1 \perp\!\!\!\perp Y_4|Y_2$ in the undirected graph in Figure 10.4(b). To achieve the same Markov properties between a DAG and an undirected graph, we need to construct a moral graph of a DAG. Recall that if the probability distribution on Y is faithful to DAG G, then the resulting moral graph is the conditional independence graph of Y.

As we will see, to recover a DAG, in particular a Gaussian DAG, since (conditional) independence can be inferred from the (partial) correlation, we need to start by testing the marginal correlation and then gradually increase the adjustment or conditioning set.

Example 23 cont. *To decide whether there is an edge between Y_2 and Y_3, one needs to sequentially test $\rho_{23} = 0$, then $\rho_{23|1} = 0$ and $\rho_{23|41} = 0$, where $\rho_{ij|k}$ indicates that we adjusted for variable Y_k. If all tests are rejected, then an edge is drawn between Y_2 and Y_3. If at any point, the test is not rejected, then there is no edge between Y_2 and Y_3. Note that this procedure only identifies the existence of an edge, not its direction. We provide more details on this later.*

Below, we discuss the PC algorithm for the multivariate Gaussian data generation process described in Example 23. The PC algorithm is a two-phase process: skeleton estimation and edge orientation. In the first phase, the PC strives to search and find a pattern representing the Markov equivalence class. This phase uses partial correlations to test conditional independencies, i.e., d-separations. Skeleton search relies on the fact that two nodes Y_1 and Y_2 are adjacent if and only if there is no subset $\mathbf{Z} \subset V/\{Y_1, Y_2\}$ that d-separates them.

Example 23 cont. (PC in action) *Figure 10.6 summarizes the steps of the PC algorithm. It works as follows*

1. *The algorithm starts from a complete undirected graph, i.e., all nodes are adjacent to each other. See Figure 10.6(b).*

2. *(Edge Elimination) In the first phase, PC implements edge elimination by constructing so called separation set (sepset) and testing for vanishing partial correlation of two nodes adjusted for sepset. At the start, the sepset is an empty set, so that PC eliminates edges between two variables if the marginal correlation is zero. In our example, ρ_{41} and ρ_{43} are equal to zero, thus the edges between the corresponding variables are eliminated. In subsequent steps, cardinality of the sepset increases by one. For example, at the kth step the cardinality of the sepset is $k-1$, and to test whether Y_i is d-separated from Y_j given a sepset, the algorithm searches over all possible subsets of size $k - 1$ that are either neighbors of Y_i or are neighbors of Y_j. In our example, in the second step, for Y_1 and Y_4, PC tests whether $\rho_{14|2} = 0$ and $\rho_{14|3} = 0$. Here, the phase 1 terminates after the third step with the sepset size equal to two. See Figure 10.6(c).*

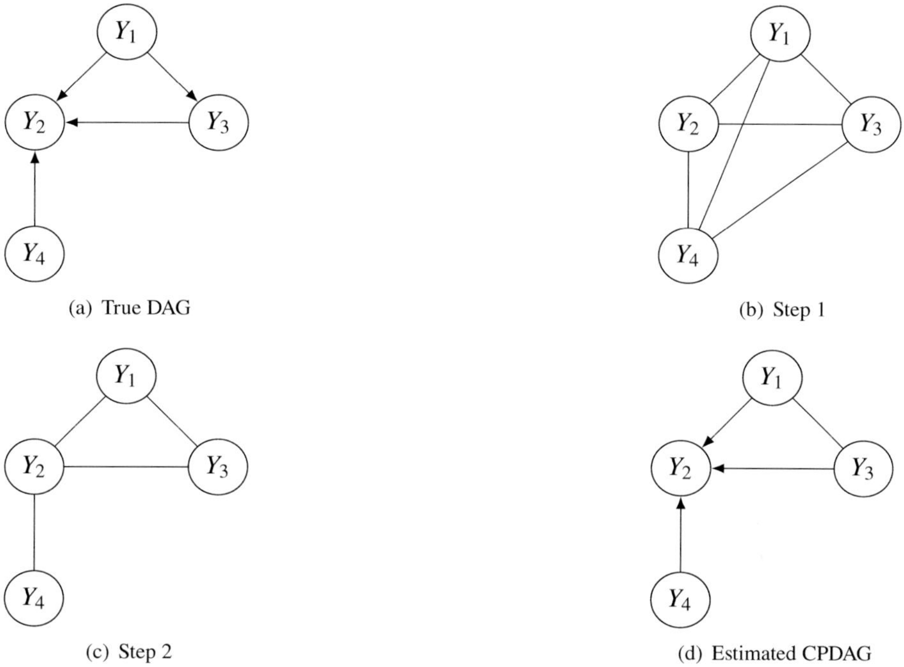

FIGURE 10.6
Steps for PC algorithm

3. *(Edge Orientation) In the second phase, PC implements edge orientation. Intuitively, since DAGs in the same Markov equivalence class share the same v-structure, then identifying and orienting edges in v-structures is relatively easy. In particular, for the structure Y_i—Y_k—Y_j, where Y_i and Y_j are not adjacent and $Y_k \notin$ sepset is a v-structure, it can be oriented as $Y_i \rightarrow Y_k \leftarrow Y_j$. In our example, Y_1—Y_2—Y_4 and Y_3—Y_2—Y_4 can be oriented as v-structures. Further orientation rules, known as Meek's orientation rules [Meek, 1995], can be applied to the obtained CPDAG. Unfortunately, for our example, no further improvement can be made. Figure 10.6(d) illustrates the final estimated CPDAG. As expected, the unoriented edge Y_1—Y_3 indicates that the Markov equivalence class contains two DAGs, where in one DAG $Y_1 \rightarrow Y_3$, as in Figure 10.6(a) and in the other DAG $Y_1 \leftarrow Y_3$.*

Notably, in Phase 2, the PC algorithm relies on the faithfulness assumption to guarantee that the edges eliminated in the early steps will not be drawn again in the subsequent steps. In particular, under faithfulness, if the sepset $C_1 \subset C_2$ then $\rho_{ij|C_1} = 0 \implies \rho_{ij|C_2} = 0$, for $i, j \notin C_2$. For details, see Bühlmann and van de Geer [2011, Chapter 13]. Another observation, that significantly decreases the computational time of the PC algorithm, is that to test d-separation between Y_i and Y_j it is sufficient to search only over the possible subsets of neighbors of Y_i and Y_j, and not all possible subsets of $V/\{Y_i, Y_j\}$. This is a consequence of the local Markov property, since if Y_i, Y_j are not adjacent then they are d-separated by either pa_i or pa_j.

In practice, in the skeleton estimation phase, we rely on a finite sample size for the estimation of the vanishing partial correlation. Consequently, PC heavily relies on the significance level to make conclusions. The latter can sometimes yield a contradictory results.

The Fast Causal Inference (FCI) algorithm and its variants [Spirtes et al., 2001] are generalizations of the PC algorithm that allow and sometimes detect unobserved confounders. For details, see Glymour et al. [2019].

10.2.2 Score-Based Methods

In contrast to constraint-based methods, the score-based approach treats structure learning as a combinatorial optimization problem. In particular, they search and test various graph structures by assigning a score $S(\boldsymbol{Y}, G)$ to each graph and selecting the one that best fits the data. Mathematically,

$$\hat{G} = \max_{G \in \mathbf{DAG}} S(\boldsymbol{Y}, G) \tag{10.8}$$

In practice, $S(\boldsymbol{Y}, G)$ is typically derived from the likelihood or penalized likelihood function, such as BIC. A score function is *decomposable* if it can be expressed as a sum of local scores for each node. In many models, such as those for Gaussian or categorical data, the score function is also score-equivalent, meaning it assigns the same value to all DAGs within the same Markov equivalence class. That is for any two DAGs G, G' in the same Markov equivalence class $S(\boldsymbol{Y}, G) = S(\boldsymbol{Y}, G')$. The BIC score is an example of a score-equivalent function. From a Bayesian perspective, the score function represents the log posterior over DAGs.

Unfortunately, the search space of all DAGs grows superexponentially, and solving (10.8) is NP-hard [Chickering, 1996]. Consequently, it is challenging to develop an exact method that finds the best scoring graph for a given finite observational data. The recent research exploits the decomposability of the score function and relies on dynamic or integer linear programming framework to scale the structure learning problem to higher dimensions [de Campos and Ji, 2011, Silander and Myllymäki, 2012].

For large graphs, the exact search is computationally infeasible and another line of research relies on greedy search algorithms to solve (10.8). A typical greedy search traverses the search space in such a way that at each step it evaluates each neighbor of the current state and moves to the next state if the current score is improved, otherwise, the search terminates. The connectivity of the search space is determined by the set of neighbors as the algorithm moves through the states. One such successful algorithm is Greedy Equivalence Search (GES)[Chickering, 2002], which relies on the BIC score and searches for CPDAG over the space of Markov equivalence classes of DAGs. GES is a two-phase procedure in which the state is represented by a CPDAG or a Markov equivalence class, and depending on the phase, the neighbors of the state are the Markov equivalent classes obtained by applying edge insertion or deletion operators to the CPDAG. The first phase starts with an empty graph, and the edge insert operator is applied to the CPDAG in each state until a local maximum is achieved. The second phase starts from the previous local maximum. Then the edge deletion operator is applied, so the DAG search travels through the Markov equivalence classes until another local maximum of the score is achieved. The next example summarizes the steps.

Example 23 cont. (GES in action) *Figures 10.7 and 10.8 illustrate the connectivity of the search space for one possible state in each phase. For each figure, we assume that the algorithm is in the state represented by (a), and (b) – (e) depicts the corresponding neighbor CPDAGs, which represent the Markov equivalence class, after edge insertion (Figure 10.7) or deletion (Figure 10.8), respectively. At this stage, for each neighbor CPDAG, GES will evaluate the corresponding score function and decide whether to move to a neighboring CPDAG or terminate Phase 1. In the first phase, insertion of edges* $Y_1 \rightarrow Y_4$ *and* $Y_4 \rightarrow Y_1$*, see cases (b) and (c) in Figure 10.7, leads to the same equivalence class and cases (d) and (e) depict CPDAGs after inserting* $Y_4 \rightarrow Y_3$ *and* $Y_3 \rightarrow Y_4$*, respectively. Figure 10.8 illustrates similar results for*

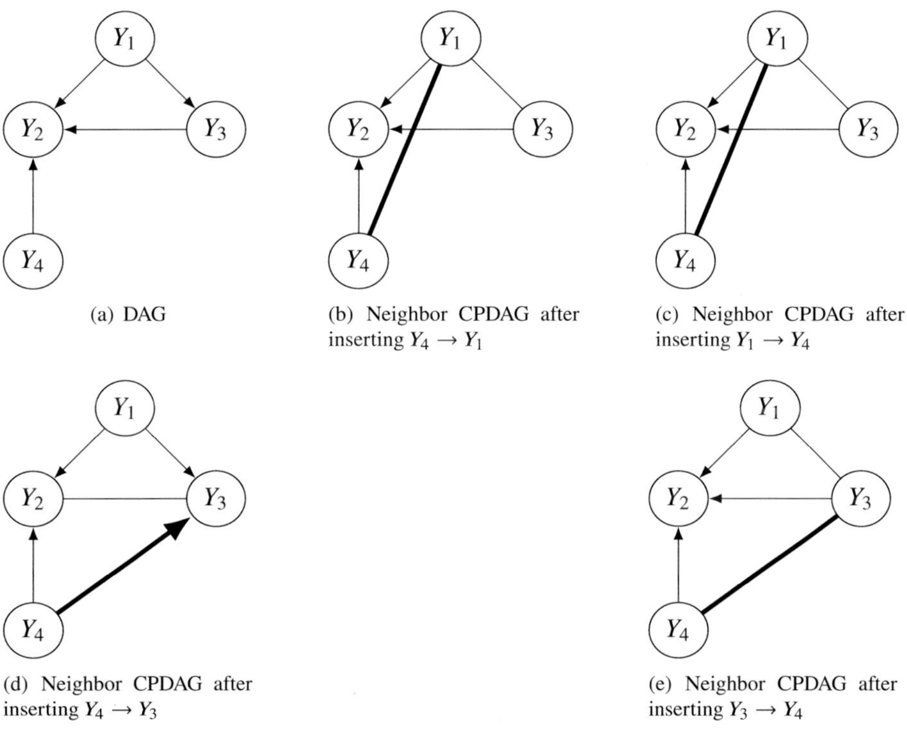

(a) DAG

(b) Neighbor CPDAG after inserting $Y_4 \rightarrow Y_1$

(c) Neighbor CPDAG after inserting $Y_1 \rightarrow Y_4$

(d) Neighbor CPDAG after inserting $Y_4 \rightarrow Y_3$

(e) Neighbor CPDAG after inserting $Y_3 \rightarrow Y_4$

FIGURE 10.7
Phase 1 neighbors

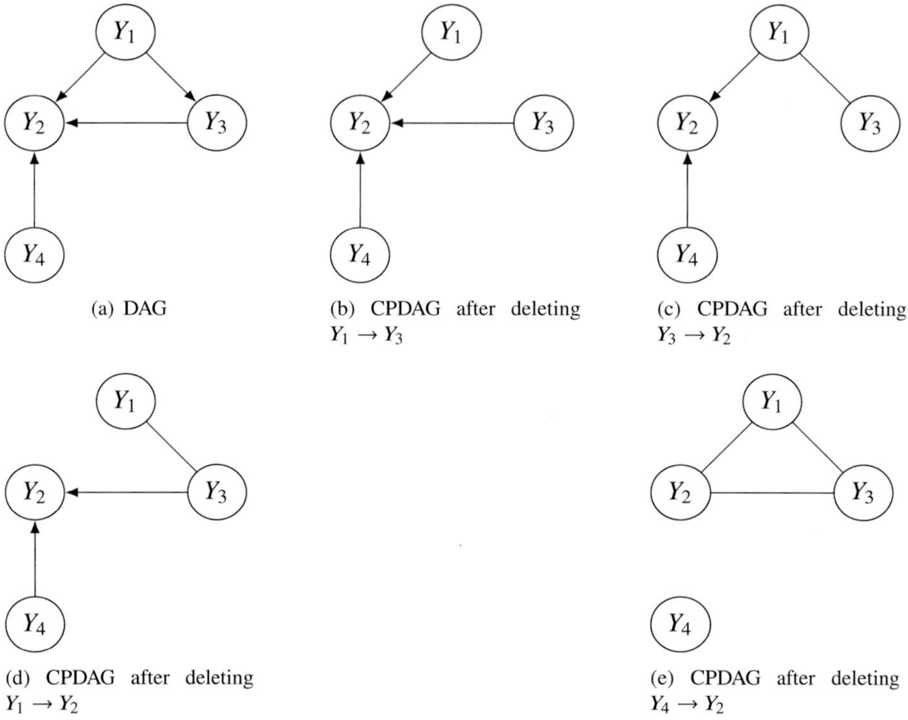

FIGURE 10.8
Phase 2 neighbors

the delete operator in the second phase. In practice, it is not necessary to evaluate the score for each DAG in the neighbor equivalence class. As Chickering [2002] showed, for the score-equivalent decomposable scores, after applying Insert(·) and Delete(·) operators, one needs to evaluate the score for the local change in the CPDAG.

The GES algorithm has good asymptotic properties. In particular, under the faithfulness assumption, when the number of observations is large, at the end of the first phase, the estimated CPDAG contains the true generating distribution P.

10.2.3 Functional-Based Methods

Both constraint-based and score-based methods require faithfulness assumption and the solution of these methods usually returns a Markov Equivalence class which generally contains more than one DAG. As a result, the DAG structure is not identifiable. Moreover, these methods by design cannot learn the structure when there are only two variables [Glymour et al., 2019].

By contrast, SEMs possess specific properties that enable the full determination of the DAG structure by introducing asymmetry into the generation process. Specifically, it is generally expected that the complexity of a physical process generating an effect from a cause should, in some way, be lower than the complexity of the reverse process. In other words, the relationship between two variables in such a model is inherently asymmetrical rather than symmetrical.

We start motivating this section with a question: given two variables, can we differentiate a parent from its child? Let's suppose $Y_1 \to Y_2$, then naturally, we expect some asymmetry between processes that generate a child from the parent

$$\begin{aligned} Y_1 &:= f_1(\varepsilon_1) \\ Y_2 &:= f_2(Y_1, \varepsilon_2) \end{aligned} \tag{10.9}$$

and the backward process

$$\begin{aligned} Y_2 &:= g_1(\tilde{\varepsilon}_1) \\ Y_1 &:= g_2(Y_2, \tilde{\varepsilon}_2) \end{aligned} \tag{10.10}$$

By assumption in (10.9), the error term ε_2 is independent of Y_1. If for the backward direction (10.10), we cannot find a function $g_2(\cdot)$ such that the error term $\tilde{\varepsilon}_2$ is independent of Y_2, then this will help determine the true data-generating direction. However, without any restrictions on the model class of $f(\cdot)$ and $g(\cdot)$ such a direction is not identifiable. The backbone of functional-based methods is to impose restrictions on the functional class or the error distributions to uniquely recover the DAG structure.

Linear Non-Gaussian methods

Consider the data generated from the following linear SEM: $Y_2 = aY_1 + \varepsilon_{Y_2}$ and $Y_1 \perp\!\!\!\perp \varepsilon_{Y_2}$. In Figure 10.9(a) and (c), we plot the regression line of Y_2 on Y_1 and the backward Y_1 on Y_2, when both error terms and covariates are Gaussian. The corresponding residuals are plotted in Figure 10.9(b) and (d). In both plots, the residual is independent from its covariate, meaning that, for the Gaussian case, the true direction of $Y_1 \to Y_2$ cannot be identified. This is a simple consequence of the Darmois-Skitovic theorem [Peters et al., 2017, Theorem 4.3] .

However, the situation changes drastically when at most one of the Y's and errors ε is Gaussian. In such case, the direction of $Y_1 \to Y_2$ can be identified. This is illustrated in Figure 10.10. Here we plot errors generated from the regression of Y_2 given Y_1 and the backward regression when Y_1 and ε_{Y_2} follow a standard uniform distribution. In the case of the true direction (Figure 10.10(b)), the error term is independent of Y_1, but this is not the case for the backward direction, displayed in Figure 10.10(d). This key observation, that error term independence can reveal causal direction, motivated the LINGAM algorithm introduced in Shimizu et al. [2006]. For the data with more than two variables, the LINGAM algorithm relies on independent component analysis (ICA)[Hyvärinen et al., 2004] to estimate the DAG structure. DIRECTLINGAM, proposed in Shimizu et al. [2011] is a further improvement to the initial LINGAM algorithm.

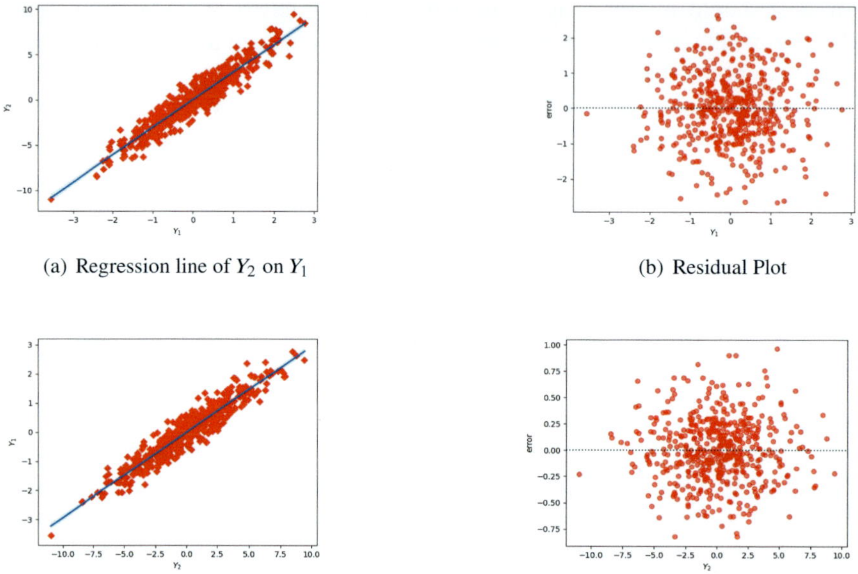

(a) Regression line of Y_2 on Y_1

(b) Residual Plot

(c) Regression line of Y_1 on Y_2

(d) Residual Plot

FIGURE 10.9
Illustration of symmetry between regression of Y_1 on Y_2 and Y_2 on Y_1 when the error terms and covariates follow Gaussian distribution.

Non-Linear methods

The (Non-) Gaussian and linear models impose strict restrictions on a data-generating process. If the assumed model is wrong and does not represent the true generating process, then the learned structure can be misleading. Therefore, in practice, it is preferable that the assumed model be general enough and identifiable. Zhang and Hyvärinen [2009] examined a large class of post-nonlinear models represented as

$$Y_i = f_{i,2}(f_{i,1}(\mathrm{pa}_i) + \varepsilon_i) \tag{10.11}$$

where ε_i's are independent, $f_{i,1}$ captures the nonlinear effect of the parent set, and the function $f_{i,2}$ is continuous and invertible. The authors show that, for the bivariate case of (10.11), i.e., $Y_2 = f_2(f_1(Y_1) + \varepsilon_2)$, the backward process $Y_1 = g_2(g_1(Y_2) + \varepsilon_1)$ is feasible under a very specific and limited condition involving second and third derivatives of the distribution of variables and error terms. For most of the models used in practice, these conditions are violated.

To learn the structure from the observational data, Zhang and Hyvärinen [2009] rely on nonlinear ICA. In particular, from (10.11) it follows that

$$\varepsilon_2 = f_2^{-1}(Y_2) - f_1(Y_1),$$

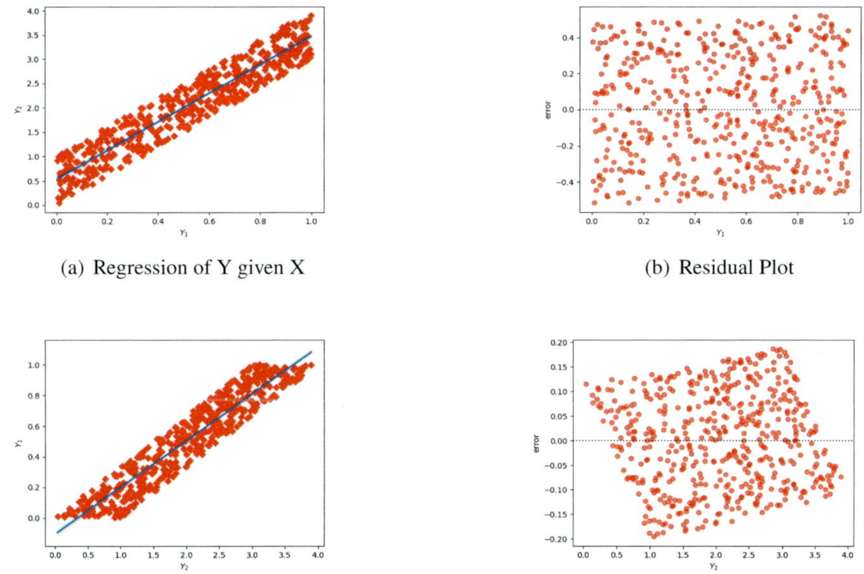

(a) Regression of Y given X

(b) Residual Plot

(c) Regression of X given Y

(d) Residual Plot

FIGURE 10.10
Uniform Case

which should be independent of Y_1. Thus, the structure $Y_1 \rightarrow Y_2$ can be learned by testing whether $\hat{\varepsilon}_2 = \hat{f}_2^{-1}(Y_2) - \hat{f}_1(Y_1)$ is independent of Y_1. Finding such functions (f_1 and f_2) is conceptually a constrained nonlinear ICA problem, and after some algebra, the mutual information between Y_1 and $\hat{\varepsilon}_2$ can be written as

$$I(Y_1, \hat{\varepsilon}_2) = -E[\log P(\hat{\varepsilon}_2)] - E[\log |\hat{f}_2(Y_2)|] + \text{const.}$$

The authors use multi-layer perceptrons to represent f_1 and f_2, and maximize $I(Y_1, \hat{\varepsilon}_2)$ to learn the corresponding parameters.

As a special case of (10.11), Peters et al. [2014] proposed the class of continuous additive noise models (ANM) $Y_i = f_{i,1}(\text{pa}_i) + \varepsilon_i$. They proposed a two-phase regression with continuous additive noise models (RESET) algorithm for the structure learning. Another special case of (10.11) is the so-called causal additive models (CAMs) defined as $Y_i = \sum_{j \in \text{pa}_i} f_{i,j}(Y_j) + \varepsilon_i$. For details, see Bühlmann et al. [2014].

Equal conditional variance and Generalization

Different from restricting a functional class, several studies [Rajaratnam and Salzman, 2013, Peters and Bühlmann, 2014, Park, 2020] and references therein, have employed (conditional) variance as a "measure of asymmetry" between the effect and the cause to establish identifiability results. Notably, Peters and Bühlmann [2014] demonstrate the identifiability of linear structural equation models (SEMs) with equal variances. Chen et al. [2019] note that the arrangement of specific conditional variances implies the identifiability of DAGs. Furthermore, Park [2020] introduce more comprehensive criteria based on conditional variances for nonlinear ANMs with unknown heterogeneous error variances. Recently Dallakyan and Ni [2025] established the identifiability of ANMs by leveraging the concept of majorization, demonstrating that their approach generalizes the methods above.

The notion of majorization arises in a variety of branches of mathematics and statistics. Informally, majorization describes the notion that the components of a vector $\boldsymbol{x} \in \mathbb{R}^p$ are less spread out or more nearly equal than the components of a vector $\boldsymbol{y} \in \mathbb{R}^p$. Formally, let $\boldsymbol{x}_{[\cdot]}$ denote the vector $\boldsymbol{x}$ in descending order, then we say $\boldsymbol{x}$ is **majorized** by $\boldsymbol{y}$, denoted $\boldsymbol{x} \prec \boldsymbol{y}$, if $\sum_{i=1}^{k} x_{[i]} \leq \sum_{i=1}^{k} y_{[i]}$, for $k = 1, \ldots, p-1$, and $\sum_{i=1}^{p} x_{[i]} = \sum_{i=1}^{p} y_{[i]}$. If $\sum_{i=1}^{p} x_{[i]} = \sum_{i=1}^{p} y_{[i]}$ does not hold, but $\sum_{i=1}^{p} x_{[i]} \leq \sum_{i=1}^{p} y_{[i]}$, then we say $\boldsymbol{y}$ **weakly majorizes** $\boldsymbol{x}$, i.e., $\boldsymbol{x} \prec_w \boldsymbol{y}$. Majorization can be exploited to learn interesting characteristics of certain classes of functions. In particular, let f be a real-valued function defined on $\boldsymbol{a} \in \mathbb{R}^p$ and $\boldsymbol{x}$ is not a permutation of $\boldsymbol{y}$. Then

$$\boldsymbol{x} \prec_w \boldsymbol{y} \implies f(\boldsymbol{x}) < f(\boldsymbol{y}), \tag{10.12}$$

if and only if $f(\boldsymbol{a})$ is strictly increasing, symmetric, and strictly convex in $\boldsymbol{a}$ [Marshall et al., 2011, Theorems 2.A.8a and 3.C.2.D] .

Let's denote a topological ordering of a DAG G by ρ_0. Without loss of generality, we assume $\rho_0 = \{1, \ldots, p\}$ and use ρ to represent any other ordering. This notation simplification helps avoid the need for longer notation such as $\rho_0(1), \ldots, \rho_0(p)$.

In the context of structure learning, let data be generated from the ANM $Y_i = f_{i,1}(\text{pa}_i) + \varepsilon_i$ and vector $\boldsymbol{x}$ contains the conditional variances using a true topological ordering

$$\boldsymbol{x} = \{\text{var}(Y_1), \ldots, E[\text{var}(Y_p|Y_1, \ldots, Y_{p-1})]\}' \tag{10.13}$$

and $\boldsymbol{y}_\rho$ contains conditional variances using any other ordering

$$\boldsymbol{y}_\rho = \{\text{var}(Y_{\rho(1)}), \ldots, E[\text{var}(Y_{\rho(p)}|Y_{\rho(1)}, \ldots, Y_{\rho(p-1)})]\}'. \tag{10.14}$$

Dallakyan and Ni [2025] showed that under Assumption 1 the topological ordering can be uniquely recovered.

Assumption 1 *Let* $\boldsymbol{Y}$ *be generated from an ANM with DAG* G*, then assume* $\boldsymbol{x} \prec_w \boldsymbol{y}_\rho$ *for all permutations* $\rho \in \mathfrak{S}_p$*, where* $\boldsymbol{x}$ *and* $\boldsymbol{y}_\rho$ *are defined in (10.13) and (10.14), respectively.*

Note that once an ordering ρ is determined, existing variable selection methods can be used to learn the parent set pa and hence the graph G. Therefore, learning a DAG G reduces to the task of learning its topological ordering. For more details, see Gao et al. [2020, Appendix A] .

To provide an intuition of how the Assumption 1 implies identification, consider bivariate ANMs generated from the following three DAGs in Figure 10.11.

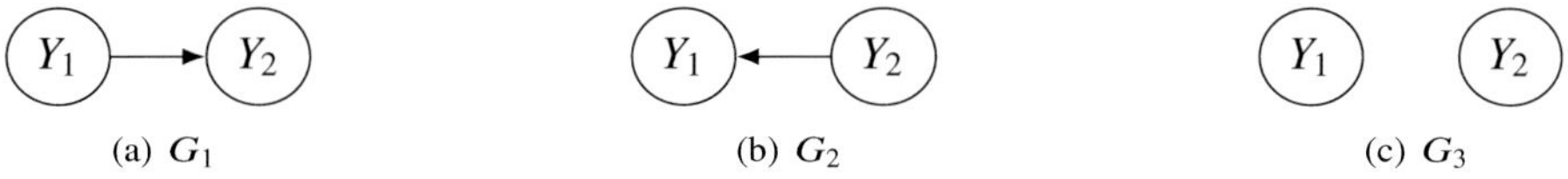

FIGURE 10.11
Three DAGs for motivation example.

From (10.12), the following identification result holds:

$$\text{DAG} := \begin{cases} G_1, & \text{if } f(\boldsymbol{x}) < f(\boldsymbol{y}_\rho) \\ G_2, & \text{if } f(\boldsymbol{x}) > f(\boldsymbol{y}_\rho)\,, \\ G_3, & \text{otherwise} \end{cases} \tag{10.15}$$

where $\boldsymbol{x} = [\text{var}(Y_1), E[\text{var}(Y_2|Y_1)]]'$, $\boldsymbol{y}_\rho = [\text{var}(Y_2), E[\text{var}(Y_1|Y_2)]]'$ and $f(\boldsymbol{x}) = \|\boldsymbol{x}\|_2$, is the Euclidean norm of $\boldsymbol{x}$.

In practice, validating Assumption 1, like many other causal assumptions, using observational data is challenging.

10.3 Structure Learning With Continuous Optimization

Continuous optimization methods are the core of most machine learning and deep learning algorithms where gradient descent is used to learn the parameters of layers. In structure learning, the continuous optimization problem recasts the hard combinatorial problem of searching DAGs in score-based methods (left-hand side of (10.16)) into a continuous optimization problem in the right-hand side of (10.16). The latter solves the constrained optimization problem to find the optimal weighted adjacency matrix that minimizes some score function $S(\boldsymbol{Y}, B)$. The constraint $h(B) = 0$ enforces acyclicity constraint in the inferred graph and $G(B)$ represents a DAG based on the weighted adjacency matrix.

$$\begin{array}{llll} \min\limits_{G(B)} S(\boldsymbol{Y}, B) & & \min\limits_{B} S(\boldsymbol{Y}, B) & \\ \text{s. t.} & G(B) \in \text{DAG} & \text{s. t.} & h(B) = 0 \end{array} \tag{10.16}$$

For the linear SEMs (10.5), pioneering work by Zheng et al. [2018], referred to as NOTEARS, used least squares scores with ℓ_1 penalty to encourage sparsity and $h(B) = \text{tr}[\exp(B \odot B)] - p$ to impose acyclicity constraint. The authors relied on the augmented Lagrangian algorithm with penalty parameters α and ρ to solve

$$S(\boldsymbol{Y}, B) + \alpha h(B) + \frac{\rho}{2} h(B)^2 \tag{10.17}$$

Wei et al. [2020] showed that the NOTEARS formulation does not satisfy Karush-Kuhn-Tucker (KKT) conditions, resulting in the algorithm failing to converge to an exact acyclic solution even when α and ρ are very large. As a remedy, they turn to the absolute value definition of the constraint with a general form $h(B) = \sum_{i=1}^{p} c_i \text{tr}(|B|^p)$ with $c_i > 0$.

Ng et al. [2020] showed that for the linear Gaussian case, the hard constraint $h(B) = 0$ in NOTEARS can be substituted with soft constraints (explained below) if the likelihood-based objective ($\mathcal{L}(\boldsymbol{Y}, B)$) is used instead of the least squares score. In particular, they cast the constrained optimization (10.16) as the unconstrained optimization problem with soft ℓ_1 and DAG constraint

$$\min_{B \in \mathbb{R}^{p \times p}} \mathcal{L}(\boldsymbol{Y}, B) + \lambda_1 \|B\|_1 + \lambda_2 h(B), \tag{10.18}$$

where λ_1 and λ_2 are the penalty coefficients. The above formulation is referred to as GOLEM.

Recently, Bello et al. [2022] rely on M-matrix (a matrix of the form $sI - A$, where $A \geq 0$ and s is greater than the spectral radius of A) to propose a novel acyclicity constraint $h_M(B) = -\log\det(sI - B \odot B) + p\log s$. The authors show that the proposed acyclicity constraint h_M is preferable to the existing constraint because

1. It does not diminish cycles of any length,
2. It has a better-behaved gradient which does not vanish.
3. Finally, computing h_M and its gradient is empirically faster.

There exists a rich literature on different versions of NOTEARS and its variants. For a recent review, we refer the reader to Vowels et al. [2022].

10.3.1 The Role of Variance, Sparsity, and Thresholding

One of the key assumptions in NOTEARS and its variants is equal error variance. However, several empirical analyses [Ng et al., 2024] have shown that the performance improvement of these algorithms diminishes when this assumption is violated. Although the GOLEM formulation in (10.18) can accommodate non-equal variance, the algorithm remains highly sensitive to the initialization value. The authors attribute the primary obstacle to performance deterioration to the nonconvexity of the optimization problem.

The same authors also investigate the role of thresholding and the sparsity penalty in continuous structure learning approaches. Their findings align with the results discussed in Chapter 2. Specifically, due to the bias introduced by the ℓ_1 penalty, MCP and SCAD penalties are found to be preferable.

A common practice in continuous structure learning is to threshold the resulting adjacency matrix at a fixed value, typically 0.3. However, Ng et al. [2024] found that using a relatively large threshold, such as 0.3, can be detrimental, as it may remove many true positives. Nonetheless, they acknowledge that thresholding can be beneficial in certain scenarios to reduce false discoveries. As such, it should be applied cautiously.

10.4 Hands-on R: Structure Learning

In this section, we utilize flow cytometry data, introduced in Example 25, to perform structure learning using the PC, GES, LINGAM, NOTEARS, and equal variance (EqVar) algorithms. For the PC, GES, and LINGAM algorithms, we employ the R package pcalg. For NOTEARS and the EqVar (proposed in Chen et al. [2019]), we use the R packages gnlearn and EqVarDAG, respectively.

We begin by importing the dataset and the necessary packages. The dataset and the true graph can be obtained from the recode package. The true graph is used as a benchmark to evaluate the performance of the different methods, with comparisons made using the Structural Hamming Distance (SHD).

```
## install additional packages
## devtools::install_github('rlebron-bioinfo/gnlearn')
## devtools::install_github("WY-Chen/EqVarDAG")
require(recode)
require(pcalg)
require(gnlearn)
require(eqVarDAG)

## obtain data
data = sachs$data

## True DAG based on "gold standard"
trueDAG <- sachs$graph

## Convert graph to graphNEL object
trueDAG = as_graphnel(trueDAG)

## CPDAG of the true DAG
trueCPDAG <- dag2cpdag(trueDAG)
```

To be able to use the true graph, we first convert it into a graphNEL object using the as_graphnel() function. Since the PC and GES methods return a CPDAG, rather than a DAG, we converted the true DAG into its corresponding CPDAG using the dag2cpdag() function for a fair comparison. The true DAG and its CPDAG are visualized in Figure 10.13(left) and Figure 10.12(left), respectively.

Next, we implement the PC algorithm and calculate the SHD score using the shd() command to evaluate its performance.

```
## Run PC algorithm
n <- nrow (data)
labels <- colnames(data)
### Use partial correlation for the indep. test
suffStat = list(C = cor(data), n = n)
pc.fit <- pcalg::pc(suffStat = suffStat,
                    indepTest = gaussCItest,
                    alpha=0.05, labels = labels,
                    verbose = TRUE)
## Estimate SHD
shd(pc.fit,trueCPDAG)
[1] 27
```

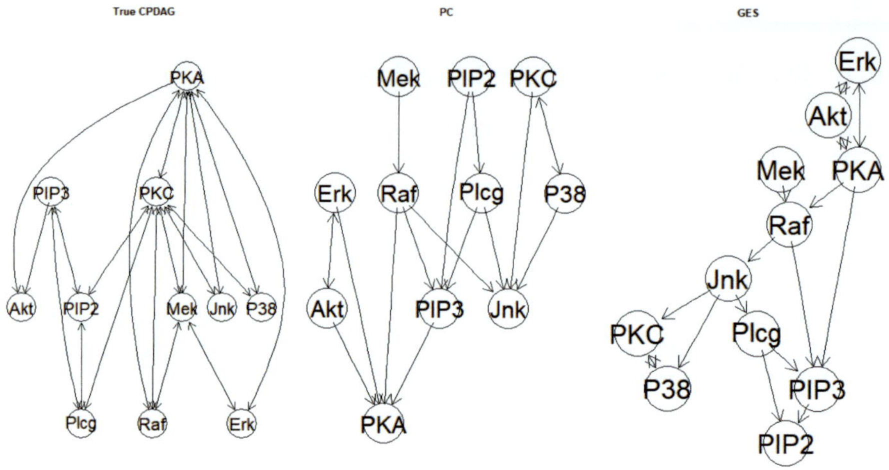

FIGURE 10.12
Compare CPDAGs for PC and GES algorithms.

Here, we set the significance level of the conditional independence test to $\alpha = 0.05$. It is important to note that α should not be interpreted as an overall significance level. Instead, in practice, it can be treated as a tuning parameter, where smaller values tend to produce sparser graphs. The SHD score for the PC algorithm is 27, and the estimated CPDAG is shown in Figure 10.12(middle).

For GES, we first define the BIC score and use the as() function to change the resulting class to the appropriate object so that shd function can be used. For details, see documentation of pcalg [Kalisch et al., 2012].

```
## Run GES algorithm
## Define the score (BIC)
score <- new("GaussL0penObsScore", data)

## Estimate the CPDAG
ges.fit <- pcalg::ges(score)

## Convert to GraphNEL class
ges.graphnel = as(ges.fit$essgraph,"graphNEL")
shd(ges.graphnel,trueCPDAG)
[1] 27
```

The estimated SHD score for GES is also equal to 27. Figure 10.12(right) shows the estimated CPDAG.

The LINGAM algorithm is implemented using the lingam() function. Unlike the PC and GES algorithms, which return a CPDAG, LINGAM directly outputs a DAG. After estimating the DAG, we construct a graph from the resulting adjacency matrix to ensure compatibility with the shd() function. The true and estimated DAGs are visualized in Figure 10.13.

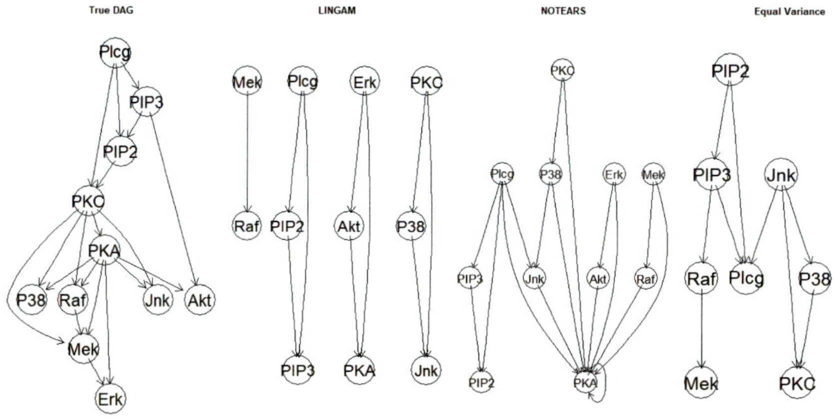

FIGURE 10.13
Compare DAGs for LINGAM, NOTEARS and EqVar algorithms.

```
## Run LINGAM algorithm
lingam.fit = lingam(data)

## Extract adjacency matrix
lingam.adj = (abs(lingam.fit$Bpruned) > 0) * 1
colnames(lingam.adj) = labels

## Convert to graphNEL class
lingam.graphnel = as(t(lingam.adj), "graphNEL")

## Estimate SHD
shd(lingam.graphnel,trueDAG)
[1] 25
```

The estimated SHD for LINGAM is 25, which is lower than for PC and GES.

Next, we apply the NOTEARS and EqVar algorithms using the notears and EqVarDAG_BU functions, respectively, and report their corresponding SHD scores. For NOTEARS, we set the Lasso penalization parameter to a small value ($\lambda = 0.05$) and use a threshold of 0.3, which is consistent with common practices in the literature [Ng et al., 2024]. For the EqVar algorithm, we employ a Z-test to recover the DAG from the ordering, using a significance level of $\alpha = 0.1$.

```
## Run NOTEARS
notears_graph = notears(data, lambda1= 0.05, max.iter = 200, w.
    threshold = 0.3)
## Convert to graphNEL
notears.graphnel = as_graphnel(notears_graph)
shd(notears.graphnel,trueDAG)
[1] 25.5

## Run EqVar
```

```
eqvardag.adj = EqVarDAG_BU(as.matrix(data), mtd = "ztest", alpha =
    0.1)$adj
colnames(eqvardag.adj) = labels
## Convert to graphNEL class
eqvar.graphnel = as(t(eqvardag.adj), "graphNEL")
## Estimate SHD
shd(eqvar.graphnel,trueDAG)
[1] 26
```

Based on the above results, the LINGAM algorithm provides the lowest SHD.

10.5 Problems

10.1 Properties of the adjacency matrix of a DAG.

(a) A matrix with all eigenvalues equal to zero is called a nilpotent matrix . Show that if A is a weighted adjacency matrix of a DAG, then A is nilpotent.

(b) Show that for the weighted adjacency matrix A, $A^k = 0$ for some power k.

(c) Finally, show that a directed graph is acyclic if and only if its weighted adjacency matrix satisfies $h(A) = \text{tr}(\exp(A \odot A)) - p = 0$.

10.2 Let $h(B) = \sum_{i=1}^{p} c_i \text{tr}(B^i)$ with $c_i > 0$ and B is non-negative weighted adjacency matrix for the DAG G. Show that if $(\nabla h(B))_{ij} > 0$ for $i \neq j$ then there exists a directed walk from node j to i in G. A directed walk is a sequence of nodes and edges in a directed graph, where each edge connects the preceding node to the following node, following the direction of the edge. Hint: first show that $(\nabla h(B))_{ij} = \sum_{k=1}^{p-1} (k+1) c_{k+1} (B^k)_{ji}$ for $i \neq j$, then recall that if $(B^k)_{ji} > 0$ for at least one k, then there exists a directed walk of length k from j to i.

10.3 Show that a directed graph G is acyclic if and only if $\text{tr}(\nabla h(B)' B) = 0$.

10.4 Suppose a random vector $\boldsymbol{Y}$ is generated from the linear SEM (10.5) with an ordering ρ. Show that for $1 \leq j \leq p$

$$L^2_{\rho(j),\rho(j)} = \text{var}(Y_{\rho(j)} | Y_{\rho(j-1)}, \dots, Y_{\rho(1)}), \tag{10.19}$$

where $\boldsymbol{L}$ is the Cholesky factor of the covariance matrix $\boldsymbol{\Sigma}_\rho$ after permuting rows and columns of $\boldsymbol{\Sigma}$ simultaneously.

10.5 Consider sachs dataset from Section 10.4.

(a) Use the pc() function to run PC algorithm, but this time set the significance level of the conditional independence test to $\alpha = 0.01$ and $\alpha = 0.1$.

(b) Estimate SHDs for both graphs.

(c) Use the plot() function to visualize the estimated CPDAGs. Describe the differences.

(d) Use the notears() function to estimate the DAG but vary the Lasso penalization parameter $\lambda = \{0.01, 0.1\}$.

(e) Report SHDs for both graphs. Which value of penalization parameter provides a better SHD result?

11

Time Series and Temporal Dependence

Data collected over time is referred to as time series where nearby observations are more (temporally) correlated than those far apart. The world of time series data is vast and growing, just think of the constant stream of data generated on the social media and digital recording devices. Historically, time series data were collected and studied in meteorology, business and economics. The problems of interest in time series analysis are forecasting and capturing the nature of *temporal dependence* through the *autocovariance function* of the series. In this chapter, the focus is on covariance stationary time series. We present the basic concepts for univariate and multivariate stationary time series in the time- and spectral-domains. For multivariate time series with the additional *cross-sectional dependence*, its Fourier transform effectively removes the temporal dependence. We highlight the similarity of the roles of the spectral density and inverse spectral density matrices to those of the covariance and precision matrices of independent multivariate data. In the spectral-domain, the analogs of marginal and partial correlations, namely coherence and partial coherence, are of central importance for studying connectivity in neuroscience and signal processing. We present time series graphical Lasso (TS-GLasso) as the natural extension of GLasso for i.i.d. multivariate data in Chapter 9. For sparsity in high-dimensional time series, there is also the option of time-domain approach within vector autoregressive (VAR) models where zeros in the VAR coefficient matrices are closely related to the notion of Granger causality. Similarly, dimension reduction techniques like PCA and factor models, which play important roles in high-dimensional time series data analysis, are presented as natural extensions of such techniques introduced in Chapter 7.

11.1 Univariate Time Series

For a time series or stochastic process $\{X_t, t \in \mathbb{Z}\}$ indexed by the set of integers, the simple regression-like difference equation connecting its present value to the immediate past via

$$X_t = \beta_0 + \beta_1 X_{t-1} + Z_t, \tag{11.1}$$

DOI: 10.1201/9781003638896-11

where Z_t is the error or noise, plays a central role in appreciating the deeper properties of stochastic processes, and more generally the finite parameter autoregressive and moving average (ARMA) models for time series data. Following the tradition in time series analysis, in this chapter we use slightly different notation like X_t, Z_t to denote the variable of interest and the error term, etc.

Heuristically, measurements closer in time are more correlated than those far apart. Thus, quantifying dependence among successive values of a series matters greatly in the statistical analysis of time series data. To further highlight this point, we consider random variables $X_1, \ldots, X_n$ sharing the same mean μ where the sample mean $\bar{X}$ is always an *unbiased* estimator of μ regardless of the dependence among the variables. However, its variance does depend on the dependence of the data, so that setting $\text{Cov}(X_i, X_j) = \sigma_{i,j}$, we have

$$n^2\text{Var}(\bar{X}) = \text{Var}(\sum_{i=1}^{n} X_i) = \sum_{i,j} \text{Cov}(X_i, X_j) = \sum_{i,j} \sigma_{i,j}. \tag{11.2}$$

This is quite different from the more familiar formula σ^2/n for i.i.d. random variables with variance σ^2. At the other extreme, when the observations are equi-correlated, i.e., $\text{Cov}(X_i, X_j) = \sigma^2\rho$ for $i \neq j$, and σ^2 for $i = j$, then the sample variance simplifies to

$$\text{Var}(\bar{X}) = \frac{\sigma^2}{n}\left[1 + (n-1)\rho\right], \tag{11.3}$$

which can be much bigger (smaller) than σ^2/n depending on whether ρ is positive (negative).

Judging by the variance of the sample mean, the *(effective) sample size* in this case is not really n, but given by the expression

$$n_e = \frac{n}{1 + (n-1)\rho}. \tag{11.4}$$

Understanding the implication of this variance inflation (deflation) factor or the effective sample size is quite important when constructing confidence intervals or testing hypothesis about the mean μ or other important parameters in the presence of dependence in the data. Fortunately, the variance inflation factor for the sample mean takes particularly manageable form for the class of covariance stationary processes introduced next.

11.1.1 Covariance Stationary Time Series

The *stationarity* property of time series data is a natural extension of observations being i.i.d. in the classical statistics. In fact, the time-invariance of the mean and variance of a covariance stationary time series ensures that its observed values hover around a fixed mean level.

We say a time series $\{X_t\}$, indexed by integers, is a *covariance stationary* time series (stochastic process) if

1. $E(X_t) = \mu$, for all integers t,

2. $\mathrm{Cov}(X_s, X_t)$, depends only on the *lag or time-separation* $|s-t|$ between the two variables.

In this case, the function $\gamma(\cdot)$ defined by

$$\gamma(h) = \mathrm{Cov}(X_{t+h}, X_t) = E(X_{t+h} - \mu)(X_t - \mu), h = 0, \pm 1, \ldots, \tag{11.5}$$

is called its *autocovariance function (ACF)*, and enjoys the following important properties for all integers $t, h = 0, \pm 1, \ldots$:

(a) Constant Variance: $\mathrm{Var}(X_t) = \gamma(0)$,

(b) Symmetry: $\gamma(h) = \gamma(-h)$,

(c) Boundedness: $|\gamma(h)| \leq \gamma(0)$.

In view of the symmetry property in (b), it is common to focus on $\gamma(h), h \geq 0$. The *autocorrelation function* (also abbreviated as ACF) of a stationary process is

$$\rho(h) = \frac{\mathrm{Cov}(X_{t+h}, X_t)}{\sqrt{\mathrm{Var}(X_{t+h})\mathrm{Var}(X_t)}} = \frac{\gamma(h)}{\gamma(0)}, \quad h = 0, 1, \ldots,$$

and the plot of $\rho(h)$ vs h, called the *theoretical correlogram* of $\{X_t\}$ reveals useful dependence (correlation) features of a time series. The central roles of the ACF in various aspects of time series analysis are summarized below.

ACF and the Accuracy of the Sample Mean: It follows from (11.2) and the properties of the ACF of a stationary time series that

$$\mathrm{Var}(\bar{X}) = \frac{\gamma(0) + 2\sum_{h=1}^{n-1}(1-\frac{h}{n})\gamma(h)}{n} = \frac{\gamma(0)}{n}\left[1 + 2\sum_{h=1}^{n-1}(1 - \frac{h}{n})\rho(h)\right]. \tag{11.6}$$

Note that for n large, the numerator converges to

$$\sigma_L^2 = \gamma(0) + 2\sum_{h=1}^{\infty}\gamma(h) = \sum_{h=-\infty}^{+\infty}\gamma(h) = \gamma(0)\left(1 + 2\sum_{h=1}^{\infty}\rho(h)\right), \tag{11.7}$$

provided that the ACF is absolutely summable, i.e.,

$$\sum_{h=-\infty}^{+\infty}|\gamma(h)| < \infty.$$

In this case, the σ_L^2 is known as the *long-run variance* of the process $\{X_t\}$. It effectively plays the role of σ^2 in the i.i.d. case, in the sense that

$$\mathrm{Var}(\bar{X}) \approx \frac{\sigma_L^2}{n},$$

see also (11.3). It turns out that the asymptotic distribution and other optimal properties of $\bar{X}$, the sample mean of a stationary time series, depend primarily on the whole covariance function $\{\gamma(h)\}$ and the rate at which it goes to zero as the lag $h \to \infty$. This is in sharp contrast with the classical theory of statistics where the asymptotic properties of the estimators depend on the shape, smoothness and other properties of a single (density) function.

ACF and Forecasting: For the moment, assuming that $X_1, \ldots, X_n$ are observed and $\{\gamma(h)\}$ is known, consider the task of predicting the next future value X_{n+1}. Let

$$X_{n+1,n} = \sum_{k=1}^{n} a_{k,n} X_{n+1-k}, \tag{11.8}$$

be the optimal linear predictor where the coefficients $a_{k,n}$'s are found by minimizing the mean square error (MSE)

$$E(X_{n+1} - \sum_{k=1}^{n} a_{k,n} X_{n+1-k})^2. \tag{11.9}$$

It follows that the finite predictor coefficients $a_{k,n}$'s and the finite prediction error variance, in symbols σ_{n+1}^2, satisfy the *normal equations*, where here for simplicity we use γ_h instead of $\gamma(h)$:

$$\begin{bmatrix} \gamma_0 & \gamma_1 & \cdots & \gamma_{n-1} \\ \gamma_1 & \gamma_0 & \cdots & \gamma_{n-2} \\ \vdots & \vdots & \vdots & \vdots \\ \gamma_{n-1} & \gamma_{n-2} & \cdots & \gamma_0 \end{bmatrix} \begin{bmatrix} a_{1,n} \\ a_{2,n} \\ \vdots \\ a_{n,n} \end{bmatrix} = \begin{bmatrix} \gamma_1 \\ \gamma_2 \\ \vdots \\ \gamma_n \end{bmatrix}, \tag{11.10}$$

$$\sigma_{n+1}^2 = E(X_{n+1} - X_{n+1,n})^2 = \gamma_0 - \sum_{k=1}^{n} a_{k,n} \gamma_k. \tag{11.11}$$

Computing these predictor coefficients for n large through matrix-inversion is fairly expensive computationally, but the Durbin-Levinson recursive algorithm [Brockwell and Davis, 1991, Chapter 5] provides a faster method of solving the system.

Interestingly, the sequence $\{a_{n,n}\}$, the last or the far away coefficients, in (11.10) is known as the *partial autocorrelation function* (PACF) of the stationary process. In fact, for each n, $a_{n,n}$ is the partial correlation between the two variables X_t and X_{t+n} adjusted for linear effects of the intervening variables $X_{t+1}, \ldots, X_{t+n-1}$. Note that at $n = 1$, $a_{1,1} = \rho(1)$ since there is no intervening variable between X_t and X_{t+1}.

ACF and the Spectral Density: The *spectral density function* of a stationary process $\{X_t\}$ with absolutely summable ACF $\gamma(\cdot)$ is defined by the Fourier series

$$f(\lambda) = \frac{1}{2\pi} \sum_{h=-\infty}^{+\infty} \gamma(h) e^{ih\lambda}, \ \lambda \in (-\pi, \pi]. \tag{11.12}$$

Note that the value of the spectral function at $\lambda = 0$ is related to the long-run variance of the time series, i.e., $\sigma_L^2 = 2\pi f(0)$; see (11.7). This already reveals the close connection between the values of the spectral density function, the ACF and the asymptotic distribution of the sample mean and other parameters of a time series.

Evidently, estimating the mean, spectral density function, and forecasting the next future value based on the past $X_1, \ldots, X_n$ require the knowledge of the ACF $\gamma(h)$'s. In practice, the ACF is estimated by averaging over the pairs (X_{t+h}, X_t) of available data leading to the *sample autocovariance function*:

$$\hat{\gamma}(h) = \frac{1}{n} \sum_{t=1}^{n-h} (X_{t+h} - \bar{X})(X_t - \bar{X}), \quad h = 0, 1, \ldots, H, \tag{11.13}$$

and the *sample autocorrelation function* is defined by

$$\widehat{\rho}(h) = \frac{\widehat{\gamma}(h)}{\widehat{\gamma}(0)}, h = 0, 1, \ldots, H.$$

A popular choice of H in practice is around $n/4$. The plot of $\widehat{\rho}(h)$ versus h is usually referred to as the *sample correlogram* of the time series data.

11.1.2 Examples of Stationary Time Series

In this section, many examples of time series models including the very important class of finite-parameter autoregressive and moving average (ARMA) processes are constructed as linear combination (filtering) of the simplest sequence of uncorrelated random variables defined next.

Example 26 White Noise: *A sequence of uncorrelated random variables Z_t, with the mean zero and common variance σ^2 is stationary, since its covariance function,*

$$Cov(Z_s, Z_t) = \delta_{s,t}\sigma^2 = \delta_{s-t,0}\sigma^2,$$

depends only on $s-t$, where $\delta_{s,t}$ is the Kronecker's delta. Its ACF is simple and takes only two distinct values zero and σ^2:

$$\gamma(h) = \delta_{h,0}\sigma^2.$$

Such a time series is called a white noise (WN) process and written as

$$\{Z_t\} \sim WN(0, \sigma^2).$$

This simplest stationary time series or the WN turns out to be the building block of the whole class of stationary processes.

Example 27 Moving Average (MA) Models: *A moving average (MA) model of order q is simply a linear combination of values of a WN* $\{Z_t\}$*:*

$$X_t = Z_t + \theta_1 Z_{t-1} + \cdots + \theta_q Z_{t-q} = \sum_{i=0}^{q} \theta_i Z_{t-i}, \tag{11.14}$$

where $\theta_0 = 1, \theta_1, \ldots, \theta_q$ *are the parameters and* q*, a nonnegative integer, is called the order of the MA process.*
Covariance stationarity of the MA process follows by noting that $E(X_t) = 0$*, for all* t*, and*

$$Cov(X_{t+h}, X_t) = E(\sum_{i=0}^{q} \theta_i Z_{t+h-i}, \sum_{j=0}^{q} \theta_j Z_{t-j}) = \sum_{i=0}^{q}\sum_{j=0}^{q} \theta_i \theta_j E(Z_{t+h-i} Z_{t-j}).$$

It is advisable to compute this double-sum separately for each $h = 0, 1, \ldots, q$*, and* $h > q$*, and use the fact that for a white noise* $E(Z_{t+h-i} Z_{t-j}) = \sigma^2 \delta_{h-i+j,0}$ *which does not depend on* t*. Thus, the ACF of the process is*

$$\gamma(h) = \begin{cases} \sigma^2 \sum_{i=0}^{q} \theta_i^2, & \text{if} \quad h = 0, \\ \sigma^2 \sum_{i=0}^{q-h} \theta_i \theta_{i+h}, & \text{if} \quad 1 \le h \le q, \\ 0, & \text{if} \quad h > q, \end{cases} \tag{11.15}$$

which vanishes at all lags $h > q$*. In particular, the theoretical correlogram of an MA(q) process is flat zero after the lag* q*, and one expects the same for the sample correlogram for* n *large. This is the key in relying on the sample correlogram to identify the order* q *if the true underlying data-generating model is of finite-order moving average type.*
The spectral density function of an MA(q) process is a trigonometric polynomial:

$$2\pi f(\lambda) = \gamma(0) + 2\sum_{h=1}^{q} \gamma(h) cos(h\lambda) = \sigma^2 |\sum_{k=0}^{q} \theta_k e^{ik\lambda}|^2. \tag{11.16}$$

The first row in Figure 11.1 displays the ACF and PACF for an MA(1) process with $\theta_1 \in \{-0.25, 0.25, 0.5\}$.

So far q was taken to be finite, what happens when $q = \infty$? It turns out that the above derivations and formulae for MA(q) hold even when $q = \infty$, provided that the sequence of the coefficients is square summable. Then, the process is referred to as *infinite order (one-sided) moving averages* or MA(∞), for short. They play a central role in time series analysis and prediction theory as most of the theoretically interesting time series models admit MA(∞) representations.

Example 28 Infinite Order one-sided MA: *An MA(*∞*) process is defined by*

$$X_t = \sum_{j=0}^{\infty} \psi_j Z_{t-j}, \quad \psi_0 = 1, \quad \sum_{j=1}^{\infty} \psi_j^2 < \infty. \tag{11.17}$$

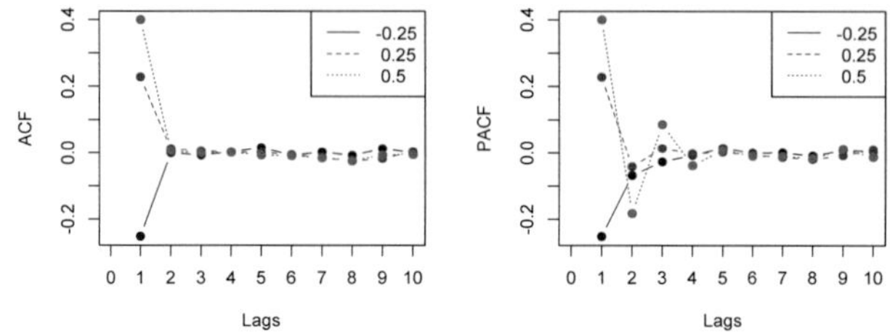

(a) MA(1) with $\theta_1 \in \{-0.25, 0.25, 0.5\}$

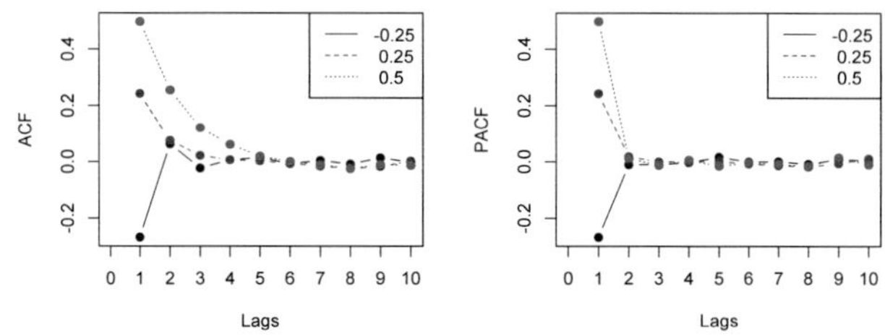

(b) AR(1) with $\phi_1 \in \{-0.25, 0.25, 0.5\}$

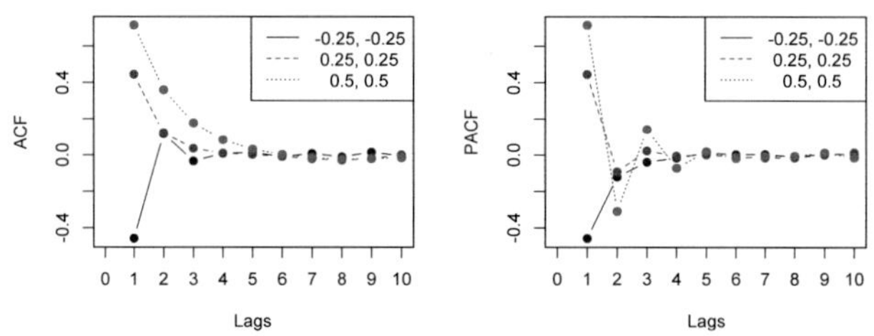

(c) ARMA(1,1) with $\theta_1, \phi_1 \in \{-0.25, 0.25, 0.5\}$

FIGURE 11.1
Correlogram and PACF of various ARMA models.

Such a process is a zero-mean stationary with the ACF

$$\gamma(h) = \sigma^2 \sum_{j=0}^{\infty} \psi_j \psi_{j+h}, \quad h = 0, 1, \ldots. \tag{11.18}$$

A more general version of MA(∞), presented next, is useful in expressing solutions of stochastic difference equations as linear combination of the input noise or shock. A *general (two-sided) linear process* is defined by

$$X_t = \sum_{j=-\infty}^{\infty} \psi_j Z_{t-j}, \qquad \sum_{j=-\infty}^{\infty} \psi_j^2 < \infty. \tag{11.19}$$

It is zero-mean stationary with the ACF

$$\gamma(h) = \sigma^2 \sum_{j=-\infty}^{\infty} \psi_j \psi_{j+h}, \quad h = 0, 1, \ldots. \tag{11.20}$$

As an important example of the MA(∞) model, for a given scalar ϕ with $|\phi| < 1$ and $\psi_j = \phi^j$, consider the process

$$X_t = \sum_{j=0}^{\infty} \phi^j Z_{t-j},$$

which is well-defined and is covariance stationary with

$$\gamma(h) = \sigma^2 \sum_{j=0}^{\infty} \phi^j \phi^{j+h} = \frac{\sigma^2}{1-\phi^2}\phi^h, \quad \rho(h) = \phi^h, \quad h = 0, 1, \ldots. \tag{11.21}$$

Note that both $\gamma(h), \rho(h)$ converge to zero exponentially fast when h goes to infinity. Evidently, this process X_t satisfies the recursion (stochastic difference equation):

$$X_t = \phi X_{t-1} + Z_t, \tag{11.22}$$

which motivates the important class of finite-parameter AR models.

Example 29 Autoregressive (AR) Models: *An autoregressive model expresses the future value of a time series in terms of its immediate past values. More precisely, for a positive integer p an autoregressive model of order p, AR(p) for short, is defined by*

$$X_t = \phi_1 X_{t-1} + \phi_2 X_{t-2} + \ldots + \phi_p X_{t-p} + Z_t, \tag{11.23}$$

where p is the order and $\phi_1, \ldots, \phi_p$ are the parameters of the model. Defining the backward shift operator B by $BX_t = X_{t-1}$, the model can be rewritten more compactly as

$$(1 - \phi_1 B - \phi_2 B^2 - \cdots - \phi_p B^p)X_t = Z_t, \quad or \quad \phi(B)X_t = Z_t, \tag{11.24}$$

where $\phi(B) = 1 - \phi_1 B - \phi_2 B^2 - \cdots - \phi_p B^p$ is called the characteristic polynomial of the process.

Unlike the coefficients of moving average models, those of AR models are restricted so that the stochastic difference equation has a stationary solution. In addition, it is desirable to choose the parameters so that for a stationary solution, the future shocks or the future values of the white noise $\{Z_t\}$ are uncorrelated with the past of the process $\{X_t\}$, i.e.,

$$\text{Cov}(X_t, Z_{t+h}) = 0, \quad \text{for} \quad h > 0. \tag{11.25}$$

Such processes are called *causal*, since they have the desirable property that *their past and present are not affected by the unknown future shocks or noises.*

To appreciate the constraint on the AR coefficients and the fundamental roles of the MA representation of the solution of the difference equations, we consider the simple AR(1) model $X_t = \phi X_{t-1} + Z_t$. Back-substituting for X_{t-1} we get

$$X_t = \phi[\phi X_{t-2} + Z_{t-1}] + Z_t = Z_t + \phi Z_{t-1} + \phi^2 X_{t-2},$$

and repeating the back-substitution step m times, one arrives at

$$X_t = \sum_{j=0}^{m-1} \phi^j Z_{t-j} + \phi^m X_{t-m}. \tag{11.26}$$

Next, for m large we study the limits of the two terms in the right-hand-side, by considering the following three distinct cases depending on the magnitude of ϕ:

Case I: $|\phi| < 1$.

The limit of the second term as $m \to \infty$ is zero, since $\mathrm{Var}(\phi^m X_{t-m}) = \phi^{2m}\gamma(0)$, and hence the solution

$$X_t = \sum_{j=0}^{\infty} \phi^j Z_{t-j},$$

is stationary and causal solution in the sense that X_t depends only on the present and past shocks.

Case II: $|\phi| = 1$.

In particular, focusing on $\phi = 1$, the AR(1) model reduces to

$$X_t = X_{t-1} + Z_t,$$

which is known as a *random walk*. For the initial value $X_0 = 0$, it has a causal solution of the form

$$X_t = Z_1 + \ldots + Z_t, \tag{11.27}$$

which turns out to be nonstationary with the autocovariance function,

$$\mathrm{Cov}(X_s, X_t) = \sigma^2 \min(s, t),$$

see Problem 11.27.

Case III: $|\phi| > 1$.

Then, both terms in (11.26) are *explosive* and a stationary solution does not seem possible. However, by turning time around and rewriting the AR(1) model as

$$X_t = \phi^{-1} X_{t+1} - \phi^{-1} Z_{t+1},$$

its coefficient now satisfies $|\phi^{-1}| < 1$ as in the Case I. Thus, we may try the alternative idea of forward-substitution, and after m forward-substitution steps we arrive at

$$X_t = \phi^{-m} X_{t+m} - \sum_{j=1}^{m-1} \phi^{-j} Z_{t+j}.$$

Now, letting $m \to \infty$, its stationary solution given by

$$X_t = -\sum_{j=1}^{\infty} \phi^{-j} Z_{t+j},$$

is *noncausal*, in the sense that the present X_t depends on the future shocks or noises like $Z_{t+1}, \ldots$.

The three cases above can be summarized succinctly in terms of the magnitude or the location of the root of the AR characteristic polynomial

$$\phi(z) = 1 - \phi z = 0, \quad z = \frac{1}{\phi},$$

in the complex plane. Indeed, the three cases correspond, respectively, to the root being outside, on, and inside the unit circle in the complex plane. Of course, for AR(p) models some of the p roots could be in any of these three locations. The second row in Figure 11.1 displays the spectral density and the ACF of AR(1) processes corresponding to the coefficients $\phi \in \{-0.25, 0.25, 0.5\}$.

Moving average representations of the potential solutions of an AR(1) model was crucial to relating the nature of its solutions (causality, nonstationarity,...) to the magnitude of its coefficient ϕ. For MA(q) models, there is also the need to express the shocks Z_t's in terms of the past observations $X_t, X_{t-1}, \ldots$, usually called the *invertibility* or the AR(∞) representation of the MA models. This is motivated by noting that, for example in $X_t = Z_t + \theta Z_{t-1}$, the observable X_t is expressed in terms of the unobservable (shocks, innovations, noises,...) Z_t, Z_{t-1}. Thus, to correct this statistical heresy, it is plausible to "invert" the model and express Z_t in terms of the past X_t's. This can be done easily by rewriting the MA(1) as

$$Z_t = -\theta Z_{t-1} + X_t, \tag{11.28}$$

and viewing it as an AR(1) model driven by the process $\{X_t\}$. Then, following the development in Case I above with $|\theta| < 1$, one obtains

$$Z_t = \sum_{i=0}^{\infty} (-\theta)^i X_{t-i}, \tag{11.29}$$

as the AR(∞) representation of an *invertible* MA(1).

The insights and results from these two simple cases carry over to the following general class of models.

Example 30 ARMA Models: *The (mixed) autoregressive and moving average models of orders p and q, ARMA (p, q) for short, is defined by*

$$X_t = \phi_1 X_{t-1} + \cdots + \phi_p X_{t-p} + Z_t + \theta_1 Z_{t-1} + \cdots + \theta_q Z_{t-q}, \tag{11.30}$$

where the ϕ_i*'s and* θ_i*'s are the AR and MA coefficients of the model. Using the backward shift operator B defined by* $BX_t = X_{t-1}$, *such models can be written compactly as*

$$\phi_p(B) X_t = \theta_q(B) Z_t,$$

where

$$\phi_p(B) = 1 - \phi_1 B - \cdots - \phi_p B^p, \ \theta_q(B) = 1 + \theta_1 B + \cdots + \theta_q B^q, \tag{11.31}$$

are, respectively, its AR and MA characteristic polynomials. The model has a causal stationary solution if all roots of $\phi_p(z) = 0$ *lie outside the unit circle in the complex plane, likewise the model is invertible if all roots of* $\theta_q(z) = 0$ *lie outside the unit circle.*

The operator B is a great book-keeping device in time series analysis, particularly for finding conditions on the coefficients so that the (mixed) ARMA models have solutions that are causal stationary with MA(∞) or AR(∞) representations. In particular, the condition for an ARMA(p, q) model to have a causal stationary solution is that all roots of $\phi_p(z) = 0$ lie outside the unit circle. And the condition for it to be invertible or have an infinite-order AR representation is that all roots of $\theta_q(z) = 0$ lie outside the unit circle. The last row in Figure 11.1 displays the correlogram and the PACF of ARMA(1,1) processes corresponding to combinations of $\theta_1, \phi_1 \in \{-0.25, 0.25, 0.5\}$.

11.1.3 Fitting ARMA Models to Time Series Data

Time series data are usually far from the ideal image of stationarity and exhibit patterns like trend, seasonality and time varying variances. Simple transformations like differencing

$$X_t - X_{t-1} = (1 - B)X_t,$$

log and Box-Cox are helpful in reducing the data to stationarity and near normality.

Figure 11.2 displays the closing daily stock prices of three major players in the technology industry (Nvidia (NVDA), IBM, and AMD) from 1 January 2019 to 1 January 2021, with noticeable upward trend in the first and third series. Time series plots of their first differences remove the trends, but highlights some local bursts of volatility or time-varying variances of the differenced series. Modeling the new feature of time-varying variances using *autoregressive conditional heteroscedastic (ARCH)* models is a topic of fundamental importance in modern finance and financial engineering.

For stationary time series data, identifying and fitting an ARMA(p, q) model is the next important goal. Just like regression analysis, there is a well-established iterative process of *model identification, estimation and diagnostics* for this important task. In this section we highlight only the principles for time series model fitting process, for further tutorial the reader may consult Shumway and Stoffer [2025] with a wealth of genuine time series data from diverse areas of applications and analyzed using different perspectives.

For model identification, the plot of the autocorrelation coefficients $\rho(h)$ versus the lag h, called the *population correlogram*, enjoys the property that it is flat zero after lag q for an MA(q) model. This property carries over to its *sample correlogram* for large samples, and can be used to identify the order q, see Figure 11.1. Traditionally, *model identification* was based on comparing the sample correlogram of the data with the theoretical correlograms of ARMA (p, q) processes to select appropriate values for p and q. However, there has been great progress in formulating this phase as a decision-theoretic problem using AIC (Akaike Information Criterion) and BIC for optimal choice of p and q, see Brockwell and Davis [1991, Chapter 9] .

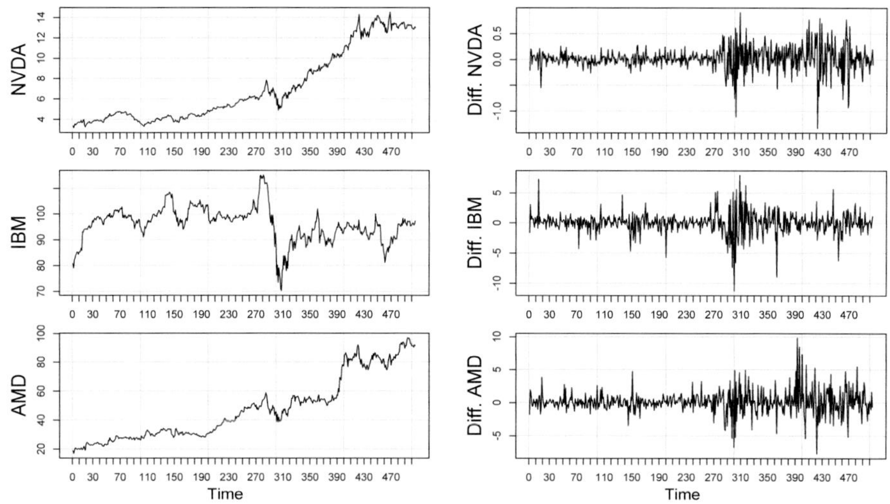

(a) Daily stock prices. (b) First difference of daily stock prices

FIGURE 11.2
Time series plots of daily stock prices and their first differences.

For model estimation, once the ARMA orders p and q are identified, maximum likelihood estimator (MLE) of the parameter vector $\theta = (\alpha_1, \ldots, \alpha_p, \beta_1, \ldots, \beta_q, \sigma^2)$ and establishing the asymptotic normality of the MLE of $\hat{\theta}$ are studied in detail in Brockwell and Davis [1991, Chapter 8] . Model diagnostics in time series is usually based on testing whether the residuals from a model are white noise and the class of *portmanteau tests* is the natural starting point.

In the statistical inference in time series, sometimes one needs to make stronger assumptions like the Z_t's are independent and identically distributed (iid) random variables or that they are normally distributed. In the first case, the shorthand $\{Z_t\} \sim$ iid $(0, \sigma^2)$, and in the second case $\{Z_t\} \sim$ iid $N(0, \sigma^2)$ are commonly used.

11.1.4 The Spectral Representation

Spectral analysis of time series is widely used in physical, natural, engineering and medical sciences where usually problems of interest lend themselves to formulation in terms of frequency, cycle, period, rhythm,..., and longer time series data are available. By contrast, in economics and social sciences where usually not enough data is available a time-domain approach based on ARMA models in the spirit of correlation and regression analysis is desirable.

Historically, the concept of periodogram and its early statistical applications were developed by Schuster [1906] in the search for an unknown period p of the data generated from the signal plus noise model

$$X_t = \alpha \cos \lambda t + \beta \sin \lambda t + Z_t, \ t = 1, \ldots, n, \tag{11.32}$$

where $\lambda = \frac{2\pi}{p}$ is the frequency cycle of the fluctuations in the data, α and β are the parameters and $\{Z_t\}$ is the measurement error.

The spectral representation of stationary processes provides a natural bridge connecting the spectral- and time-domain approaches in time series analysis. It expresses any stationary process as the Fourier transform of a stochastic process with simpler dependence structures [Cramer, 1940]. More precisely,

$$X_t = \int_{-\pi}^{\pi} e^{it\lambda} dZ(\lambda), t = 0, \pm 1, \ldots, \tag{11.33}$$

where $\{Z(\lambda); -\pi < \lambda \leq \pi\}$ is a complex-valued process with *orthogonal increments*, that is

$$\text{Cov}\,(dZ(\lambda_1), dZ(\lambda_2)) = 0, \quad \text{for} \quad \lambda_1 \neq \lambda_2.$$

Recall that the covariance of two mean zero complex-valued random variables X and Y is defined by

$$\text{Cov}(X, Y) = EXY^*,$$

where Y^* stands for the complex conjugate of Y. It is helpful to think of the integral as a limit of a partial sum or alternatively as a limit of the infinite sum appearing in the definition of an MA(∞) process, and to think of $Z(\lambda)$ as an analog of a white noise defined on an interval rather than on the set of integers. With this interpretation of the stochastic integral, the spectral representation (11.33) of a stationary processes formally leads to a similar (Fourier) representation of its *autocovariance function* $\{\gamma(h)\}$:

$$\begin{cases} \gamma(h) & = Cov(X_{t+h}, X_t) = \int_{-\pi}^{\pi} e^{ik\lambda} dF(\lambda), h = 0, \pm 1, \ldots, \\ dF(\lambda) & = E|d\,Z\,(\lambda)|^2. \end{cases} \tag{11.34}$$

That the function $F(\lambda)$ is nondecreasing on $(-\pi, \pi]$ follows from the fact that $Z(\cdot)$ has orthogonal increments; see Problem 11.8. This monotone function and $f(\lambda)$, the derivative of its absolutely continuous part, are called the *spectral distribution* and the *spectral density* of the stationary process $\{X_t\}$, respectively.

Setting $h = 0$ in (11.33) one obtains

$$\gamma(0) = Var(X_t) = \int_{-\pi}^{\pi} dF(\lambda), \tag{11.35}$$

which shows directly the role of the spectral distribution in dividing up the *total power*, $Var(X_t)$, among the frequencies λ on $(-\pi, \pi]$. The mystery of how the unobservable (complex-valued) random process $\{Z(\lambda); -\pi < \lambda \leq \pi\}$ holds the key to such distribution of the power over the frequencies raises this natural question: How to express $Z(\lambda)$ in (11.33) in terms of the original process $\{X_t\}$? This question is reminiscent of invertibility of moving average models discussed in the previous section. An affirmative answer is provided through the Fourier inversion formula and then truncating the resulting infinite series $\sum_{t=-\infty}^{\infty} X_t e^{it\lambda}$, to the observed segment of the process for $t = 1, \ldots, n$. Thus, the finite-segment analogs of (11.34) are

$$\begin{cases} dZ(\lambda) & \approx \frac{1}{\sqrt{n}} \sum_{t=1}^{n} X_t e^{-it\lambda} = d(\lambda), \\ dF(\lambda) & \approx \frac{1}{n} |\sum_{t=1}^{n} X_t e^{-it\lambda}|^2 = I(\lambda). \end{cases} \tag{11.36}$$

These are referred to, respectively, as the *discrete Fourier transform* (DFT) and *periodogram* of the time series data $X_1, \ldots, X_n$. The critical importance of (11.36) in time series analysis is due to its ability to transform dependent random variables $\{X_t\}$ into approximately uncorrelated variables $\{d(\lambda),\ \lambda \in (-\pi, \pi)\}$, see Brockwell and Davis [1991, Section 10.3]. More precisely, for large n and a broad class of stationary processes the random variables $d(\lambda_k)$ and $I(\lambda_k)$, corresponding to the values of $d(\cdot)$ and $I(\cdot)$ at the *Fourier frequencies* $F_n = \{\lambda_k = \pi k/n,\ k = 0, 1, \ldots, n-1\}$, are approximately independent with normal and a multiple of chi-square distributions, respectively.

As an estimator, the periodogram $I(\lambda)$ turns out to be a naive and notoriously poor estimator of the spectral density function $f(\lambda)$ of a stationary process because periodogram ordinates are highly irregular and erratic, see Figure 11.3. But, since periodogram ordinates at different Fourier frequencies are approximately independent, a much better and smoother estimator of $f(\lambda_j)$ is obtained by averaging (smoothing) a few periodogram ordinates in its neighborhood

$$\widehat{f}(\lambda_j) = \frac{1}{2\pi(2m+1)} \sum_{k=-m}^{m} I(\lambda_{j+k}), \quad j \in F_n, \tag{11.37}$$

for a positive integer m growing slowly with n. The size of this neighborhood m, called the *window length* or *bandwidth* or *smoothing parameter*, is important in controlling the level of smoothness of the estimator and establishing consistency of the estimator, see Brockwell and Davis [1991, Section 10.4]. Figure 11.3 plots periodogram of MA(1) process superimposed with smoothed density estimator with $m = 10, n = 1000$ and the true spectral density function. This asymptotic independence of ordinates of $d(\lambda)$ at the Fourier frequencies $\lambda_k \in F_n$, is crucial in formulating spectral-domain analogs of the classical statistical methods such as the analysis of variance [Tukey, 1962] over orthogonal components indexed by frequencies, regression, principal components, canonical correlation analysis, and dynamic factor models, see Brillinger [1975] and Forni et al. [2000].

11.2 Multivariate Stationary Time Series

Multivariate time series data provide the opportunity and challenge of dealing simultaneously with *temporal and cross-sectional* correlations through matrix-valued

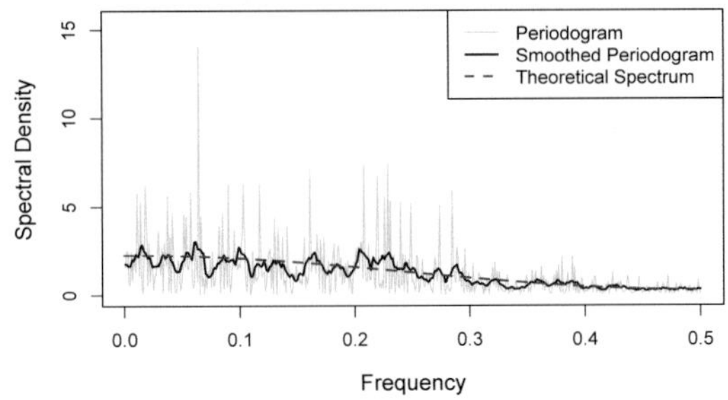

FIGURE 11.3
Periodogram and smoothed periodogram ($m = 10$) of and MA(1) with $\theta = 0.5$.

autocovariance and spectral density functions. We begin with the definition of multivariate covariance stationarity, and note that in this chapter we use d in place of p to denote the dimension of a vector, and boldface capital X is used for a vector time series.

A d-variate stochastic process $\{\boldsymbol{X}_t = (X_{1,t}, \ldots, X_{d,t})'\}$ is *covariance stationary* if

1. its mean vector $E(\boldsymbol{X}_t) = \boldsymbol{\mu}$, does not depend on t,
2. the covariance matrix $\text{Cov}(\boldsymbol{X}_s, \boldsymbol{X}_t) = E(\boldsymbol{X}_s\boldsymbol{X}_t') - \boldsymbol{\mu}\boldsymbol{\mu}' = \Gamma(s-t)$, depends only on the *lag or the time-separation* of the time-indexed random vectors.

The matrix-valued function $\Gamma(\cdot)$ is called its *autocovariance function (ACF)*. As in the univariate case the ACF turns out to be the Fourier transform of its matrix-valued spectral density function $f(\lambda)$. In analogy with the multivariate data in the previous chapters, the eigendecompositions of the lag-0 matrix $\Gamma(0) = \text{Cov}(\boldsymbol{X}_t)$ and the spectral matrix $f(\lambda)$ are expected to play fundamental roles in time series PCA, factor models and graphical models.

After introducing vector AR (VAR) models and discussing Granger's causality in the time-domain, we focus on the spectral density matrix of multivariate stationary time series and highlight the similarity of its roles with those of the covariance matrix of i.i.d. multivariate data. This enables a sense of parallelism between concepts in multivariate statistics and multivariate time series analysis centered around the concepts of coherence and partial coherence, the spectral-domain analogs of marginal and partial correlations. These concepts are important in the search for sparsity of $f(\lambda)$ and its inverse and developing analogs of Gaussian graphical models (GGM) for multivariate time series data.

A multivariate white noise is the building block of multivariate stationary processes just like its univariate counterpart. A sequence of uncorrelated random vectors $\boldsymbol{Z}_t, t \in \mathbb{Z}$, with mean zero and covariance matrix $\boldsymbol{\Sigma}$ is called a (vector) *white noise (WN) process*. Evidently, it is a multivariate stationary time series since

$$\text{Cov}(\boldsymbol{Z}_s, \boldsymbol{Z}_t) = \delta_{s,t}\boldsymbol{\Sigma} = \delta_{s-t,0}\boldsymbol{\Sigma},$$

depends only on $s - t$. Its ACF and spectral density matrices are given by

$$\Gamma(h) = \delta_{h,0}\boldsymbol{\Sigma}, \quad f(\lambda) = \frac{1}{2\pi}\boldsymbol{\Sigma}, \quad \lambda \in (-\pi, \pi].$$

We note that for a multivariate white noise, its cross-sectional or contemporaneous dependence is governed by the covariance matrix $\boldsymbol{\Sigma}$. Temporal or serial dependence is introduced through linear combinations of a white noise. In particular, vector MA models, as linear combinations of vector white noise, continue to play important theoretical roles in multivariate time series (Problem 11.9). However, vector autoregressive (VAR) models have proved more useful in the analysis of multivariate time series data for business and economics, engineering and neurosciences.

Example 31 Vector Autoregressive (VAR) Processes: *A vector autoregression of order p or VAR(p) is defined by*

$$\boldsymbol{X}_t = \Phi_1 \boldsymbol{X}_{t-1} + \ldots + \Phi_p \boldsymbol{X}_{t-p} + \boldsymbol{Z}_t, \ \boldsymbol{Z}_t \sim WN(0, \boldsymbol{\Sigma}),$$

where $\Phi_1, \ldots, \Phi_p$ are the AR coefficient matrices, and $\{\boldsymbol{Z}_t\}$ is a white noise process. For the model to have a causal stationary solution all the roots of the polynomial equation

$$det(I_d - \sum_{i=1}^{p} \Phi_i z^i) = 0, \ z \in C,$$

must lie outside the unit circle in the complex plane, see Example 30 and Brockwell and Davis [1991].

Given multivariate time series data $\boldsymbol{X}_1, \ldots, \boldsymbol{X}_n$, fitting VAR models amounts to estimating its order p and the coefficient matrices $\Phi_1, \ldots, \Phi_p$. Unlike the univariate case, the sample correlogram and partial correlogram are not that helpful in the multivariate case, and it is more common to rely on AIC and BIC-type criteria for order selection. In addition, when the number of time series d is large or even moderate, a VAR model is over-parameterized in that the number of parameters d^2p, grows quadratically in the dimension d.

There is a growing interest in developing sparse methods to overcome the computational problem and the interpretation of the model parameters; see Song and Bickel [2011], Davis et al. [2016], Nicholson et al. [2016], Dallakyan et al. [2022]. Most popular methods for sparse VAR estimation rely on a suitable mix of time- and spectral-domain techniques.

11.2.1 Spectral Representation and Analysis

Multivariate extensions of (11.33)–(11.34) hold for any d-variate zero-mean stationary process and can be realized as the Fourier coefficients of a multivariate orthogonal increment stochastic process.

More precisely, a zero-mean multivariate stationary time series admits the representation

$$\boldsymbol{X}_t = \int_{-\pi}^{\pi} e^{-it\lambda} d\boldsymbol{Z}(\lambda), t = 0, \pm 1, \ldots, \tag{11.38}$$

where $\{\mathbf{Z}(\lambda); -\pi < \lambda \leq \pi\}$ is a d-variate *orthogonal increments* process, namely

$$\text{Cov}\,(d\mathbf{Z}(\lambda_1), d\mathbf{Z}(\lambda_2)) = \mathbf{0}, \quad \text{for} \quad \lambda_1 \neq \lambda_2.$$

With the proper interpretation of the stochastic integral, the spectral representation (11.38) of the stationary processes leads to similar (Fourier) representation of its matrix-valued *autocovariance function* $\{\Gamma(h)\}$:

$$\begin{cases} \Gamma(h) = \int_{-\pi}^{\pi} e^{-ik\lambda} dF(\lambda), & h = 0, \pm 1, \ldots, \\ dF(\lambda) = Ed\,\mathbf{Z}\,(\lambda) d\,\mathbf{Z}\,(\lambda)^*, \end{cases} \tag{11.39}$$

where for a matrix M, M^* stands for its complex-conjugate transpose. The $d \times d$ matrix-valued function $F(\lambda)$ is nondecreasing on $(-\pi, \pi]$ in the sense that for any $\lambda_2 > \lambda_1$, the matrix $F(\lambda_2) - F(\lambda_1)$ is positive semi-definite, a consequence of $\mathbf{Z}(\cdot)$ having orthogonal increments. The monotone (matrix-valued) function $F(\lambda)$ and $f(\lambda) = (f_{ij}(\lambda))$, the derivative of its absolutely continuous part, are called the *spectral distribution* and the *spectral density* of the multivariate stationary process $\{\boldsymbol{X}_t\}$, respectively. For $i \neq j$, the (i, j)-th entry of the spectral density matrix is referred to as the *cross-spectral density* of the two individual (marginal) series $\{X_{t,i}\}$ and $\{X_{t,j}\}$, while for $i = j$, it reduces to the spectral density of a single series.

The dependence terminologies above are reminiscent of $\text{Cov}(X, Y)$ and $\text{Var}(X)$ for two random variables. Thus, in the spectral domain and in analogy with computing correlation coefficients from covariances, given a spectral density matrix, the *spectral coherence (SC) or coherency* at the frequency λ between two marginal series [Brillinger, 1981] is defined as

$$\kappa_{ij}(\lambda) := \frac{f_{ij}(\lambda)}{\sqrt{f_{ii}(\lambda) f_{jj}(\lambda)}}, \quad \lambda \in [-\pi, \pi]. \tag{11.40}$$

It is a complex-valued analog of the correlation coefficient between the two random increments $dZ_i(\lambda)$ and $dZ_j(\lambda)$, and hence satisfies $|\kappa_{ij}(\lambda)| \leq 1$.

Example 32 *For the bivariate ARMA(1,1) process*

$$\boldsymbol{X}_t = \begin{bmatrix} 0.4 & 0 \\ 0 & 0.8 \end{bmatrix} \boldsymbol{X}_{t-1} + \begin{bmatrix} 0.3 & -0.5 \\ 0.6 & 0.8 \end{bmatrix} \mathbf{Z}_{t-1} + \mathbf{Z}_t, \tag{11.41}$$

with the identity covariance matrix for the white noise, the components of its 2×2 *spectral density matrix are plotted in Figure 11.4. Note that since the cross-spectral density function* f_{12} *is complex-valued, its real and imagery parts are plotted separately.*

As in the univariate case, the discrete Fourier transform of a time series plays a fundamental role in providing the ingredients for estimating the spectral density matrix. The multivariate analogs of the discrete Fourier transform and periodogram of the data $\boldsymbol{X}_t, t = 1, \ldots, n$, are:

$$\begin{cases} d(\lambda) = \frac{1}{\sqrt{n}} \sum_{t=1}^{n} \boldsymbol{X}_t e^{-it\lambda}, \\ I(\lambda) = \frac{1}{n} d(\lambda) d(\lambda)^*, \end{cases} \tag{11.42}$$

It follows from (11.42) that for large n and a broad class of stationary processes the random vectors and matrices $d(\lambda_k)$ and $I(\lambda_k)$, corresponding to the values of $d(\cdot)$ and $I(\cdot)$ at the *Fourier frequencies* are approximately independent with complex multivariate normal and a multiple of Wishart distributions [Brockwell and Davis, 1991], respectively.

In addition to nonparametric methods of estimating the spectral density matrix as the multivariate analog of (11.37), there is a class of parametric estimators obtained by fitting vector autoregressive (VAR) models of increasing orders p_n to the time series data. Such estimators are known to be asymptotically mean squared consistent [Brockwell and Davis, 1991] when the dimension d is fixed and p_n and n grow to infinity. A recent review of nonparametric estimation of spectral density matrices, with a focus on Ledoit-Wolf type shrinkage, can be found in von Sachs [2020].

11.2.2 Linear Filters

Linear filters or linear combinations of lagged values of a stationary process play important roles in time series analysis, signal processing and statistical estimation.

Let $\boldsymbol{X}_t$ be a zero-mean d-variate stationary process with the spectral representation (11.38) and consider the r-variate process $\boldsymbol{Y}_t$ obtained from $\boldsymbol{X}_t$ using a linear time-invariant filter of the form

$$\boldsymbol{Y}_t = \sum_{k=-\infty}^{\infty} \Psi_k \boldsymbol{X}_{t-k}, \tag{11.43}$$

with suitable summability condition on the $r \times d$ filter coefficient matrices Ψ_k's. The filter is time-invariant in the sense that Ψ_k's do not depend on the time t. The function

$$\Psi(\lambda) = \sum_{k=-\infty}^{\infty} \Psi_k e^{-ik\lambda}, \tag{11.44}$$

is called the *transfer function* of the filter. Note that this filter setup is a more general version of the infinite moving average process discussed earlier in that the input process $\boldsymbol{X}_t$ in not required to be white noise.

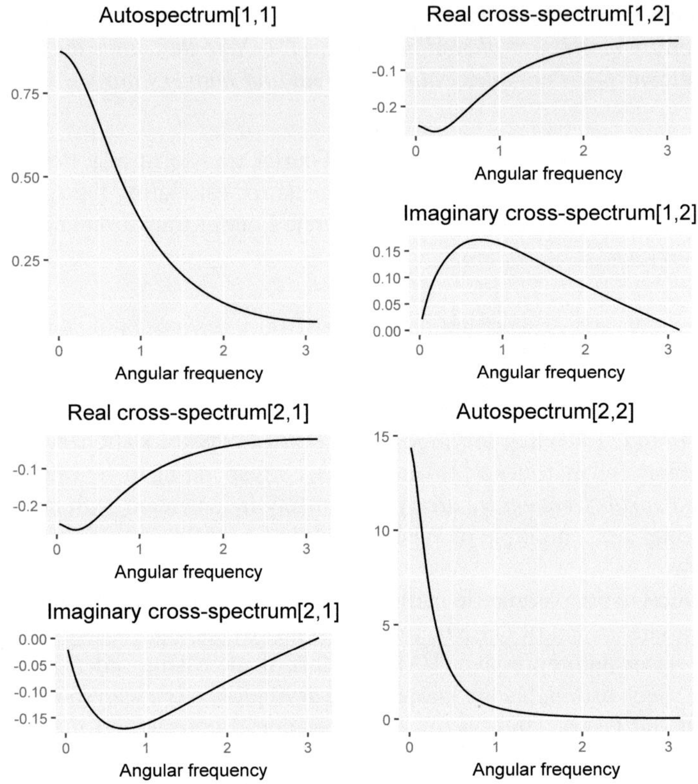

FIGURE 11.4
Plots of components of the spectral density matrix of a bivariate ARMA(1,1).

Linear filters coupled with the spectral representation of stationary processes provide a powerful framework for deriving or proving deeper results on prediction and estimation of stationary processes. For example, stationarity of the output process $\boldsymbol{Y}_t$ above and its spectral representation are established simultaneously using the spectral representation of the stationary process $\boldsymbol{X}_t$ and a formal exchange of the order of integral and the infinite sum:

$$\boldsymbol{Y}_t = \sum_{k=-\infty}^{\infty} \Psi_k \int_{-\pi}^{\pi} e^{i(t-k)\lambda} \mathbf{Z}_X(d\lambda) = \int_{-\pi}^{\pi} e^{it\lambda} \Psi(\lambda) \mathbf{Z}_X(d\lambda) = \int_{-\pi}^{\pi} e^{it\lambda} \mathbf{Z}_Y(d\lambda).$$

Consequently, the output $\boldsymbol{Y}_t$ is stationary with the random spectral measure

$$\mathbf{Z}_Y(d\lambda) = \Psi(\lambda)\mathbf{Z}_X(d\lambda),$$

and the spectral density matrix

$$f_Y(\lambda) = \Psi(\lambda) E\left(\mathbf{Z}_X(d\lambda)\mathbf{Z}_X(d\lambda)^*\right) \Psi(\lambda)^* = \Psi(\lambda) f_X(\lambda) \Psi(\lambda)^*, \tag{11.45}$$

which involves the transfer function of the filter and the spectral density matrix of the input process. We note that the derivation and the end result are reminiscent of computing covariance of linear transformation of random vectors:

$$\mathrm{Cov}(AX) = A\mathrm{Cov}(X)A^*.$$

Some important consequences of the identity involving the two spectral densities in (11.45) are given next. The key idea is to view VAR models as linear filters transforming $\{X_t\}$ to a white noise process.

Example 33 Spectral density matrix of VAR models:
As an immediate consequence of (11.45) when a VAR model is causal or has a stationary solution, see Example 31, its spectral density matrix is given by

$$f(\lambda) = \frac{1}{2\pi}\Phi(\lambda)^{-1}\Sigma\Phi^*(\lambda)^{-1},$$

where

$$\Phi(\lambda) = I_d - \Phi_1 e^{-i\lambda} - \ldots - \Phi_p e^{-ip\lambda}$$

is the characteristic polynomial of the VAR model. Its inverse spectral matrix given by

$$f^{-1}(\lambda) = 2\pi\Phi^*(\lambda)\Sigma^{-1}\Phi(\lambda),$$

plays a role similar to the precision matrices for i.i.d. data so far as graphical models and conditional independence is concerned. The identity (11.45) is also useful for deriving the spectral density of ARMA models; see Problem 11.10.

The next example provides a much more general and fundamental role of a linear filter.

Example 34 Whitening Filters: *Another important consequence of (11.45), for $r = d$, is that of finding a time-invariant linear filter that transforms $\boldsymbol{X}_t$ into d uncorrelated univariate stationary time series. This kind of filters is of great interest in principal component analysis and factor analysis of multivariate stationary time series in the spectral domain. Indeed, we consider the following two examples of such transfer functions:*

(a) For a fixed frequency λ take the transfer function $\Psi(\lambda)$ to be the orthogonal matrix $P(\lambda)$ diagonalizing the spectral density matrix $f_X(\lambda)$, namely

$$\Psi(\lambda)f_X(\lambda)\Psi(\lambda)^* = diag(\omega_1(\lambda), \ldots, \omega_d(\lambda)),$$

where

$$\omega_1(\lambda) \leq \ldots \leq \omega_d(\lambda),$$

are the ordered eigenvalues. Then, it follows that

$$f_Y(\lambda) = diag(\omega_1(\lambda), \ldots, \omega_d(\lambda)),$$

so that the components of the filtered series $\boldsymbol{Y}_t$ *are uncorrelated univariate stationary time series with spectral density functions* $\omega_i(\lambda),\ i = 1, \ldots, d$.

(b) For a fixed λ *take* $\Psi(\lambda) = \Phi(\lambda)^{-1}$, *where* $\Phi(\lambda)$ *is obtained from factorizing an invertible spectral density matrix of* $\boldsymbol{X}_t$ *as*

$$f_X(\lambda) = \Phi(\lambda)\Phi(\lambda)^*. \tag{11.46}$$

Then, from (11.45) it is evident that

$$f_Y(\lambda) = \frac{1}{2\pi} I_d,$$

so that $\boldsymbol{Y}_t$ *is a multivariate white noise and all its components are uncorrelated univariate processes. Popular choices of* $\Phi(\lambda)$ *are* $f_X(\lambda)^{1/2}$, *the symmetric square-root of the spectral density matrix. This leads to a two-sided filter (involving the entire past and future of the input process), which is not too interesting for forecasting purposes. The alternative choice of the optimal factor of* $f_X(\lambda)$ *from prediction theory of multivariate stationary processes leads to a one-sided linear filter involving past and present of the input process. This optimal factor has been playing increasingly important roles in generalized dynamic factor models; see Section 11.7.2 and Hallin and Lippi [2013].*

As the final important example of the use of the notion of filter we present the solution to a prediction problem in the spectral domain. Given a $(d+r)$-variate stationary processes $(\boldsymbol{X}_t', \boldsymbol{Y}_t')'$ with spectral density matrices f_{XX}, f_{YY}, and the cross-spectral density matrix f_{XY}, consider the problem of finding a general *optimal linear filter*, that is finding $r \times d$ matrices Ψ_k's so that $\sum_{k=-\infty}^{\infty} \Psi_k \boldsymbol{X}_{t-k}$ is close to $\boldsymbol{Y}_t$ in the sense that it minimizes the trace of the covariance matrix of the errors:

$$E(\boldsymbol{Y}_t - \sum_{k=-\infty}^{\infty} \Psi_k \boldsymbol{X}_{t-k})(\boldsymbol{Y}_t - \sum_{k=-\infty}^{\infty} \Psi_k \boldsymbol{X}_{t-k})'. \tag{11.47}$$

This problem and its solution are the time series analogs of the multivariate regression problem in Section 4.1.2, with similar corollaries as far as regression interpretation of the entries of the inverse spectral density matrix is concerned. The following theorem provides properties of its solution, for a proof see Brillinger [1981].

Theorem 14 *For the* $(d+r)$*-variate stationary process* $(\boldsymbol{X}_t', \boldsymbol{Y}_t')'$ *assume that* $f_{XX}(\lambda)$ *is invertible for all* $\lambda \in [-\pi, \pi]$. *Then,*

(a) the $r \times d$ *optimal transfer function or the solution of (11.47) is given by*

$$\Psi(\lambda) = f_{YX}(\lambda) f_{XX}(\lambda)^{-1},$$

and referred to as the complex regression coefficient of $\boldsymbol{Y}_t$ *on* $\boldsymbol{X}_t$.

(b) the r*-variate regression residual series*

$$\boldsymbol{\epsilon}_t = \boldsymbol{Y}_t - \sum_{k=-\infty}^{\infty} \Psi_k \boldsymbol{X}_{t-k},$$

is a stationary process with the spectral density matrix

$$f_{\epsilon\epsilon}(\lambda) = f_{YY}(\lambda) - f_{YX}(\lambda) f_{XX}^{-1}(\lambda) f_{XY}(\lambda),$$

having the Schur complement form.

(c) *the minimum of the criterion (11.47) is the $r \times r$ matrix*

$$\int_{-\pi}^{\pi} [f_{YY}(\lambda) - f_{YX}(\lambda) f_{XX}^{-1}(\lambda) f_{XY}(\lambda)] d\lambda.$$

The results in (a) and (b) are the time series analogs of (4.12) and (4.13). The regression residual series are particularly useful in motivating and defining the time series analogs of partial correlations in the spectral domain. More precisely, let $\{X_{t,i}\}$ and $\{X_{t,j}\}$ be two distinct marginal series of a d-variate stationary process $\{\boldsymbol{X}_t\}$, and $\{\boldsymbol{X}_{t,-ij}\}$ denotes the remaining $(d-2)$ marginal processes. With $\boldsymbol{Y}_t = (X_{t,i}, X_{t,j})'$, the conditional correlation between these two time series is computed by adjusting for the linear effect of the remaining series $\{\boldsymbol{X}_{t,-ij}\}$ using the optimal linear filter in (11.43). The two series $\{X_{t,i}\}$ and $\{X_{t,j}\}$ are said to be conditionally uncorrelated, if their regression residual series $\{\varepsilon_{t,i}\}$ and $\{\varepsilon_{t,j}\}$ are uncorrelated at all lags; i.e., $\mathrm{Cor}(\varepsilon_{t+k,i}, \varepsilon_{t,j}) = 0$ for all $t, k \in \mathbb{Z}$. In the frequency domain, this is equivalent to their cross-spectral density being zero at all frequencies $\lambda \in [0, \pi]$.

The *partial spectral coherence* (PSC) [Brillinger, 1981] between two distinct time series is defined by

$$\kappa_{ij}(\lambda) := \frac{f_{\epsilon\epsilon,ij}(\lambda)}{\sqrt{f_{\epsilon\epsilon,ii}(\lambda) f_{\epsilon\epsilon,jj}(\lambda)}}, \tag{11.48}$$

where its magnitude is less or equal to one in absolute value. As an important early application of this concept, Brillinger [1981, Figure 7.8.9] provides a comprehensive analysis of a multivariate monthly mean temperature time series of dimension $d = 14$ using PSC, where the interest is to identify the frequencies or frequency bands that drive the linear relationship between two components of the multivariate time series.

From the computational perspective, it should be noted that computing the residual series involves inverting $(d-2) \times (d-2)$ matrices for $\binom{d}{2}$ pairs of bivariate series, which is challenging for high dimensional data. Fortunately, it is known that [Dahlhaus, 2000] one can simultaneously compute PSC for all the $\binom{d}{2}$ pairs by inverting the $d \times d$ spectral density matrix only once. More precisely, setting $\Theta(\lambda) = (\theta_{ij}(\lambda)) := f(\lambda)^{-1}$, the PSC is computed using

$$\kappa_{ij}(\lambda) = -\frac{\theta_{ij}(\lambda)}{\sqrt{\theta_{ii}(\lambda)\theta_{jj}(\lambda)}}, \tag{11.49}$$

where $\theta_{ii}(\lambda)$, $\theta_{jj}(\lambda)$ are the ith and jth diagonal entries and $\theta_{ij}(\lambda)$ the (i, j)th entry of $\Theta(\lambda)$, the inverse spectral density matrix; see Section 4.2. This development shows that the PSC and the inverse spectral density matrix are the complete analogs of the familiar notions of partial correlations and inverse covariance matrices in multivariate statistics discussed in Chapter 9. However, regularization of spectral density matrix with complex entries (Hermitian matrices) could bring fresh new challenges on both theoretical and computational fronts as discussed in Section 11.4.

11.3 Sparse Vector Autoregressions

A vector autoregressive (VAR) model is a powerful tool for analyzing multivariate time series data arising from business and economics, engineering and medical imaging. The recent increase in availability of such time series data pivots the interest of researchers toward sparse high-dimensional VAR models which goes far beyond the customary and simpler notion of Granger causality [Granger, 1969].

Using regularization there are three different approaches to sparse VAR estimation:

1. Regularized least square estimators via Lasso-type penalties [Song and Bickel, 2011, Basu and Michailidis, 2015, Skripnikov and Michailidis, 2019, Safikhani and Shojaie, 2022],

2. Regularized maximum-likelihood estimators [Basu and Michailidis, 2015, Davis et al., 2016, Yuen et al., 2018],

3. Regularized Yule-Walker estimators using the CLIME or Dantzig estimators [Han et al., 2015, Wu and Wu, 2016, Ding et al., 2017].

The more popular regularized least square VAR methods ignore the contemporaneous dependence in the time series since the loss function does not include the covariance matrix of the error terms. Song and Bickel [2011] discuss the possible impact in fitting a VAR model while ignoring the contemporaneous dependence. Empirical results in Davis et al. [2016] show that the forecast performance of the VAR model improves when the covariance of the errors are incorporated at the estimation stage.

11.3.1 VAR Models and Granger Causality

Granger causality is usually studied for two univariate series. Here we present it for a multivariate time series partitioned in two subseries.

Let $\boldsymbol{X}_t$ be a zero-mean d-variate stationary process with the spectral density matrix $f(\lambda)$ and partitioned as $\boldsymbol{X}_t = (\boldsymbol{X}'_{A,t}, \boldsymbol{X}'_{B,t})'$, where A, B are a partition of the index set $\{1, 2, \ldots, d\}$. The key question of Granger causality is whether the past of the first series $\boldsymbol{X}_A$ improves the one-step ahead prediction of the second series $\boldsymbol{X}_B$. More precisely, we say that $\boldsymbol{X}_A$ *does not Granger cause* $\boldsymbol{X}_B$ if

$$\mathrm{Cov}(\boldsymbol{X}_{B,t}|\boldsymbol{X}_{A,s}, \boldsymbol{X}_{B,s}, s < t) = \mathrm{Cov}(\boldsymbol{X}_{B,t}|\boldsymbol{X}_{B,s}, s < t). \tag{11.50}$$

Otherwise, it is said that $\boldsymbol{X}_A$ *Granger causes* $\boldsymbol{X}_B$. Granger [1969] also introduced the notion of *instantaneous causality* which means that the approximation of $\boldsymbol{X}_{B,t}$ is more accurate if $\boldsymbol{X}_{A,t}$ is known at time t. In other words, adding the present $\boldsymbol{X}_{A,t}$ to the conditioning set in the left-hand-side of (11.50) will tip the balance.

The following characterization links Granger causality with sparsity of the AR coefficients and the innovation covariance matrix. For a proof of the theorem and more discussions on causality, see Deistler and Scherrer [2022, Chapter 9].

Theorem 15 *(A characterization of Granger causality) Consider the possibly infinite autoregressive representation*

$$X_t = \sum_{k=1}^{\infty} \Phi_k X_{t-k} + Z_t, \ Z_t \sim N_d(0, \Sigma),$$

where Φ_k's are the AR coefficient matrices, and $\{Z_t\}$ is a white noise process. Suppose that the Φ_k's and Σ are partitioned compatibly with $X_t = (X'_{A,t}, X'_{B,t})'$. Then,

(i) X_A does not Granger cause X_B, if and only if the $(2, 1)$ block of every Φ_k is zero.

(ii) X_A neither Granger causes X_B, nor does it cause X_B instantaneously, if and only if the $(2, 1)$ block for every Φ_k is zero and the covariance matrix Σ is block diagonal.

Thus, in the AR models, $\Phi_{k,ij} \neq 0$, for some $k \neq 0$, implies that the jth component Granger causes the ith component (edge $j \longrightarrow i$ in the graph), conditional on the past effects from all the components of X_t. In addition, residual contemporaneous effects among the d components can be captured by $\Theta = \Sigma^{-1}$, where $\theta_{ij} \neq 0$ represents an undirected edge $i - j$ in a graph or network. For an illustration of Granger causality with a multivariate time series data; see Section 11.5.3.

Since the dimension of the parameter space is quadratic in the dimension d of the time series, the problem of searching for zeros is inherently high-dimensional. A more systematic search and estimation of the nonzero entries are often conducted assuming the presence of a lower-dimensional structure on these matrices, and employing a penalized likelihood method like the GLasso algorithm.

Remark 6 *A novel reparameterization of the Φ_k's in an infinite-order VAR model as $\sum_{j=1}^{q} \ell_{kj}(\theta) G_j$ in the spirit of LCM, in Chapter 6, is given by Zheng [2025] where G_j's are unknown coefficient matrices and $\ell_{kj}(\theta)$ are q sequences of scalar functions characterizing the exponential decay to zero of the VAR coefficient matrices with θ containing a finite number of parameters. The $\ell_{kj}(\theta)$'s are chosen to control the temporal dependence, while the cross-sectional dependence is captured by the G_j's. The two directions of dependencies in the model can be interpreted separately, since they are characterized by different sets of parameters. This separation naturally allows sparsity on the G_j's determining the cross-sectional dependence. For example, a Granger causal network of the components of the series is directly linked to the aggregate sparsity pattern of G_j's.*

11.3.2 Sparse Vector Autoregressions (sVAR, msVAR)

For sparse estimation of VAR models there is a tendency to rely on a mix of time- and spectral-domain approaches. In this section we present Davis et al. [2016]'s two-stage approach where in the first stage it relies on the partial spectral coherences (PSC) (11.49) to identify a set of non-zero autoregressive coefficients. Then, constrained maximum likelihood estimation is used to re-estimate the AR coefficient matrices.

First, we present the two-stage sparse VAR (sVAR) algorithm in Davis et al. [2016] followed by more details and descriptions of its stages and the key ingredients.

Stage 1: Model Selection:

In the first stage the PSC (11.49) is used to screen out zero entries in the autoregressive (AR) coefficient matrices. Specifically, if two marginal time series are conditionally uncorrelated at all frequencies, then all corresponding AR coefficients are set to zero, i.e.,

$$PSC_{ij}(\lambda) = 0,\ \lambda \in [0, \pi] \Longrightarrow \Phi_k(i, j) = \Phi_k(j, i) = 0,\ k = 1, \ldots, p. \tag{11.51}$$

We note that this implication used in the algorithm may not hold in general (Problem 11.12). In practice, due to sampling variability, estimated PSCs are rarely exactly zero even if the two series were conditionally uncorrelated. This issue is addressed [Davis et al., 2016] by ranking the supremum over all frequencies of the squared modulus of the PSC values

$$\mathcal{S}_{ij} := \sup |PSC_{ij}(\lambda)|^2, \tag{11.52}$$

from largest to smallest and searching for a gap/threshold that separates non-zero PSCs.

A larger value of $\mathcal{S}_{ij}$ suggests stronger conditional dependence between the ith and jth series. The ranking of these summary statistics forms the basis for selecting candidate AR coefficients. The output of Stage 1 is a constrained VAR model with $(d + 2M^*)p^*$ non-zero AR coefficients. If the proportion of selected pairs M^* is small, then the number of non-zero parameters is much smaller than that for the fully-parameterized VAR(p^*) which is d^2p^*, for more on (p^*, M^*); see (11.54). In Step 5 of the Algorithm 4, the parameter estimation under the zero constraints is performed using constrained log-likelihood estimation as described in Lütkepohl [2007, Chapter 5] .

Stage 2: Refinement:

While Stage 1 is effective at eliminating many irrelevant AR coefficients, it may also introduce spurious non-zero estimates due to noise or indirect effects. In particular, PSC can only assess relationships between pairs of series and does not account for diagonal terms of the coefficient matrices or grouped coefficients associated with the lag structure. To refine the selection, Stage 2 evaluates the statistical significance of each nonzero AR coefficient obtained from Stage 1. For each coefficient $\Phi_k(i, j)$, a t-statistic is computed as:

$$t_{i,j,k} := \frac{\Phi_k(i, j)}{\mathrm{se}(\Phi_k(i, j))}, \tag{11.53}$$

where $se(\Phi_k(i, j))$ is the standard error derived from the asymptotic distribution of the constrained maximum likelihood estimator [Lütkepohl, 2007]. These statistics are ranked in absolute value, and a final model is selected by minimizing the BIC over

Algorithm 4 sVAR algorithm

1: **Input** $X_1, \ldots, X_n \in \mathbb{R}^d$

Stage 1

2: Invert estimated spectral density matrix and compute the PSC for all $d(d-1)/2$ pairs of distinct marginal series.

3: Construct a sequence Q_1 by ranking summary statistics S_{ij}'s in (11.52) from highest to lowest.

4: For each $(p, M) \in \mathbb{P} \times \mathbb{M}$, set the order of autoregression to p and feed the top M pairs in the sequence Q_1 to the VAR model. Estimate parameters under this constraint and compute the corresponding $BIC(p, M)$:

$$BIC(p, M) = -2 \log L(\Phi_1, \ldots, \Phi_p) + \log n \cdot (d + 2M)p. \tag{11.54}$$

5: Choose the number of lags and non-zero pairs (p^*, M^*) that gives the minimum BIC value over $\mathbb{P} \times \mathbb{M}$.

Stage 2

6: For each of the non-zero AR coefficient compute the t-statistic in (11.53).

7: Construct the sequence Q_2 of the $(d+2M^*)p^*$ triplets (i, j, k) by ranking $|t_{i,j,k}|$ from highest to lowest.

8: For each $m \in \{0, 1, \ldots, (d + 2M^*)p^*\}$ select the m non-zero AR coefficients corresponding to the top triplets in the sequence Q_2 and compute

$$\text{BIC}(m) = -2 \log L + m \log n.$$

Choose m^* that gives the minimum BIC value.

9: **Output** $\Phi_1, \ldots, \Phi_{p^*}$

nested models of increasing size. The result is a sparse VAR model with m^* nonzero coefficients, denoted by sVAR(p^*, m^*).

Additional modifications proposed in Dallakyan et al. [2022] improve the computational time of the Algorithm 4. Instead of relying on PSC values and extensive grid search, the proposed msVAR begins by estimating the inverse spectral density matrix using the TS-GLasso algorithm introduced in Section 11.4. This approach identifies conditional independencies between time series components in the frequency domain by fitting a sparse graphical model to the inverse spectrum. The TS-GLasso avoids inversion of high-dimensional spectral density matrices, which can be computationally demanding for larger d. In particular, Steps 1–3 of Algorithm 4 are substituted by the TS-GLasso algorithm, with the ensuing M^* non-zero elements, using BIC only once to choose the order p^* compared to sVAR. Additionally, in the second stage, rather than using t-statistics of the AR coefficients, msVAR employs the FDR [Benjamini and Hochberg, 1995] procedure for further refinement. The advantage of FDR adaptation is twofold: First, instead of pairwise t-statistic, it implements multiple hypothesis testing. Second, it eliminates the need for Step 8 in Stage two of the sVAR proposal.

Figure 11.8 presents sVAR and msVAR estimates for a VAR(1) coefficient matrix with $d = 20$ and the true coefficient matrix as in the left plot. The generated VAR model satisfies condition (11.51). For details, see Section 11.5.4. It can be seen that sVAR provides a sparser result and on average is 10 times slower than msVAR.

We compare the performance of this two-stage algorithm with the TS-GLasso algorithm concerned with sparse estimation of the inverse spectral density matrix introduced next.

11.4 Time Series GLasso (TS-GLasso)

Displaying the interrelationships among multiple time series via graphs or networks is a topic of great importance in neuroscience and economics under the heading of *connectivity*. There are two distinct approaches in the time- and spectral-domains based on the sparsity of VAR models and the inverse spectral density matrices, respectively. They both aim to reduce the impact of temporal dependence while searching for sparse networks.

In time series graphical model the focus is on understanding marginal and conditional relationships between the entire (all leads and lags) components of the multivariate process. More precisely, for a d-dimensional (stationary) process $\boldsymbol{X}_t$, $t = 1, \ldots, n$, let $G = (V, E)$ be a graph with nodes $V = \{1, \ldots, d\}$ where each node $v \in V$ corresponds to the whole times series $\{X_{v,t}\}$, instead of a single random variable as in GGM. Absence of an edge between two nodes (two entire time series) is characterized by the conditional independence of the marginal series $\{X_{t,i}\}$ and $\{X_{t,j}\}$, given the rest $\boldsymbol{X}_{t,-ij}$. In other words, $\{X_{t,i}\} \perp \{X_{t,j}\}$ conditionally, if and only if $\theta_{ij}(\lambda) = 0$, or $\text{PSC}_{ij}(\lambda) = 0$ for all $\lambda \in [-\pi, \pi]$. As in GGM, it is instructive to view this equivalence in terms of zeros of the inverse spectral density matrix $\Theta(\lambda) = f(\lambda)^{-1} = (f^{ij}(\lambda))$.

Remark 7 *In the spectral domain, the discrete Fourier transform (DFT) effectively sends a multivariate stationary process to a sequence of uncorrelated random vectors at Fourier frequencies. It serves as a two-way bridge*

$$\boldsymbol{\Sigma}^{-1} \longleftrightarrow f(\lambda)^{-1}, \tag{11.55}$$

to link and emphasize the nearly parallel roles of these matrices in the construction of sparse networks for multivariate independent data and temporally dependent multivariate stationary time series data. However, there are notable differences demanding more conceptual attention and computational efforts. The most notable differences in the time series setup are:

1. *instead of sparse estimation of a single precision matrix the problem becomes that of sparse estimation of a group of precision matrices,*

$$f(\lambda_j)^{-1},\ j = 1, \ldots, M,$$

at the selected Fourier frequencies. This essentially amounts to solving the so-called joint graphical Lasso problem discussed in Guo et al. [2011] and Pourahmadi [2013, Section 5.7].

2. *the entries of the spectral density matrix $f(\lambda)$ are complex-valued scalar functions and due attention must be paid to optimization of the ensuing objective functions.*

These differences are discussed in detail in the next section.

11.4.1 Stationary TS-GLasso

The genesis of GGM for multivariate time series can be traced to Brillinger [1996] and Dahlhaus [2000] where the idea of hypothesis testing based on the partial spectral coherence and its asymptotic distribution was employed for edge selection. However, when the number of components in the multivariate time series is large or when the components are highly correlated, the estimates of the PSC are unstable [Fiecas et al., 2010] and directly inverting the spectral density matrix is not advisable.

Motivated by the phenomenal success of GLasso for i.i.d. multivariate data, a similar penalized log-likelihood framework has been developed to learn a multivariate time series graph [Jung et al., 2015, Tugnait, 2021, Dallakyan et al., 2022]. Recall from Chapter 9 that given the sample data, a regularized Gaussian graphical model estimation amounts to solving the following optimization problem:

$$\min_{\Theta} - \log\det(\Theta) + \mathrm{tr}(\mathbf{S}\Theta) + P(\Theta, \lambda), \tag{11.56}$$

where $\det(\cdot)$ is determinant of the matrix, $\mathbf{S}$ is the sample covariance matrix, and λ is a tuning parameter in $P(\Theta, \lambda) = \lambda\|\Theta\|_1 = \lambda \sum_{ij} |\theta_{ij}|$. For stationary processes, a key role is played by the fact that under some regularity conditions, the discrete Fourier transforms at neighboring Fourier frequencies are asymptotically independent and nearly identically distributed multivariate complex-valued normal. This asymptotic independence is used to develop an approximate likelihood function and hence a time series extension of the GLasso algorithm. More precisely, one replaces the likelihood function of the Gaussian time series data $\boldsymbol{X}_1, \ldots, \boldsymbol{X}_n$ by its Whittle approximation involving the DFT's $\boldsymbol{y}_k = d(\lambda_k)$ and $f_k = f(\lambda_k)$ at the Fourier frequencies:

$$\mathrm{Whittle}(f_0, \ldots, f_{n-1}) \approx \prod_{k=0}^{n-1} |f_k^{-1}| \exp(-\boldsymbol{y}_k^* f_k^{-1} \boldsymbol{y}_k). \tag{11.57}$$

Considering the product around a desired frequency λ_j leads to the following penalized Whittle log-likelihood:

$$\min_{\Theta_j} - \log\det(\Theta_j) + \mathrm{tr}(\hat{f}_j \Theta_j) + P(\Theta_j, \lambda), \tag{11.58}$$

where $\Theta_j = \Theta(\lambda_j) = f^{-1}(\lambda_j)$ and $\hat{f}_j$ is the smoothed spectral density estimate at the Fourier frequency λ_j as in (11.37) averaged over $2m + 1$ consecutive frequencies. It is instructive to note that such an $\hat{f}_j$ is effectively replacing the sample covariance matrix $\mathbf{S}$ in the GLasso objective function (11.56). In practice, it is reasonable to expect that spectral precision matrices at neighboring Fourier frequencies exhibit similar sparsity structures. This assumption is incorporated in the TS-GLasso algorithm in Dallakyan et al. [2022] via the following joint graphical model formulation:

$$\min_{\Theta_j, 1 \leq j \leq M} \sum_{j=1}^{M} \Big[-\log\det(\Theta_j) + \mathrm{tr}(\hat{f}_j \Theta_j) \Big] + P(\Theta[\cdot], \lambda), \tag{11.59}$$

where the penalty

$$P(\Theta[\cdot], \lambda) = \lambda \sum_{i \neq l} \sqrt{\sum_{j=1}^{M} |\theta_{il}(\lambda_j)|^2}$$

encourages a shared sparsity structure across the M Fourier frequencies λ_j, $1 \leq j \leq M$.

Compared to the standard GLasso algorithm, the main challenge in solving (11.59) lies in the fact that the entries of the spectral precision matrix Θ are complex-valued. This issue is dealt with in Dallakyan et al. [2022] by resorting to Wirtinger calculus [Wirtinger, 1927] coupled with the notion of Wirtinger subgradients [Bouboulis et al., 2012] and the alternating direction method of multipliers (ADMM) [Boyd et al., 2011] to solve the ensuing optimization problem. Fortunately, the more recent complex graphical Lasso (CGLasso) algorithm in Deb et al. [2024] addresses this challenge more effectively by leveraging the well-known field isomorphism that maps complex numbers to 2×2 matrices:

$$\phi : C \to R^{2\times 2}, \; z = a + bi \to \begin{bmatrix} a & -b \\ b & a \end{bmatrix}.$$

This mapping allows the optimization in (11.59) to be implemented in field of real numbers, reducing it to a familiar GLasso algorithm from Chapter 9. The key modification is that the analogous Lasso regression for each column becomes a group Lasso with orthogonal predictors within each group, for which a closed-form solution exists [Hastie et al., 2015]. Note that even though the map $\phi(\cdot)$ doubles the dimensionality of the input, the computational cost remain comparable to directly handling complex values objects.

Analogous to CLIME and node-wise regression algorithms for multivariate data, given a smooth estimator of the density matrix at a Fourier frequency λ_j, a sparse spectral precision matrix can be obtained using the *sparse inverse periodogram estimator* (SIPE) algorithm in Fiecas et al. [2019] which employs the idea of CLIME in Cai and Liu [2011], or the regression approach adopted in Krampe and Paparoditis [2025] using the node-wise regression in Meinshausen and Bühlmann [2006].

As for analysis of nonstationary data, a large body of research in neuroscience points to dynamic connectivity in human brain where the interactions among brain regions evolve over time. While some recent work in statistics use the change point idea within VAR models to capture these patterns, a systematic nonparametric framework that allows smooth evolution of connectivity has remained elusive. In Basu and Subba Rao [2023] a bona fide graphical modeling framework for nonstationary time series is proposed which captures the time-varying dynamics. They further show that the conditional relationships (independence and nonstationarity) are encoded in the Fourier transform of an infinite-dimensional covariance operator, which can be learned from data using node-wise regression of discrete Fourier transforms.

For an illustration of TS-GLasso and comparing its performance with sVAR and msVAR; see Section 11.5.4.

11.5 Hands-on R: Time Series Analysis

In this section we illustrate a few key time series tools and concepts using R, simulated and real datasets. Time series data analysis usually starts with examining various scatterplots of the data and searching for linear patterns as in regression analysis. For example, a linear pattern in the scatterplot of X_t vs X_{t-1} supports a model like (11.1).

11.5.1 Time Series Plot and Lagplots

Here, we use the astsa package from Shumway and Stoffer [2025], to present a *time series plot* of 100 observations, namely the plot of X_t versus t, and the lagplot, namely plots of X_t versus X_{t-h}, $h = 1, 2, \dots, H$, for a realization of MA(1) process

$$X_t = Z_t + 0.5Z_{t-1}$$

```
require(astsa)
set.seed(1234)
## Generate time series
x = sarima.sim(ma = 0.5, n = 100)

par(mfrow = c(1,3))
## Plot time series
tsplot(x, ylab =  expression(X[t]))
abline(h= 0)
## Plot X[t] against X[t-1]
tsplot(stats::lag(x, -1),x , ylab =  expression(X[t]),
       xlab = expression(X[t-1]),
       type = "p", xy.labels = FALSE,)
abline(a = 0, b = 0.4)
## Plot X[t] against X[t-2]
tsplot(stats::lag(x, -2),x , ylab =  expression(X[t]),
       xlab = expression(X[t-2]),
       type = "p", xy.labels = FALSE)
abline(h = 0)
par(mfrow = c(1,1))
```

The time series plot of generated data in the left of Figure 11.5 indicates possible positive correlation between the adjacent observations. This point is emphasized in

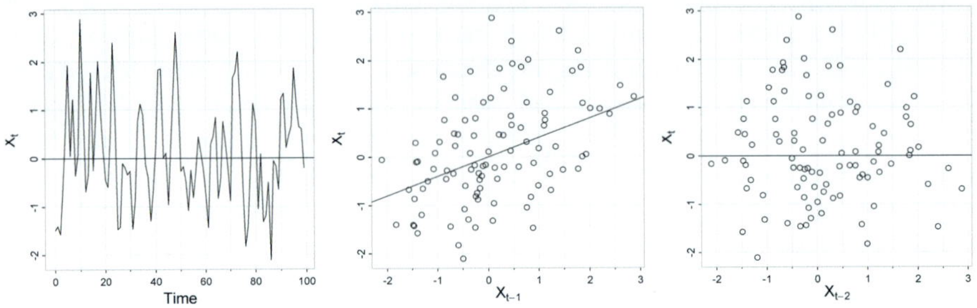

(a) MA(1) process (b) X_t vs X_{t-1} (c) X_t vs X_{t-2}

FIGURE 11.5
(a) Time series plot of data from an MA(1) process, (b) plot of X_t against X_{t-1}, (c) plot of X_t against X_{t-2}.

the middle plot in Figure 11.5 where we plot X_t against X_{t-1}. The slope of the line in the figure represents the correlation, which is 0.4. However, since X_t is independent from X_{t-2}, in the right of Figure 11.5, the true correlation coefficient between X_t and X_{t-h} is zero for any $h \geq 2$.

The spectral density of the time series can be obtained by using arma.spec() function.

```
arma.spec(ma = 0.5)
```

11.5.2 Univariate Time Series: Periodogram

To get a better feel for the role of periodograms in detecting periodicity, we continue utilizing the astsa package to generate stationary time series data and plot its periodogram.

Consider a time series of length $n = 100$, generated as the sum of two periodic time series

$$X_t = X_{t1} + X_{t2},$$

where

$$X_{t1} = 2\cos(2\pi 1/10) + 4\sin(2\pi 1/10),$$
$$X_{t2} = 3\cos(2\pi 1/5) + 6\sin(2\pi 1/5).$$

We can plot and estimate the spectral density using mvspec() function.

```
## Generate time series
X_t1 = 2*cos(2*pi*1:100*1/10) + 4*sin(2*pi*1:100*1/10)
X_t2 = 3*cos(2*pi*1:100*1/5) + 6*sin(2*pi*1:100*1/5)

X_t = X_t1 + X_t2
```

```
## Plot time series and periodogram
par(mfrow = c(2,2))
tsplot(X_t1, ylab = expression(X[t1]))
tsplot(X_t2, ylab = expression(X[t2]))
tsplot(X_t, ylab = expression(X[t]))
## Estimate and plot periodogram
mvspec(X_t, main = "")
par(mfrow = c(1,1))
```

From the bottom right plot in Figure 11.6 it can be seen that the periodogram is nonzero and peaked at the expected frequency points 0.1 and 0.2.

11.5.3 VAR and Granger Causality

We use the vars package to illustrate and implement VAR and Granger causality in R for airpollute data available in recode package. The series were recorded hourly during the year 2006 at Azusa, California. The $d = 5$ variables in the data include CO and NO (pollutants mainly emitted from the cars and industry), NO 2 and O3 (generated from different reactions in the atmosphere), and the global solar radiation intensity R. Similar datasets were analyzed in Dahlhaus and Eichler [2003].

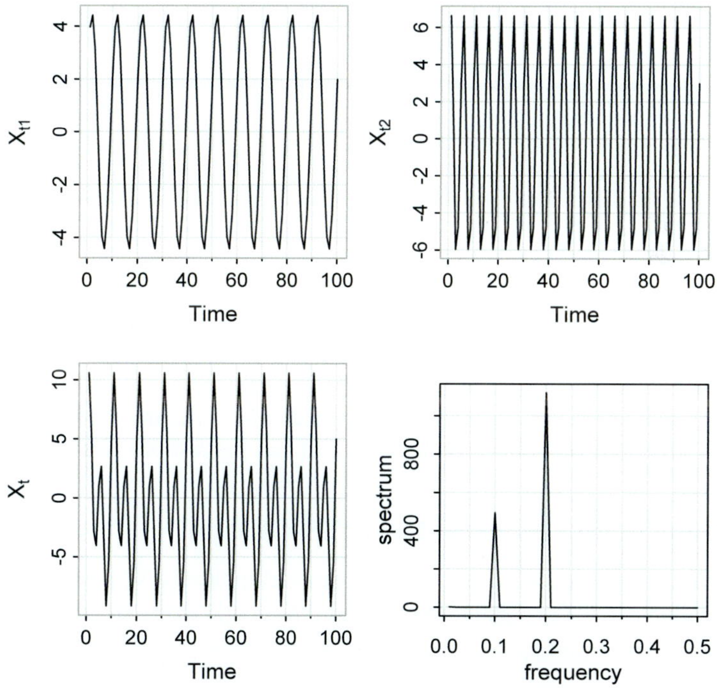

FIGURE 11.6
Plots of two periodic components, their sum and the periodogram.

Figure 11.7 displays the average of daily measurements for the above five pollutants.

```
## VAR and Granger Causality
require(vars)
require(recode)
data(airpollute)
air.var = VAR(y = airpollute, lag.max = 10, ic = "SC" )
A = coef(air.var)
```

We have specified the maximum lag as 10 for VAR models and use the Schwarz criterion (SC) or the BIC criterion to select the order p which turns out to be 5. To summarize the results, we use the summary() function.

```
summary(air.var)
```

To save the estimated coefficient matrix in variable A, type

```
A = coef(air.var)
```

Next, to test whether the variable CO Granger causes the rest of variables we use the causality() function.

```
causality(air.var, cause = "CO")
$Granger

  Granger causality H0: CO do not Granger-cause NO NO2 OZONE R

data:  VAR object air.var
F-Test = 11.179, df1 = 20, df2 = 43645, p-value < 2.2e-16


$Instant

  H0: No instantaneous causality between: CO and NO NO2 OZONE R

data:  VAR object air.var
Chi-squared = 3450, df = 4, p-value < 2.2e-16
```

The small p-value in the output indicates that we should reject the null hypothesis and conclude that CO does Granger-cause the rest of the variables. The command also tests for the instantaneous causality by testing for nonzero correlation between the error processes of the cause and effect variables. Here, the small p-value indicates that there is instantaneous causality between CO and the rest of variables.

11.5.4 Time Series Graphical Lasso (TS-GLasso)

In this section, we illustrate the use of time series graphical lasso using the tsglasso package, available in https://github.com/adallak/tsgl. To install the package, type

```
devtools::install_github("adallak/tsgl")
```

For illustration purposes, we simulate multivariate time series data of dimension $d = 20$ from a VAR(1) process with block diagonal inverse spectral density matrix by using the simulateVAR() function.

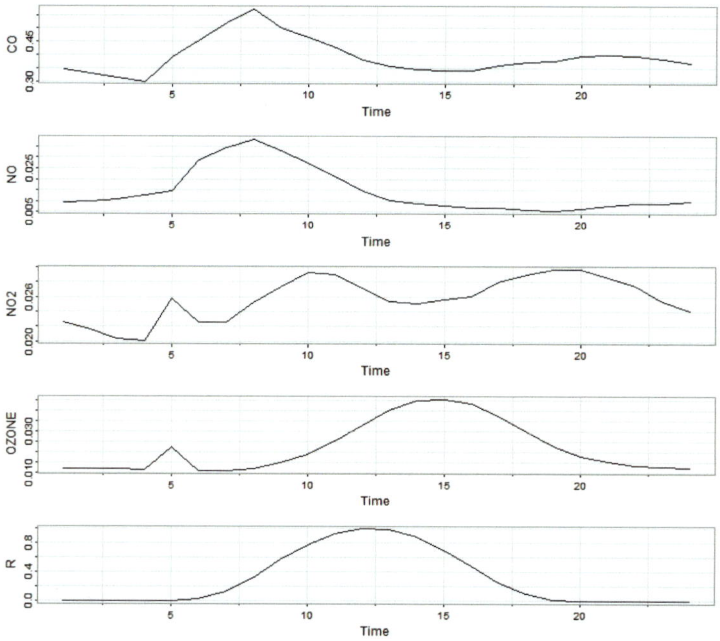

FIGURE 11.7
Average of daily measurements for the five pollutants.

```
require(tsglasso)
set.seed(12)
K = 20    ## Data dimension
N = 1026   ## Number of observations
block = 4 ## Number of blocks
p = 1     ## Number of lags
burn = 300
Sigma = diag(K)
## Generate block diagonal coefficient matrix
A = gendiagAR(K = K, p = p, block = block)
## Simulate time series
dta = simulateVAR(coefMatr=A, intercept =rep(0,K),
                  size=N, burn=burn,
                  Sigma=Sigma, error.dist="normal")$simData
```

Then we use the tsglasso() function to implement TS-GLasso. The threshold is selected such that the number of estimated edges is equal to the true number of edges.

```
halfWindowLength = 6
ispd = tsglasso(dta, lambda = 1, method = "strict",
    halfWindowLength = halfWindowLength)$X
ispd_post = postprocess(ispd, threshold = 0.017)
est_adj=      adjacency_from_tsgl(ispd_post$X, diag= FALSE)
```

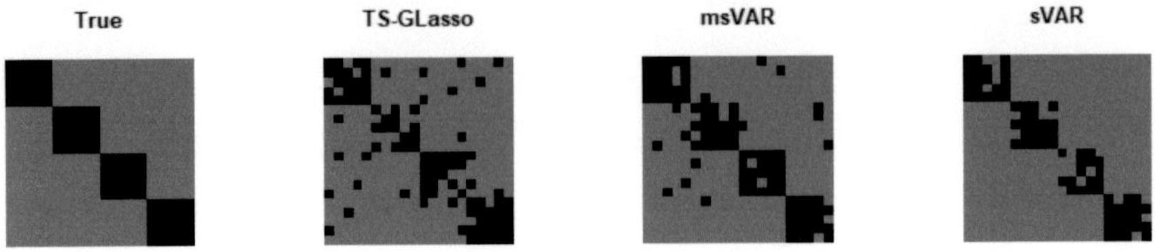

FIGURE 11.8
Example of tsglasso algorithm, msVAR and sVAR.

These three methods perform differently, but at least visually the msVAR seems to do better than the TS-GLasso and sVAR in revealing the true pattern.

11.6 Advanced Topics: Time Series PCA and Factor Models

Modeling multivariate time series data is a nontrivial task, especially in high dimensions, due to the presence of cross-sectional and temporal correlations and hence a large number of parameters. Dimension-reduction and regularization methods discussed in earlier chapters provide means of reducing the number of parameters. This section is concerned with the conceptual and mathematical underpinnings of developing genuine analogs of time series PCA and factor models for multivariate stationary time series. We highlight some of the challenges in handling simultaneously the temporal and cross-sectional dependencies. For deeper understanding the readers are encouraged to consult the cited original references and then dive into the growing literature of dimension-reduction in high-dimensional time series.

11.6.1 Time Series PCA

In this section, the focus is on dimension-reduction in the spirit of classical PCA and the goal is to transform a multivariate stationary time series linearly and contemporaneously into another multivariate series of the same dimension composed of a number of decorrelated subseries.

More precisely, given a d-dimensional zero-mean stationary time series $\{\boldsymbol{Y}_t\}$, the aim is to find a matrix $\mathbf{A}$ which transforms the series linearly into q serially and contemporaneously uncorrelated subseries of $\{\boldsymbol{X}_t\}$:

$$\boldsymbol{Y}_t = \mathbf{A}\boldsymbol{X}_t, \quad \boldsymbol{X}_t = \begin{pmatrix} \boldsymbol{X}_t^{(1)} \\ \boldsymbol{X}_t^{(2)} \\ \cdots \\ \boldsymbol{X}_t^{(q)} \end{pmatrix}, \quad \mathrm{Cov}(\boldsymbol{X}_t^{(i)}, \boldsymbol{X}_s^{(j)}) = 0, \quad \text{for } i \neq j,\ s, t \in \mathbb{Z}. \tag{11.60}$$

Thus, the autocovariance matrices of the transformed series $\boldsymbol{X}_t$ have a block diagonal structure with q blocks of length d_j for the subseries $\boldsymbol{X}_t^{(j)}$ and $\sum_{j=1}^{q} d_j = d$. The key question is existence of a matrix $\mathbf{A}$ which does not depend on t. A related, more general and desirable solution from the orthogonalization perspective in the frequency domain using linear filters is given in Example 34.

Our presentation here follows closely that of Chang et al. [2018]. From data analysis and dimension-reduction perspectives, we note that existence of a time-invariant matrix $\mathbf{A}$ is tantamount to reducing the more difficult task of modeling a high-dimensional time series to that of modeling a number of uncorrelated subseries in lower dimensions. In fact, when such a transformation exists, and once the subseries are identified and modeled separately, the results can be transformed back to the original series. However, there is the possibility of non-existence of such a matrix $\mathbf{A}$ in which case one should be content with approximate or weakly decorrelated subseries.

11.6.2 The Methodology

Like the standard PCA for i.i.d. multivariate data, one expects the existence and search for a matrix $\mathbf{A}$ to boil down to computing the eigenvalues and normalized eigenvectors of a certain positive-definite matrix involving possibly the autocovariance matrices of a zero-mean multivariate stationary process. A novelty of the method presented here is its use of the autocovariance matrices at lags beyond zero.

Without loss of generality, one may assume that

$$\mathrm{Var}(\boldsymbol{Y}_t) = \mathrm{Var}(\boldsymbol{X}_t) = I_d. \tag{11.61}$$

Then, from linearity of the transformation in (11.60) and computing

$$\mathrm{Cov}(\boldsymbol{Y}_t) = \mathrm{Cov}(\mathbf{A}\boldsymbol{X}_t) = \mathbf{A}\mathrm{Cov}(\boldsymbol{X}_t)\mathbf{A}',$$

it follows that $\mathbf{AA}' = I_p$, so that $\mathbf{A}$ must be an orthogonal matrix and $\boldsymbol{X}_t = \mathbf{A}'\boldsymbol{Y}_t$. Corresponding to the segmentation of $\boldsymbol{X}_t$, we may partition the columns of the matrix $\mathbf{A}$ as $(\mathbf{A}_1, ..., \mathbf{A}_q)$ and note that

$$\boldsymbol{X}_t^{(j)} = \mathbf{A}_j'\boldsymbol{Y}_t, \quad j = 1, \ldots, q. \tag{11.62}$$

For $j = 1, \ldots, q$, let H_j be a $d_j \times d_j$ orthogonal matrix and define the block-diagonal matrix H = block diag$(H_1, \ldots, H_q)$. Note that both $(\mathbf{A}, \boldsymbol{X}_t)$ and $(\mathbf{A}H, H'\boldsymbol{X}_t)$ satisfy (11.60), so that $\mathbf{A}$ and $\boldsymbol{X}_t$ are not uniquely identified, see also Section 7.2.2, but the column spaces of $\mathbf{A}$ and $(\mathbf{A}_1, ..., \mathbf{A}_q)$ namely $C(\mathbf{A}_1), ..., C(\mathbf{A}_q)$ can be uniquely identified.

It turns out that the search for $\mathbf{A}$ involves the autocovariance functions of the two time series, namely $\Gamma_Y(k) = \mathrm{Cov}(\boldsymbol{Y}_{t-k}, \boldsymbol{Y}_t)$ and $\Gamma_X(k) = \mathrm{Cov}(\boldsymbol{X}_{t-k}, \boldsymbol{X}_t)$, through the eigenanalysis of the following two positive-definite matrices:

$$M_Y = \sum_{k=0}^{k_0} \Gamma_Y(k)\Gamma_Y(k)' = I_p + \sum_{k=1}^{k_0} \Gamma_Y(k)\Gamma_Y(k)', \qquad (11.63)$$

$$M_X = \sum_{k=0}^{k_0} \Gamma_X(k)\Gamma_X(k)' = I_p + \sum_{k=1}^{k_0} \Gamma_X(k)\Gamma_X(k)', \qquad (11.64)$$

where k_0 is a preselected integer. Moreover, they satisfy

$$M_Y = \mathbf{A} M_X \mathbf{A}', \qquad (11.65)$$

where due to the segmentation of the series $\boldsymbol{X}_t$ the matrices $\Gamma_X(k)$ and M_X are block-diagonal.

The connection between the eigenvalues and eigenvectors of the matrices M_X and M_Y are explored next. Let D and P_X be the diagonal and orthogonal matrices of eigenvalues and eigenvectors, respectively, of the matrix M_X which satisfy

$$M_X P_X = P_X D. \qquad (11.66)$$

Pre-multiplying this equation by $\mathbf{A}$ and using (11.65), one obtains

$$M_Y \mathbf{A} P_X = \mathbf{A} P_X D, \qquad (11.67)$$

showing that the orthonormal eigenvectors of M_Y and M_X satisfy

$$P_Y = \mathbf{A} P_X \quad \text{and} \quad P_Y' \boldsymbol{Y}_t = P_X' \boldsymbol{X}_t. \qquad (11.68)$$

This already implies that $\mathbf{A} = P_Y P_X'$, giving the matrix $\mathbf{A}$ which we are looking for, up to a rotation of the columns of P_Y. A more precise description of the process is given next.

The matrix M_X being block-diagonal can be written as

$$M_X = \text{block diag}(M_{X_1}, M_{X_2}, ..., M_{X_q}),$$

with M_{X_j} as its j-th $d_j \times d_j$, $1 \le j \le q$, diagonal block. Assuming that M_{X_i} and M_{X_j} have no common eigenvalues for any $i \neq j$, we arrange these eigenvalues in the matrix D according to the order of the q block matrices listed above. The matrix of orthogonal eigenvectors P_X can be shown to be a block-diagonal orthogonal matrix with the same block-diagonal structure as M_X; see Problem 11.14. Since the eigenvalues of M_X are latent/unknown, any P_X from (11.66) is a column permutation of such a block-orthogonal matrix. This in turn implies that each component of $P_X' \boldsymbol{X}_t$ is a linear combination of elements from exactly one of the subseries, so that in view of the second identity in (11.68) the d components of $P_Y' \boldsymbol{Y}_t$ can be partitioned into q groups such that there exist neither contemporaneous nor serial correlations across different groups. Thus, $P_Y' \boldsymbol{Y}_t$ can be seen as the desired permutation of $\boldsymbol{X}_t$, and P_Y as a column-permutation of $\mathbf{A}$. These observations lead to the following two-step estimation for $\mathbf{A}$ and $\boldsymbol{X}_t$:

1. Compute $\hat{M}_Y$, a consistent estimator for M_Y and let $\hat{P}_Y$ be the corresponding orthonormal matrix of eigenvectors.
2. The columns of the required transformation matrix $\hat{\mathbf{A}}$ is a permutation of the columns of $\hat{P}_Y$ such that the series $\hat{\boldsymbol{X}}_t = \hat{\mathbf{A}}' \boldsymbol{Y}_t$ is segmented into q subseries which are serially uncorrelated; see (11.60).

From (11.68) the series $P_Y' \boldsymbol{Y}_t$ is "almost" the needed transformed series. The only remaining task is to rearrange the columns of P_Y so that the transformed series $P_Y' \boldsymbol{Y}_t$ are serially uncorrelated. Indeed, Step 1 provides an estimator of $\mathbf{A}$ up to permuting its columns. In principle, the permutation in Step 2 can be done by visual inspection of the cross-correlation plots which works well when d the number of series is small to moderate. The transformed series obtained right after Step 1, $\hat{\boldsymbol{Z}}_t = \hat{P}_Y' \boldsymbol{Y}_t$, is a starting point where permuting the columns of $\hat{P}_Y$ is equivalent to permuting the components of the series $\hat{\boldsymbol{Z}}_t$. Step 2 is carried out by grouping together components in $\hat{\boldsymbol{Z}}_t$ so that serial correlation may exist within groups but not across groups. This is further elaborated in the next section.

Remark 8 *The matrix $M_Y = GG'$ where the matrix*

$$G = (\Gamma_Y(1), \ldots, \Gamma_Y(k_0)), \tag{11.69}$$

appears in computing the canonical correlations between $\boldsymbol{Y}_t$ and $\boldsymbol{Y}_{(k_0)} = (\boldsymbol{Y}_{t-1}', \ldots, \boldsymbol{Y}_{t-k_0}')'$. In fact, determining the rank of the matrix

$$Cov(\boldsymbol{Y}_{(k_0)}, (\boldsymbol{Y}_{t-1}', \ldots, \boldsymbol{Y}_{t-k_0}')' = G, \tag{11.70}$$

plays a central role in the scalar component analysis of multivariate stationary processes in Tiao and Tsay [1989].

11.6.3 Permutation by Maximum Cross-Correlations

In this section, components of the estimated series $\hat{\boldsymbol{Z}}_t$ are re-ordered so that highly cross-correlated components will be assigned to the same group. For the alternative method of identifying the connected pair of components of $\hat{\boldsymbol{Z}}_t$ by a false discovery rate (FDR) procedure based on the multiple tests for cross-correlations of each pair series, see Chang et al. [2018].

More formally, groups are formed based on hypothesis tests for significant cross-correlations between pairs of components. For any pair (i, j), the following multiple hypothesis test is carried out,

$$H_0 : \rho_{i,j}(h) = 0 \;\text{ for any }\; h = 0, \pm 1, \pm 2, ..., \pm m, \tag{11.71}$$

for a pre-chosen value of m and where $\rho_{i,j}(h)$ is the cross-correlation between the selected components (i, j) at lag h. We say that two components are *connected* if there is a significant cross-correlation between them. The multiple tests are performed for every pair of components of $\hat{Z}_t$. The following grouping or permutation scheme is used which also selects the number of groups q:

(P1.) Start with the d groups where each group contains one component of $\hat{\boldsymbol{Z}}_t$ only.

(P2.) Combine two groups together if one connected pair is split over the two groups.

(P3.) Repeat Step 2 until all connected pairs are within one group.

Given the series $\hat{\boldsymbol{Z}}_t = (\hat{z}_{t,1}, \ldots, \hat{z}_{t,d})'$, for any pair $1 \leq i < j \leq d$, the sample cross-correlation estimate is given by

$$\hat{\rho}_{i,j}(h) = I(h \geq 0)\frac{\sum_{t=1}^{n-h}(\hat{z}_{i,t+h} - \bar{z}_{i,.})(\hat{z}_{j,t} - \bar{z}_{j,.})}{(\sum_{t=1}^{n}(\hat{z}_{i,t} - \bar{z}_{i,.})^2 \sum_{t=1}^{n}(\hat{z}_{j,t} - \bar{z}_{j,.})^2)^{1/2}} + I(h < 0)\frac{\sum_{t=1-h}^{n}(\hat{z}_{i,t+h} - \bar{z}_{i,.})(\hat{z}_{j,t} - \bar{z}_{j,.})}{(\sum_{t=1}^{n}(\hat{z}_{i,t} - \bar{z}_{i,.})^2 \sum_{t=1}^{n}(\hat{z}_{j,t} - \bar{z}_{j,.})^2)^{1/2}}$$

where $\bar{z}_{i,.} = n^{-1}\sum_{t=1}^{n}\hat{z}_{i,t}$. The maximum cross-correlation statistic is then given by

$$\hat{L}_n(i, j) = \max_{|h| \leq m}|\hat{\rho}_{i,j}(h)|, \tag{11.72}$$

and the null hypothesis of significant cross-correlation would be rejected for the (i, j) pair if this statistic is more than a specified threshold.

Instead of multiple testing, a ratio-based statistic involving the maximum cross-correlation statistic is more prudent. Next, for $d_0 = d(d-1)/2$ the d_0 maximum cross-correlation statistics above are arranged in descending order,

$$\hat{L}_1 \geq \hat{L}_2 \geq \ldots \geq \hat{L}_{d_0},$$

and $\hat{r}$, the estimated number of groups or clusters, is given by

$$\hat{r} = \underset{1 \leq j \leq c d_0}{\text{argmin}} \frac{\hat{L}_j}{\hat{L}_{j+1}}. \tag{11.73}$$

This ratio statistic is motivated by observing that if there were actually r connected pairs among the d_0 pairs of the components of $\boldsymbol{X}_t$, by arranging the true maximum cross-correlations in the descending order as above, we expect to have $L_r > 0$ and $L_{r+1} = 0$, where then the ratio L_j/L_{j+1} takes a value like ∞ for $j = r$. This is the intuition behind defining the estimator r in (11.73). For a similar ratio-based idea for selecting the number of factors in factor models; see (7.6).

We reject H_0, for pairs corresponding to $\hat{L}_1, \ldots, \hat{L}_{\hat{r}}$. The $c \in (0, 1)$ is a preselected multiplier. It is essential for guarding against possible encounters with 0 or very small values in one of the $\hat{L}_j$, which may give rise to highly volatile values or

∞ as the ratio. Once the $\hat{r}$ is determined, the pairs corresponding to the first $\hat{r}$ maximum cross-correlations are declared connected or significantly cross-correlated and groups are formed based on these pairs.

Not surprisingly, the notion of connectedness touched upon here can be framed in a graph language. For this, let a graph $(V, \widehat{E})$ have d vertices $V = \{1, 2, \ldots, d\}$ corresponding to the d components series of $\hat{\boldsymbol{Z}}_t$. We define an edge connecting vertices i and j if H_0 is rejected by the ratio test based on (11.73). Then, for

$$\boldsymbol{Z}_t = P_Y' \boldsymbol{Y}_t = (z_{1,t}, \ldots, z_{d,t})',$$

define

$$E = \{(i, j) : \max_{|h| \leq m} |\mathrm{Corr}(z_{i,t+h}, z_{j,t})| > 0, 1 \leq i < j \leq d\},$$

and view it as the set of edges for the population graph. Under mild conditions, Proposition 2 in Chang et al. [2018] shows that the graph $(V, \widehat{E}_n)$ is a consistent estimator of the population graph (V, E).

11.6.4 Hands-on R: Time Series PCA

In Section 7.2.4, we used the globalbank dataset, available in the complementary R package recode, to illustrate factor analysis and PCA and it was found that the latent factors capture cross-correlations both within and across countries.

In this section, we apply time series PCA, introduced in Section 11.6, to further study the connectedness of banks across different countries. For this, we use the PCA_TS() function from the R package HDTSA. Since the time series PCA procedure returns groups of series, we order the variables by country so that banks from the same country appear consecutively in a group. This action improves interpretability, as it is expected that banks from the same country to belong to the same group.

Figure 11.9 shows the cross-correlograms of the first 4 series for the first 15 lags. The maximum cross-correlations between pairs at each lag are displayed in Figure 11.10. We observe that the maximum cross-correlation tapers off to around 0.4 as the number of lags increases. These plots are produced using the cross_correlogram() and plot_max_ccf_all_pairs() functions from the R package recode.

```
## Import required packages
require(HDTSA)
require(recode)
## Import data
data(globalbank)
## Countries of each bank
countries = c("Great Britain", "Japan", "USA", "Great Britain", "USA",
              "USA", "Japan", "Great Britain", "Japan",
              "USA", "Great Britain", "Italy", "USA", "Italy", "USA",
              "Canada", "Canada", "Australia",
              "Canada", "Australia", "Great Britain", "China", "
   Australia",
```

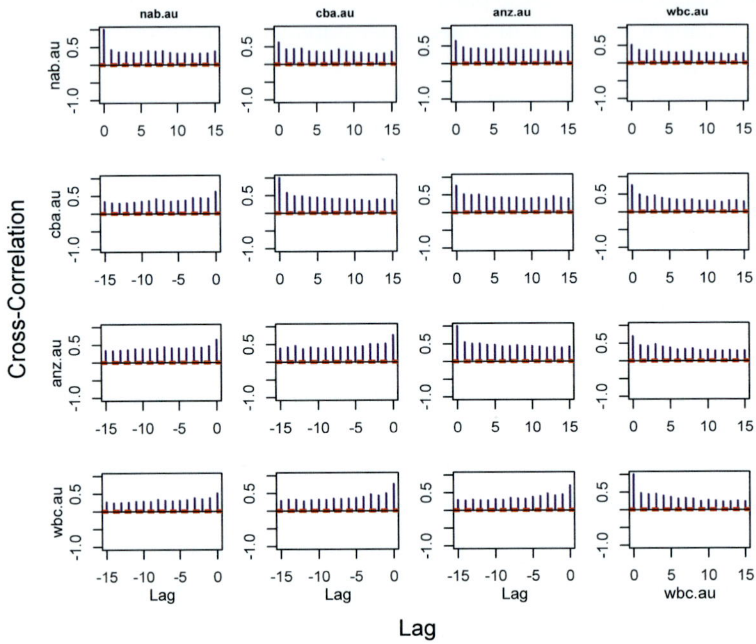

FIGURE 11.9
Cross-correlograms of the first 4 series.

```
              "Australia", "China", "China", "Canada",
              "Japan", "Japan", "Japan", "Canada", "USA", "USA", "USA"
    ,
              "China", "USA", "China", "Italy", "USA", "USA", "Canada"
    ,
              "USA", "Italy", "Australia", "Japan",
              "Japan", "USA", "USA", "Japan", "Italy", "Japan", "Japan
    ",
              "Italy", "Japan")
## Order column names based on the countries
data_mat = as.matrix(globalbank)
colnames(data_mat) = colnames(globalbank)
data_mat = data_mat[,order(countries)]
```

Next, we implement time series PCA using the permutation method of max with the number of lags equal to 2 and 3, respectively.

```
## Use max for permutation
res_max_lag2 <- PCA_TS(data_mat, lag.k =2,permutation = "max")
res_max_lag2$Groups
## Use fdr for permutation with beta = 0.01
```

```
res_max_lag3 <- PCA_TS(data_mat, lag.k =3, permutation = "max")
res_max_lag3$Groups
```

For the number of lags equal 2, Group 1 consists of series $\{1, 2, 3, 4, 5, 6, 8\}$, while the remaining series each form their own group. When the number of lags is equal to 3, Group 1 consists of series $\{1, 2, 3, 4, 5, 6, 7, 8, 9, 15\}$ and the remaining series each form their own group. They yield slightly different results, indicating mild sensitivity to the number of lags. Problem 11.15 asks to estimate the groups with number of lags equal to 5.

11.6.5 Time Series PCA: Spectral Domain

There are two methods of developing PCA in the spectral domain for a zero-mean multivariate stationary time series with the spectral density matrix $f(\lambda)$ and the ordered eigenvalues

$$\omega_1(\lambda) \geq \ldots \geq \omega_d(\lambda),\ \lambda \in (-\pi, \pi], \tag{11.74}$$

and the corresponding normalized eigenvectors $\boldsymbol{e}_i(\lambda)$'s with complex-valued entries. The two methods are equivalent: the first relies on classical PCA for each frequency λ (frequency-by-frequency), and the second relies on linear filters to produce d uncorrelated univariate time series with certain properties.

As for i.i.d. multivariate data in Section 7.1, for a fixed frequency λ one searches for a unit length vector $\boldsymbol{c}(\lambda)$ with complex entries to maximize $\text{Var}(\boldsymbol{c}(\lambda)^* Y_t) = \boldsymbol{c}^* f(\lambda)\boldsymbol{c}$. It follows that the maximum is achieved at $\boldsymbol{c}(\lambda) = \boldsymbol{e}_1(\lambda)$, such that the maximum value for this choice is $\omega_1(\lambda)$. The process $Z_t(\omega) = \boldsymbol{e}_1(\lambda)^* Y_t$, which is

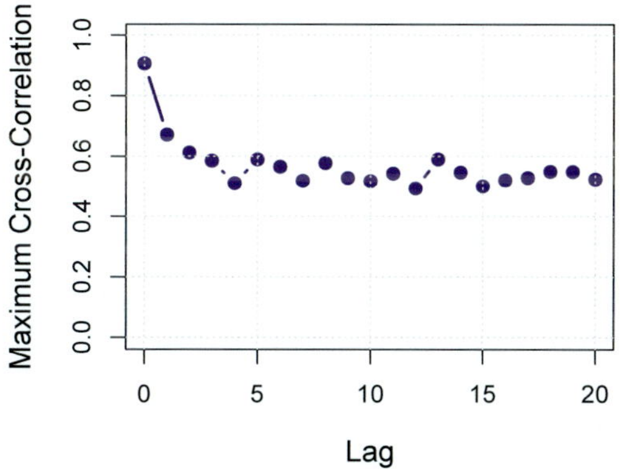

FIGURE 11.10
Maximum cross-correlations between pairs.

stationary with the spectral density function $f_Z(\lambda) = \boldsymbol{c}^* f(\lambda)\boldsymbol{c} = \omega_1(\lambda)$, is called the *first principal component* at frequency λ. Similarly, the kth PC at frequency λ and their spectral density functions are

$$Z_{t,k}(\omega) = \boldsymbol{e}_k(\lambda)^* Y_t), \quad f_k(\lambda) = \omega_k(\lambda), \quad k = 1, \ldots, d.$$

As for the second method, consider the problem of finding a linear filter, that is finding $1 \times d$ matrices Ψ_k's so that $\sum_{k=-\infty}^{\infty} \Psi_k \boldsymbol{X}_{t-k}$ is close to $\boldsymbol{Y}_t$ in the sense that it minimizes the trace of covariance matrix of the approximation error series:

$$E(\boldsymbol{Y}_t - \sum_{k=-\infty}^{\infty} \Psi_k \boldsymbol{X}_{t-k})(\boldsymbol{Y}_t - \sum_{k=-\infty}^{\infty} \Psi_k \boldsymbol{X}_{t-k})'. \tag{11.75}$$

This problem and its solution are the time series analogs of the multivariate regression problem, for its solution see Example 34 and Theorem 14. For more detailed motivation and applications see Brillinger [1981].

11.7 Time Series Factor Models

As in i.i.d. multivariate data, factor models for time series are concerned with using a low-dimensional latent process to describe the dynamics of high-dimensional observations. They start with writing a multivariate time series as a sum of two orthogonal components usually referred to as *common and idiosyncratic components*. However, handling simultaneously the cross-sectional and temporal dependencies in the time series data offers unprecedented challenges and opportunities for presenting variety of novel models and ideas. There is a vast and rapidly growing literature of time series factor models which can be divided into two broad categories of *static* and *dynamic* models. The features of time series factor models so far as departures from the standard factor models in terms of the structures of the common, idiosyncratic components and high-dimensionality are presented in this section,

11.7.1 Dynamic Factor Models

In finance and economics one usually deals with the analysis of the so-called panel data collected from a doubly indexed stationary stochastic process $\{X_{it}; i \in \mathbb{N}, t \in \mathbb{Z}\}$. Since the underlying panel data has infinite cross-section, it is common to consider the situation where data is generated from the finite cross-section $\{X_{it}; i \in \{1, \ldots, d\}, t \in \mathbb{Z}\}$ or the d-variate time series $\{X_t^{(d)}, t \in \mathbb{Z}\}$. This naturally presents the opportunity for asymptotic analysis when the cross-sectional dimension $d \to \infty$, so that one must deal with high-dimensional time series under some regularity assumptions on the two components of the factor model and temporal dependence.

We consider a decomposition of the form

$$X_{it} = \chi_{it} + \epsilon_{it}, \ \{\chi_{it}\} \perp \{\epsilon_{is}\}, \ s, t \in \mathbb{Z} \tag{11.76}$$

where χ_{it} is the common components, and ϵ_{it} the idiosyncratic components. Compared to the factor model in Chapter 7, here both the common and idiosyncratic components could be temporally correlated. This presents a unique opportunity to consider dynamic models for the common component process leading to the novel idea of *dynamic factor models* for time series in Geweke [1977] and Sargent and Sims [1977] as

$$\chi_{it} = L_i(B)F_t, \quad i = 1, \ldots, d, \ t = 1, \ldots, n$$

where F_t is the r-vector of latent factors and the operator $L_i(B)$ the ith row of the $d \times r$ matrix $L(B)$, with B as the backshift operator. This formulation allows the r factors in F_t to load differently (have different coefficients) for different time series at different time points. For example, a common choice of the dynamic loading matrix $L(B)$ corresponds to VAR models of the form

$$L(B)F_t = (I - \Phi_1 B - \Phi_2 B^2 - \cdots - \Phi_p B^p)F_t = \boldsymbol{u}_t, \tag{11.77}$$

where Φ_j's are $r \times r$ matrices and $\boldsymbol{u}_t$ is a vector white noise series referred to as the *common shocks*. The dependence structure of $\{\epsilon_{it}\}$ in (11.76) is the same as the cross-sectional orthogonality of the idiosyncratic components in the i.i.d. factor models in Chapter 7. However, this cannot be expected to hold in most high-dimensional econometric datasets. Indeed, Chamberlain [1983] and Chamberlain and Rothschild [1983] recognized that the idiosyncratic component could be (serially and cross-sectionally) correlated with sparse covariance matrix instead of being diagonal. Their class of static *approximate factor models* kept the common components χ_{it} intact as in i.i.d. factor models, but the idiosyncratic covariance non-diagonal and general. A consequence of this bold move and generality of the idiosyncratic process is that the model was no longer identifiable for a fixed d. Another novelty of Chamberlain and Rothschild [1983] was the use of high-dimensional asymptotics, long before it became popular in the first decade of the 2000s in the context of i.i.d. high-dimensional data where both d and n are allowed to go to infinity. Notably, $d \longrightarrow \infty$ plays a central role for the asymptotic identification of the summands, while $n \longrightarrow \infty$ allows for consistent estimation of the model parameters.

In practice, analysis of approximate factor models along with estimation of the model parameters have been developed mainly in the time-domain applying the PCA method of estimation to the sample (instantaneous, lag-0) covariance matrix of data, as covered extensively in Chapter 7. Understandably, these methods seem fairly straightforward and quite popular in the econometrics literature as compared to the alternative and more sophisticated *general dynamic factor model* (GDFM) methodologies rooted in the spectral-domain presented briefly in the next section.

11.7.2 General Dynamic Factor Models (GDFM)

For the next novel model, it seems plausible to combine the dynamic models for the common components χ_{it} and a sparse idiosyncratic covariance matrix of approximate factor models while allowing $d \to \infty$. This is precisely the idea of GDFM proposed in Forni et al. [2000] and Forni and Lippi [2001] and initially studied entirely in the spectral-domain. The class of GDFMs is broad enough to subsume all the earlier factor models. Extensive subsequent theoretical and empirical research over the last two decades, as reviewed and summarized in Barigozzi and Hallin [2026], all seem to point to the conceptual advantages of the GDFM framework over the widely used and popular static approximate factor models in Chamberlain and Rothschild [1983].

As one might expect there is a sense of duality between the time- and spectral-domain methodologies for factor models. In fact, the key working assumption for GDFM is the spectral-domain analog of the pervasive (strong) factors that a finite number of the leading eigenvalues of the spectral density matrix diverge to infinity on the frequency domain (almost everywhere).

Spectral Pervasive Factors: For $f(\lambda)$ the $d \times d$ spectral density matrix of the time series $\{X_{it}; i \in \{1, \ldots, d\}, t \in \{1, \ldots, n\}$ with the ordered eigenvalues

$$\omega_1^{(d)}(\lambda) \geq \ldots \geq \omega_d^{(d)}(\lambda),\ \lambda \in (-\pi, \pi], \tag{11.78}$$

there exists a finite $q > 0$, not depending on d, such that

$$\lim_{d \longrightarrow \infty} \omega_q^{(d)}(\lambda) = \infty, \quad \text{and} \quad \lim_{d \longrightarrow \infty} \omega_{q+1}^{(d)}(\lambda) < \infty,\ \lambda - \text{a.e. in } (-\pi, \pi]. \tag{11.79}$$

Note that here to prevent confusion, we have used ω to denote an eigenvalue instead of the traditional λ.

Under the spectral pervasiveness, it is known [Forni et al., 2000] that the time series of the common component admits the representation

$$\chi_{it} = \ell_{i1}(B)u_{1t} + \ell_{i2}(B)u_{2t} + \cdots + \ell_{iq}(B)u_{qt} = L_i(B)\boldsymbol{u}_t, \tag{11.80}$$

where $\mathbf{u}_t = (u_{1t} \cdots u_{qt})'$ is a q-variate orthogonal white noise and the filters $\ell_{ij}(B)$, with B the backshift operator, are square-summable. Of course, it is natural to ask: Is there any possible connection between GDFM and the static approximate factor models? It turns out that the answer is affirmative if the dimension r of the linear space $H_t(\chi) = \text{span closure}\{\chi_{it}, i \in \mathbb{N}\}$ is finite for any t. This is the space constructed from finite linear combinations of χ_{it} and their limits in the mean-square sense. In fact, when r is finite with $r \geq q$, then the common component series can be written [Forni et al., 2009] alternatively as

$$\chi_{it} = c_{i1}F_{1t} + \cdots + c_{ir}F_{rt}, \tag{11.81}$$

where $F_t = (F_{1t}, \ldots, F_{rt})' = L_i(B)\mathbf{u}_t$ and c_{ik}'s are scalars. Furthermore, if $L_i(B) =$ ${}_i$ is a constant matrix, then $F_t = \boldsymbol{u}_t$ is a white noise process, and (11.80) reduces to the so-called *static factor model* of Chapter 7. Then, the series $\{F_{jt}\}, j = 1, \ldots, r$, are called the *static factors* as opposed to the dynamic factors.

The estimation problem here is that of recovering the unobserved common and idiosyncratic components χ_{it} and ϵ_{it}, the common shocks $\boldsymbol{u}_t$ and the filters $\ell_{ij}(B)$'s from a finite realization $X_{it}, i = 1, \ldots, d; t = 1, \ldots, n$. While there are methods of estimating these, it is fruitful to note that the models (11.76)–(11.77) together form a state-space model where the factors F_t are the states. For its estimation, Barigozzi and Luciani [2024] rely on the popular Expectation-Maximization (EM) algorithm and the Kalman smoother to estimate factor loadings and factor scores in a quasi-likelihood setup. They show that when both the cross-sectional dimension and the sample size diverge to infinity, then

1. the estimated loadings are $\sqrt{n}$-consistent and asymptotically normal,
2. the estimated factor scores are $\sqrt{d}$-consistent and asymptotically normal, and equivalent to their weighted least-squares estimates,
3. the estimated loadings are asymptotically as efficient as those obtained by the PCA method, and the estimated factor scores are more efficient if the idiosyncratic covariance is sparse enough.

Historically the setup for theoretical analysis of GDFM has been in the frequency domain using the eigenvalues and eigenvectors of the spectral density matrix at each frequency [Forni et al., 2000]. But practitioners seem to prefer time-domain methods and there are some recent studies dealing with the estimation problem in the time-domain [Hallin and Lippi, 2013] which completely avoids the spectral analysis. Its key starting assumption is the existence of the orthogonal decomposition (11.76).

A comprehensive coverage of high-dimensional time series analysis with a focus on the topic of GDFM and its connections with the static factor models can be found in the book by Hallin et al. [2020].

11.7.3 Factor Models and CCA

In this section we present a new class of factor models due to Gao and Tsay [2022] which is a non-singular linear transformation of the combined vector of common factors and the idiosyncratic components. It can be described best within the following framework with slightly different notation than before.

Let $\{\boldsymbol{Y}_t\}$ be a d-variate mean-zero stationary time series admitting a generic factor model of the form

$$\boldsymbol{Y}_t = A\boldsymbol{X}_t + \boldsymbol{\epsilon}_t, \tag{11.82}$$

where $\{\boldsymbol{X}_t\}$ is a serially correlated (dynamic) latent r-vector process, A is a $d \times r$ full-rank factor loading matrix, $\boldsymbol{\epsilon}_t$ is the idiosyncratic error and $\{\boldsymbol{X}_t\}$ and $\{\boldsymbol{\epsilon}_t\}$ are independent. In this section, $\boldsymbol{Y}_t$ and $\boldsymbol{X}_t$ are assumed to be stationary and r is treated as known.

Thus far, we have seen two broad schools of thought for specifying factor models and estimating the loading matrix and the factor scores under different assumptions on the nature and extent of dependence of the common and idiosyncratic processes. The first and most popular procedure relies on the principal component analysis (PCA) of $\Gamma_Y(0) = \boldsymbol{\Sigma}_Y$, the lag-zero or the instantaneous covariance matrix of the data. In other words, it focuses on the cross-sectional dependence at the cost of ignoring the temporal dependence at higher lags, as discussed extensively in Chapter 7. The second procedure focuses on incorporating temporal dependence through eigen-analysis or PCA of positive-definite matrices involving autocovariances $\Gamma_Y(k) = \text{Cov}(\boldsymbol{Y}_t, \boldsymbol{Y}_{t-k})$ for $k \geq 1$, going beyond lag-zero, and can be divided in the following two categories:

(TD) relies on the sum of "squares" of auto-covariance matrices at finite number of lags; see (11.87), Lam et al. [2011] and Lam and Yao [2012]. It assumes that $\{\boldsymbol{X}_t\}$ is serially correlated and $\{\boldsymbol{\epsilon}_t\}$ is a vector white noise process.

(SD) relies on the Fourier transform of the entire autocovariance function and the PCA operation is performed in the frequency domain involving the eigenvalues and eigenvectors of spectral density matrices; see Section 11.7.2.

The new class of factor models in Gao and Tsay [2022] is of the form

$$\boldsymbol{Y}_t = L \begin{pmatrix} F_t \\ \boldsymbol{\epsilon}_t \end{pmatrix} = \begin{pmatrix} L_1 & L_2 \end{pmatrix} \begin{pmatrix} F_t \\ \boldsymbol{\epsilon}_t \end{pmatrix} = L_1 F_t + L_2 \boldsymbol{\epsilon}_t, \tag{11.83}$$

where L is a $d \times d$ invertible matrix, L_1 and L_2 are $d \times r$ and $d \times v$ matrices with $d = r + v$. It is a special case of (11.60) and a non-singular linear transformation of a r-dimensional common factor process F_t, and $\boldsymbol{\epsilon}_t$ a $v = (d - r)$- dimensional idiosyncratic component. The factor process F_t is temporally dependent, $\boldsymbol{\epsilon}_t$ is a white noise series, and the pair $(F_t, \boldsymbol{\epsilon}_t)$ satisfies the basic assumptions of orthogonal factor models (Section 7.2.1), namely,

$$\text{Cov}(F_t) = I_r, \quad \text{Cov}(\boldsymbol{\epsilon}_t) = I_v, \quad \text{Cov}(F_t, \boldsymbol{\epsilon}_t) = \boldsymbol{0}.$$

Moreover, no linear combinations of F_t is serially uncorrelated, because otherwise the dimension of F_t can be reduced.

Decompositions of the form (11.83) arise from canonical correlation analysis of $\boldsymbol{Y}_t$ and its lagged values, as described in Remark 8. This approach belongs to the time-domain (TD) class of methods where factors capture the full dynamics of the data, while cross-sectional dependence is explained by both the factors and the idiosyncratic components. This is quite different from the approximate factor models where the factors capture most of the cross-sectional dependence, while the idiosyncratic terms may contain some non-trivial temporal dependence. So far as extracting the common dynamic components of the data is concerned, the new approach is closer to the dynamic factor models in Section 11.7.2, for more discussions see Gao and Tsay [2022].

We note that $v = 0$ simply means that no linear combination of $\boldsymbol{Y}_t$ is white noise, so that dimension reduction is impossible. Quite the opposite holds when $r = 0$. In particular, for $r > 0$, the dependence or serial correlation in $\boldsymbol{Y}_t$ is solely driven by the latent factor process $\{F_t\}$. In what follows, to exclude trivial cases, we assume that both $r > 0$ and $v > 0$.

Further insights into the model can be gained using the QR or the singular value decomposition of L_1 and L_2 as

$$L_1 = A_1 Q_1, \quad L_2 = A_2 Q_2, \quad \text{with} \quad A_1' A_1 = I_r, \; A_2' A_2 = I_v, \tag{11.84}$$

where A_1 and A_2 are half orthogonal matrices. Then, setting $\boldsymbol{X}_t = Q_1 F_t$ and $e_t = Q_2 \boldsymbol{\epsilon}_t$, the model (11.83) can be rewritten as

$$\boldsymbol{Y}_t = A_1 \boldsymbol{X}_t + A_2 e_t, \tag{11.85}$$

which appears like the traditional factor model (11.82). As expected, A_1 and X_t are not uniquely identified, but the linear space spanned by the column vectors of A_1, denoted by $C(A_1)$ and called the factor loading space, is uniquely defined. With similar definition for the column space of A_2.

For the statistical meaning of A_1 and A_2 and their roles in estimation of the factor loadings and the latent factors, it is instructive to introduce two half orthonormal matrices $B_1 \in \mathbb{R}^{d \times v}$ and $B_2 \in \mathbb{R}^{d \times r}$ by

$$B_1' A_1 = 0, \quad B_2' A_2 = 0,$$

so that their column spaces are orthogonal to those of A_1 and A_2. We partition

$$(A_1, B_1) = (\mathbf{a}_1, \ldots, \mathbf{a}_r, \mathbf{b}_1, \ldots, \mathbf{b}_v), \quad (A_2, B_2) = (\mathbf{a}_{r+1}, \ldots, \mathbf{a}_d, \mathbf{b}_{v+1}, \ldots, \mathbf{b}_d),$$

in terms of their columns. Then, pre-multiplying (11.85) by B_1', one obtains

$$B_1' \boldsymbol{Y}_t = B_1' A_2 e_t, \tag{11.86}$$

showing that $\{B_1' \boldsymbol{Y}_t\}$ is a white noise process.

The next key step is to identify the columns of B_1 and B_2 as the eigenvectors corresponding to the zero eigenvalues of the following two positive semi-definite matrices

$$M_Y = \sum_{k=1}^{k_0} \Gamma_Y(k) \Gamma_Y(k)', \quad Z_Y = \Gamma_Y B_1 B_1' \Gamma_Y, \tag{11.87}$$

where k_0 is a prespecified nonnegative integer , $\Gamma_Y(k) = \mathrm{Cov}(\boldsymbol{Y}_t, \boldsymbol{Y}_{t-k})$ is the lag-k autocovariance and $\Gamma_Y(0) = \boldsymbol{\Sigma}_Y$ is the lag-zero or instantaneous covariance of the time series. Similarly, we define

$$\Gamma_X(k) = \mathrm{Cov}(\boldsymbol{X}_t, \boldsymbol{X}_{t-k}), \quad \Gamma_{Xe}(k) = \mathrm{Cov}(\boldsymbol{X}_t, e_{t-k}),$$

and note that in the traditional factor models it is usually assumed that $\boldsymbol{X}_t$ and e_s are uncorrelated for all leads and lags, namely for any t and s. However, here we assume less and only require that

$$\mathrm{Cov}(\boldsymbol{X}_t, e_{t+j}) = \mathbf{0}, \quad \text{for } j \geq 0.$$

In other words, the past latent values are required to be uncorrelated with the future idiosyncratic errors, a sort of causality in the latent space. Then, from (11.83) and routine covariance computation it follows that for $k \geq 1$,

$$\Gamma_Y(k) = A_1 \Gamma_X(k) A_1' + A_1 \Gamma_{Xe}(k) A_2',$$

and for $k = 0$,

$$\Gamma_Y = A_1 \Gamma_X A_1' + A_2 \Gamma_{Xe} A_2'.$$

From $B_1' A_1 = 0$, it follows easily that $M_Y B_1 = 0$, so that the columns of B_1 are the eigenvectors corresponding to the zero eigenvalues of M_Y, and the factor loading space $C(A_1)$ is spanned by eigenvectors corresponding to its nonzero eigenvalues.

As for the relevance of Z_Y, B_2 and estimation of the common factors, we note that premultiplying (11.85) by B_2' one obtains

$$B_2' \boldsymbol{Y}_t = B_2' A_1 \boldsymbol{X}_t, \tag{11.88}$$

which is uncorrelated with $B_1' \boldsymbol{Y}_t$; see (11.86). Namely,

$$\text{Cov}(B_2' \boldsymbol{Y}_t, B_1' \boldsymbol{Y}_t) = B_2' \text{Cov}(\boldsymbol{Y}_t, \boldsymbol{Y}_t) B_1' = B_2' \boldsymbol{\Sigma}_Y B_1' = \boldsymbol{0} \Longrightarrow B_2' \boldsymbol{\Sigma}_Y B_1 B_1' \boldsymbol{\Sigma}_Y B_2 = \boldsymbol{0},$$

and hence the columns of B_2 are the eigenvectors corresponding to the zero eigenvalues of the matrix Z_Y in the second equation in (11.87). The $r \times r$ matrix $B_2' A_1$ appearing in (11.88) is known to be invertible [Gao and Tsay, 2022], and therefore the latent process is estimated by

$$\boldsymbol{X}_t = (B_2' A_1)^{-1} B_2' \boldsymbol{Y}_t, \tag{11.89}$$

even though B_2 is not unique.

Thus far, the methodology has been developed assuming the number of latent factors r is known. Estimation of r based on time series data is discussed next, where (11.86) is used in conjunction with a white noise test [Gao and Tsay, 2022].

11.7.4 Model Fitting Procedure and Prediction

This section provides a brief outline of estimating r, the factor loading matrix A_1, and the common factor process $\boldsymbol{X}_t$ in the model (11.85) from time series data $\boldsymbol{Y}_t, t = 1, \ldots, n$, when d, the dimension of the data, is not too large.

There are a number of methods for estimating the unknown number of factors r, see for example the discussion around (7.6). An alternative method in Gao and Tsay [2022] is based on performing white noise tests, see (11.86), to determine the number of white noise components $\hat{v}$ first and then setting $\hat{r} = d - \hat{v}$.

Let $\widehat{G}$ be the matrix of eigenvectors of the sample matrix $\widehat{M}_Y$, arranged according to the decreasing order of eigenvalues and define

$$\widehat{U}_t = \widehat{G} \boldsymbol{Y}_t = (\widehat{U}_{1t}, \ldots, \widehat{U}_{dt})',$$

as the transformed data series. The plan is to estimate v by sequentially testing for the number of white noise series in $\widehat{U}_t$. If the dimension d is small or moderate, a bottom-up procedure is used to determine if $\widehat{U}_{it}$ is a white noise series starting with $i = d$ or the last column. The test used is the well-known Ljung-Box portmanteau test, which tests for white noise based on the sum of squares of the first few values of the sample autocorrelation function of a univariate time series. If the null hypothesis is rejected, then $\hat{v} = 0$ and $\hat{r} = d$; otherwise, reduce i by one and repeat the testing process. Evidently, testing for the number of white noise components will continue until $i = 1$. If all transformed series $\widehat{U}_{it}$ are white noise, then $\hat{v} = d$ and $\hat{r} = 0$. In general, if $\widehat{U}_{it}$ is not a white noise series but $\widehat{U}_{jt}$ are for $j = i + 1, \dots, d$, then $\hat{v} = d - i$ and $\hat{r} = i$, with $\widehat{G} = (\widehat{A}_1, \widehat{B}_1)$ where $\widehat{A}_1 \in \mathbb{R}^{d\times\hat{r}}$ and $\widehat{B}_1 \in \mathbb{R}^{d\times\hat{v}}$. An eigenanalysis of $\widehat{Z}_Y$ leads to $\widehat{B}_2 = (\hat{b}_{\hat{v}+1}, \dots, \hat{b}_d)$ where $\hat{b}_{\hat{v}+1}, \dots, \hat{b}_d$ are the eigenvectors corresponding to the r smallest eigenvalues of $\hat{Z}_Y$.

Substituting $\widehat{A}_1$ and $\widehat{B}_2$ in (11.89) leads to the estimated factor process $\widehat{\boldsymbol{X}}_t$ which will be used to compute the h-step ahead prediction of the $\boldsymbol{Y}_t$ series using

$$\widehat{\boldsymbol{Y}}_{n+h} = \widehat{A}_1 \widehat{\boldsymbol{X}}_{n+h}, \tag{11.90}$$

where $\widehat{\boldsymbol{X}}_{n+h}$ is an h-step ahead forecast for $\boldsymbol{X}_t$ based on the estimated past values $\widehat{\boldsymbol{X}}_1, \dots, \widehat{\boldsymbol{X}}_n$. This is usually computed by fitting a VAR model to $\widehat{\boldsymbol{X}}_1, \dots, \widehat{\boldsymbol{X}}_n$.

11.8 Problems

11.1 Let $X_i = X + e_i, i = 1, \dots, n$, where X and $e_1, \dots, e_n$, are mutually independent mean zero random variables with $\text{Var}(X) = \sigma_1^2$ and $\text{Var}(e_i) = \sigma_e^2, i = 1, \dots, n$.

(a) Compute the covariance matrix of $\mathbf{X} = (X_1, \dots, X_n)'$ and decide if it is stationary.

(b) Find the inverse of the covariance matrix. Is it sparse?

11.2 Compute the autocovariance function of the random walk in (11.27) and write the $n \times n$ covariance matrix of $\mathbf{X} = (X_1, \dots, X_n)'$.

11.3 Find σ_L^2 for the following stationary (Toeplitz) covariance matrices:

(a) AR(1): $\Sigma = \sigma^2(\rho^{|i-j|}), |\rho| < 1$.

(b) MA(1): $\gamma(0) = \sigma^2(1+\theta^2), \gamma(1) = \sigma^2\theta, \gamma(h) = 0, h > 1$, where the parameter θ is a real number.

(c) (CS)(ρ, σ^2) : With the entries $\text{Cov}(X_i, X_j) = \sigma^2\rho$ for $i \neq j$ and σ^2 for $i = j$.

11.4 (a) Starting from the definition of least-squares predictor in (12.2), derive the normal equations (11.10) and the formula for the prediction error variance.

(b) Solve the normal equations (11.10) for the MA(1) model $X_t = Z_t + \theta Z_{t-1}$, for $n = 2$.

(c) For the MA(1) model $X_t = Z_t - Z_{t-1}$, directly verify that $a_{k,n} = -(1 - \frac{k}{n+1})$, $k = 1, \dots, n$, solves the normal equations (11.10). Hence, its best linear predictor is given by

$$X_{n+1,n} = -\frac{1}{n+1}\sum_{t=1}^{n}(n+1-t)X_t,$$

with the prediction error variance

$$\sigma_{n+1}^2 = \frac{n+2}{n+1}\sigma^2.$$

(d) Show that the solution of the normal equations (11.10) for the AR(1) model $X_t = \phi X_{t-1} + Z_t, |\phi| < 1$, is given by

$$a_{1,n} = \phi, \; a_{k,n} = 0, \; k > 1.$$

What are the best linear predictor and the corresponding prediction error variance?

(e) Solve the normal equations for any n when the ACF is the $CS(\rho, \sigma^2)$ or equi-correlated matrix. Write the formulae for the best linear predictor and the corresponding prediction error variance.

(f) Compare the formulae in (e) with those in (b), (c) and (d).

11.5 Suppose $X_1, \ldots, X_n$ are the observed segment of a stationary time series and its ACF $\{\gamma(h)\}$ is known. Consider the task of predicting X_{n+2} or finding the two-step ahead linear predictor. Let

$$X_{n+2,n} = \sum_{k=1}^{n}\tilde{a}_{k,n}X_{n+1-k}, \tag{11.91}$$

be the optimal linear predictor where the coefficients are found by minimizing the mean square error (MSE)

$$E(X_{n+2} - \sum_{k=1}^{n}\tilde{a}_{k,n}X_{n+1-k})^2. \tag{11.92}$$

(a) Derive the normal equations for the predictor coefficients, and compare it to (11.10). What are the similarities? Differences?

(b) Repeat (b)–(f) of the previous problem this time for the two-step ahead predictors of the three models.

(c) Write the normal equations for computing the h-step ahead predictor coefficients.

11.6 Recall that AR(p) process is causal if and only if all roots of its AR polynomial $\phi_p(z) = 0$ lie outside the unit circle in the complex plane. Consider the following AR(2) process.

$$\boldsymbol{X}_t = 1.5\boldsymbol{X}_{t-1} - 0.7\boldsymbol{X}_{t-2} + \boldsymbol{\epsilon}_t$$

(a) Find its AR polynomial $\phi_p(z)$

(b) Solve $\phi_p(z) = 0$ and find its roots. Is this process causal?

(c) Verify your answer by plotting the roots in the complex plane using the following R code:

```
# AR polynomial coefficients: 1 - 1.5z + 0.7z^2
 phi <- c(1.5, -0.7)

# Roots of the AR polynomial (we flip sign)
roots <- polyroot(c(1, -phi))
modulus <- Mod(roots)

# Print modulus
print(modulus)
[1] 1.195229 1.195229
# Plot roots on complex plane
plot(roots, xlim=c(-2,2), ylim=c(-2,2), xlab="Real", ylab="
    Imaginary",
  main="Roots of AR Polynomial", pch=19, col="blue")
  symbols(0, 0, circles=1, inches=FALSE, add=TRUE, lty=2) #
    unit circle
  abline(h=0, v=0, lty=3)
```

11.7 Use the acf1() function from the astsa package to plot the ACF and PACF of the following models:

(a) MA(2), AR(2), ARMA(2,1) by choosing the coefficients randomly in the interval $(-1, 1)$. Hint: use the sarima.sim() function for data generation, where ar() and ma() options specify coefficients for AR and MA components, respectively. .

(b) Compute the roots of their characteristic polynomials. Explain any (computer) error messages you may get in terms of causality and invertibility of the underlying models.

11.8 Show that the function $F(\lambda)$ defined in (11.34) is a monotone function.

11.9 Starting with $\{\boldsymbol{Z}_t\} \sim WN(0, \boldsymbol{\Sigma})$, a positive integer q and matrices $\Theta_k, k = 1, \ldots, q$, a vector MA($q$) process is defined by

$$\boldsymbol{X}_t = \boldsymbol{Z}_t + \sum_{k=1}^{q} \Theta_k \boldsymbol{Z}_{t-k}, \quad t \in \mathbb{Z},$$

which is a linear filter as in (11.43).

(a) Find the transfer function of the filter and the spectral density matrix of $\boldsymbol{X}_t$.

(b) Compute the autocovariance function $\mathrm{Cov}(\boldsymbol{X}_t, \boldsymbol{X}_{t-h})$.

(c) What happens in (a)–(b) when $q = \infty$?

11.10 A vector autoregressive and moving average (VARMA) model of order p, q or VARMA(p, q) is defined by

$$\boldsymbol{X}_t = \Phi_1 \boldsymbol{X}_{t-1} + \ldots + \Phi_p \boldsymbol{X}_{t-p} + \boldsymbol{Z}_t + \sum_{k=1}^{q} \Theta_k \boldsymbol{Z}_{t-k}, \ \boldsymbol{Z}_t \sim \mathrm{WN}(0, \Sigma),$$

where Φ_i's and Θ_i's are the AR and the MA coefficient matrices.

(a) Assuming that the model has a causal stationary solution, find its spectral density matrix in terms of the AR and MA characteristic polynomials $\Phi(\lambda)$ and $\Theta(\lambda)$.

(b) State the conditions on the AR coefficient matrices for the model to have a causal stationary solution.

(c) State the conditions on the MA coefficient matrices for the model to be invertible or to have an AR(∞) representation.

11.11 For the bivariate ARMA(1,1) model in (32),

(a) find the explicit form of its 2×2 spectral density matrix.

(b) find the spectral coherence (SC) or the coherency function between the two series and plot it.

(c) find the inverse of the spectral density matrix in (a).

11.12 For a causal VAR(p) model, set $\Gamma_k = \sum_{i=0}^{p-k} \Phi_i' \Theta \Phi_{i+k}$ with $\Theta = \Sigma^{-1}$ and $\Phi_0 = I$. Show that

$$\mathrm{PSC}_{ij}(\lambda) = 0 \iff (\Gamma_k)_{ij} = 0,$$

for $k = 1, \ldots, p$.

11.13 Consider the airpollute dataset in Section 11.5.3.

(a) Fit a VAR model to the data, but this time use AIC criterion for the lag selection. Summarize and report the estimation results and the selected number of lags.

(b) Use the normality.test() function from the R package vars to test for multivariate Jarque-Bera test. What is your conclusion?

(c) In time series data analysis impulse response functions (irfs) are used to analyze how the system responds to an impulse or shock. Use the irf() function to compute the response of NO to the shock of CO.

(d) Use the plot() function to visualize estimated irfs.

11.14 For the spectral decomposition of the block-diagonal matrix M_X in (11.66) show that

(a) The matrix P_X can be taken to be a block-diagonal orthogonal matrix with the same block structure as M_X.

(b) An orthogonal matrix P_X satisfies (11.66), if and only if its columns are a permutation of the columns of the block-diagonal orthogonal matrix in (a) provided that any two different blocks $M_{X,i}$ and $M_{X,j}$ do not share the same eigenvalues.

(c) When different blocks in (11.66) share the same eigenvalue(s), a version of (a)–(b) still holds, but with M_X replaced by a smaller $\tilde{M}_X$ which is also a block diagonal matrix with fewer than q blocks obtained by combining together those blocks sharing at least one common eigenvalue into one larger block. In particular, in this case the proposed time series PCA method cannot separate $\boldsymbol{X}_t^{(i)}$ and $\boldsymbol{X}_t^{(j)}$ sharing at least one common eigenvalue.

11.15 Repeat the analysis of Section 11.6.4 when the number of lags is equal to 5. Report the groups and the members in each group.

12

Spatial Dependence and Cholesky Factor

While time series data consists of measuring a variable over time, spatial data (aka random field) is concerned with measuring a variable like temperature, rainfall, CO2, over the plane, sphere or other domains. A key difference between the two areas is whether their index sets are *ordered* or not, with the index sets of spatial data being unordered subsets of $\mathbb{R}^d$ for which there is the opportunity for partial ordering along a coordinate, for example. In time series one is usually interested in predicting the future values using the past observations, but in spatial data situations there is no natural notion of "past and future". Fortunately, "order", "past and future" play no roles in defining the central notion of *stationarity*, though it does play a role in reducing the computational burden in maximum likelihood estimation and the prediction (interpolation) of unobserved values. In fact, since the introduction of Vecchia [1988] approximation of the likelihood function for Gaussian data, there have been growing activities in using certain (partial) ordering and the ensuing Cholesky factor of the inverse covariance matrix. This chapter provides only a brief overview of the emerging strong connections between time series and spatial data involving Cholesky factors. It does not touch the vast literature on spatial data analysis and geostatistics for which there are many excellent book-length coverage such as Cressie [1993], Stein [1999] and Diggle and Ribeiro [2007]. Analysis of multivariate spatio-temporal data encounters even more challenges in dealing simultaneously with the *cross-sectional, temporal and spatial* dependencies.

12.1 Spatial Data and Covariance Functions

One usually observes a random field or spatial process $\{Y(x); x \in \mathbb{R}^d\}$ at locations $x_1, \ldots, x_n$ in a bounded region $\mathcal{D}$ of $\mathbb{R}^d$, where the goal is to predict (interpolate) an unobserved value $Y(x_0)$, $x_0 \in \mathcal{D}$, using the observed data and the knowledge of the first two moments or the mean and covariance functions of the spatial process.

DOI: 10.1201/9781003638896-12

12.1.1 Linear Prediction and the Covariance Function

It is well-known that for the task of linear prediction (interpolation), the mean and covariance functions of the underlying process defined by

$$\mu(x) = E(Y(x)),\ \gamma(x, y) = \text{Cov}(Y(x), Y(y)), \tag{12.1}$$

play a central role. For the time being we assume that these functions exist and are known.

In analogy with time series forecasting in Section 11.1.1, arranging the observations at the n locations in the n-vector $Y = (Y(x_1), \ldots, Y(x_n))'$, and using a linear predictor of the form $a_0 + \mathbf{a}'Y$, its mean squared error can be decomposed as

$$\begin{aligned} E\{Y(x_0) - a_0 - \mathbf{a}'Y\}^2 = \{\mu(x_0) - a_0 - \boldsymbol{a}'\mu\}^2 + \gamma(0, 0) \\ - 2\boldsymbol{a}'\boldsymbol{\gamma}_n + \boldsymbol{a}'\Gamma_n\boldsymbol{a}, \end{aligned} \tag{12.2}$$

where

$$\boldsymbol{\mu} = E(Y),\ \boldsymbol{\gamma}_n = \text{Cov}\{Y, Y(x_0)\}, \Gamma_n = \text{Cov}(Y),$$

and a_0, the n-vector $\boldsymbol{a}$ are the predictor coefficients. Evidently, for any given $\boldsymbol{a}$ the first term in the decomposition can be made zero by choosing a_0 suitably, but the second term is minimized by

$$\boldsymbol{a} = \Gamma_n^{-1}\gamma_n, \tag{12.3}$$

provided that Γ_n is invertible. Thus, just as in time series prediction, see (11.10), computing the linear predictor coefficients requires inverting an $n \times n$ covariance matrix. This is a computationally demanding task n is large, like 10, 000, which is seen quite common in spatial data analysis. It also reveals the centrality of the role of the covariance function/matrix of a random field in its linear prediction. Of course, as in time series computation, a way to reduce the computational burden is to impose further reasonable structures on the covariance structures such as stationarity, isotropic, etc. so that they are parsimonious or sparse and hence less expensive to store with scalable computational advantage.

In what follows, we restrict attention to Gaussian random fields where, for all $n \geq 1, x_1, \ldots, x_n$, the n-vectors Y have multivariate normal distributions completely determined by their mean and covariance functions or their first two moments. A Gaussian random fields is said to be (weakly, second-order) *stationary* if its mean function is constant and its covariance function $\gamma(x_1, x_2) = \text{Cov}(Y(x_1), Y(x_2))$ depends on $x_1 - x_2$, or the location separation. A stationary random field is said to be *isotropic* if its covariance function depends on $||x_1 - x_2||$, the Euclidean distance between the two locations. Consequently, the covariance function of an isotropic stationary random field is a function of a single variable taking values in $[0, \infty)$.

Since the covariance function of a stationary random field is a positive-definite function, it can be written as the Fourier transform of a unique positive measure on $\mathbb{R}^d$ with the spectral density function $f(\lambda)$, see Example 13. A few important examples of such covariance and spectral density functions are given next, for more examples and details see Stein [1999].

Example 35 Gaussian Covariance: *A smooth and familiar isotopic covariance and its spectral density functions on $\mathbb{R}$ are given by:*

$$\gamma(t) = ce^{-\alpha t^2}, \quad f(\lambda) = \frac{1}{2}c(\pi\alpha)^{-1/2}e^{-\lambda^2/(4\alpha)},$$

where c, α are positive scalars.

Note that both functions are infinitely differentiable, which has important consequences for inference and prediction.

Example 36 Triangular Covariance Function: *The triangular class of nonsmooth covariance functions with oscillating spectral density functions is given by*

$$\gamma(t) = c(a - |t|)^+, \quad f(\lambda) = c\pi^{-1}\{1 - \cos(a\lambda)\}/\lambda^2,$$

where a and c are positive scalars.

Again, the oscillatory behavior of the density function has important consequences for inference and prediction.

The next class of covariance functions is perhaps the single most important one in the whole area of spatial data analysis.

Example 37 Matérn Class: *Consists of covariance functions of the form*

$$\gamma(t) = \frac{\pi^{1/2}\phi}{2^{\nu-1}\Gamma(\nu + 1/2)\alpha^{2\nu}}(\alpha|t|)^{\nu}\mathcal{K}_{\nu}(\alpha|t|), \tag{12.4}$$

where all three parameters ν, ϕ, α are positive, and $\mathcal{K}_\nu$ is a modified Bessel function. Its corresponding spectral density function has the more familiar and simpler form

$$f(\lambda) = \phi(\alpha^2 + \lambda^2)^{-\nu-1/2}. \tag{12.5}$$

Evidently, ν is the more consequential parameter as far as smoothness of the random field is concerned. In fact, the random field is m times mean square differentiable if and only if $\nu > m$. For ν of the form $\nu = m + \frac{1}{2}$ with m a nonnegative integer, the spectral density is a rational function and the covariance function is of the form $e^{-\alpha|t|}$ times a polynomial of degree m in $|t|$. For example, the Matérn covariances for $\nu = 1/2, 3/2$ are:

$$\gamma(t) = \pi\phi\alpha^{-1}e^{-\alpha|t|}, \quad \frac{1}{2}\pi\phi\alpha^{-3}e^{-\alpha|t|}(1 + \alpha|t|).$$

Intuitively, the interpolation (Kriging) is more precise and its error is smaller if the underlying process is locally smooth. This latter point can be made more clear and precise by choosing to work with random fields with a continuous index set $\mathcal{D}$ in $\mathbb{R}^d$.

12.2 MLE of the Covariance Parameters

While in some applications the measurements (variables) in the vector Y are *ordered* naturally as in time series and longitudinal data, this is not the case for spatial data. In the first case, it is common to take advantage of the regression interpretation of the entries of the Cholesky factor of the precision matrix and induce sparsity in it through penalized likelihood or other means (Section 4.3).

For spatial data, in spite of lack of a natural order among the measurements, there has been growing interest in introducing pseudo-order among the measurement sites and using the circle of ideas linking the Cholesky factor, Directed Acyclic Graphs (DAGs) models and Vecchia [1988] approximation of the likelihood function for Gaussian data. The simplest way to get to the root of these connections is to note that with the Cholesky factorization $\Gamma_n = CC'$, the computational cost of

$$\Gamma_n^{-1} = C'^{-1}C^{-1},$$

for prediction in (12.3) and for MLE is considerably lower when the inverse Cholesky factor is sparse, see Katzfuss and Guinness [2021].

12.2.1 Vecchia Approximation

The idea of Vecchia [1988] approximation is to approximate the likelihood function of a Gaussian distribution as a product of univariate conditional densities, each of which is conditioned on a suitably chosen subset of the "previous" variables.

Recall that a given random vector $Y = (Y_1, \ldots, Y_n)'$, its density can be factorized for any $y \in \mathbb{R}^n$ as

$$f_Y(y) = \prod f_{Y_i|Y_{[i-1]}}(y_i|y_{[i-1]}), \quad [i-1] = \{1, 2, \ldots, i-1\},$$

and the computational burden can be reduced by replacing the conditioning set $[i-1]$ for each i by a suitably chosen smaller subset $c(i)$ with cardinality $m < p$. The likelihood or the joint density is then approximated by

$$\tilde{f}(Y) = \prod f_{Y_i|Y_{[i-1]}}(y_i|y_{c(i)}). \tag{12.6}$$

An early indication of the potential application of such approximation is Cressie and Davidson [1998] who showed that Vecchia's approximation of density implies a Markov random field model with sparse precision matrix.

Numerous suggestions and rationales for various choices of $c(i)$ have been proposed in the last three decades. For example, Stein et al. [2004] recommended including in it the indices of some close and some far-away observations, and grouping observations for computational advantages in the likelihood approximations. Further extensions are proposed in Sun and Stein [2016] and Huang and Sun [2018].

In addition, the recent surge of interest in implementing Vecchia [1988] likelihood approximation has infused new energy into the question of "ordering" measurement sites in spatial data, variables in multivariate data, and columns of matrices.

1. *Coord ordering*: Vecchia [1988]'s original approximation of the likelihood is based on ordering the locations by a spatial coordinate (lexicographical ordering), referred to commonly as *coord ordering*.

2. *Maxmin-Ordering*: Guinness [2018] proposed an approximate maximum-minimum distance *(maxmin) ordering*, which sequentially picks each location in the ordering by aiming to maximize the distance to the nearest of the previous locations. Heuristically, it ensures that each location is surrounded by previous locations in the ordering. He showed that the maxmin ordering can lead to substantial improvements over coord ordering in settings without any nugget or noise.

3. In the context of space-time data, Zhu and Liu [2009] rely on the Cholesky decomposition of the precision matrix, based on several ordering schemes using the spatial locations of the observations. Their theoretical and simulation studies show that the regression-based penalized normal likelihood method performs competitively.

For more discussions and examples of ordering the variables, see Zhu and Liu [2009] and Guinness [2018].

12.2.2 Sparse Inverse Cholesky Factor: Known Order

It is known that Vecchia's likelihood approximation amounts to implicitly computing a sparse approximate inverse Cholesky factor of the underlying covariance matrix [Datta et al., 2016, Katzfuss and Guinness, 2021] .

Recently, Schäfer et al. [2021] has shown that sparse inverse Cholesky factors are optimal in the Kullback-Leibler (KL) divergence for two multivariate normal distributions; see Problem 12.6. For the role of KL-divergence in statistical inference we note that if f_2 is an approximation of the true density function f_1, then the KL-divergence can be viewed as the expected difference between the true and approximate log-densities. Consequently, its minimization is directly related to the accurate approximations for likelihood-based inference for the model parameters. From here on, to emphasize the central role of covariance matrices, we replace the expression $D_{KL}(f_1, f_2)$ for multivariate normal densities by $D_{KL}(\boldsymbol{\Sigma}_1, \boldsymbol{\Sigma}_2)$.

For a positive-definite matrix $\boldsymbol{\Sigma}$, consider the problem of finding an approximate Cholesky factor of its inverse by solving

$$C = \underset{\hat{C} \in S}{\operatorname{argmin}}\, D_{KL}(\boldsymbol{\Sigma}, (\hat{C}\hat{C}')^{-1}), \tag{12.7}$$

where S is a known sparsity set for the Cholesky factor. From (4.38) it follows that

$$C = \underset{\hat{C} \in S}{\operatorname{argmin}} \left(\operatorname{tr}(\hat{C}\boldsymbol{\Sigma}\hat{C}') - \log |\hat{C}\hat{C}'| \right). \tag{12.8}$$

After some manipulation and using the given sparsity pattern (Problem 12.6), it reduces to the quadratic optimization problem:

$$C = \underset{\hat{C} \in S}{\text{argmin}} \sum_{k=1}^{p} \left(\hat{C}_{s_k,k} \boldsymbol{\Sigma}_{s_k,s_k} \hat{C}'_{s_k,k} - 2 \log \hat{C}_{k,k} \right), \tag{12.9}$$

which involves only the (smaller) vectors $\hat{C}_{s_k,k}$'s with $s_i = \{j : (j,i) \in S\}$, composed of the nonzero entries of the columns of $\hat{C}$, and $\boldsymbol{\Sigma}_{s_i,s_i}$ is the restriction of $\boldsymbol{\Sigma}$ corresponding to the set of indices s_i. Then, for each k taking the derivative of the kth sum with respect to $\hat{C}_{s_k,k}$ and setting it to zero, one obtains the closed-form solution

$$\boldsymbol{\Sigma}_{s_k,s_k} \hat{C}_{s_k,k} = \frac{\boldsymbol{e}_1}{C_{k.k}}, \tag{12.10}$$

where $\boldsymbol{e}_1$ is a vector of dimension s_i whose first entry is one and the rest is zero.

There are two alternative characterizations of the closed-form solution above with a given sparsity pattern as the solution of two distinct optimization problems coming from different application areas. These include Kaporin [1990], who arrived at a closed-form expression of the approximate inverse Cholesky factor of a positive-definite matrix as the minimizer of the K-condition number of a preconditioned system subject to a sparsity constraint. Later, Yeremin et al. [2000] showed that Kaporin's solution is also the minimizer of a Frobenius-norm of the error subject to a diagonal scaling constraint.

What to do when the sparsity pattern S is unknown?

12.2.3 Unknown Sparsity Pattern: Penalized Likelihood

Regularization is the method of choice when the sparsity pattern is unknown, and there is a sort of known order among the variables or measurement sites. We provide a brief review of existing regularization methods in the literature.

For unknown pattern of sparsity of C, regularization methods can set some of its entries to zero. In one of the first attempts at inducing sparsity in the precision matrix, Huang et al. [2006] inadvertently considered penalizing the off-diagonal entries of its inverse Cholesky factor by minimizing

$$Q(T,D) = n \log |\boldsymbol{\Sigma}| + \sum_{i=1}^{n} \boldsymbol{Y}_i' T' D^{-1} T \boldsymbol{Y}_i + \lambda ||\phi_t||_1 = \sum_{t=1}^{p} Q_t(\phi_t, \sigma_t^2), \tag{12.11}$$

with

$$Q_t(\phi_t, \sigma_t^2) = \frac{\phi_t' S_{tt} \phi_t + 2\phi_t' S_{.t} + S_{tt}}{\sigma_t^2} + \log \sigma_t^2 + \lambda ||\phi_t||_1,$$

where S_{tt} and other vectors are the sample analog of $\boldsymbol{\Sigma}_t$ in (4.26) and $\phi_t = (\phi_{t1}, \ldots, \phi_{t,t-1})'$ is the t-th row of the modified Cholesky factor T. A cyclic block coordinate descent algorithm was used where each iteration consists of minimizing $Q_t(\phi_t, \sigma_t^2)$ with respect to ϕ_t (fixing σ_t^2 at its current value), and then with respect to σ^2 (fixing ϕ_t at its current value). Using a local quadratic approximation to the penalty function, it was shown that the method is computationally tractable, see also Huang et al. [2007].

As pointed out by Khare et al. [2019, Lemma 1.1], the objective function is neither jointly convex nor bi-convex in (T, D), and the convergence of this iterative method is not guaranteed. Interestingly, Shojaie and Michailidis [2010] note that choosing a sparsity pattern in T is equivalent to choosing an underlying Directed Acyclic Graph (DAG) model, and set their goal to minimize the objective function $Q(T, I_p)$, which is convex in T since it is quadratic in T plus an ℓ_1 penalty term; see Problem 12.5. It turns out that whether the true D is the identity matrix, this procedure leads to a valid approach for model/DAG selection.

The Lasso or ℓ_1-penalty usually results in placing zeros in T with no regular patterns so that the sparsity of the precision matrix is not guaranteed. Nevertheless, sparsity of an inverse Cholesky factor can reduce the computational cost of computing the forecasts and MLE of the covariance parameters. However, a suitably structured sparse Cholesky factor, like a banded one, leads to a sparse precision matrix. Subsequent research has focused on various penalty functions to band lower triangular matrices so that simultaneous sparsity of estimators of T and $\boldsymbol{\Theta}$ can be guaranteed:

1. **Variable Banding (Nested Lasso):** The patterns of zeros in T is controlled and pushed farther away from the main diagonal using the *nested* Lasso penalty [Levina et al., 2008]. It replaces the penalty in (12.11) by

$$P(\phi_t) = |\phi_{t,t-1}| + \frac{|\phi_{t,t-2}|}{|\phi_{t,t-1}|} + \cdots + \frac{|\phi_{t,1}|}{|\phi_{t,2}|}, \tag{12.12}$$

   where $0/0$ is defined to be zero which forces a contiguous pattern of zeros in T right below the main diagonal, in the sense that if the jth variable is not included in the tth regression ($\phi_{tj} = 0$), then all the preceding variables are also excluded. Hence, the tth regression only uses the k_t immediate predecessors of the tth variable, and each regression could have a different order k_t; see (12.13). The scaling of coefficients could be an issue; see Levina et al. [2008] for various alternatives. Since this formulation is nonconvex and nonlinear, an iterative local quadratic approximation algorithm was used to fit the model.

2. **Adaptive Banding:** An alternative to the nested banding structure [Leng and Li, 2011] is the *forward adaptive banding* approach for estimating T using variable band lengths $\{k_j; j = 1, \ldots, p\}$. For selecting the band lengths, consider the modified BIC:

$$\text{BIC}(k_1, \ldots, k_p) = n \log |\widehat{\Sigma}| + \sum_{i=1}^{n} \boldsymbol{Y}_i' \widehat{T} \widehat{D}^{-1} \widehat{T} \boldsymbol{Y}_i + C_n \log(n) \sum_{j=1}^{p} k_j, \tag{12.13}$$

   where $\widehat{T}$ and $\widehat{D}$ are constructed from estimated autoregressive coefficients and prediction variances by fitting AR(k_j) models to the successive variables. Then, the k_j's are the minimizers of (12.13) over

$$k_j \leq \min\{n/(\log n)^2, j-1\}, \quad j = 1, \ldots, p,$$

with a sequence C_n diverging to infinity. Fortunately, (12.13) decouples as

$$BIC = \sum_{j=1}^{p} \left\{ n \log \widehat{\sigma}_j^2 + \sum_{i=1}^{n} \widehat{\varepsilon}_{ij}^2 / \widehat{\sigma}_j^2 + C_n \log(n) k_j \right\},$$

so that each k_j can be chosen separately.

3. **Uniform or k-Banding:** When $k_t = k$ for all t, the Cholesky factor T is said to be uniform or k-banded. In this case, Bickel and Levina [2008a] provide conditions ensuring consistency in the operator norm for precision matrix estimates based on banded Cholesky factors.

4. **Convex Sparse Cholesky Selection (CSCS):** As noted earlier the objective function $Q(T, D)$ in (12.11) is not jointly convex in its arguments, so that there is no theoretical guarantees of convergence of the iterative methods. Fortunately, it was noted later that rewriting it in terms of the standard Cholesky factor $L = D^{-1}T$ leads to a convex objective function, see Yu and Bien [2017], Khare et al. [2019], Aragam and Zhou [2015].

   The CSCS algorithm in Khare et al. [2019] deals with situations where there is a known domain-specific ordering of the variables and the estimation problem boils down to estimating the sparsity pattern in L, the lower triangular with positive diagonal entries such that $\mathbf{\Theta} = L'L$. The objective function for CSCS is

$$Q_{CSCS}(L) = \text{tr}(L'L\mathbf{S}) - 2\log|L| + \lambda \sum_{i<j} L_{ij}, \tag{12.14}$$

   where $\mathbf{S}$ is the sample covariance matrix, the first two terms correspond to the Gaussian log-likelihood, and the third term is an 'ℓ_1-penalty term to induce sparsity in the lower triangular matrix L. By contrast, rewriting (12.11) in terms of L it reduces to

$$Q = \text{tr}(LL'\mathbf{S}) - 2\log|L| + \lambda \sum_{i<j} |L_{ij}|, \tag{12.15}$$

   where the only difference with (12.14) is seen in its third term (Problem 12.7).

12.2.4 Sparse Cholesky Factor: Unknown Order

In many applications the ordering among the variables is unknown. In such settings, a possible solution is to consider the ordering as an additional parameter in the objective function and optimizing over it.

This problem has been studied in the DAG literature covered in Section 10.2 where a DAG G admits a topological ordering of its nodes, that is there exists a permutation order π such that $\pi(j) < \pi(k)$ only if k is not an ancestor of j. For linear SEM (10.5, the existence of a topological ordering leads to the permutation-similarity of B to a strictly lower triangular matrix $B_\pi = P_\pi B P'_\pi$ by permuting rows and columns of B simultaneously through a permutation matrix P_π. For Gaussian DAGs, where $Y \sim N_p(0, \Sigma)$, the DAG learning problem can be formulated as

$$\arg\min_{P_\pi, L} \operatorname{tr}(P_\pi \mathbf{S} P'_\pi L'_\pi L_\pi) - \log|L_\pi| + \lambda \|L_\pi\|_0^-, \tag{12.16}$$

where $L_\pi = P_\pi L P'_\pi$ and L is the Cholesky factor of the precision matrix $\mathbf{\Theta} = L'L$, and $\|L_\pi\|_0^-$ denotes the ℓ_0 penalty of the nondiagonal entries of the lower triangular matrix. In other words, the goal is to select a permutation matrix P_π that results in the smallest number of edges, i.e., the sparsest Cholesky factor. For further details and discussions, see Raskutti and Uhler [2018], Ye et al. [2020] and Dallakyan and Pourahmadi [2021].

Alternatively, the Concave penalized Coordinate Descent with reparameterization (CCDr) algorithm in Aragam and Zhou [2015] considers the factorization $\mathbf{\Theta} = BB'$ where B varies over the space of $p \times p$ matrices such that by permuting its rows and columns it can be reduced to a lower triangular matrix with positive diagonal entries. The corresponding optimization problem is significantly more complicated, and the resulting estimate can only be shown to be a local minimum of the objective functions.

12.3 Problems

12.1 Prove the identity in (12.2). Show that a_0 and the MSE of the linear predictor at x_0 in (12.2) are given by

$$a_0 = \mu(x_0) - \mathbf{a}'\boldsymbol{\mu}, \quad \sigma^2(x_0) = \gamma(0,0) - \boldsymbol{\gamma}'_n \Gamma_n^{-1} \boldsymbol{\gamma}_n.$$

12.2 (a) Plot the covariance and spectral density functions in Example 36 for $a = c = 1$.

(b) Plot the covariance and spectral density functions of the Mattern class for $\phi == \alpha = 1$ and $\nu = 3/2$.

12.3 Consider defining a random field as a collection of random variables on the lattice points $(s,t) \in \mathbb{Z}^2$ by

$$Y_{s,t} = X + \varepsilon_{s,t}, \tag{12.17}$$

where X and $\varepsilon_{s,t}$'s are independent random variables with means zero and $\operatorname{Var}(X) = \sigma_x^2$ with $\operatorname{Var}(\epsilon_{s,t}) = \sigma_e^2$.

(a) Compute the covariance function of the random field. Is it stationary? Isotropic?

(b) for n observations at sites $x_1, \ldots, x_n$, write the $n \times n$ covariance matrix Γ_n, and find its inverse.

(c) For a lattice point x_0 distinct from the observations sites, write the form of linear least-squares predictor of Y_{x_9} in terms of the observed values of random field $Y(x_1), \ldots, Y(x_n)$. Find its MSE.

12.4 Use basic matrix algebra and show that the penalized likelihood in (12.11) can be decoupled as

$$Q(T, D) = \left\{ \frac{S_{11}}{\sigma_1^2} + \log \sigma_1^2 \right\} + \sum_{t=2}^{p} \left\{ \frac{\phi_t' S_{tt} \phi_t + 2\phi_t' S_{.t} + S_{tt}}{\sigma_t^2} + \log \sigma_t^2 + \lambda ||\phi_t||_1 \right\},$$

12.5 (a) Starting from (12.11) show that

$$Q(T, I_p) = \text{tr}(T'\mathbf{S}T) + \lambda \sum_{i<j} |\phi_{ij}|.$$

(b) Discuss the impact of the assumption $D = I_p$ when the goal is to estimate $\mathbf{\Sigma} = T^{-1} D (T')^{-1}$.

(c) As an alternative to (b), discuss the pros and cons of estimating T by minimizing $Q(T, I_p)$, but estimating D directly from the modified Cholesky decomposition of the sample covariance matrix $\mathbf{S}$. What can go wrong when $n < p$?

12.6 Starting with the KL objective function (12.7), complete the derivations of the identities (12.8)–(12.10).

12.7 Show that the objective function in (12.14) is convex in L, but the one in (12.15) is not. Discuss the convexity of the first two terms and the third term in each objective function.

12.8 Show that for a $p \times p$ positive-definite matrix A, $c \geq 0$ and $\boldsymbol{x} = (x_1, \ldots, x_{p-1}, x_p) \in \mathbb{R}^{p-1} \times \mathbb{R}_+$ the function

$$f(\boldsymbol{x}) = -2c \log x_p + \boldsymbol{x}' A \boldsymbol{x} + \lambda \sum_{i=1}^{p-1} |x_i|,$$

where $\lambda > 0$ is a tuning parameter. It is interesting to note that with $c = 0$ and without the restriction that $x_p > 0$, the minimization problem for this function would have been equivalent to the Lasso regression problem in Chapter 2. With the logarithmic term and positivity constraint the corresponding optimization problem is important in covariance estimation [Khare et al., 2019].

A

Basics of the R Programming Language

At the end of each chapter, we include a Hands-on in R section where we demonstrate practical applications of the methods discussed in that chapter. The purpose of such sections is to teach the reader R basics, discuss the pros and cons of the methods from the practical point of view, and familiarize the reader with the necessary syntax. The code script for each chapter is available at https://rcdg-book.cv/code.html.

A.1 Quick Introduction to R

The best way to use any programming language is to use it often and regularly. To use R, the reader needs to obtain their own copy. R can be downloaded from https://cran.r-project.org/ web page. The top of the web page contains three links for downloading different flavors of R: Windows, Linux, and Mac.

RStudio or Positron are code editors that help you write commands and different functions in R. They can be downloaded from https://posit.co/. We will use RStudio throughout the book. Note that even if we use RStudio, we still need to download R.

At a basic level, R and RStudio can be used as a calculator. For example, if we type in the console

```
> 1 + 1
[1] 2
```

Not surprisingly, the output `[1] 2` reassures that the sum is 2 and `[1]` indicates that the observation number at the beginning of the line is the first observation. R and RStudio support a variety of mathematical functions, such as

```
# square root
> sqrt(9)
[1] 3

# exponential
> exp(1)
[1] 2.718282
```

DOI: 10.1201/9781003638896-A

```
# logarithm
> log(1)
[1] 0
```

We recommend using comments # throughout your script to improve its clarity and reproducibility.

```
# This is a comment
# Use comments to explain your code

# Save your work in a script file (.R)
# Run code with Ctrl+Enter (Windows) or Cmd+Enter (Mac)
```

So far, we have displayed the results of our code directly in the console. However, for practical work, you should store results in objects for later use. To make your code reusable and reproducible, we recommend writing your commands in a script file that you can save and reload as needed.

A.2 Objects and Data Types

The objects are the backbone of most programming languages. These objects can be anything: numbers, strings, functions, plots, etc. To create an object, we need to give it a name and assign a value to it. In R, this can be done using the assignment operator `<-` or the equal operator `=`.

In the next script, we assign the number 2 to the object `first_object` and assign the result of the expression `2 * 2` to the object `second_object`.

```
first_object = 2

second_object = 2 * 2
```

Once we created the objects, we can use them for different purposes

```
third_object = first_object + second_object
```

To display the values of an object, we can utilize the `print()` function. We discuss R functions in the next section.

```
> print(third_object)
[1] 6
```

R has several basic data types that are important to understand:

```
# Numeric (double)
x <- 5.5
class(x)
@[1] "numeric"

# Integer
y <- 5L
class(y)
@[1] "integer"
```

```
# Character (string)
name <- "Statistics"
class(name)
[1] "character"

# Logical (boolean)
is_true <- TRUE
class(is_true)
@[1] "logical"

# Factor (categorical data)
categories <- factor(c("low", "medium", "high", "low"))
print(categories)
@[1] low    medium high    low
Levels: high low medium@
```

A.3 Functions and Packages

To create a complex model in R, it is helpful to utilize the power of R by using the base functions. One of the most common functions is c(), which joins together the entered inputs and stores them as a vector. A vector is considered a type of data structure in R. The discussion of data structures is out of the scope of this book.

```
vec <- c(11, 22, 33, 44, 55)

print(vec)
[1] 11 22 33 44 55
```

Given the vector vec, we can use other R functions to analyze it. For example, we may want to find the mean and standard deviation of the entries and store them as objects.

```
# Calculate the mean
mean_vec <- mean(vec)
sd_vec <- sd(vec)
```

To build a matrix in R, we can use the function matrix().

```
my_mat <- matrix(c(1,2,3,
                   4,5,6), nrow = 2, ncol = 3, byrow = TRUE)
print(my_mat)
```

To index the elements of a vector or a matrix, we use the square bracket [].

```
# print the third value of vec
print(vec[3])

# print the second and third column of my_mat
print(my_mat[,2:3])

# sum the second row of my_mat and the first
```

```
# three elements of vec and assign it to object
my_sum = my_mat[2,] + vec[1:3]
print(my_sum)
```

Data frames are the most common data structure for statistical analysis in R. Unlike matrices, data frames can contain different data types in different columns.

```
# Create a data frame
students <- data.frame(
  name = c("Alice", "Bob", "Charlie"),
  age = c(20, 22, 21),
  grade = c(85.5, 92.0, 88.5),
  passed = c(TRUE, TRUE, TRUE)
)
# Access columns
students$name
# Summary statistics
summary(students)
```

Sometimes, the base R functions are not enough to solve our problem. In such cases, we can use R to write our own function. In the next script, we provide a script for a simple function that takes two numbers as inputs and returns the maximum.

```
my_max <- function(a, b){
    if (a > b) {
        return(a)
    } else {
        return(b)
    }
}
```

Now we can test our function

```
print(my_max(-2, 6))
[1] 6
```

Note that it is the job of the programmer to make sure that the function works properly and does not produce unexpected results.

```
print(my_max("as", "sd"))
[1] "sd"
```

This result is problematic because our function should compare numbers, not strings. R doesn't automatically know our intent, so we need to add input validation to our code. We can do this by checking that both inputs are numeric using `is.numeric()`.

```
my_max <- function(a, b){
    if (!is.numeric(a) | !is.numeric(b)) {
        stop("The inputs should be numeric")
    }
    if (a > b) {
        return(a)
    } else {
        return(b)
    }
}
```

```
print(my_max("as", "sd"))
Error in my_max("as", "sd") : The inputs should be numeric
```

Suppose we save my_max() function in R script named max.R. If we want to use it later in a different R script, we can source it

```
source("max.R")
print(my_max(2, 3))
[1] 3
```

It is important to note that max.R should be located in the working directory; otherwise, R will complain that the file is not found. To check or set your working directory use getwd() or setwd(), respectively.

So far, we have discussed how to source custom functions from R scripts. While sourcing is useful for individual scripts, R packages provide a much more powerful and organized way to extend R's functionality.

A.4 R Packages

What are R Packages?

Packages are collections of functions, datasets, and documentation bundled together for easy distribution and use. For example, there are packages for statistical modeling, data visualization, machine learning, bioinformatics, and much more.

Where to Find Packages

R packages can be downloaded from several repositories:

- **CRAN** (Comprehensive R Archive Network): The main repository with over 18,000 packages
- **Bioconductor**: Specialized repository for bioinformatics packages
- **GitHub**: Development versions and packages not yet on CRAN

Most packages are installed from CRAN using the install.packages() function. For example:

```
# Install a package from CRAN
install.packages("ggplot2")
```

Installing the recode *Package*

This book uses a custom R package called recode, which stands for **Re**gressions in **Co**variances, **De**pendencies, and Graphs. The recode package contains all the datasets and specialized functions used throughout this book, making it easy to reproduce the examples and complete the exercises.

Since recode is hosted on GitHub (for details, see https://github.com/adallak/recode) rather than CRAN, we need to use the devtools package to install it. Here's how:

```
# Step 1: Install devtools (only needed once)
install.packages("devtools")

# Step 2: Install recode from GitHub
devtools::install_github("adallak/recode")
```

The first line installs devtools, which provides tools for installing packages from GitHub. The second line uses the install_github() function from devtools to download and install recode. Note that you only need to install a package once, though you may want to reinstall it occasionally to get updates.

Loading and Using Packages

After installing a package, you need to load it in each R session before using its functions:

```
# Load the recode package
library(recode)

# Now you can access datasets and functions from recode
# For example:
data(globalbank)
```

Getting Help with Packages

Once a package is loaded, you can access its documentation:

```
# View package documentation
help(package = "recode")

# Get help on a specific function
?function_name

# See all available datasets in a package
data(package = "recode")
```

A.5 Importing and Exploring Data

In the previous section, we showed an example of importing data from the R package. In practice, you will often need to import your own data from external files. The most common formats are CSV (comma-separated values) and text files.

The primary function for importing data into R is read.table(), which is flexible enough to handle various file formats. To see all available options, type ?read.table() in the console.

Let's import a hypothetical dataset. First, it's important to understand file paths. R looks for files relative to your working directory. You can check your current working directory with getwd() and change it with setwd().

```
# Check current working directory
getwd()

# Import the data
data <- read.table("Data/toydata.csv", sep = ",", header = TRUE)
```

If R cannot find the file, it will return an error. Make sure the file path is correct and the file exists in the specified location. The first argument, which is expected to be specified as a string, shows the directory of the toydata.csv file, sep = "," indicates that the imported file is comma-separated, and finally header = TRUE specified that the first row of the file contains the variable names.

After importing data, the first step is to examine its structure. The str() function provides a compact summary of the data frame:

```
str(data)
'data.frame': 2676 obs. of  54 variables:
 $ hsba.ln    : num  8.05e-05 2.79e-05 7.03e-05 ...
 $ X8306.to   : num  0.000258 0.000258 0.00179  ...
 $ jpm        : num  1.30e-04 1.03e-04 1.11e-04 ...
##Output is omitted
```

This output tells us that:

- The object is a data frame with 2,676 observations (rows)
- There are 54 variables (columns)
- Each variable is listed with its type (num for numeric) and first few values

Other useful functions for examining data structure include:

```
# View first 6 rows
head(data)

# View last 6 rows
tail(data)

# Get dimensions
dim(data)       # Returns: rows, columns
nrow(data)      # Number of rows
ncol(data)      # Number of columns

# Get variable names
names(data)
colnames(data)
```

Finally, we can produce a quick summary of statistics of the variables using summary() function.

```
# Summary of entire dataset
summary(data)
```

```
# Summary of a single variable
summary(data$hsba.ln)
@      Min.    1st Qu.     Median       Mean    3rd Qu.       Max.
3.520e-06 4.138e-05 8.720e-05 2.615e-04 2.080e-04 1.620e-02@
```

For more specific statistics, you can use individual functions:

```
mean(data$hsba.ln)
sd(data$hsba.ln)              # Standard deviation
var(data$hsba.ln)             # Variance
quantile(data$hsba.ln)        # Quantiles
range(data$hsba.ln)           # Min and max
```

One of the most challenging aspects of any data science project is exploratory data analysis (EDA), which helps you understand the patterns, distributions, and relationships in your data. Graphics are an essential part of EDA.

R has three main graphical systems:

- **Base R graphics**: Simple, fast, and included by default
- **ggplot2**: More sophisticated and customizable (part of tidyverse)
- **lattice**: Good for multi-panel plots

To display any variable in the base R graphics, for example jpm, we use plot() function.

```
# Plot a variable over its index (observation number)
plot(data$jpm,
     main = "JPM Values Over Observations",
     xlab = "Observation Number",
     ylab = "JPM Value",
     pch = 20,
     col = "blue")
```

This type of plot is useful for detecting trends, outliers, or patterns over the sequence of observations.

For examining distributions, other plot types are more appropriate:

```
# Histogram to show distribution
hist(data$jpm,
     main = "Distribution of JPM Values",
     xlab = "JPM Value",
     col = "lightblue",
     breaks = 30)

# Boxplot to identify outliers and quartiles
boxplot(data$jpm,
        main = "Boxplot of JPM Values",
        ylab = "JPM Value",
        col = "lightgreen")

# Scatter plot between two variables
plot(data$hsba.ln, data$jpm,
     main = "Relationship between HSBA and JPM",
     xlab = "HSBA",
```

```
    ylab = "JPM",
    pch = 20,
    col = "darkred")
```

Working with Subsets of Data

You can access specific parts of your data using indexing:

```
# Access a single variable
data$jpm

# Access first 10 rows of a variable
data$jpm[1:10]

# Access multiple variables
data[, c("jpm", "hsba.ln")]

# Filter rows based on conditions
high_values <- data[data$jpm > 0.001, ]

# Count observations meeting a condition
sum(data$jpm > 0.001)
```

Saving Your Work

After cleaning or transforming your data, you may want to save it:

```
# Save as CSV
write.csv(data, "cleaned_data.csv", row.names = FALSE)

# Save as RDS (R's native format, preserves data types)
saveRDS(data, "cleaned_data.rds")

# Save entire workspace
save.image("workspace.RData")
```

Bibliography

K. M. Abadir and J. R. Magnus. *Matrix Algebra*. Econometric Exercises. Cambridge University Press, 2005. ISBN 9781139443647.

K. M. Abadir, W. Distaso, and F. Žikeš. Design-free estimation of variance matrices. *Journal of Econometrics*, 181(2):165–180, 2014. ISSN 0304-4076. doi: https://doi.org/10.1016/j.jeconom.2014.03.010.

S. C. Ahn and A. R. Horenstein. Eigenvalue ratio test for the number of factors. *Econometrica*, 81(3):1203–1227, 2013. doi: https://doi.org/10.3982/ECTA8968.

C. Alakus, D. Larocque, and A. Labbe. Covariance regression with random forests. *BMC Bioinformatics*, 24, 06 2023. doi: 10.1186/s12859-023-05377-y.

M. Aldrin. Multivariate prediction using softly shrunk reduced-rank regression. *The American Statistician*, 54(1):29–34, 2000. doi: 10.1080/00031305.2000.10474504.

T. W. Anderson. Estimating linear restrictions on regression coefficients for multivariate normal distributions. *Annals of Mathematical Statistics*, 22:327–351, 1951.

T. W. Anderson. Asymptotically efficient estimation of covariance matrices with linear structure. *The Annals of Statistics*, 1:135–141, 1973.

T. W. Anderson. *An Introduction to Multivariate Statistics*. Wiley, New York, 2003.

B. Aragam and Q. Zhou. Concave penalized estimation of sparse gaussian bayesian networks. *Journal of Machine Learning Research*, 16(69):2273–2328, 2015. URL http://jmlr.org/papers/v16/aragam15a.html.

I. Archakov and P. R. Hansen. A new parametrization of correlation matrices. *Econometrica*, 89(4):1699–1715, 2021. doi: https://doi.org/10.3982/ECTA16910.

M. Aure, S. Jernström, M. Krohn, H. Vollan, E. Due, E. Rødland, R. Kåresen, P. Ram, Y. Lu, G. Mills, K. Kleivi, A. L. Børresen-Dale, O. Lingjærde, and V. Kristensen. Integrated analysis reveals microrna networks coordinately expressed with key proteins in breast cancer. *Genome Medicine*, 7:21, 2015. doi: 10.1186/s13073-015-0135-5.

J. Bai and K. Li. Maximum likelihood estimation and inference for approximate factor models of high dimension. *The Review of Economics and Statistics*, 98(2): 298–309, 2016.

J. Bai and Y. Liao. Efficient estimation of approximate factor models via penalized maximum likelihood. *Journal of Econometrics*, 191(1):1–18, 2016. ISSN 0304-4076. doi: https://doi.org/10.1016/j.jeconom.2015.10.003. URL https://www.sciencedirect.com/science/article/pii/S0304407615002535.

J. Bai and S. Ng. Determining the number of factors in approximate factor models. *Econometrica*, 70(1):191–221, 2002.

Pierre Baldi and Kurt Hornik. Neural networks and principal component analysis: Learning from examples without local minima. *Neural Networks*, 2(1):53–58, 1989. ISSN 0893-6080. doi: https://doi.org/10.1016/0893-6080(89)90014-2. URL https://www.sciencedirect.com/science/article/pii/0893608089900142.

O. Banerjee, L. E. Ghaoui, and A. d'Aspermont. Model selection through sparse maximum likelihood estimation for multivariate gaussian or binary data. *Journal of Machine Learning Research*, 9:485–516, 2008.

Matteo Barigozzi and Marc Hallin. The dynamic, the static, and the weak: Factor models and the analysis of high-dimensional time series. *Journal of Time Series Analysis*, 47(1):201–219, 2026. doi: https://doi.org/10.1111/jtsa.12837. URL https://onlinelibrary.wiley.com/doi/abs/10.1111/jtsa.12837.

Matteo Barigozzi and Matteo Luciani. Quasi maximum likelihood estimation and inference of large approximate dynamic factor models via the em algorithm, 2024. URL https://arxiv.org/abs/1910.03821.

S. Basu and G. Michailidis. Regularized estimation in sparse high-dimensional time series models. *The Annals of Statistics*, 43(4):1535–1567, 08 2015.

S. Basu and S. Subba Rao. Graphical models for nonstationary time series. *The Annals of Statistics*, 51(4):1453 – 1483, 2023. doi: 10.1214/22-AOS2205. URL https://doi.org/10.1214/22-AOS2205.

H. Battey. Eigen structure of a new class of covariance and inverse covariance matrices. *Bernoulli*, 23(4B):3166 – 3177, 2017. doi: 10.3150/16-BEJ840. URL https://doi.org/10.3150/16-BEJ840.

H. Battey. Eigen structure of a new class of covariance and inverse covariance matrices. *Bernoulli*, 23(4B):3166–3177, 2018. ISSN 13507265.

T. Bedford and R. Cooke. *Probabilistic Risk Analysis: Foundations and Methods*. Cambridge University Press, 2001. ISBN 9780521773201.

T. Bedford and R. M. Cooke. Vines–a new graphical model for dependent random variables. *The Annals of Statistics*, 30(4):1031 – 1068, 2002. doi: 10.1214/aos/1031689016. URL https://doi.org/10.1214/aos/1031689016.

H. Begleiter. EEG Database. UCI Machine Learning Repository, 1999. DOI: https://doi.org/10.24432/C5TS3D.

K. Bello, B. Aragam, and P. Ravikumar. Dagma: Learning dags via m-matrices and a log-determinant acyclicity characterization. In S. Koyejo, S. Mohamed, A. Agarwal, D. Belgrave, K. Cho, and A. Oh, editors, *Advances in Neural Information Processing Systems*, volume 35, pages 8226–8239. Curran Associates, Inc., 2022.

A. Belloni, V. Chernozhukov, and L. Wang. Square-root lasso: pivotal recovery of sparse signals via conic programming. *Biometrika*, 98(4):791–806, 2011. ISSN 00063444, 14643510.

Y. Benjamini and Y. Hochberg. Controlling the false discovery rate: a practical and powerful approach to multiple testing. *Journal of the Royal Statistical Society. Series B*, pages 289–300, 1995.

J. Besag. Statistical analysis of non-lattice data. *Journal of the Royal Statistical Society. Series D (The Statistician)*, 24(3):179–195, 1975. ISSN 00390526, 14679884.

H. Bhatia. Infinitely divisible matrices. *American Math Monthly*, pages 221–235, 2006.

P. J. Bickel and E. Levina. Covariance regularization by thresholding. *Annals of Statistics*, 36:2577–2604, 2008b.

P. J. Bickel and L. Levina. Some theory for Fisher's linear discriminant function, naive Bayes and some alternatives when there are many more variables than observations. *Bernoulli*, 10:989–1010, 2004.

P. J. Bickel and L. Levina. Regularized estimation of large covariance matrices. *The Annals of Statistics*, 36:199–227, 2008a.

J. Bien. Graph-guided banding of the covariance matrix. *Journal of the American Statistical Association*, 114(526):782–792, April 2019. doi: 10.1080/01621459.2018.1442720.

J. Bien and R. J. Tibshirani. Sparse estimation of a covariance matrix. *Biometrika*, 98:807–820, 2011.

R. J. Boik. Spectral models for covariance matrices. *Biometrika*, 89:159–182, 2002.

K. A. Bollen. *Structural Equations with Latent Variables*. John Wiley and Sons, New York, 1989.

W. H. Bonat and B. Jørgensen. Multivariate covariance generalized linear models. *Journal of the Royal Statistical Society. Series C (Applied Statistics)*, 65(5):649–675, 2016. ISSN 00359254, 14679876.

P. Bouboulis, K. Slavakis, and S. Theodoridis. Adaptive learning in complex reproducing kernel hilbert spaces employing wirtinger's subgradients. *IEEE Transactions on Neural Networks and Learning Systems*, 23(3):425–438, 2012.

S. Boyd and L. Vandenberghe. *Convex Optimization*. Cambridge University Press, 2004.

S. Boyd, N. Parikh, E. Chu, B. Peleato, and J. Eckstein. Distributed optimization and statistical learning via the alternating direction method of multipliers. *Foundations and Trends® in Machine Learning*, 3(1):1–122, 2011.

L. Breiman. Heuristics of instability and stabilization in model selection. *The Annals of Statistics*, 24:2350–2383, 1996.

L. Breiman and J. H. Friedman. Predicting multivariate responses in multiple linear regression. *Journal of the Royal Statistical Society. Series B*, 59:3–37, 1997.

D. R. Brillinger. The Identification of Point Process Systems. *The Annals of Probability*, 3(6):909 – 924, 1975. doi: 10.1214/aop/1176996218. URL https://doi.org/10.1214/aop/1176996218.

D. R. Brillinger. *Time Series: Data Analysis and Theory*. Society for Industrial and Applied Mathematics, Philadelphia, PA, USA, 1981.

D. R. Brillinger. Remarks concerning graphical models for time series and point processes. *Brazilian Review of Econometrics*, 16(1), 1996.

P. J. Brockwell and R. A. Davis. *Time Series: Theory and Methods: Theory and Methods*. Springer Series in Statistics. Springer New York, 1991. ISBN 9780387974293.

P. J. Brown, N. D. Le, and J. V. Zidek. *Inference for a Covariance Matrix. In Aspects of Uncertainty (P.R. Freeman and A.F.M. Smith, ed.)*. Wiley, Chichester, U.K, 1994.

M. W. Browne. Generalized least squares estimators in the analysis of covariance structures. *South African Statistical Journal*, 8(1):1–24, 1974. doi: 10.10520/AJA0038271X_175.

P. Bühlmann and S. van de Geer. *Statistics for High-Dimensional Data: Methods, Theory and Applications*. Springer, Springer Heidelberg, 2011.

A. J. Butte, P. Tamayo, D. Slonim, T. R. Golub, and I. S. Kohane. Discovering functional relationships between rna expression and chemotherapeutic susceptibility using relevance networks. *Proceedings of the National Academy of Sciences*, 97(22):12182–12186, 2000. doi: 10.1073/pnas.220392197. URL https://www.pnas.org/doi/abs/10.1073/pnas.220392197.

P. Bühlmann, J. Peters, and J. Ernest. CAM: Causal additive models, high-dimensional order search and penalized regression. *The Annals of Statistics*, 42 (6):2526 – 2556, 2014. doi: 10.1214/14-AOS1260.

T. T. Cai and W. Liu. Adaptive thresholding for sparse covariance estimation. *Journal of the American Statistical Association*, 106:672–684, 2011.

T. T. Cai, W. Liu, and X. Luo. A constrained ℓ_1 minimization approach to sparse precision matrix estimation. *Journal of the American Statistical Association*, 106: 594–607, 2011.

T. T. Cai, H. Li, W. Liu, and J. Xie. Covariate-adjusted precision matrix estimation with an application in genetical genomics. *Biometrika*, 100(1):139–156, 11 2012. ISSN 0006-3444. doi: 10.1093/biomet/ass058.

T. T. Cai, Z. Ren, and H. H. Zhou. Estimating structured high-dimensional covariance and precision matrices: Optimal rates and adaptive estimation. *Electronic Journal of Statistics*, 10:1–59, 2016.

E. J. Candes and J. Romberg. Quantitative robust uncertainty principles and optimally sparse decompositions. *Foundations of Computational Mathematics*, 6(2):227–254, April 2006. ISSN 1615-3375.

R. J. Carroll and D. Ruppert. *Transformation and Weighting in Regression*. Chapman and Hall, London, 1988.

R. B. Cattell. The scree test for the number of factors. *Multivariate Behavioral Research*, 1(2):245–276, 1966.

G. Chamberlain. Funds, factors, and diversification in arbitrage pricing models. *Econometrica*, 51(5):1305–1323, 1983. ISSN 00129682, 14680262.

G. Chamberlain and M. Rothschild. Arbitrage, factor structure and meanvariance analysis in large asset markets. *Econometrica*, 51:1305–1324, 1983.

V. Chandrasekaran, P. A. Parrilo, and A. S. Willsky. Latent variable graphical model selection via convex optimization. *The Annals of Statistics*, 40(4):1935 – 1967, 2012. doi: 10.1214/11-AOS949. URL https://doi.org/10.1214/11-AOS949.

J. Chang, B. Guo, and Q. Yao. Principal component analysis for second-order stationary vector time series. *The Annals of Statistics*, 46(5):2094 – 2124, 2018. doi: 10.1214/17-AOS1613. URL https://doi.org/10.1214/17-AOS1613.

C. Chapman and E. Feit. *R For Marketing Research and Analytics*. 01 2019. ISBN 978-3-030-14315-2. doi: 10.1007/978-3-030-14316-9.

A. Chattopadhyay, D. Hoh, D. Kramer, T. Maiti, and S. Sinha. Cmple: Correlation modeling to decode photosynthesis using the minorize–maximize algorithm. *Journal of Agricultural, Biological and Environmental Statistics*, 2024. doi: 10.1007/s13253-024-00627-9.

S. Chaudhuri, M. Drton, and T. S. Richardson. Estimation of a covariance matrix with zeros. *Biometrika*, 94:199–216, 2007.

J. Chen and Z. Chen. Extended Bayesian information criteria for model selection with large model spaces. *Biometrika*, 95(3):759–771, 2008. ISSN 0006-3444. doi: 10.1093/biomet/asn034.

M. Chen, Z. Ren, H. Zhao, and H. Zhou. Asymptotically normal and efficient estimation of covariate-adjusted gaussian graphical model. *Journal of the American Statistical Association*, 111(513):394–406, 2016. doi: 10.1080/01621459.2015.1010039.

Mengjie Chen, Chao Gao, Zhao Ren, and Harrison H. Zhou. Sparse cca via precision adjusted iterative thresholding. *arXiv: Statistics Theory*, 2013.

W. Chen, M. Drton, and Y. S. Wang. On causal discovery with an equal-variance assumption. *Biometrika*, 106(4):973–980, 2019. ISSN 0006-3444. doi: 10.1093/biomet/asz049.

Z. Chen and D. Dunson. Random effects selection in linear mixed models. *Biometrics*, 59:762–769, 2003.

J. Cheng, E. Levina, P. Wang, and J. Zhu. A sparse ising model with covariates. *Biometrics*, 70(4):943–953, 2014.

D. M. Chickering. *Learning Bayesian Networks is NP-Complete*, pages 121–130. Springer-Verlag, learning from data: artificial intelligence and statistics v edition, 1996.

D. M. Chickering. Optimal structure identification with greedy search. *Journal of Machine Learning Research*, 3:507–554, 2002. ISSN 1532-4435. doi: 10.1162/153244303321897717. URL https://doi.org/10.1162/153244303321897717.

T. Y. M. Chiu, T. Leonard, and K. W. Tsui. The matrix-logarithm covariance model. *Journal of the American Statistical Association*, 91:198–210, 1996.

E. S. Christensen. Statistical properties of i-projections within exponential families. *Scandinavian Journal of Statistics*, 16(4):307–318, 1989.

L. Cibinel, A. Roverato, and V. Vinciotti. A unified approach to penalized likelihood estimation of covariance matrices in high dimensions, 2024. URL https://arxiv.org/abs/2410.02403.

G. Connor and A. R. Korajczyk. Performance measurement with the arbitrage pricing theory: A new framework for analysis. *Journal of Financial Economics*, 15(3):373–394, 1986. ISSN 0304-405X. doi: https://doi.org/10.1016/0304-405X(86)90027-9. URL https://www.sciencedirect.com/science/article/pii/0304405X86900279.

G. Connor and O. Linton. Semiparametric estimation of a characteristic-based factor model of common stock returns. *Journal of Empirical Finance*, 14(5):694–717, 2007. ISSN 0927-5398. doi: https://doi.org/10.1016/j.jempfin.2006.10.001.

G. Connor, M. Hagmann, and O. Linton. Efficient semiparametric estimation of the fama–french model and extensions. *Econometrica*, 80(2):713–754, 2012. doi: https://doi.org/10.3982/ECTA7432.

R. D. Cook and L. Forzani. Principal Fitted Components for Dimension Reduction in Regression. *Statistical Science*, 23(4):485 – 501, 2008. doi: 10.1214/08-STS275. URL https://doi.org/10.1214/08-STS275.

D. R. Cox and N. Wermuth. *Multivariate Dependencies: Models, Analysis and Interpretation.* Chapman & Hall/CRC Monographs on Statistics & Applied Probability. Taylor & Francis, 1996. ISBN 9780412754104.

H. Cramer. On the theory of stationary random processes. *Annals of Mathematics*, 41(1):215–230, 1940.

N. Cressie and J. L. Davidson. Image analysis with partially ordered markov models. *Computational Statistics and Data Analysis*, 29(1):1–26, 1998. ISSN 0167-9473. doi: https://doi.org/10.1016/S0167-9473(98)00052-8. URL https://www.sciencedirect.com/science/article/pii/S0167947398000528.

N. A. C. Cressie. *Statistics for Spatial Data.* John Wiley and Sons, Ltd, 1993. ISBN 9781119115151. doi: https://doi.org/10.1002/9781119115151.

M. Crowder. On the use of a working correlation matrix in using generalised linear models for repeated measures. *Biometrika*, 82(2):407–410, 1995. ISSN 00063444, 14643510.

C. Czado. *Analyzing Dependent Data with Vine Copulas: A Practical Guide With R (Lecture Notes in Statistics, 222).* Springer, paperback edition, May 2019. ISBN 978-3-030-13784-7.

C. Czado and T. Nagler. Vine copula based modeling. *Annual Review of Statistics and Its Application*, 9(Volume 9, 2022):453–477, 2022. doi: https://doi.org/10.1146/annurev-statistics-040220-101153.

R. Dahlhaus. Graphical interaction models for multivariate time series. *Metrika*, 51 (2):157–172, 2000.

R. Dahlhaus and M. Eichler. Causality and graphical models in time series analysis. In Green P., Hjort N., and Richardson S., editors, *Highly Structured Stochastic Systems*, chapter 4. Oxford University Press, Oxford, 2003.

A. Dallakyan and Y. Ni. Generalized criterion for identifiability of additive noise models using majorization. In *Proceedings of The 28th International Conference on Artificial Intelligence and Statistics*, volume 258 of *Proceedings of Machine Learning Research*, pages 37–45. PMLR, 03–05 May 2025. URL https://proceedings.mlr.press/v258/dallakyan25a.html.

A. Dallakyan and M. Pourahmadi. Learning bayesian networks through birkhoff polytope: A relaxation method. *ArXiv*, abs/2107.01658, 2021.

A. Dallakyan, R. Kim, and M. Pourahmadi. Time series graphical lasso and sparse var estimation. *Computational Statistics and Data Analysis*, 176(C), 2022. ISSN 0167-9473. URL https://doi.org/10.1016/j.csda.2022.107557.

M. Daniele, W. Pohlmeier, and A. Zagidullina. A sparse approximate factor model for high-dimensional covariance matrix estimation and portfolio selection. *Journal of Financial Econometrics*, 23(1):nbae017, 07 2024. ISSN 1479-8409. doi: 10.1093/jjfinec/nbae017. URL https://doi.org/10.1093/jjfinec/nbae017.

J. Das and H. You. Hint: High-quality protein interactomes and their applications in understanding human disease. *BMC Systems Biology*, 6(92), 2012. doi: https://doi.org/10.1186/1752-0509-6-92.

A. Datta, S. Banerjee, A. O. Finley, and A. E. Gelfand. Hierarchical nearest-neighbor gaussian process models for large geostatistical datasets. *Journal of the American Statistical Association*, 111(514):800–812, 2016. doi: 10.1080/01621459.2015.1044091.

C. Davis and W. M. Kahan. The rotation of eigenvectors by a perturbation. iii. *SIAM Journal on Numerical Analysis*, 7(1):1–46, 1970. ISSN 00361429.

R. A. Davis, P. Zang, and T. Zheng. Sparse vector autoregressive modeling. *Journal of Computational and Graphical Statistics*, 25(4):1077–1096, 2016.

T. A. Davis. *Direct Methods for Sparse Linear Systems*. Society for Industrial and Applied Mathematics, Philadelphia, PA, 2006.

C. P. de Campos and Q. Ji. Efficient structure learning of bayesian networks using constraints. *Journal of Machine Learning Research*, 12:663–689, March 2011. ISSN 1532-4435.

N. Deb, A. Kuceyeski, and S. Basu. Regularized estimation of sparse spectral precision matrices, 2024.

M. Deistler and W. Scherrer. *Time Series Models*. Lecture Notes in Statistics. 2022. ISBN 9783031132131. URL https://books.google.com/books?id=L7CWEAAAQBAJ.

M. Demirer, F. X. Diebold, L. Liu, and K. Yilmaz. Estimating global bank network connectedness. *Journal of Applied Econometrics*, 33(1):1–15, 2018.

P. Diggle and P. J. Ribeiro. *Model-based Geostatistics*. Springer Series in Statistics. Springer New York, 2007. ISBN 9780387485362.

P. Diggle, K. Y. Liang, S. L. Zeger, and P. J. Heagerty. *Analysis of Longitudinal Data, Second Edition*. Clarendon Press, Oxford, 2002.

X. Ding, Z. Qiu, and X. Chen. Sparse transition matrix estimation for high-dimensional and locally stationary vector autoregressive models. *Electronic Journal of Statistics*, 11(2):3871–3902, 2017.

P. Domingos and M. Pazzani. On the optimality of the simple bayesian classifier under zero-one loss. *Machine Learning*, 29:103–130, 1997.

D. Donoho, M. Gavish, and I. Johnstone. Optimal shrinkage of eigenvalues in the spiked covariance model. *The Annals of Statistics*, 46, 2016. doi: 10.1214/17-AOS1601.

D. Donoho, M. Gavish, and E. Romanov. ScreeNOT: Exact MSE-optimal singular value thresholding in correlated noise. *The Annals of Statistics*, 51(1):122 – 148, 2023. doi: 10.1214/22-AOS2232. URL https://doi.org/10.1214/22-AOS2232.

M. Drton and T. S. Richardson. Graphical methods for efficient likelihood inference in gaussian covariance models. *Journal of Machine Learning Research*, 9:893–914, jun 2008. ISSN 1532-4435.

S. Dudoit, J. Fridlyand, and T. P. Speed. Comparison of discrimination methods for the classification of tumors using gene expression data. *Journal of the American Statistical Association*, 97:77–87, 2002.

L. Dumitrescu, W. Qian, and J. N. K. Rao. A Weighted Composite Likelihood Approach to Inference from Clustered Survey Data Under a Two-Level Model. *Sankhya A: The Indian Journal of Statistics*, 83(2):814–843, 2021. doi: 10.1007/s13171-020-00234-.

B. Efron. *Large-Scale Inference: Empirical Bayes Methods for Estimation, Testing, and Prediction*. Cambridge University Press, Cambridge, 2010.

B. Efron, T. Hastie, I. Johnstone, and R. Tibshirani. Least angle regression. *The Annals of Statistics*, 32:407–499, 2004.

R. F. Engle. Dynamic conditional correlation: a simple class of multivariate generalized autoregressive conditional heteroscedasticity models. *Journal of Business and Economic Statistics*, 20:339–350, 2002.

R. F. Engle and K. F. Kroner. Multivariate simultaneous generalized arch. *Econometric Theory*, 11:122 – 150, 1995.

R. F. Engle, O. Ledoit, and M. Wolf. Large dynamic covariance matrices. *Journal of Business & Economic Statistics*, 37(2):363–375, 2019.

I. G. Evans. Bayesian estimation of parameters of a multivariate normal distribution. *Journal of the Royal Statistical Society: Series B (Methodological)*, 27(2):279–283, 1965. ISSN 0035-9246. doi: 10.1111/j.2517-6161.1965.tb01494.x. URL https://doi.org/10.1111/j.2517-6161.1965.tb01494.x.

E. Fama and K. R. French. The cross-section of expected stock returns. *Journal of Finance*, 47:427–465, 1992.

J. Fan and R. Li. Variable selection via nonconcave penalized likelihood and its oracle properties. *Journal of the American Statistical Association*, 96:1348–1360, 2001.

J. Fan and J. Lv. Sure independence screening for ultrahigh dimensional feature space. *Journal of Royal Statistical Society, B*, 70:849–911, 2008.

J. Fan, Y. Liao, and M. Mincheva. Large covariance estimation by thresholding principal orthogonal complements. *Preprint, Princeton University*, 2011.

J. Fan, Y. Liao, and M. Mincheva. Large Covariance Estimation by Thresholding Principal Orthogonal Complements. *Journal of the Royal Statistical Society Series B: Statistical Methodology*, 75(4):603–680, 08 2013. ISSN 1369-7412. doi: 10.1111/rssb.12016. URL https://doi.org/10.1111/rssb.12016.

J. Fan, Y. Liao, and H. Liu. An overview of the estimation of large covariance and precision matrices. *The Econometrics Journal*, 19(1):C1–C32, 2016a. doi: https://doi.org/10.1111/ectj.12061. URL https://onlinelibrary.wiley.com/doi/abs/10.1111/ectj.12061.

J. Fan, Y. Liao, and W. Wang. Projected principal component analysis in factor models. *The Annals of Statistics*, 44(1):219 – 254, 2016b. doi: 10.1214/15-AOS1364.

J. Fan, Y. Ke, and K. Wang. Factor-adjusted regularized model selection. *Journal of Econometrics*, 216(1):71–85, 2020a. ISSN 0304-4076. doi: https://doi.org/10.1016/j.jeconom.2020.01.006. Annals Issue in honor of George Tiao: Statistical Learning for Dependent Data.

J. Fan, Y. Ke, and K. Wang. Factor-adjusted regularized model selection. *Journal of Econometrics*, 216(1):71–85, 2020b. ISSN 0304-4076. doi: https://doi.org/10.1016/j.jeconom.2020.01.006. URL https://www.sciencedirect.com/science/article/pii/S0304407620300117. Annals Issue in honor of George Tiao: Statistical Learning for Dependent Data.

J. Fan, R. Li, C. H. Zhang, and H. Zou. *Statistical Foundations of Data Science*. Chapman and Hall/CRC, New York, 2020c.

S. Fan and R. J. Hyndman. Forecasting electricity demand in australian national electricity market. In *2012 IEEE Power and Energy Society General Meeting*, pages 1–4, 2012. doi: 10.1109/PESGM.2012.6345304.

J. Felsenstein. Maximum likelihood and minimum-steps methods for estimating evolutionary trees from data on discrete characters. *Systematic Zoology*, 22(3):240–249, 1973. ISSN 00397989.

J. Felsenstein. Evolutionary trees from dna sequences: A maximum likelihood approach. *Journal of Molecular Evolution*, 17:368–376, 1981.

M. Fiecas, H. Ombao, C. Linkletter, W. Thompson, and J. Sanes. Functional connectivity: Shrinkage estimation and randomization test. *NeuroImage*, 49(4):3005–3014, 2010.

M. Fiecas, C. Leng, W. Liu, and Y. Yu. Spectral analysis of high-dimensional time series. *Electronic Journal of Statistics*, 13:4079–4101, 01 2019. doi: 10.1214/19-EJS1621.

G. M. Fitzmaurice, N. M. Laird, and J. H. Ware. *Applied Longitudinal Analysis*. Wiley Series in Probability and Statistics. Wiley, 2011. ISBN 9780470380277.

B. Flury. *Common Principal Components and Related Multivariate Models*. Wiley, New York, 1988.

M. Forni and M. Lippi. The generalized dynamic factor model: Representation theory. *Econometric Theory*, 17(6):1113–1141, 2001. ISSN 02664666, 14694360.

M. Forni, M. Hallin, M. Lippi, and L. Reichlin. The generalized dynamic-factor model: Identification and estimation. *The Review of Economics and Statistics*, 82 (4):540–554, 2000. ISSN 00346535, 15309142. URL http://www.jstor.org/stable/2646650.

M. Forni, D. Giannone, M. Lippi, and L. Reichlin. Opening the black box: Structural factor models with large cross sections. *Econometric Theory*, 25(5):1319–1347, 2009. ISSN 02664666, 14694360.

E. B. Fox and D. B. Dunson. Bayesian nonparametric covariance regression. *Journal of Machine Learning Research*, 16(77):2501–2542, 2015. URL http://jmlr.org/papers/v16/fox15a.html.

R. Foygel and M. Drton. Extended bayesian information criteria for gaussian graphical models. In *NIPS*, pages 604–612, 2010.

A. M. Franks. Reducing subspace models for large-scale covariance regression. *Biometrics*, 78(4):1604–1613, 2022. doi: https://doi.org/10.1111/biom.13531.

A. M. Franks and P. Hoff. Shared subspace models for multi-group covariance estimation. *Journal of Machine Learning Research*, 20(171):1–37, 2019. URL http://jmlr.org/papers/v20/18-484.html.

J. Friedman, T. Hastie, H. Höfling, and R. Tibshirani. Pathwise coordinate optimization. *The Annals of Applied Statistics*, 1(2):302 – 332, 2007. doi: 10.1214/07-AOAS131.

J. Friedman, T. Hastie, and R. Tibshirani. Sparse inverse covariance estimation with the graphical lasso. *Biostatistics*, 9:432–441, 2008.

K. Friston. Functional and effective connectivity: A review. *Brain Connectivity*, 1: 13–36, 2011. doi: 10.1089/brain.2011.0008.

R. Furrer and T. Bengtsson. Estimation of high-dimensional prior and posterior covariance matrices in kalman filter variants. *Journal of Multivariate Analysis*, 98: 227–255, 2007.

Connor G. and Korajczyk A. R. Risk and return in an equilibrium apt: Application of a new test methodology. *Journal of Financial Economics*, 21(2):255–289, 1988. ISSN 0304-405X. doi: https://doi.org/10.1016/0304-405X(88)90062-1. URL https://www.sciencedirect.com/science/article/pii/0304405X88900621.

M. Gao, Y. Ding, and B. Aragam. A polynomial-time algorithm for learning nonparametric causal graphs. In H. Larochelle, M. Ranzato, R. Hadsell, M. F. Balcan, and H. Lin, editors, *Advances in Neural Information Processing Systems*, volume 33, pages 11599–11611. Curran Associates, Inc., 2020.

Z. Gao and R. S. Tsay. Modeling high-dimensional time series: A factor model with dynamically dependent factors and diverging eigenvalues. *Journal of the American Statistical Association*, 117(539):1398–1414, 2022. doi: 10.1080/01621459.2020.1862668.

J. Geweke. *The Dynamic Factor Analysis of Economic Time Series Models*. SSRI workshop series. Social Systems Research Institute, University of Wisconsin-Madison, 1977.

V. Gioia, M. Fasiolo, J. Browell, and R. Bellio. Additive covariance matrix models: Modeling regional electricity net-demand in great britain. *Journal of the American Statistical Association*, 0(0):1–13, 2024. doi: 10.1080/01621459.2024.2412361.

C. Glymour, K. Zhang, and P. Spirtes. Review of causal discovery methods based on graphical models. *Frontiers in Genetics*, 10, 2019.

G. H. Golub and C. F. Van Loan. *Matrix Computations*. The Johns Hopkins University Press, Baltimore, 1996.

C. W. J. Granger. Investigating causal relations by econometric models and cross-spectral methods. *Econometrica*, 37(3):424–438, 1969. ISSN 00129682, 14680262.

R. M. Gray. Toeplitz and circulant matrices: A review. *Foundations and Trends® in Communications and Information Theory*, 2(3):155–239, 2006. ISSN 1567-2190. doi: 10.1561/0100000006. URL http://dx.doi.org/10.1561/0100000006.

J. Guinness. Permutation and grouping methods for sharpening gaussian process approximations. *Technometrics*, 60(4):415–429, 2018. doi: 10.1080/00401706.2018.1437476.

J. Guo, L. Levina, G. Michailidis, and J. Zhu. Joint estimation of multiple graphical models. *Biometrika*, 98:432–441, 2011.

L. R. Haff. Empirical bayes estimation of the multivariate normal covariance matrix. *The Annals of Statistics*, 8:586–597, 1980.

M. Hallin and M. Lippi. Factor models in high-dimensional time series—a time-domain approach. *Stochastic Processes and their Applications*, 123(7):2678–2695, 2013. ISSN 0304-4149. doi: https://doi.org/10.1016/j.spa.2013.04.001. A Special Issue on the Occasion of the 2013 International Year of Statistics.

M. Hallin, M. Lippi, M. Barigozzi, M. Forni, and P. Zaffaroni. *Time Series in High Dimensions*. WORLD SCIENTIFIC, 2020. doi: 10.1142/11204.

J. M. Hammersley and P. Clifford. Markov fields on finite graphs and lattices. 1971.

F. Han, H. Lu, and H. Liu. A direct estimation of high dimensional stationary vector autoregressions. *Journal of Machine Learning Research*, 16(97):3115–3150, 2015.

P. Han and P. X. K. Song. A note on improving quadratic inference functions using a linear shrinkage approach. *Statistics and Probability Letters*, 81(3):438–445, 2011. ISSN 0167-7152. doi: https://doi.org/10.1016/j.spl.2010.12.010.

N. Hans, N. Klein, F. Faschingbauer, M. Schneider, and A. Mayr. Boosting distributional copula regression. *Biometrics*, 79(3):2298–2310, 2023. doi: https://doi.org/10.1111/biom.13765.

L. P. Hansen. Large sample properties of generalized method of moments estimators. *Econometrica*, 50(4):1029–1054, 1982. ISSN 00129682, 14680262.

A. Harvey, E. Ruiz, and N. Shephard. Multivariate stochastic variance models. *The Review of Economic Studies*, 61(2):247–264, 1994. ISSN 00346527, 1467937X.

T. Hastie, R. Tibshirani, and J. Friedman. *The Elements of Statistical Learning: Data Mining, Inference, and Prediction, Second Edition*, Springer, New York, 2009.

T. Hastie, R. Tibshirani, and M. Wainwright. *Statistical Learning with Sparsity: The Lasso and Generalizations*. Chapman & Hall/CRC, 2015. ISBN 1498712169.

T. J. Hastie and R. J. Tibshirani. *Generalized Additive Models*. Chapman & HALL, CRC, 1990.

M. Heide, B. R. Auer, and F. Schuhmacher. Optimal versus naive diversification in commodity futures markets. *Journal of Futures Markets*, 2024. doi: https://doi.org/10.1002/fut.22550.

H. Hoefling. A path algorithm for the fused lasso signal approximator. *Journal of Computational and Graphical Statistics*, 19(4):984–1006, 2010. doi: 10.1198/jcgs.2010.09208.

P. Hoff. A hierarchical eigenmodel for pooled covariance estimation. *Journal of the Royal Statistical Society Series B*, 71:971–992, 2009.

P. Hoff and Niu. A covariance regression model. Technical Report arXiv:1102.5721v1 [stat.ME], University of Washington, 2011.

H. Höfling and R. Tibshirani. Estimation of sparse binary pairwise markov networks using pseudo-likelihoods. *Journal of Machine Learning Research*, 10(32):883–906, 2009.

H. Hotelling. The most predictable criterion. *The Journal of Educational Psychology*, 26:139–143, 1935.

T. Hothorn, L. Most, and P. Bühlmann. Most Likely Transformations. *Scandinavian Journal of Statistics*, 45(1):110–134, March 2018. doi: 10.1111/sjos.12291.

J. Hu, Y. Chen, C. Leng, and C. Y. Tang. Applied regression analysis of correlations for correlated data. *The Annals of Applied Statistics*, 18(1):184 – 198, 2024. doi: 10.1214/23-AOAS1785.

H. Huang and Y. Sun. Hierarchical low rank approximation of likelihoods for large spatial datasets. *Journal of Computational and Graphical Statistics*, 27(1):110–118, 2018. doi: 10.1080/10618600.2017.1356324.

J. Z. Huang, N. Liu, M. Pourahmadi, and L. Liu. Covariance matrix selection and estimation via penalized normal likelihood. *Biometrika*, 93:85–98, 2006.

J. Z. Huang, L. Liu, and N. Liu. Estimation of large covariance matrices of longitudinal data with basis function approximations. *Journal of Statistical Computation and Graphics*, 16:189–209, 2007.

R. M. Hutchison, T. Womelsdorf, E. Allen, P. Bandettini, V. Calhoun, M. Corbetta, S. Della Penna, J. Duyn, G. Glover, J. Gonzalez-Castillo, D. Handwerker, S. Keilholz, V. Kiviniemi, D. Leopold, F. Pasquale, O. Sporns, M. Walter, and C. Chang. Dynamic functional connectivity: Promise, issues, and interpretations. *NeuroImage*, 80, 05 2013. doi: 10.1016/j.neuroimage.2013.05.079.

A. Hyvärinen, J. Karhunen, and E. Oja. *Independent Component Analysis*. Wiley, 2004. ISBN 9780471464198.

E. Ising. Beitrag zur Theorie des Ferromagnetismus. *Zeitschrift für Physik*, 31:253 – 258, 1925. doi: 10.1007/BF02980577.

W. James and C. Stein. Estimation with quadratic loss. *Proceedings of the Fourth Berkeley Sympossium on Mathematical and Statistical Probability (Jerzy Neyman, ed.)*, 1:361–379, 1961.

G. Jiang, S. K. Sarkar, and F. Hsuan. A likelihood ratio test and its modifications for the homogeneity of the covariance matrices of dependent multivariate normals. *Journal of Statistical Planning and Inference*, 81:95–111, 1999.

H. Joe. Families of m-variate distributions with given margins and m(m-1)/2 bivariate dependence parameters. *Lecture Notes-Monograph Series*, 28:120–141, 1996. ISSN 07492170.

H. Joe. *Dependence Modeling with Copulas*. Chapman & Hall/CRC Monographs on Statistics & Applied Probability. Taylor & Francis, 2014. ISBN 9781466583221. URL https://books.google.com/books?id=09ThAwAAQBAJ.

N. A. Johnson. A dynamic programming algorithm for the fused lasso and l 0-segmentation. *Journal of Computational and Graphical Statistics*, 22(2):246–260, 2013. doi: 10.1080/10618600.2012.681238.

R. A. Johnson and D. W. Wichern. *Applied Multivariate Statistical Analysis*. Pearson Prentice Hall, New Jersey, 2008.

I. M. Johnstone. On the distribution of the largest eigenvalue in principal component analysis. *Annals of Statistics*, 29:295–327, 2001.

I. M. Johnstone and A. Y. Lu. Sparse principal components analysis. *Journal of the American Statistical Association*, 104:682–693, 2009a.

I. M. Johnstone and A. Y. Lu. On the consistency and sparsity for principal component analysis in high dimensions. *Journal of the American Statistical Association*, 104: 682–693, 2009b.

I. T. Jolliffe. *Principal Component Analysis*. Springer Series in Statistics. Springer, 2002. ISBN 9780387954424. URL https://books.google.com/books?id=TtVF-ao4fI8C.

I. T. Jolliffe, N. T. Trendafilov, and M. Uddin. A modified principal component technique based on the lasso. *Journal of Computational and Graphical Statistics*, 12: 531–547, 2003.

A. Jung, G. Hannak, and N. Goertz. Graphical lasso based model selection for time series. *IEEE Signal Processing Letters*, 22:1781–1785, 2015.

H. Kaiser. The varimax criterion for analytic rotation in factor analysis. *Psychometrika*, 23(3):187–200, September 1958. doi: 10.1007/BF02289233. URL https://ideas.repec.org/a/spr/psycho/v23y1958i3p187-200.html.

M. Kalisch and P. Bühlmann. Estimating high-dimensional directed acyclic graphs with the pc-algorithm. *Journal of Machine Learning Research*, 8(22):613–636, 2007.

M. Kalisch, M. Mächler, D. Colombo, M. H. Maathuis, and P. Bühlmann. Causal inference using graphical models with the R package pcalg. *Journal of Statistical Software*, 47(11):1–26, 2012. doi: 10.18637/jss.v047.i11.

I. E. Kaporin. An alternative approach to estimating the iteration number of the conjugate gradient method. *Numerical Methods and Software*, pages 53–72, 1990.

T. Kariya and H. Kurata. *Generalized Least Squares*. Wiley Series in Probability and Statistics. Wiley, 2004. ISBN 9780470866986. URL https://books.google.com/books?id=KtDkJe8srP4C.

M. Katzfuss and J. Guinness. A General Framework for Vecchia Approximations of Gaussian Processes. *Statistical Science*, 36(1):124 – 141, 2021. doi: 10.1214/19-STS755. URL https://doi.org/10.1214/19-STS755.

G. Kauermann. On a dualization of graphical gaussian models. *Scandinavian Journal of Statistics*, 23(1):105–116, 1996.

J. L. Kazdan. A short note on tridiagonal matrices. https://www2.math.upenn.edu/~kazdan/AMCS602/tridiag-short.pdf, 2006. Accessed: 2025-07-31.

K. Khare, S. Y. Oh, and B. Rajaratnam. A convex pseudolikelihood framework for high dimensional partial correlation estimation with convergence guarantees. *Journal of the Royal Statistical Society Series B: Statistical Methodology*, 77(4): 803–825, 09 2014. ISSN 1369-7412. doi: 10.1111/rssb.12088. URL https://doi.org/10.1111/rssb.12088.

K. Khare, S. Y. Oh, S. Rahman, and B. Rajaratnam. A scalable sparse cholesky based approach for learning high-dimensional covariance matrices in ordered data. *Machine Learning*, 108(12):2061–2086, 2019. doi: 10.1007/s10994-019-05810-5.

K. Khare, S. Rahman, B. Rajaratnam, and J. Zhou. Scalable and non-iterative graphical model estimation, 2024. URL https://arxiv.org/abs/2408.11718.

R. Kim and I. Gaynanova. A sparse linear model for positive definite estimation of covariance matrices, 2025. URL https://arxiv.org/abs/2503.09026.

R. Kim and J. Zhang. High-dimensional covariance regression with application to co-expression qtl detection, 2024. URL https://arxiv.org/abs/2404.02093.

S. J. Kim, K. Koh, S. Boyd, and D. Gorinevsky. ℓ_1 trend filtering. *SIAM Review*, 51 (2):339–360, 2009. doi: 10.1137/070690274.

N. Klein. Distributional regression for data analysis. *Annual Review of Statistics and Its Application*, 11(Volume 11, 2024):321–346, 2024. ISSN 2326-831X. doi: https://doi.org/10.1146/annurev-statistics-040722-053607. URL https://www.annualreviews.org/content/journals/10.1146/annurev-statistics-040722-053607.

N. Klein, T. Kneib, S. Klasen, and S. Lang. Bayesian Structured Additive Distributional Regression for Multivariate Responses. *Journal of the Royal Statistical Society Series C: Applied Statistics*, 64(4):569–591, 12 2015. doi: 10.1111/rssc.12090.

N. Klein, T. Hothorn, L. Barbanti, and T. Kneib. Multivariate conditional transformation models. *Scandinavian Journal of Statistics*, 49(1):116–142, March 2022. doi: 10.1111/sjos.12501.

T. Kneib and L. Fahrmeir. A mixed model approach for geoadditive hazard regression. *Scandinavian Journal of Statistics*, 34(1):207–228, 2007. doi: https://doi.org/10.1111/j.1467-9469.2006.00524.x. URL https://onlinelibrary.wiley.com/doi/abs/10.1111/j.1467-9469.2006.00524.x.

T. Kneib, A. Silbersdorff, and B. Säfken. Rage against the mean – a review of distributional regression approaches. *Econometrics and Statistics*, 26:99–123, 2023. ISSN 2452-3062. doi: https://doi.org/10.1016/j.ecosta.2021.07.006. URL https://www.sciencedirect.com/science/article/pii/S2452306221000824.

D. Koller and N. Friedman. *Probabilistic Graphical Models: Principles and Techniques - Adaptive Computation and Machine Learning*. The MIT Press, 2009. ISBN 0262013193.

J. Krampe and E. Paparoditis. Frequency domain statistical inference for high-dimensional time series. *Journal of the American Statistical Association*, 0(0): 1–13, 2025. doi: 10.1080/01621459.2025.2479244.

Pavel Krupskii and Harry Joe. Factor copula models for multivariate data. *Journal of Multivariate Analysis*, 120:85–101, 2013. ISSN 0047-259X. doi: https://doi.org/10.1016/j.jmva.2013.05.001. URL https://www.sciencedirect.com/science/article/pii/S0047259X13000870.

G. Lafit, F. Tuerlinckx, I. Myin-Germeys, and E. Ceulemans. A partial correlation screening approach for controlling the false positive rate in sparse gaussian graphical models. *Scientific Reports*, 9, 11 2019. doi: 10.1038/s41598-019-53795-x.

P. Lai, W. Liang, F. Wang, and Q. Zhang. Feature screening of quadratic inference functions for ultra high dimensional longitudinal data. *Journal of Statistical Computation and Simulation*, 90:2614–2630, 2020.

C. Lam and J. Fan. Sparsistency and rates of convergence in large covariance matrices estimation. *Annals of Statistics*, 37:4254–4278, 2009.

C. Lam and Q. Yao. Factor modeling for high-dimensional time series: Inference for the number of factors. *The Annals of Statistics*, 40(2):694–726, 2012. doi: 10.1214/12-AOS970. URL https://doi.org/10.1214/12-AOS970.

C. Lam, Q. Yao, and N. Bathia. Estimation of latent factors for high-dimensional time series. *Biometrika*, 98(4):901–918, 2011.

S. L. Lauritzen and P. Zwiernik. Locally associated graphical models and mixed convex exponential families. *The Annals of Statistics*, 2022.

O. Ledoit and M. Wolf. A well-conditioned estimator for large-dimensional covariance matrices. *Journal of Multivariate Analysis*, 88:365–411, 2004.

O. Ledoit and M. Wolf. The power of (non-)linear shrinking: A review and guide to covariance matrix estimation. *Journal of Financial Econometrics*, 20(1):187–218, 06 2022. ISSN 1479-8409. doi: 10.1093/jjfinec/nbaa007. URL https://doi.org/10.1093/jjfinec/nbaa007.

W. Lee and Y. Liu. Simultaneous multiple response regression and inverse covariance matrix estimation via penalized gaussian maximum likelihood. *Journal of Multivariate Analysis*, 111:241–255, 10 2012. doi: 10.1016/j.jmva.2012.03.013.

C. Leng and B. Li. Forward adaptive banding for estimating large covariance matrices. *Biometrika*, 98:821–830, 2011.

T. Leonard and J. S. J. Hsu. Bayesian inference for a covariance matrix. *Annals of Statistics*, 20:1669–1696, 1992.

J. P. LeSage and R. K. Pace. A matrix exponential spatial specification. *Journal of Econometrics*, 140:198–214, 2007.

E. Levina, A. J. Rothman, and J. Zhu. Sparse estimation of large covariance matrices via a nested lasso penalty. *Annals of Applied Statistics*, 2:245–263, 2008.

S. Leyder, J. Raymaekers, and P. J. Rousseeuw. Robust distance covariance. *International Statistical Review*, 2025. doi: https://doi.org/10.1111/insr.70005.

X. Li, H. Joe, and C. Genest. A factor-copula latent-vine time series model for extreme flood insurance losses. *Journal of the American Statistical Association*, pages 1–20, 10 2025. doi: 10.1080/01621459.2025.2579953.

K. Y. Liang and S. L. Zeger. Longitudinal data analysis using generalized linear models. *Biometrika*, 73:13–22, 1986a.

K. Y. Liang and S. L. Zeger. Longitudinal data analysis using generalized linear models. *Biometrika*, 73(1):13–22, 04 1986b. ISSN 0006-3444. doi: 10.1093/biomet/73.1.13.

S. P. Lin and M. D. Perlman. *A Monte Carlo comparison of four estimators of a covariance matrix. In Multivariate Analysis, 6, Ed. P.R.Krishnaiah*. North-Holland, Amsterdam, 1985.

H. Liu and L. Wang. Tiger: A tuning-insensitive approach for optimally estimating gaussian graphical models. *Preprint, MIT*, 2012.

K. Lounici. Sup-norm convergence rate and sign concentration property of lasso and dantzig estimators. *Electronic Journal of Statistics*, 2:90–102, 2008.

L. Luo and P. X. K. Song. Renewable Estimation and Incremental Inference in Generalized Linear Models with Streaming Data Sets. *Journal of the Royal Statistical Society Series B: Statistical Methodology*, 82(1):69–97, 12 2019. ISSN 1369-7412. doi: 10.1111/rssb.12352.

L. Luo, L. Zhou, and P. X. K. Song. Real-time regression analysis of streaming clustered data with possible abnormal data batches. *Journal of the American Statistical Association*, 118(543):2029–2044, 2023. doi: 10.1080/01621459.2022.2026778.

H. Lütkepohl. *New Introduction to Multiple Time Series Analysis*. Springer, New York, 2007.

J. Lv, H. Yang, and C. Guo. An efficient and robust variable selection method for longitudinal generalized linear models. *Computational Statistics and Data Analysis*, 82:74–88, 2015. ISSN 0167-9473.

H. Markowitz. Portfolio selection. *The Journal of Finance*, 7(1):77–91, 1952. URL http://www.jstor.org/stable/2975974.

G. Marra and R. Radice. Bivariate copula additive models for location, scale and shape. *Computational Statistics and Data Analysis*, 112(C):99–113, 2017.

A. W. Marshall, I. Olkin, and B. C. Arnold. *Inequalities: Theory of Majorization and its Applications*, volume 143. Springer, second edition, 2011. doi: 10.1007/978-0-387-68276-1.

A. Maurya. A well-conditioned and sparse estimation of covariance and inverse covariance matrices using a joint penalty. *Journal of Machine Learning Research*, 17(130):1–28, 2016. URL http://jmlr.org/papers/v17/15-345.html.

R. Mazumder and T. J. Hastie. The graphical lasso: New insights and alternatives. *Electronic Journal of Statistics*, 6:2125–2149, 2011.

G. P. McCabe. Principal variables. *Technometrics*, 26(2):137–144, 1984. ISSN 00401706.

P. McCullagh and J. A. Nelder. *Generalized Linear Models. Second Edition.* Chapman and Hall, London, 1989.

C. Meek. Causal inference and causal explanation with background knowledge. In *UAI*, 1995.

N. Meinshausen and P. Bühlmann. High-dimensional graphs and variable selection with the lasso. *The Annals of Statistics*, 34:1436–1462, 2006.

N. Meinshausen and B. Yu. Lasso-type recovery of sparse representations for high-dimensional data. *The Annals of Statistics*, 37(1):246–270, 2009. doi: 10.1214/07-AOS582. URL https://doi.org/10.1214/07-AOS582.

S. Muller, S. Wang, and A. H. Welsh. The effect of the working correlation on fitting models to longitudinal data. *Scandinavian Journal of Statistics*, 51(2):891–912, 2024. doi: https://doi.org/10.1111/sjos.12704.

T. Muschinski, M. N. Lang, G. J. Mayr, J. W. Messner, A. Zeileis, and T. Simon. Predicting power ramps from joint distributions of future wind speeds. *Wind Energy Science*, 7(6):2393–2405, 2022. doi: 10.5194/wes-7-2393-2022. URL https://wes.copernicus.org/articles/7/2393/2022/.

I. Ng, A. Ghassami, and K. Zhang. On the role of sparsity and dag constraints for learning linear dags. In *Proceedings of the 34th International Conference on Neural Information Processing Systems*, NIPS '20, 2020. ISBN 9781713829546.

I. Ng, B. Huang, and K. Zhang. Structure learning with continuous optimization: A sober look and beyond. In F. Locatello and V. Didelez, editors, *Proceedings of the Third Conference on Causal Learning and Reasoning*, volume 236 of *Proceedings of Machine Learning Research*, pages 71–105. PMLR, 01–03 Apr 2024.

Y. Ni, F. C. Stingo, and V. Baladandayuthapani. Bayesian graphical regression. *Journal of the American Statistical Association*, 114(525):184–197, 2019. doi: 10.1080/01621459.2017.1389739.

W. B. Nicholson, J. Bien, and D. S. Matteson. Hierarchical vector autoregression. *Arxiv preprint arXiv:1412.5250v2*, 2016.

X. Niu and P. D. Hoff. Joint mean and covariance modeling of multiple health outcome measures. *The Annals of Applied Statistics*, 13(1):321 – 339, 2019. doi: 10.1214/18-AOAS1187. URL https://doi.org/10.1214/18-AOAS1187.

Y. Niu, Y. Ni, D. Pati, and B. K. Mallick. Covariate-assisted bayesian graph learning for heterogeneous data. *Journal of the American Statistical Association*, 0(0):1–15, 2023. doi: 10.1080/01621459.2023.2233744.

A. Onatski. Determining the number of factors from empirical distribution of eigenvalues. *The Review of Economics and Statistics*, 92(4):1004–1016, 2010.

R. Opgen-Rhein and K. Strimmer. Accurate ranking of differentially expressed genes by a distribution-free shrinkage approach. *Statistical Applications in Genetics and Molecular Biology*, 6(1), 2007. doi: doi:10.2202/1544-6115.1252. URL https://doi.org/10.2202/1544-6115.1252.

J. Pan and G. MacKenzie. On modelling mean-covariance structure in longitudinal studies. *Biometrika*, 90:239–244, 2003.

G. Park. Identifiability of additive noise models using conditional variances. *Journal of Machine Learning Research*, 21(75):1–34, 2020.

G. Park and Y. Kim. Identifiability of gaussian linear structural equation models with homogeneous and heterogeneous error variances. *Journal of the Korean Statistical Society*, 49:276–292, 03 2020. doi: 10.1007/s42952-019-00019-7.

H. G. Park. Bayesian estimation of covariate assisted principal regression for brain functional connectivity, 2023.

D. Paul. Asymptotics of sample eigenstructure for a large dimensional spiked covariance model. *Statistica Sinica*, 17(4):1617–1642, 2007. ISSN 10170405, 19968507.

J. Pearl. *Causality: Models, Reasoning and Inference*. Cambridge University Press, USA, 2nd edition, 2009.

J. Peng, P. Wang, N. Zhou, and J. Zhu. Partial correlation estimation by joint sparse regression models. *Journal of the American Statistical Association*, 104:735–746, 2009.

K. W. Penrose, A. G. Nelson, and A. G. Fisher. Generalized body composition prediction equations for men using simple measurement techniques. *Medicine and Science in Sports and Exercise*, 17:189, 1985.

R. Perry, S. Panigrahi, J. Bien, and D. Witten. Inference on the proportion of variance explained in principal component analysis. *Journal of the American Statistical Association*, 0(0):1–11, 2025. doi: 10.1080/01621459.2025.2538895.

J. Peters and P. Bühlmann. Identifiability of gaussian structural equation models with equal error variances. *Biometrika*, 101:219–228, 2014.

J. Peters, J. M. Mooij, D. Janzing, and B. Schölkopf. Causal discovery with continuous additive noise models. *Journal of Machine Learning Research*, 15(58):2009–2053, 2014. URL http://jmlr.org/papers/v15/peters14a.html.

J. Peters, D. Janzing, and B. Scholkopf. *Elements of Causal Inference: Foundations and Learning Algorithms*. The MIT Press, 2017. ISBN 0262037319.

K. B. Petersen and M. S. Pedersen. The matrix cookbook, 2008. URL http://www2.imm.dtu.dk/pubdb/p.php?3274.

J. Pinheiro and D. Bates. *Mixed-Effects Models in S and S-PLUS*. Statistics and Computing. Springer New York, 2000. ISBN 9780387989570.

R. A. Poldrack, A. Kittur, D. J. Kalar, E. Miller, C. Seppa, Y. Gil, D. S. Parker, F. W. Sabb, and R. M. Bilder. The cognitive atlas: Toward a knowledge foundation for cognitive neuroscience. *Frontiers in Neuroinformatics*, 5, 2011. URL https://api.semanticscholar.org/CorpusID:14629685.

M. Pourahmadi. Joint mean-covariance models with applications to longitudinal data: Unconstrained parameterisation. *Biometrika*, 86:677–690, 1999.

M. Pourahmadi. Maximum likelihood estimation of generalized linear models for multivariate normal covariance matrix. *Biometrika*, 87:425–435, 2000.

M. Pourahmadi. *Foundations of Time Series Analysis and Prediction Theory*. Wiley, New York, 2001.

M. Pourahmadi. Simultaneous modeling of covariance matrices : Glm, bayesian and nonparametric perspective. *Correlated Data Modelling 2004, D. Gregori et al. (eds)*, FrancoAngeli, Milan, Italy, 2004.

M. Pourahmadi. Cholesky decompositions and estimation of a multivariate normal covariance matrix: Parameter orthogonality. *Biometrika*, 94:1006–1013, 2007a.

M. Pourahmadi. *High-Dimensional Covariance Estimation*. Wiley, New York, 2013.

A. Qu and R. Li. Quadratic Inference Functions for Varying-Coefficient Models with Longitudinal Data. *Biometrics*, 62:379–391, 2006.

A. Qu, B. G. Lindsay, and B. Li. Improving generalised estimating equations using quadratic inference functions. *Biometrika*, 87(4):823–836, 12 2000. ISSN 0006-3444. doi: 10.1093/biomet/87.4.823. URL https://doi.org/10.1093/biomet/87.4.823.

B. Rajaratnam and J. Salzman. Best permutation analysis. *Journal of Multivariate Analysis*, 121:193–223, October 2013. doi: 10.1016/j.jmva.2013.03.001.

G. Raskutti and C. Uhler. Learning directed acyclic graph models based on sparsest permutations. *Stat*, (1), 2018.

P. Ravikumar, M. J. Wainwright, and B. Yu. High-dimensional covariance estimation by minimizing ℓ_1-penalized log-determinant divergence. *Electronic Journal of Statistics*, 5:935–980, 2011.

S. Reid, R. Tibshirani, and J. Friedman. A study of error variance estimation in lasso regression. *Statistica Sinica*, 26(1):35–67, 2016. ISSN 10170405, 19968507.

Z. Ren, T. Sun, C. H. Zhang, and H. H. Zhou. Asymptotic normality and optimalities in estimation of large gaussian graphical models. *The Annals of Statistics*, 43(3): 991–1026, 2015.

R. A. Rigby and D. M. Stasinopoulos. Generalized additive models for location, scale and shape. *Journal of the Royal Statistical Society: Series C (Applied Statistics)*, 54(3):507–554, 2005.

K. Rohe and M. Zeng. Vintage factor analysis with Varimax performs statistical inference. *Journal of the Royal Statistical Society Series B: Statistical Methodology*, 85(4):1037–1060, 07 2023. ISSN 1369-7412. doi: 10.1093/jrsssb/qkad029.

A. J. Rothman, P. J. Bickel, E. Levina, and J. Zhu. Sparse permutation invariant covariance estimation. *Electronic Journal of Staistics*, 2:494–515, 2008.

A. J. Rothman, E. Levina, and J. Zhu. Generalized thresholding of large covariance matrices. *Journal of the American Statistical Association (Theory and Methods)*, 104:177–186, 2009.

A. J. Rothman, E. Levina, and J. Zhu. A new approach to cholesky-based estimation of high-dimensional covariance matrices. *Biometrika*, 97:539–550, 2010a.

A. J. Rothman, E. Levina, and J. Zhu. Sparse multivariate regression with covariance estimation. *Journal of Computational and Graphical Statistics*, 19:947–962, 2010b.

K. Sachs, O. Perez, D. Pe'er, D. A. Lauffenburger, and G. P. Nolan. Causal protein-signaling networks derived from multiparameter single-cell data. *Science*, 308 (5721):523–529, 2005.

A. Safikhani and A. Shojaie. Joint structural break detection and parameter estimation in high-dimensional nonstationary var models. *Journal of the American Statistical Association*, 117(537):251–264, 2022. URL https://doi.org/10.1080/01621459.2020.1770097

T. Sargent and C. Sims. Business cycle modeling without pretending to have too much a priori economic theory. Working Papers 55, Federal Reserve Bank of Minneapolis, 1977.

SAS Institute Inc. *SAS/STAT 9.2 User's Guide*. SAS Institute Inc., Second edition, 2009. SAS Institute Inc.

F. Schäfer, M. Katzfuss, and H. Owhadi. Sparse cholesky factorization by kullback–leibler minimization. *SIAM Journal on Scientific Computing*, 43(3):A2019–A2046, 2021. doi: 10.1137/20M1336254.

J. Schafer and K. Strimmer. A shrinkage approach to large-scale covariance matrix estimation and implications for functional genomics. *Statistical Applications in Genetics and Molecular Biology*, 4(1), 2005. doi: doi:10.2202/1544-6115.1175. URL https://doi.org/10.2202/1544-6115.1175.

A. Schuster. The periodogram and its optical analogy. *Proceedings of the Royal Society of London. Series A, Containing Papers of a Mathematical and Physical Character*, 77(515):136–140, 1906. doi: 10.1098/rspa.1906.0011.

L. Seidel. Uber ein verfahren, die gleichungen, auf welche die methode der kleinsten quadrate fuhrt, sowie lineare gleichungen uberhaupt, durch successive annaherung aufzulosen. *Abhandlungen der Mathematisch-Physikalischen Klasse der Königlich Bayerischen Akademie der Wissenschaften*, 1874.

C. Seiler and S. Holmes. Multivariate heteroscedasticity models for functional brain connectivity. *Frontiers in Neuroscience*, 11, 2017. doi: 10.3389/fnins.2017.00696.

W. Sharpe. *Portfolio Theory and Capital Markets*. McGraw-Hill, New York, 1970.

S. Shimizu, P. O. Hoyer, A. Hyvärinen, and A. Kerminen. A linear non-gaussian acyclic model for causal discovery. *Journal of Machine Learning Research*, 7(72): 2003–2030, 2006.

S. Shimizu, T. Inazumi, Y. Sogawa, A. Hyvärinen, Y. Kawahara, T. Washio, P. O. Hoyer, and K. Bollen. Directlingam: A direct method for learning a linear non-gaussian structural equation model. *Journal of Machine Learning Research*, 12 (33):1225–1248, 2011. URL http://jmlr.org/papers/v12/shimizu11a.html.

A. Shojaie and G. Michailidis. Penalized likelihood methods for estimation of sparse high-dimensional directed acyclic graphs. *Biometrika*, 97(3):519–538, 2010. ISSN 00063444, 14643510.

R. H. Shumway and D. S. Stoffer. *Time Series Analysis and Its Applications, Fifth Edition*. Springer-Verlag, Berlin, Heidelberg, 2025. ISBN 9783031705830.

T. Silander and P. Myllymäki. A simple approach for finding the globally optimal bayesian network structure. *Proceedings of the 22nd Annual Conference on Uncertainty in Artificial Intelligence (UAI-06)*, 06 2012.

B. Skagerberg, J. F. MacGregor, and C. Kiparissides. Multivariate data analysis applied to low-density polyethylene reactors. *Chemometrics and Intelligent Laboratory Systems*, 14(1):341–356, 1992. ISSN 0169-7439. doi: https://doi.org/10.

1016/0169-7439(92)80117-M. URL https://www.sciencedirect.com/science/article/pii/016974399280117M. Proceedings of the 2nd Scandinavian Symposium on Chemometrics.

M. Sklar. Fonctions de repartition an dimensions et leurs marges. *Publications de l'Institut Statistique de l'Université de Paris*, 8:229–231, 1959.

A. Skripnikov and G. Michailidis. Joint estimation of multiple network granger causal models. *Econometrics and Statistics*, 10:120–133, 2019.

S. M. Smith, P. T. Fox, K. L. Miller, D. C. Glahn, P. M. Fox, C. E. Mackay, N. Filippini, K. E. Watkins, R. Toro, A. R. Laird, and C. F. Beckmann. Correspondence of the brain's functional architecture during activation and rest. *Proceedings of the National Academy of Sciences*, 106(31):13040–13045, 2009. doi: 10.1073/pnas.0905267106.

P. X. K. Song, Z. Jiang, E. Park, and A. Qu. Quadratic inference functions in marginal models for longitudinal data. *Statistics in Medicine*, 28, 2009.

S. Song and P. J. Bickel. Large vector auto regressions. *Arxiv preprint arXiv:1106.3915*, 2011.

P. Spirtes and C. Glymour. An algorithm for fast recovery of sparse causal graphs. *Social Science Computer Review*, 9(1):62–72, 1991.

P. Spirtes, C. Glymour, and R. Scheines. *Causation, Prediction, and Search.* MIT Press, 2nd edition, 2001.

N. Stadler, P. Bühlmann, and S. Geer. l1-penalization for mixture regression models. *TEST*, 19(2):209–256, August 2010. doi: 10.1007/s11749-010-0197-z.

C. Stein. Estimation of a covariance matrix. *In Reitz lecture, Atlanta, Georgia, 1975. 39th Annual Meeting IMS*, 1975.

M. L. Stein. *Interpolation of Spatial Data: Some Theory for Kriging*. Springer Series in Statistics. Springer New York, 1999. ISBN 9781461214946. URL https://books.google.com/books?id=aZXwBwAAQBAJ.

M. L. Stein, Z. Chi, and L. J. Welty. Approximating likelihoods for large spatial data sets. *Journal of the Royal Statistical Society Series B: Statistical Methodology*, 66 (2):275–296, 04 2004. ISSN 1369-7412. doi: 10.1046/j.1369-7412.2003.05512.x. URL https://doi.org/10.1046/j.1369-7412.2003.05512.x.

J. H. Stock and M. W. Watson. Macroeconomic forecasting using diffusion indexes. *Journal of Business & Economic Statistics*, 20(2):147–162, 2002a. doi: 10.1198/073500102317351921.

J. H. Stock and M. W. Watson. Forecasting using principal components from a large number of predictors. *Journal of the American Statistical Association*, 97(460): 1167–1179, 2002b. doi: 10.1198/016214502388618960.

W. Su, M. Bogdan, and E. Candès. False discoveries occur early on the lasso path. *The Annals of Statistics*, 45(5):2133–2150, 2017.

T. Sun and C. H. Zhang. Scaled sparse linear regression. *Biometrika*, 99(4):879–898, 2012. ISSN 00063444, 14643510.

Y. Sun and M. L. Stein. Statistically and computationally efficient estimating equations for large spatial datasets. *Journal of Computational and Graphical Statistics*, 25(1):187–208, 2016. doi: 10.1080/10618600.2014.975230.

T. H. Szatrowski. Necessary and sufficient conditions for explicit solutions in the multivariate normal estimation problem for patterned means and covariances. *The Annals of Statistics*, 8:802–810, 1980.

G. J. Székely, M. L. Rizzo, and N. K. Bakirov. Measuring and testing dependence by correlation of distances. *The Annals of Statistics*, 35(6):2769 – 2794, 2007. doi: 10.1214/009053607000000505. URL https://doi.org/10.1214/009053607000000505.

Z. Y. Tho, F. K. C. Hui, and T. Zou. An ising similarity regression model for modeling multivariate binary data, 2024.

L. L. Thurstone. *Multiple Factor Analysis*. University of Chicago Press, Chicago, 1947.

G. C. Tiao and R. S. Tsay. Model specification in multivariate time series. *Journal of the Royal Statistical Society: Series B (Methodological)*, 51(2):157–195, 1989. doi: https://doi.org/10.1111/j.2517-6161.1989.tb01756.x.

R. Tibshirani. Regression shrinkage and selection via the lasso. *Journal of the Royal Statistical Society. Series B*, 58:267–288, 1996.

R. J. Tibshirani and J. Taylor. Degrees of freedom in lasso problems. *The Annals of Statistics*, 40(2):1198–1232, 2012.

M. E. Tipping and C. M. Bishop. Probabilistic principal component analysis. *Journal of the Royal Statistical Society: Series B (Statistical Methodology)*, 61(3):611–622, 1999. doi: https://doi.org/10.1111/1467-9868.00196.

C. Tran and G. Yu. A completely tuning-free and robust approach to sparse precision matrix estimation. In *Proceedings of the 39th International Conference on Machine Learning*, volume 162 of *Proceedings of Machine Learning Research*, pages 21733–21750. PMLR, 17–23 Jul 2022.

W. F. Trench. Explicit inversion formulas for toeplitz band matrices. *SIAM Journal on Algebraic Discrete Methods*, 6(4):546–554, 1985. doi: 10.1137/0606054.

W. F. Trench. An unnoticed consequence of szegö's distribution theorem. *The American Mathematical Monthly*, 121(10):pp. 917–921, 2014. ISSN 00029890, 19300972.

Y. Tsang, M. Coates, and R. Nowak. Passive network tomography using em algorithms. In *2001 IEEE International Conference on Acoustics, Speech, and Signal Processing. Proceedings (Cat. No.01CH37221)*, volume 3, pages 1469–1472 vol.3, 2001. doi: 10.1109/ICASSP.2001.941208.

P. Tseng. Convergence of a block coordinate descent method for nondifferentiable minimization. *Journal of Optimization Theory and Applications*, 109(3):475–494, jun 2001. ISSN 0022-3239. doi: 10.1023/A:1017501703105.

J. K. Tugnait. On sparse high-dimensional graphical model learning for dependent time series, 2021.

J. W. Tukey. The Future of Data Analysis. *The Annals of Mathematical Statistics*, 33 (1):1–67, 1962. doi: 10.1214/aoms/1177704711. URL https://doi.org/10.1214/aoms/1177704711.

B. A. Turlach, W. N. Venables, and S. J. Wright. Simultaneous variable selection. *Technometrics*, 47(3):349–363, 2005.

C. Uhler, G. Raskutti, P. Bühlmann, and B. Yu. Geometry of the faithfulness assumption in causal inference. *The Annals of Statistics*, 41(2):436–463, 2013. ISSN 00905364, 21688966.

S. van de Geer, P. Bühlmann, and S. Zhou. The adaptive and the thresholded Lasso for potentially misspecified models (and a lower bound for the Lasso). *Electronic Journal of Statistics*, 5:688–749, 2011. doi: 10.1214/11-EJS624.

S. A. van de Geer and P. Bühlmann. On the conditions used to prove oracle results for the Lasso. *Electronic Journal of Statistics*, 3:1360–1392, 2009. doi: 10.1214/09-EJS506. URL https://doi.org/10.1214/09-EJS506.

M. G. P. van der Wijst, H. Brugge, D. H. de Vries, P. Deelen, M. A. Swertz, and L. Franke. Single-cell rna sequencing identifies cell type-specific cis-eqtls and co-expression qtls. *Nature Genetics*, 50:493 – 497, 2018.

D. C. Van Essen, S. M. Smith, D. M. Barch, T. E. J. Behrens, E. Yacoub, and K. Ugurbil. The wu-minn human connectome project: An overview. *NeuroImage*, 80:62–79, 2013. ISSN 1053-8119. doi: https://doi.org/10.1016/j.neuroimage.2013.05.041. Mapping the Connectome.

W. N. van Wieringen. Lecture notes on ridge regression, 2023. URL https://arxiv.org/abs/1509.09169.

L.H. Vanegas, L.M. Rondón, and G.A. Paula. Generalized estimating equations using the new r package glmtoolbox. *The R Journal*, 15:105–133, 11 2023. doi: 10.32614/RJ-2023-056.

T. Vatter and T. Nagler. Generalized additive models for pair-copula constructions. *Journal of Computational and Graphical Statistics*, 27(4):715–727, 2018. doi: 10.1080/10618600.2018.1451338.

A. V. Vecchia. Estimation and model identification for continuous spatial processes. *Journal of the Royal Statistical Society. Series B*, 50:297–312, 1988.

R. von Sachs. Nonparametric spectral analysis of multivariate time series. *Annual Review of Statistics and Its Application*, 7(Volume 7, 2020):361–386, 2020. doi: https://doi.org/10.1146/annurev-statistics-031219-041138.

M. J. Vowels, N. C. Camgoz, and R. Bowden. D'ya like dags? a survey on structure learning and causal discovery. *ACM Computing Surveys*, 55(4), November 2022. ISSN 0360-0300. doi: 10.1145/3527154. URL https://doi.org/10.1145/3527154.

O. Wahltinez, M. Lee, A. Erlinger, M. Daswani, P. Yawalkar, K. Murphy, and M. Brenner. Covid-19 open-data: curating a fine-grained, global-scale data repository for sars-cov-2. 2020. URL https://goo.gle/covid-19-open-data. Work in progress.

H. Wang. Coordinate descent algorithm for covariance graphical lasso. *Statistics and Computing*, 24:521–529, 2014.

J. Wang. Joint estimation of sparse multivariate regression and conditional graphical models. *Statistica Sinica*, 25(3):831–851, 2015. ISSN 10170405, 19968507.

L. Wang, J. Zhou, and A. Qu. Penalized generalized estimating equations for high-dimensional longitudinal data analysis. *Biometrics*, 68(2):353–360, 2012a. doi: https://doi.org/10.1111/j.1541-0420.2011.01678.x.

L. Wang, B. Peng, J. Bradic, R. Li, and Y. Wu. A tuning-free robust and efficient approach to high-dimensional regression. *Journal of the American Statistical Association*, 115(532):1700–1714, 2020. doi: 10.1080/01621459.2020.1840989.

Linbo Wang, Thomas Richardson, and James Robins. Causal inference: A tale of three frameworks, 2025. URL https://arxiv.org/abs/2511.21516.

M. Wang and G. I. Allen. Thresholded graphical lasso adjusts for latent variables. *Biometrika*, 2022. ISSN 1464-3510. doi: 10.1093/biomet/asac060. URL https://doi.org/10.1093/biomet/asac060.

Y. Wang, S. Joseph, X. Liu, M. Kelley, and R. Rekaya. Snpxge2: a database for human snp-coexpression associations. *Bioinformatics (Oxford, England)*, 28:403–10, 11 2012b. doi: 10.1093/bioinformatics/btr663.

Y. G. Wang and V. Carey. Working correlation structure misspecification, estimation and covariate design: Implications for generalised estimating equations performance. *Biometrika*, 90(1):29–41, 03 2003.

D. Warton. Penalized normal likelihood and ridge regularization of correlation and covariance matrices. *Journal of the American Statistical Association*, 103:340–349, 2008.

L. Wasserman and K. Roeder. High-dimensional variable selection. *The Annals of Statistics*, 37(5A):2178–2201, 2009. ISSN 00905364, 21688966.

R. W. M. Wedderburn. Quasi-likelihood functions, generalized linear models, and the gauss-newton method. *Biometrika*, 61(3):439–447, 1974. ISSN 00063444, 14643510. URL http://www.jstor.org/stable/2334725.

D. Wei, T. Gao, and Y. Yu. Dags with no fears: A closer look at continuous optimization for learning bayesian networks. In H. Larochelle, M. Ranzato, R. Hadsell, M. F. Balcan, and H. Lin, editors, *Advances in Neural Information Processing Systems*, volume 33, pages 3895–3906. Curran Associates, Inc., 2020.

A. Weinstein, W. J. Su, M. Bogdan, R. F. Barber, and E. J. Candes. A power analysis for model-x knockoffs with ℓ_p-regularized statistics, 2022. URL https://arxiv.org/abs/2007.15346.

P. Westgate and T. Braun. An improved quadratic inference function for parameter estimation in the analysis of correlated data. *Biostatistics*, 32(19):3260–3273, 2013. doi: 10.1002/sim.5715.

H. Weyl. Das asymptotische verteilungsgesetz der eigenwerte linearer partieller differentialgleichungen. *Mathematische Annalen*, (71):441–479, 1912.

J. Whittaker. *Graphical Models in Applied Multivariate Statistics*. John Wiley and Sons, Chichester, 1990.

W. Wirtinger. Zur formalen theorie der funktionen von mehr komplexen veränderlichen. *Mathematische Annalen*, 97:357–375, 1927.

D. M. Witten and R. Tibshirani. Covariance-regularized regression and classification for high-dimensional problems. *Journal of the Royal Statistical Society, Series B*, 71:615–636, 2009.

D. M. Witten, R. Tibshirani, and T. Hastie. A penalized matrix decomposition, with applications to sparse principal components and canonical correlation analysis. *Biostatistics*, 10:515–534, 2009.

R. D. Wolfinger. Heterogeneous variance: Covariance structures for repeated measures. *Journal of Agricultural, Biological, and Environmental Statistics*, 1(2):205–230, 1996. ISSN 10857117.

W. B. Wu and Y. N. Wu. Performance bounds for parameter estimates of high-dimensional linear models with correlated errors. *Electronic Journal of Statistics*, 10(1):352–379, 2016.

Y. Wu and L. Wang. A survey of tuning parameter selection for high-dimensional regression. *Annual Review of Statistics and Its Application*, 7(1):209–226, 2020.

J. Xu and K. Lange. A proximal distance algorithm for likelihood-based sparse covariance estimation. *Biometrika*, 109(4):1047–1066, 02 2022. ISSN 1464-3510. doi: 10.1093/biomet/asac011. URL https://doi.org/10.1093/biomet/asac011.

L. L. Xu, J. Q. Li, and L. Y. Fu. Analysis of gee with a mixture working correlation matrix for diverging number of covariates. *Journal of Statistical Computation and Simulation*, 92(4):748–763, 2022. doi: 10.1080/00949655.2021.1974438.

L. Xue, S. Ma, and H. Zou. Positive definite ℓ_1 penalized estimation of large covariance matrices. *Journal of the American Statistical Association*, 2012.

H. Ye and J. Pan. Modeling covariance structures in generalized estimating equations for longitudinal data. *Biometrika*, 93:911–926, 2006.

Q. Ye, A. Amini, and Q. Zhou. Optimizing regularized cholesky score for order-based learning of bayesian networks. *IEEE Transactions on Pattern Analysis and Machine Intelligence*, pages 1–1, 2020.

A. Yeremin, L. Kolotilina, and A. Nikishin. Factorized sparse approximate inverse preconditionings. iii: Iterative construction of preconditioners. *Journal of Mathematical Sciences*, 101:3237–3254, 09 2000. doi: 10.1007/BF02672769.

J. Yin and H. Li. A sparse conditional Gaussian graphical model for analysis of genetical genomics data. *The Annals of Applied Statistics*, 5(4):2630 – 2650, 2011. doi: 10.1214/11-AOAS494. URL https://doi.org/10.1214/11-AOAS494.

G. Yu and J. Bien. Learning local dependence in ordered data. *Journal of Machine Learning Research*, 18:1–60, 2017.

G. Yu and J. Bien. Estimating the error variance in a high-dimensional linear model. *Biometrika*, 106(3):pp. 533–546, 2019.

T. P. Yuen, H. Wong, and K. F. C. Yiu. On constrained estimation of graphical time series models. *Computational Statistics and Data Analysis*, 124:27–52, 2018.

A. Zellner. An efficient method of estimating seemingly unrelated regressions and tests for aggregation bias. *Journal of the American Statistical Association*, 57(298): 348–368, 1962. ISSN 01621459, 1537274X.

Bo Zhang, ZhiXiang Zhang, and Guangming Pan. Estimating the number of significant components in high-dimensional pca. *Biometrika*, page asaf092, 12 2025. ISSN 1464-3510. doi: 10.1093/biomet/asaf092. URL https://doi.org/10.1093/biomet/asaf092.

C. H. Zhang. Nearly unbiased variable selection under minimax concave penalty. *The Annals of Statistics*, 38(2):894 – 942, 2010. doi: 10.1214/09-AOS729. URL https://doi.org/10.1214/09-AOS729.

J. Zhang and Y. Li. High-dimensional gaussian graphical regression models with covariates. *Journal of the American Statistical Association*, 118(543):2088–2100, 2023. doi: 10.1080/01621459.2022.2034632.

K. Zhang and A. Hyvärinen. On the identifiability of the post-nonlinear causal model. In *Proceedings of the Twenty-Fifth Conference on Uncertainty in Artificial Intelligence*, UAI '09, page 647–655. AUAI Press, 2009. ISBN 9780974903958.

T. Zhao, H. Liu, K. Roeder, J. Lafferty, and L. Wasserman. The huge package for high-dimensional undirected graph estimation in r. *Journal of Machine Learning Research*, 2012.

Y. Zhao, B. Wang, S. H. Mostofsky, B. S. Caffo, and X. Luo. Covariate Assisted Principal regression for covariance matrix outcomes. *Biostatistics*, 22(3):629–645, 2019. ISSN 1465-4644. doi: 10.1093/biostatistics/kxz057. URL https://doi.org/10.1093/biostatistics/kxz057.

Y. Zhao, B. Caffo, and X. Luo. Principal regression for high dimensional covariance matrices. *Electronic Journal of Statistics*, 15(2):4192–4235, 2021. doi: 10.1214/21-EJS1887. URL https://doi.org/10.1214/21-EJS1887.

X. Zheng, W. Fung, and Z. Zhu. Robust estimation in joint mean–covariance regression model for longitudinal data. *Annals of the Institute of Statistical Mathematics*, 65(4):617–638, 2013. doi: 10.1007/s10463-012-0383-8.

X. Zheng, W. K. Fung, and Z. Zhu. Variable selection in robust joint mean and covariance model for longitudinal data analysis. *Statistica Sinica*, 24(2):515–531, 2014. ISSN 10170405, 19968507.

X. Zheng, B. Aragam, P. K. Ravikumar, and E. P. Xing. Dags with no tears: Continuous optimization for structure learning. In S. Bengio, H. Wallach, H. Larochelle, K. Grauman, N. Cesa-Bianchi, and R. Garnett, editors, *Advances in Neural Information Processing Systems*, volume 31. Curran Associates, Inc., 2018.

Y. Zheng. An interpretable and efficient infinite-order vector autoregressive model for high-dimensional time series. *Journal of the American Statistical Association*, 120(549):212–225, 2025. doi: 10.1080/01621459.2024.2311365.

J. Zhou and A. Qu. Informative Estimation and Selection of Correlation Structure for Longitudinal Data. *Journal of the American Statistical Association*, 107(498):701–710, 2012. doi: 10.1080/01621459.2012.682.

S. Zhou. Thresholded lasso for high dimensional variable selection and statistical estimation. *arXiv: Statistics Theory*, 2010.

Z. Zhu and Y. Liu. Estimating spatial covariance using penalized likelihood with weighted l1 penalty. *Journal of Nonparametric Statistics*, 21:925–942, 2009.

J. N. Zorich. *The History of Correlation*. CRC Press, 2024. ISBN 9781040261774. URL https://books.google.com/books?id=Q0dHEQAAQBAJ.

H. Zou. The adaptive lasso and its oracle properties. *Journal of the American Statistical Association*, 101:1418–1429, 2006.

T. Zou, W. Lan, H. Wang, and C. L. Tsai. Covariance regression analysis. *Journal of the American Statistical Association*, 112(517):266–281, 2017. doi: 10.1080/01621459.2015.1131699.

P. Zwiernik. Entropic covariance models. *The Annals of Statistics*, 53(4):1371–1405 2025. URL https://doi.org/10.1214/24-AOS2474. 2023.

Author Index

Index

For Product Safety Concerns and Information please contact our EU representative GPSR@taylorandfrancis.com Taylor & Francis Verlag GmbH, Kaufingerstraße 24, 80331 München, Germany

Batch number: 10793307

Printed by Printforce, the Netherlands